Oxford Graduate Texts in Mathematics

OXFORD GRADUATE TEXTS IN MATHEMATICS

Riemannian Holonomy Groups and Calibrated Geometry

Dominic D. Joyce

The Mathematical Institute, 24-29 St. Giles', Oxford, OX1 3LB

OXFORD
UNIVERSITY PRESS

OXFORD
UNIVERSITY PRESS

Great Clarendon Street, Oxford OX2 6DP

Oxford University Press is a department of the University of Oxford.
It furthers the University's objective of excellence in research, scholarship,
and education by publishing worldwide in

Oxford New York

Auckland Cape Town Dar es Salaam Hong Kong Karachi
Kuala Lumpur Madrid Melbourne Mexico City Nairobi
New Delhi Shanghai Taipei Toronto

With offices in

Argentina Austria Brazil Chile Czech Republic France Greece
Guatemala Hungary Italy Japan Poland Portugal Singapore
South Korea Switzerland Thailand Turkey Ukraine Vietnam

Oxford is a registered trade mark of Oxford University Press
in the UK and in certain other countries

Published in the United States
by Oxford University Press Inc., New York

© Dominic D. Joyce 2007

British Library Cataloguing in Publication Data

Data available

Library of Congress Cataloging in Publication Data

Data available

Typeset by the author using LaTex
Printed in Great Britain
on acid-free paper by
Biddles Ltd., King's Lynn Norfolk

ISBN 978–0–19–921560–7
ISBN 978–0–19–921559–1 (Pbk.)

1 3 5 7 9 10 8 6 4 2

Preface

Riemannian holonomy groups is an area of Riemannian geometry, in the field of differential geometry. The holonomy group $\mathrm{Hol}(g)$ of a Riemannian manifold (M, g) determines the geometrical structures on M compatible with g. Thus, Berger's classification of Riemannian holonomy groups gives a list of interesting geometrical structures compatible with a Riemannian metric, and the aim of the subject is to study each such structure in depth. Most of the holonomy groups on Berger's list turn out to be important in string theory in theoretical physics.

Given some class of mathematical objects, there is often a natural class of subobjects living inside them, such as groups and subgroups for instance. The natural subobjects of Riemannian manifolds (M, g) with special holonomy are *calibrated submanifolds*—lower-dimensional, volume-minimizing submanifolds N in M compatible with the geometric structures coming from the holonomy reduction. So calibrated geometry is an obvious companion subject for Riemannian holonomy groups. Calibrated submanifolds are also important in string theory, as 'supersymmetric cycles' or 'branes'.

This is a graduate textbook on Riemannian holonomy groups and calibrated geometry. It is aimed at graduates and researchers working in differential geometry, and also at physicists working in string theory, though the book is written very much from a mathematical point of view. It could be used as the basis of a graduate lecture course. The main prerequisites are a good understanding of topology, differential geometry, manifolds, and Lie groups at the advanced undergraduate or early graduate level. Some knowledge of Hilbert and Banach spaces would also be very useful, but not essential.

A little more than half this book is a revised version of parts of my monograph, *Compact Manifolds with Special Holonomy*, Oxford University Press, 2000, reference [188]. The main goal of [188] was to publish an extended research project on compact manifolds with holonomy G_2 and $\mathrm{Spin}(7)$, so Chapters 8–15 were almost wholly my own research. Chapters 1–7 of [188] have been rewritten to form Chapters 1, 2, 3, 5, 6, 7 and 10 respectively of this book, the core of the Riemannian holonomy material.

To this I have added new material on quaternionic Kähler manifolds in Chapter 10; Chapter 11 on the exceptional holonomy groups, which summarizes Chapters 10–15 of [188] and subsequent developments; and four new chapters on calibrated geometry, Chapters 4, 8, 9 and 12 below. This textbook is not intended to replace the monograph [188], and I hope my most discerning readers will want to own both. But unless you have a particular interest in compact manifolds with holonomy G_2 or $\mathrm{Spin}(7)$, this book is probably the better of the two to buy.

This book is not a vehicle for publishing my own research, and I have aimed to select material based on how I see the field and what I think it would be useful for a new researcher in the subject to know. No doubt I have overemphasized my own contributions, and I apologize for this; my excuse is that I knew them best, and they were easiest to plagiarize. Calibrated geometry is a younger field than Riemannian holonomy, and a very active area of research. I have tried in Chapters 8, 9 and 12 to discuss the frontiers of current research, and open problems I think are worth attention.

Some other books on Riemannian holonomy groups are Salamon [296] and [188], and they are also discussed in Kobayashi and Nomizu [214, 215], Besse [30, Ch. 10], Gross, Huybrechts and the author [138, Part I] and Berger [28, Ch. 13]. The only other book I know on calibrated geometry is Harvey [150].

Acknowledgements. Many people have shared their insights and ideas on these subjects with me; I would like in particular to thank Bobby Acharya, Tom Bridgeland, Robert Bryant, Simon Donaldson, Mark Haskins, Simon Salamon and Richard Thomas. I am grateful to Maximilian Kreuzer for permission to reproduce in Figure 7.1 the graph of Hodge numbers of Calabi–Yau 3-folds from Kreuzer and Skarke [225], and to the EPSRC for financial support whilst I was writing this book.

I dedicate this book to my wife Jayne and daughters Tilly and Kitty, without whom my life would have been only half as enjoyable, and this book written in half the time.

Oxford
September 2006   D.D.J.

Contents

1
Background material

In this chapter we explain some background necessary for understanding the rest of the book. We shall assume that the reader is already familiar with the basic ideas of differential and Riemannian geometry (in particular, manifolds and submanifolds, tensors, and Riemannian metrics) and of algebraic topology (in particular, fundamental group, homology and cohomology). We start in §1.1 with a short introduction to exterior forms on manifolds, de Rham cohomology, and Hodge theory. These will be essential tools later in the book, and we discuss them out of completeness, and to fix notation.

The rest of the chapter is an introduction to the analysis of elliptic operators on manifolds. Section 1.2 defines *Sobolev* and *Hölder spaces*, which are Banach spaces of functions and tensors on a manifold, and discusses their basic properties. Then §1.3–§1.5 define *elliptic operators*, a special class of partial differential operators, and explain how solutions of elliptic equations have good existence and regularity properties in Sobolev and Hölder spaces.

1.1 Exterior forms on manifolds

We introduce exterior forms on manifolds, and summarize two theories involving exterior forms—de Rham cohomology and Hodge theory. The books by Bredon [49], Bott and Tu [40] and Warner [338] are good references for the material in this section.

Let M be an n-manifold, with tangent bundle TM and cotangent bundle T^*M. The kth exterior power of the bundle T^*M is written $\Lambda^k T^*M$. It is a real vector bundle over M, with fibres of dimension $\binom{n}{k}$. Smooth sections of $\Lambda^k T^*M$ are called k-*forms*, and the vector space of k-forms is written $C^\infty(\Lambda^k T^*M)$.

Now $\Lambda^k T^*M$ is a subbundle of $\bigotimes^k T^*M$, so k-forms are tensors on M, and may be written using index notation. We shall use the common notation that a collection of tensor indices enclosed in square brackets $[\ldots]$ are to be antisymmetrized over. That is, if $T_{a_1 a_2 \ldots a_k}$ is a tensor with k indices, then

$$T_{[a_1 \ldots a_k]} = \tfrac{1}{k!} \sum_{\sigma \in S_k} \text{sign}(\sigma) T_{a_{\sigma(1)} \ldots a_{\sigma(k)}},$$

where S_k is the group of permutations of $\{1, 2, \ldots, k\}$, and $\text{sign}(\sigma)$ is 1 if σ is even, and -1 if σ is odd. Then a k-form α on M is a tensor $\alpha_{a_1 \ldots a_k}$ with k covariant indices that is antisymmetric, i.e. that satisfies $\alpha_{a_1 \ldots a_k} = \alpha_{[a_1 \ldots a_k]}$.

The *exterior product* $\wedge$ and the *exterior derivative* d are important natural operations on forms. If α is a k-form and β an l-form then $\alpha \wedge \beta$ is a $(k+l)$-form and $d\alpha$ a $(k+1)$-form, which are given in index notation by

1

$$(\alpha \wedge \beta)_{a_1 \ldots a_{k+l}} = \alpha_{[a_1 \ldots a_k} \beta_{a_{k+1} \ldots a_{k+l}]} \quad \text{and}$$

$$(\mathrm{d}\alpha)_{a_1 \ldots a_{k+1}} = T_{[a_1 \ldots a_{k+1}]}, \quad \text{where} \quad T_{a_1 \ldots a_{k+1}} = \frac{\partial \alpha_{a_2 \ldots a_{k+1}}}{\partial x^{a_1}}.$$

If α is a k-form and β an l-form then

$$\mathrm{d}(\mathrm{d}\alpha) = 0, \quad \alpha \wedge \beta = (-1)^{kl} \beta \wedge \alpha \quad \text{and} \quad \mathrm{d}(\alpha \wedge \beta) = (\mathrm{d}\alpha) \wedge \beta + (-1)^k \alpha \wedge (\mathrm{d}\beta).$$

The first of these is written $\mathrm{d}^2 = 0$, and is a fundamental property of d. If $\mathrm{d}\alpha = 0$, then α is *closed*, and if $\alpha = \mathrm{d}\beta$ for some β then α is *exact*. As $\mathrm{d}^2 = 0$, every exact form is closed. If M is a compact, oriented n-manifold and α an $(n-1)$-form, then *Stokes' Theorem* says that $\int_M \mathrm{d}\alpha = 0$.

1.1.1   De Rham cohomology

Let M be a smooth n-manifold. As $\mathrm{d}^2 = 0$, the chain of operators

$$0 \to C^\infty(\Lambda^0 T^* M) \xrightarrow{\mathrm{d}} C^\infty(\Lambda^1 T^* M) \xrightarrow{\mathrm{d}} \cdots \xrightarrow{\mathrm{d}} C^\infty(\Lambda^n T^* M) \to 0$$

forms a *complex*, and therefore we may find its cohomology groups. For $k = 0, \ldots, n$, define the *de Rham cohomology groups* $H_{DR}^k(M, \mathbb{R})$ of M by

$$H_{DR}^k(M, \mathbb{R}) = \frac{\mathrm{Ker}\big(\mathrm{d} : C^\infty(\Lambda^k T^* M) \to C^\infty(\Lambda^{k+1} T^* M)\big)}{\mathrm{Im}\big(\mathrm{d} : C^\infty(\Lambda^{k-1} T^* M) \to C^\infty(\Lambda^k T^* M)\big)}.$$

That is, $H_{DR}^k(M, \mathbb{R})$ is the quotient of the vector space of closed k-forms on M by the vector space of exact k-forms on M. If η is a closed k-form, then the *cohomology class* $[\eta]$ of η in $H_{DR}^k(M, \mathbb{R})$ is $\eta + \mathrm{Im}\,\mathrm{d}$, and η is a *representative* for $[\eta]$.

There are several different ways to define the *cohomology* of topological spaces, for example, singular, Alexander–Spanier and Čech cohomology. If the topological space is well-behaved (e.g. if it is paracompact and Hausdorff) then the corresponding cohomology groups are all isomorphic. The de Rham Theorem [338, p. 206], [40, Th. 8.9] is a result of this kind.

Theorem 1.1.1. (The de Rham Theorem) *Let M be a smooth manifold. Then the de Rham cohomology groups $H_{DR}^k(M, \mathbb{R})$ are canonically isomorphic to the singular, Alexander–Spanier and Čech cohomology groups of M over $\mathbb{R}$.*

Thus the de Rham cohomology groups are topological invariants of M. As there is usually no need to distinguish between de Rham and other sorts of cohomology, we will write $H^k(M, \mathbb{R})$ instead of $H_{DR}^k(M, \mathbb{R})$ for the de Rham cohomology groups. The kth *Betti number* b^k or $b^k(M)$ is $b^k = \dim H^k(M, \mathbb{R})$. The Betti numbers are important topological invariants of a manifold.

Theorem 1.1.2 (Poincaré duality) *Let M be a compact, oriented n-manifold. Then there is a canonical isomorphism $H^{n-k}(M, \mathbb{R}) \cong \big(H^k(M, \mathbb{R})\big)^*$, and the Betti numbers satisfy $b^k = b^{n-k}$.*

1.1.2 Exterior forms on Riemannian manifolds

Now let M be a compact, oriented Riemannian n-manifold, with metric g. The metric and the orientation combine to give a *volume form* dV_g on M, which can be used to integrate functions on M. We shall define two sorts of inner product on k-forms. Let α, β be k-forms on M, and define (α, β) by

$$(\alpha, \beta) = \alpha_{a_1 \ldots a_k} \beta_{b_1 \ldots b_k} g^{a_1 b_1} \ldots g^{a_k b_k},$$

in index notation. Then (α, β) is a function on M. We call (α, β) the *pointwise inner product of α, β*. Now for k-forms α, β, define $\langle \alpha, \beta \rangle = \int_M (\alpha, \beta) dV_g$. As M is compact, $\langle \alpha, \beta \rangle$ exists in $\mathbb{R}$ provided α, β are (for instance) continuous. We call $\langle \alpha, \beta \rangle$ the L^2 *inner product of α, β*. (This is because it is the inner product of the Hilbert space $L^2(\Lambda^k T^* M)$, which will be defined in §1.2.)

The *Hodge star* is an isomorphism of vector bundles $* : \Lambda^k T^* M \rightarrow \Lambda^{n-k} T^* M$, which is defined as follows. Let β be a k-form on M. Then $*\beta$ is the unique $(n-k)$-form that satisfies the equation $\alpha \wedge (*\beta) = (\alpha, \beta) dV_g$ for all k-forms α on M. The Hodge star is well-defined, and depends upon g and the orientation of M. It satisfies the identities $*1 = dV_g$ and $*(*\beta) = (-1)^{k(n-k)} \beta$, for β a k-form, so that $*^{-1} = (-1)^{k(n-k)} *$.

Define an operator $d^* : C^\infty(\Lambda^k T^* M) \rightarrow C^\infty(\Lambda^{k-1} T^* M)$ by

$$d^* \beta = (-1)^{kn+n+1} * d(*\beta).$$

Let α be a $(k-1)$-form and β a k-form on M. Then

$$\langle \alpha, d^* \beta \rangle = \int_M (\alpha, d^* \beta) dV_g = \int_M \alpha \wedge (*d^* \beta) = (-1)^k \int_M \alpha \wedge * \beta.$$

But $d(\alpha \wedge * \beta) = (d\alpha) \wedge * \beta + (-1)^{k-1} \alpha \wedge d * \beta$, and as M is compact $\int_M d(\alpha \wedge * \beta) = 0$ by Stokes' Theorem. Therefore

$$(-1)^k \int_M \alpha \wedge d * \beta = \int_M d\alpha \wedge * \beta = \int_M (d\alpha, \beta) dV_g = \langle d\alpha, \beta \rangle.$$

Combining the two equations shows that $\langle \alpha, d^* \beta \rangle = \langle d\alpha, \beta \rangle$. This technique is called *integration by parts*. Thus d^* has the formal properties of the adjoint of d, and is sometimes called the *formal adjoint* of d.

As $d^2 = 0$ we see that $(d^*)^2 = 0$. If a k-form α satisfies $d^* \alpha = 0$, then α is *coclosed*, and if $\alpha = d^* \beta$ for some β then α is *coexact*. The *Laplacian* Δ is $\Delta = dd^* + d^*d$. Then $\Delta : C^\infty(\Lambda^k T^* M) \rightarrow C^\infty(\Lambda^k T^* M)$ is a linear elliptic partial differential operator of order 2. By convention $d^* = 0$ on functions, so $\Delta = d^*d$ on functions.

Several different operators are called Laplacians. When we need to distinguish between them we will refer to this one as the d-*Laplacian*, and write it Δ_d. If α is a k-form and $\Delta \alpha = 0$, then α is called a *harmonic form*.

1.1.3 Hodge theory

Let M be a compact, oriented Riemannian manifold, and define

$$\mathscr{H}^k = \mathrm{Ker}\big(\Delta : C^\infty(\Lambda^k T^* M) \rightarrow C^\infty(\Lambda^k T^* M) \big),$$

so that $\mathscr{H}^k$ is the vector space of harmonic k-forms on M. Suppose $\alpha \in \mathscr{H}^k$. Then $\Delta\alpha = 0$, and thus $\langle \alpha, \Delta\alpha \rangle = 0$. But $\Delta = \mathrm{d}\mathrm{d}^* + \mathrm{d}^*\mathrm{d}$, so

$$0 = \langle \alpha, \mathrm{d}\mathrm{d}^*\alpha \rangle + \langle \alpha, \mathrm{d}^*\mathrm{d}\alpha \rangle = \langle \mathrm{d}^*\alpha, \mathrm{d}^*\alpha \rangle + \langle \mathrm{d}\alpha, \mathrm{d}\alpha \rangle = \|\mathrm{d}^*\alpha\|_{L^2}^2 + \|\mathrm{d}\alpha\|_{L^2}^2,$$

where $\| \,.\, \|_{L^2}$ is the L^2 norm defined in §1.2. Thus $\|\mathrm{d}^*\alpha\|_{L^2} = \|\mathrm{d}\alpha\|_{L^2} = 0$, so that $\mathrm{d}^*\alpha = \mathrm{d}\alpha = 0$. Conversely, if $\mathrm{d}^*\alpha = \mathrm{d}\alpha = 0$ then $\Delta\alpha = (\mathrm{d}\mathrm{d}^* + \mathrm{d}^*\mathrm{d})\alpha = 0$, so that a k-form α lies in $\mathscr{H}^k$ if and only if it is closed and coclosed. Note also that if $\alpha \in \mathscr{H}^k$, then $*\alpha \in \mathscr{H}^{n-k}$.

The next result is proved in [338, Th. 6.8].

Theorem 1.1.3. (The Hodge Decomposition Theorem) *Let M be a compact, oriented Riemannian manifold, and write d_k for d acting on k-forms and d_k^* for d^* acting on k-forms. Then*

$$C^\infty(\Lambda^k T^* M) = \mathscr{H}^k \oplus \mathrm{Im}(\mathrm{d}_{k-1}) \oplus \mathrm{Im}(\mathrm{d}_{k+1}^*).$$

Moreover, $\mathrm{Ker}(\mathrm{d}_k) = \mathscr{H}^k \oplus \mathrm{Im}(\mathrm{d}_{k-1})$ and $\mathrm{Ker}(\mathrm{d}_k^*) = \mathscr{H}^k \oplus \mathrm{Im}(\mathrm{d}_{k+1}^*)$.

Now $H^k_{DR}(M, \mathbb{R}) = \mathrm{Ker}(\mathrm{d}_k)/\mathrm{Im}(\mathrm{d}_{k-1})$, and as $\mathrm{Ker}(\mathrm{d}_k) = \mathscr{H}^k \oplus \mathrm{Im}(\mathrm{d}_{k-1})$ there is a canonical isomorphism between $\mathscr{H}^k$ and $H^k_{DR}(M, \mathbb{R})$. Thus we have:

Theorem 1.1.4. (Hodge's Theorem) *Let M be a compact, oriented Riemannian manifold. Then every de Rham cohomology class on M contains a unique harmonic representative, and $\mathscr{H}^k \cong H^k_{DR}(M, \mathbb{R})$.*

1.2 Introduction to analysis

Let M be a Riemannian manifold with metric g. In problems in analysis it is often useful to consider infinite-dimensional vector spaces of functions on M, and to equip these vector spaces with norms, making them into Banach spaces. In this book we will meet four different types of Banach spaces of this sort, written $L^q(M)$, $L^q_k(M)$, $C^k(M)$ and $C^{k,\alpha}(M)$, and they are defined below.

1.2.1 Lebesgue spaces and Sobolev spaces

Let M be a Riemannian manifold with metric g. For $q \geqslant 1$, define the *Lebesgue space* $L^q(M)$ to be the set of locally integrable functions f on M for which the norm

$$\|f\|_{L^q} = \left(\int_M |f|^q \mathrm{d}V_g \right)^{1/q}$$

is finite. Here $\mathrm{d}V_g$ is the volume form of the metric g. Suppose that $r, s, t \geqslant 1$ and that $1/r = 1/s + 1/t$. If $\phi \in L^s(M)$, $\psi \in L^t(M)$, then $\phi\psi \in L^r(M)$, and $\|\phi\psi\|_{L^r} \leqslant \|\phi\|_{L^s}\|\psi\|_{L^t}$; this is *Hölder's inequality*.

Let $q \geqslant 1$ and let k be a nonnegative integer. Define the *Sobolev space* $L^q_k(M)$ to be the set of $f \in L^q(M)$ such that f is k times weakly differentiable and $|\nabla^j f| \in L^q(M)$ for $j \leqslant k$. Define the *Sobolev norm* on $L^q_k(M)$ to be

$$\|f\|_{L^q_k} = \left(\textstyle\sum_{j=0}^k \int_M |\nabla^j f|^q \mathrm{d}V_g \right)^{1/q}.$$

Then $L^q_k(M)$ is a Banach space with respect to the Sobolev norm. Furthermore, $L^2_k(M)$ is a Hilbert space.

The spaces $L^q(M)$, $L^q_k(M)$ are vector spaces of real functions on M. We generalize this idea to vector spaces of sections of a vector bundle over M. So, let $V \to M$ be a vector bundle on M, equipped with Euclidean metrics on its fibres, and $\hat{\nabla}$ be a connection on V preserving these metrics. Then as above, for $q \geqslant 1$, define the *Lebesgue space* $L^q(V)$ to be the set of locally integrable sections v of V over M for which the norm

$$\|v\|_{L^q} = \left(\int_M |v|^q \mathrm{d}V_g \right)^{1/q}$$

is finite, and the *Sobolev space* $L^q_k(V)$ to be the set of $v \in L^q(V)$ such that v is k times weakly differentiable and $|\hat{\nabla}^j v| \in L^q(M)$ for $j \leqslant k$, with the obvious Sobolev norm.

1.2.2 C^k spaces and Hölder spaces

Let M be a Riemannian manifold with metric g. For each integer $k \geqslant 0$, define $C^k(M)$ to be the space of continuous, bounded functions f on M that have k continuous, bounded derivatives, and define the norm $\|.\|_{C^k}$ on $C^k(M)$ by $\|f\|_{C^k} = \sum_{j=0}^{k} \sup_M |\nabla^j f|$, where ∇ is the Levi-Civita connection.

The fourth class of vector spaces are the Hölder spaces $C^{k,\alpha}(M)$ for $k \geqslant 0$ an integer and $\alpha \in (0,1)$. We begin by defining $C^{0,\alpha}(M)$. Let $d(x,y)$ be the distance between $x, y \in M$ calculated using g, and let $\alpha \in (0,1)$. Then a function f on M is said to be *Hölder continuous with exponent α* if

$$[f]_\alpha = \sup_{x \neq y \in M} \frac{|f(x) - f(y)|}{d(x,y)^\alpha}$$

is finite. Any Hölder continuous function f is continuous. The vector space $C^{0,\alpha}(M)$ is the set of continuous, bounded functions on M which are Hölder continuous with exponent α, and the norm on $C^{0,\alpha}(M)$ is $\|f\|_{C^{0,\alpha}} = \|f\|_{C^0} + [f]_\alpha$.

In the same way, we shall define Hölder norms on spaces of sections v of a vector bundle V over M, equipped with Euclidean metrics in the fibres, and a connection $\hat{\nabla}$ preserving these metrics. Let $\delta(g)$ be the injectivity radius of the metric g on M, which we suppose to be positive, and set

$$[v]_\alpha = \sup_{\substack{x \neq y \in M \\ d(x,y) < \delta(g)}} \frac{|v(x) - v(y)|}{d(x,y)^\alpha}, \tag{1.1}$$

whenever the supremum exists. Now we have a problem interpreting $|v(x) - v(y)|$ in this equation, since $v(x)$ and $v(y)$ lie in different vector spaces. We make sense of it in the following way. When $x \neq y \in M$ and $d(x,y) < \delta(g)$, there is a unique geodesic γ of length $d(x,y)$ joining x and y in M. Parallel translation along γ using $\hat{\nabla}$ identifies the fibres of V over x and y, and the metrics on the fibres. With this understanding, the expression $|v(x) - v(y)|$ is well-defined.

So, define $C^{k,\alpha}(M)$ to be the set of f in $C^k(M)$ for which the supremum $[\nabla^k f]_\alpha$ defined by (1.1) exists, working in the vector bundle $\bigotimes^k T^*M$ with its natural metric and connection. The Hölder norm on $C^{k,\alpha}(M)$ is $\|f\|_{C^{k,\alpha}} = \|f\|_{C^k} + [\nabla^k f]_\alpha$. With this norm, $C^{k,\alpha}(M)$ is a Banach space, called a *Hölder space*.

Hölder continuity is analogous to a sort of fractional differentiability. To see this, observe that if $f \in C^1(M)$ and $x \neq y \in M$ then $|f(x) - f(y)| \leqslant 2\|f\|_{C^0}$, and $|f(x) - f(y)|/d(x,y) \leqslant \|\nabla f\|_{C^0}$ by the Mean Value Theorem. Hence $[f]_\alpha$ exists, and

$$[f]_\alpha \leqslant \left(2\|f\|_{C^0}\right)^{1-\alpha}\|\nabla f\|_{C^0}^\alpha.$$

Thus $[f]_\alpha$ is a sort of interpolation between the C^0 and C^1 norms of f. It can help to think of $C^{k,\alpha}(M)$ as the space of functions on M that are $(k + \alpha)$ times differentiable.

Now suppose that V is a vector bundle on M with Euclidean metrics on its fibres, and ∇^V is a connection on V preserving these metrics. As in the case of Lebesgue and Sobolev spaces, we may generalize the definitions above in an obvious way to give Banach spaces $C^k(V)$ and $C^{k,\alpha}(V)$ of sections of V, and we leave this to the reader.

1.2.3   Embedding theorems

An important tool in problems involving Sobolev spaces is the Sobolev Embedding Theorem, which includes one Sobolev space inside another. Embedding theorems are dealt with at length by Aubin in [16, §2.3–§2.9]. The following comes from [16, Th. 2.30].

Theorem 1.2.1. (Sobolev Embedding Theorem) *Suppose M is a compact Riemannian n-manifold, $k, l \in \mathbb{Z}$ with $k \geqslant l \geqslant 0$, $q, r \in \mathbb{R}$ with $q, r \geqslant 1$, and $\alpha \in (0,1)$. If*

$$\frac{1}{q} \leqslant \frac{1}{r} + \frac{k-l}{n},$$

then $L_k^q(M)$ is continuously embedded in $L_l^r(M)$ by inclusion. If

$$\frac{1}{q} \leqslant \frac{k-l-\alpha}{n},$$

then $L_k^q(M)$ is continuously embedded in $C^{l,\alpha}(M)$ by inclusion.

Next we define the idea of a compact linear map between Banach spaces.

Definition 1.2.2   Let U_1, U_2 be Banach spaces, and let $\psi : U_1 \to U_2$ be a continuous linear map. Let $B_1 = \{u \in U_1 : \|u\|_{U_1} \leqslant 1\}$ be the unit ball in U_1. We call ψ a *compact linear map* if the image $\psi(B_1)$ of B_1 is a precompact subset of U_2, that is, if its closure $\overline{\psi(B_1)}$ is a compact subset of U_2.

It turns out that some of the embeddings of Sobolev and Hölder spaces given in the Sobolev Embedding Theorem are compact linear maps in the above sense. This is called the Kondrakov Theorem, and can be found in [16, Th. 2.34].

Theorem 1.2.3. (The Kondrakov Theorem)   *Suppose M is a compact Riemannian n-manifold, $k, l \in \mathbb{Z}$ with $k \geqslant l \geqslant 0$, $q, r \in \mathbb{R}$ with $q, r \geqslant 1$, and $\alpha \in (0,1)$. If*

$$\frac{1}{q} < \frac{1}{r} + \frac{k-l}{n}$$

then the embedding $L_k^q(M) \hookrightarrow L_l^r(M)$ is compact. If

$$\frac{1}{q} < \frac{k-l-\alpha}{n}$$

then $L_k^q(M) \hookrightarrow C^{l,\alpha}(M)$ is compact. Also $C^{k,\alpha}(M) \hookrightarrow C^k(M)$ is compact.

Finally, we state two related results, the Inverse Mapping Theorem and the Implicit Mapping Theorem for Banach spaces, which can be found in Lang [230, Th. 1.2, p. 128] and [230, Th. 2.1, p. 131].

Theorem 1.2.4. (Inverse Mapping Theorem) *Let X, Y be Banach spaces, and U an open neighbourhood of x in X. Suppose the function $F : U \to Y$ is C^k for some $k \geqslant 1$, with $F(x) = y$, and the first derivative $\mathrm{d}F_x : X \to Y$ of F at x is an isomorphism of X, Y as both vector spaces and topological spaces. Then there are open neighbourhoods $U' \subset U$ of x in X and V' of y in Y, such that $F : U' \to V'$ is a C^k-isomorphism.*

Theorem 1.2.5. (Implicit Mapping Theorem) *Let X, Y and Z be Banach spaces, and U, V open neighbourhoods of 0 in X and Y. Suppose the function $F : U \times V \to Z$ is C^k for some $k \geqslant 1$, with $F(0,0) = 0$, and $\mathrm{d}F_{(0,0)}|_Y : Y \to Z$ is an isomorphism of Y, Z as vector and topological spaces. Then there exists a connected open neighbourhood $U' \subset U$ of 0 in X and a unique C^k map $G : U' \to V$ such that $G(0) = 0$ and $F(x, G(x)) = 0$ for all $x \in U'$.*

1.3 Introduction to elliptic operators

In this section we define *elliptic operators*, which are a special sort of partial differential operator on a manifold. Many of the differential operators that crop up in problems in geometry, applied mathematics and physics are elliptic. For example, consider the equation $\Delta u = f$ on a Riemannian manifold M, where Δ is the Laplacian, and u, f are real functions on M. It turns out that Δ is a linear elliptic operator.

The theory of linear elliptic operators tells us two things about the equation $\Delta u = f$. First, there is a theory about the existence of solutions u to this equation. If f is a given function, there are simple criteria to decide whether or not there exists a function u with $\Delta u = f$. Secondly, there is a theory about the regularity of solutions u, that is, how smooth u is. Roughly speaking, u is as smooth as the problem allows, so that if f is k times differentiable, then u is $k+2$ times differentiable, but this is an oversimplification. These theories of regularity and existence of solutions to elliptic equations will be explained in §1.4 and §1.5.

Here we will define elliptic operators, and give a few examples and basic facts. Although the underlying idea of ellipticity is fairly simple, there are many variations on the theme—elliptic operators can be linear, quasilinear or nonlinear, for instance, and they can operate on functions or on sections of vector bundles, and so on. Some useful references for the material in this section are the books by Gilbarg and Trudinger [126] and Morrey [267], and the appendix in Besse [30].

1.3.1 Partial differential operators on functions

Let M be a manifold, and ∇ a connection on the tangent bundle of M, for instance, the Levi-Civita connection of a Riemannian metric on M. Let u be a smooth function on M. Then the kth derivative of u using ∇ is $\nabla^k f$, or in index notation $\nabla_{a_1} \cdots \nabla_{a_k} u$. We will write $\nabla_{a_1 \ldots a_k} u$ as a shorthand for this k^{th} derivative $\nabla_{a_1} \cdots \nabla_{a_k} u$. Here is the definition of a partial differential operator on functions.

Definition 1.3.1 A *partial differential operator* or *differential operator* P on M of order k is an operator taking real functions u on M to real functions on M, that depends on u and its first k derivatives. Explicitly, if u is a real function on M such that the first k derivatives $\nabla u, \ldots, \nabla^k u$ of u exist (possibly in some weak sense), then $P(u)$ or Pu is a real function on M given by

$$(Pu)(x) = Q\big(x, u(x), \nabla u(x), \ldots, \nabla^k u(x)\big) \tag{1.2}$$

for $x \in M$, where Q is some real function of its arguments.

It is usual to require that this function Q is at least continuous in all its arguments. If Q is a smooth function of its arguments, then P is called a *smooth* differential operator. If Pu is linear in u (that is, $P(\alpha u + \beta v) = \alpha Pu + \beta Pv$ for u, v functions and $\alpha, \beta \in \mathbb{R}$) then P is called a *linear* differential operator. If P is not linear, it is called *nonlinear*.

Here is an example. Let P be a linear differential operator of order 2, and let $(x_1, \ldots, x_n)$ be coordinates on an open set in M. Then we may write

$$(Pu)(x) = \sum_{i,j=1}^{n} a^{ij}(x) \frac{\partial^2 u}{\partial x_i \partial x_j}(x) + \sum_{i=1}^{n} b^i(x) \frac{\partial u}{\partial x_i}(x) + c(x)u(x), \tag{1.3}$$

where for $i, j = 1, \ldots, n$, each of a^{ij}, b^i and c are real functions on this coordinate patch, and $a^{ij} = a^{ji}$. We call a^{ij}, b^i and c the *coefficients* of the operator P, so that, for instance, we say P has Hölder continuous coefficients if each of a^{ij}, b^i and c are Hölder continuous functions. Also, a^{ij} are called the *leading coefficients*, as they are the coefficients of the highest order derivative of u.

Now in §1.2 we defined various vector spaces of functions: $C^k(M)$, $C^\infty(M)$, Hölder spaces and Sobolev spaces. It is often useful to regard a differential operator as a mapping between two of these vector spaces. For instance, if P is a *smooth* differential operator of order k, and $u \in C^\infty(M)$, then $Pu \in C^\infty(M)$, so P maps $C^\infty(M) \to C^\infty(M)$. On the other hand, if $u \in C^{k+l}(M)$ then $Pu \in C^l(M)$, so that P also maps $C^{k+l}(M) \to C^l(M)$.

It is not necessary to assume P is a smooth operator. For instance, let P be a linear differential operator of order k. It is easy to see that if the coefficients of P are bounded, then $P : L^q_{k+l}(M) \to L^q_l(M)$ is a linear map, and if the coefficients of P are at least $C^{l,\alpha}$, then $P : C^{k+l,\alpha}(M) \to C^{l,\alpha}(M)$ is also a linear map, and so on. In this way we can consider an operator P to act on several different vector spaces of functions.

Definition 1.3.2 Let P be a (nonlinear) differential operator of order k, that is defined as in (1.2) by a function Q that is at least C^1 in the arguments $u, \nabla u, \ldots, \nabla^k u$. Let u be a real function with k derivatives. We define the *linearization* $L_u P$ of P at u to be the derivative of $P(v)$ with respect to v at u, that is,

$$L_u Pv = \lim_{\alpha \to 0} \left(\frac{P(u + \alpha v) - P(u)}{\alpha} \right). \tag{1.4}$$

Then $L_u P$ is a *linear* differential operator of order k. If P is linear then $L_u P = P$. Note that even if P is a smooth operator, the linearization $L_u P$ need not be smooth if u

is not smooth. For instance, if P is of order k and $u \in C^{k+l}(M)$, then $L_u P$ will have C^l coefficients in general, as it depends on the kth derivatives of u.

Many properties of a linear differential operator P depend only on the highest order derivatives occurring in P. The *symbol* of P is a convenient way to isolate these highest order terms.

Definition 1.3.3 Let P be a linear differential operator on functions of order k. Then in index notation, we may write

$$Pu = A^{i_1 \cdots i_k} \nabla_{i_1 \dots i_k} u + B^{i_1 \cdots i_{k-1}} \nabla_{i_1 \dots i_{k-1}} u + \cdots + K^{i_1} \nabla_{i_1} u + Lu,$$

where $A, B, \dots, K$ are symmetric tensors and L a real function on M. For each point $x \in M$ and each $\xi \in T_x^* M$, define $\sigma_\xi(P; x) = A^{i_1 \cdots i_k} \xi_{i_1} \xi_{i_2} \dots \xi_{i_k}$. Let $\sigma(P) : T^* M \to \mathbb{R}$ be the function with value $\sigma_\xi(P; x)$ at each $\xi \in T_x^* M$. Then $\sigma(P)$ is called the *symbol* or *principal symbol* of P. It is a homogeneous polynomial of degree k on each cotangent space.

1.3.2 Elliptic operators on functions

Now we can define linear elliptic operators on functions.

Definition 1.3.4 Let P be a linear differential operator of degree k on M. We say P is an *elliptic operator* if for each $x \in M$ and each nonzero $\xi \in T_x^* M$, we have $\sigma_\xi(P; x) \neq 0$, where $\sigma(P)$ is the principal symbol of P.

Thus, $\sigma(P)$ must be nonzero on each $T_x^* M \setminus \{0\}$, that is, on the complement of the zero section in $T^* M$. Suppose $\dim M > 1$. Then $T_x^* M \setminus \{0\}$ is connected, and as $\sigma(P)$ is continuous on $T_x^* M$, either $\sigma_\xi(P; x) > 0$ for all $\xi \in T_x^* M \setminus \{0\}$, or $\sigma_\xi(P; x) < 0$ for all $\xi \in T_x^* M \setminus \{0\}$. However, $\sigma_{-\xi}(P; x) = (-1)^k \sigma_\xi(P; x)$. It follows that if $\dim M > 1$, then the degree k of an elliptic operator P must be *even*. Also, if M is connected and P has continuous leading coefficients, then $\sigma(P)$ is continuous on a connected space, so that either $\sigma(P) > 0$ or $\sigma(P) < 0$ on the whole of the complement of the zero section in $T^* M$.

For example, let P be a linear differential operator of order 2, given in a coordinate system $(x_1, \dots, x_n)$ by (1.3). At each point $x \in M$, the leading coefficients $a^{ij}(x)$ form a real symmetric $n \times n$ matrix. The condition for P to be elliptic is that $a^{ij} \xi_i \xi_j \neq 0$ whenever $\xi \neq 0$, that is, either $a^{ij} \xi_i \xi_j > 0$ for all nonzero ξ or $a^{ij} \xi_i \xi_j < 0$ for all nonzero ξ. This is equivalent to saying that the eigenvalues of the matrix $a^{ij}(x)$ must either all be positive, or all be negative.

The best known example of a linear elliptic operator is the Laplacian on a Riemannian manifold, defined by $\Delta u = -g^{ij} \nabla_{ij} u$. The symbol $\sigma(\Delta)$ is $\sigma_\xi(\Delta; x) = -g^{ij} \xi_i \xi_j = -|\xi|^2$, so that if $\xi \neq 0$ then $\sigma_\xi(\Delta; x) < 0$, and Δ is elliptic. Next we define *nonlinear* elliptic operators.

Definition 1.3.5 Let P be a (nonlinear) differential operator of degree k on M, and let u be a function with k derivatives. We say P is *elliptic at* u if the linearization $L_u P$ of P at u is elliptic. A nonlinear P may be elliptic at some functions u and not at others.

1.3.3 Differential operators on vector bundles

Now let M be a manifold, and let V, W be vector bundles over M. As above, let ∇ be some connection on TM, and let ∇^V be a connection on V. Let v be a section of V. By coupling the connections ∇ and ∇^V, one can form repeated derivatives of v. We will write $\nabla^V_{a_1 a_2 \ldots a_k} v$ for the kth derivative of v defined in this way. Here is the idea of differential operator on vector bundles.

Definition 1.3.6 A *differential operator* P *of order* k *taking sections of* V *to sections of* W is an operator taking sections v of V to sections of W, that depends on v and its first k derivatives. Explicitly, if v is a k times differentiable section of V then Pv is given by

$$\big(Pv\big)(x) = Q\big(x, v(x), \nabla^V_{a_1} v(x), \ldots, \nabla^V_{a_1 \ldots a_k} v(x)\big) \in W_x$$

for $x \in M$. If Q is a smooth function of its arguments, then P is called *smooth*, and if Pv is linear in v then P is called *linear*. If P is not linear, it is *nonlinear*. If P is a (nonlinear) differential operator defined by a function Q that is C^1 in the arguments $v, \nabla^V_{a_1} v, \ldots, \nabla^V_{a_1 \ldots a_k} v$, then we define the *linearization* $L_u P$ at u by (1.4). Although P maps sections of V to sections of W, by an abuse of notation we may also say that P is a differential operator *from V to W*.

This is a natural generalization of differential operators on functions. Since real functions are the same thing as sections of the trivial line bundle over M with fibre $\mathbb{R}$, a differential operator on functions is just the special case when $V = W = \mathbb{R}$. Here are some examples. The operators

$$\mathrm{d} : C^\infty(\Lambda^k T^* M) \to C^\infty(\Lambda^{k+1} T^* M), \quad \mathrm{d}^* : C^\infty(\Lambda^k T^* M) \to C^\infty(\Lambda^{k-1} T^* M),$$
$$\text{and} \quad \Delta : C^\infty(\Lambda^k T^* M) \to C^\infty(\Lambda^k T^* M)$$

introduced in §1.1 are all smooth linear differential operators acting on the vector bundle $\Lambda^k T^* M$, where d, d^* have order 1 and Δ has order 2. A connection ∇^V on a vector bundle V is a smooth linear differential operator of order 1, mapping from V to $V \otimes T^* M$, and so on.

As in the case of differential operators on functions, we can regard differential operators on vector bundles as mapping a vector space of sections of V to a vector space of sections of W. For instance, if P is a smooth, linear differential operator of order k from V to W, then P acts by $P : C^\infty(V) \to C^\infty(W)$, $P : C^{k+l,\alpha}(V) \to C^{l,\alpha}(W)$ and $P : L^q_{k+l}(V) \to L^q_l(W)$.

Let P be a linear differential operator of order k from V to W. Then in index notation, we write

$$Pv = A^{i_1 \ldots i_k} \nabla_{i_1 \ldots i_k} v + B^{i_1 \ldots i_{k-1}} \nabla_{i_1 \ldots i_{k-1}} v + \cdots + K^{i_1} \nabla_{i_1} v + Lv. \tag{1.5}$$

However, here $A^{i_1 \ldots i_k}, B^{i_1 \ldots i_{k_1}}, \ldots$ are not ordinary tensors, but tensors taking values in $V^* \otimes W$. For instance, if ξ_i is a 1-form at $x \in M$, then $A^{i_1 \ldots i_k}(x) \xi_{i_1} \ldots \xi_{i_k}$ is not a real number, but an element of $V^*_x \otimes W_x$, or equivalently, a linear map from V_x to W_x, the fibres of V and W at x.

One can represent this in index notation by writing $A_\beta^{\alpha\, i_1\cdots i_k}$ in place of $A^{i_1\cdots i_k}$, where $i_1,\ldots, i_k$ are indices for TM, α is an index for W, and β is an index for V^*, but we prefer to suppress the indices for V and W. We call $A^{i_1\cdots i_k},\ldots, L$ the *coefficients* of P. Next we define the *symbol* of a linear differential operator on vector bundles.

Definition 1.3.7 Let P be a linear differential operator of order k, mapping sections of V to sections of W, that is given by (1.5) in index notation. For each point $x \in M$ and each $\xi \in T_x^*M$, define $\sigma_\xi(P; x) = A^{i_1\cdots i_k}\xi_{i_1}\xi_{i_2}\ldots\xi_{i_k}$. Then $\sigma_\xi(P; x)$ is a linear map from V_x to W_x. Let $\sigma(P) : T^*M \times V \to W$ be the bundle map defined by $\sigma(P)(\xi, v) = \sigma_\xi(P; x)v \in W_x$ whenever $x \in M, \xi \in T_x^*M$ and $v \in V_x$. Then $\sigma(P)$ is called the *symbol* or *principal symbol* of P, and $\sigma(P)(\xi, v)$ is homogeneous of degree k in ξ and linear in v.

1.3.4   Elliptic operators on vector bundles

Now we define linear elliptic operators on vector bundles.

Definition 1.3.8 Let V, W be vector bundles over a manifold M, and let P be a linear differential operator of degree k from V to W. We say P is an *elliptic operator* if for each $x \in M$ and each nonzero $\xi \in T_x^*M$, the linear map $\sigma_\xi(P; x) : V_x \to W_x$ is *invertible*, where $\sigma(P)$ is the principal symbol of P.

Also, we say that P is an *underdetermined elliptic operator* if for each $x \in M$ and each $0 \neq \xi \in T_x^*M$, the map $\sigma_\xi(P; x) : V_x \to W_x$ is *surjective*, and that P is an *overdetermined elliptic operator* if for each $x \in M$ and each $0 \neq \xi \in T_x^*M$, the map $\sigma_\xi(P; x) : V_x \to W_x$ is *injective*. If P is a (nonlinear) differential operator of degree k from V to W, and v is a section of V with k derivatives, then we say P is *elliptic at* v if the linearization $L_v P$ of P at v is elliptic.

Suppose the vector bundles V, W have fibres $\mathbb{R}^l$ and $\mathbb{R}^m$ respectively. If $x \in M$ then $V_x \cong \mathbb{R}^l$ and $W_x \cong \mathbb{R}^m$, so that $\sigma_\xi(P; x) : \mathbb{R}^l \to \mathbb{R}^m$. Thus, $\sigma_\xi(P; x)$ can only be invertible if $l = m$, it can only be surjective if $l \geqslant m$, and it can only be injective if $l \geqslant m$. So, if P is *elliptic* then $\dim V = \dim W$, if P is *underdetermined elliptic* then $\dim V \geqslant \dim W$, and if P is *overdetermined elliptic* then $\dim V \leqslant \dim W$.

Consider the equation $P(v) = w$ on M. Locally we can think of v as a collection of l real functions, and the equation $P(v) = w$ as being m simultaneous equations on the l functions of v. Now, guided by elementary linear algebra, we expect that a system of m equations in l variables is likely to have many solutions if $l > m$ (underdetermined), one solution if $l = m$, and no solutions at all if $l < m$ (overdetermined). This can help in thinking about differential operators on vector bundles.

Some authors (particularly of older texts) make a distinction between *elliptic equations*, by which they mean elliptic equations in one real function, and *elliptic systems*, by which they mean systems of l real equations in l real functions for $l > 1$, which we deal with using vector bundles. We will not make this distinction, but will refer to both cases as elliptic equations.

Papers about elliptic systems often use a more general concept than we have given, in which the operators can have mixed degree. (See Morrey [267], for instance). It seems to be a general rule that results proved for elliptic equations (in one real function), can also be proved for elliptic systems (in several real functions). However, it can be difficult to locate the proof for elliptic systems in the literature, as many papers deal only with elliptic equations in one real function.

Here are some examples. Let M be a Riemannian manifold of dimension n, and consider the operators d, d^* and Δ on M defined in §1.1. Now $d : C^\infty(\Lambda^0 T^* M) \to C^\infty(\Lambda^1 T^* M)$ is a smooth linear differential operator of order 1. For $x \in M$ and $\xi \in T_x^* M$, the symbol is $\sigma_\xi(d; x)v = v\,\xi$, for $v \in \mathbb{R} = \Lambda^0 T_x^* M$. Thus, if $\xi \neq 0$, $\sigma_\xi(d; x)$ is injective, and d is overdetermined elliptic. But if $n > 1$ then $\sigma_\xi(d; x)$ is not surjective, so d is not elliptic. Similarly, $d^* : C^\infty(\Lambda^1 T^* M) \to C^\infty(\Lambda^0 T^* M)$ is underdetermined elliptic. It can also be shown that the operator

$$d + d^* : C^\infty\big(\bigoplus_{k=0}^n \Lambda^k T^* M\big) \longrightarrow C^\infty\big(\bigoplus_{k=0}^n \Lambda^k T^* M\big)$$

is a smooth linear elliptic operator of order 1, and the Laplacian $\Delta : C^\infty(\Lambda^k T^* M) \to C^\infty(\Lambda^k T^* M)$ on k-forms is smooth, linear and elliptic of order 2 for each k.

1.3.5   Elliptic operators over compact manifolds

Let M be a compact Riemannian manifold. Then from §1.2, $L^2(M)$ is a Banach space of functions on M. In fact, it is a Hilbert space, with the L^2 inner product $\langle u_1, u_2 \rangle = \int_M u_1 u_2 \, dV_g$ for $u_1, u_2 \in L^2(M)$. We can also use this inner product on any vector subspace of $L^2(M)$, such as $C^\infty(M)$. In the same way, if V is a vector bundle over M equipped with Euclidean metrics on its fibres, then $L^2(V)$ is a Hilbert space of sections of V, with inner product $\langle \, , \, \rangle_V$ given by $\langle v_1, v_2 \rangle_V = \int_M (v_1, v_2) dV_g$.

Now suppose that V, W are vector bundles over M, equipped with metrics on the fibres, and let P be a linear differential operator of order k from V to W, with coefficients at least k times differentiable. It turns out that there is a unique linear differential operator P^* of order k from W to V, with continuous coefficients, such that $\langle Pv, w \rangle_W = \langle v, P^* w \rangle_V$ whenever $v \in L_k^2(V)$ and $w \in L_k^2(W)$. This operator P^* is called the *adjoint* or *formal adjoint* of P. We have already met an example of this in §1.1.2, where the adjoint d^* of the exterior derivative d was explicitly constructed.

Here are some properties of adjoint operators. We have $(P^*)^* = P$ for any P. If P is smooth then P^* is smooth. If $V = W$ and $P = P^*$, then P is called *self-adjoint*; the Laplacian Δ on functions or k-forms is an example of a self-adjoint elliptic operator. If P is elliptic then P^* is elliptic, and if P is overdetermined elliptic then P^* is underdetermined elliptic, and vice versa.

One can write down an explicit formula for P^* in terms of the coefficients of P and the metric g. Because of this, adjoint operators are still well-defined when the manifold M is not compact, or has nonempty boundary. However, in these cases the equation $\langle Pv, w \rangle_W = \langle v, P^* w \rangle_V$ may no longer hold, and must be modified by boundary terms.

1.4   Regularity of solutions of elliptic equations

Let M be a compact manifold and V, W vector bundles over M, and suppose P is a smooth linear elliptic operator of order k from V to W. Consider the equation $Pv = w$. Clearly, if $v \in C^{k+l}(V)$ then $w \in C^l(W)$, as w is a function of v and its first k derivatives, all of which are l times differentiable. It is natural to ask whether the converse holds, that is, if $w \in C^l(W)$, is it necessarily true that $v \in C^{k+l}(V)$?

In fact this is false, and an example is given by Morrey [267, p. 54]. However, it is in general true that for $\alpha \in (0, 1)$, if $w \in C^{l,\alpha}(W)$ then $v \in C^{k+l,\alpha}(V)$, and for $p > 1$, if $w \in L_l^p(W)$ then $v \in L_{k+l}^p(V)$. One way to interpret this is that if v is the solution of a linear elliptic equation, then v must be as smooth as the problem allows it to be. This property is called *elliptic regularity*. The main reason that Hölder and Sobolev spaces are used a lot in analysis, instead of the simpler C^k spaces, is that they have this regularity property but the C^k spaces do not.

Let us begin by quoting a rather general elliptic regularity result, taken from [30, Th. 27, Th. 31, p. 463–4]. For a proof, see [267, Th. 6.4.8, p. 251].

Theorem 1.4.1 *Suppose M is a compact Riemannian manifold, V, W are vector bundles over M of the same dimension, and P is a smooth, linear, elliptic differential operator of order k from V to W. Let $\alpha \in (0, 1), p > 1$, and $l \geqslant 0$ be an integer. Suppose that $P(v) = w$ holds weakly, with $v \in L^1(V)$ and $w \in L^1(W)$. If $w \in C^\infty(W)$, then $v \in C^\infty(V)$. If $w \in L_l^p(W)$ then $v \in L_{k+l}^p(V)$, and*

$$\|v\|_{L_{k+l}^p} \leqslant C\big(\|w\|_{L_l^p} + \|v\|_{L^1}\big), \tag{1.6}$$

for some $C > 0$ independent of v, w. If $w \in C^{l,\alpha}(W)$, then $v \in C^{k+l,\alpha}(V)$, and

$$\|v\|_{C^{k+l,\alpha}} \leqslant C\big(\|w\|_{C^{l,\alpha}} + \|v\|_{C^0}\big), \tag{1.7}$$

for some $C > 0$ independent of v, w.

The estimates (1.6) and (1.7) are called the L^p *estimates* and *Schauder estimates* for P respectively. Theorem 1.4.1 is for smooth linear elliptic operators. However, when studying nonlinear problems in analysis, it is often necessary to deal with linear elliptic operators that are not smooth. Here are the Schauder estimates for operators with Hölder continuous coefficients, taken from the same references as the previous result.

Theorem 1.4.2 *Suppose M is a compact Riemannian manifold, V, W are vector bundles over M of the same dimension, and P is a linear, elliptic differential operator of order k from V to W. Let $\alpha \in (0, 1)$ and $l \geqslant 0$ be an integer. Suppose that the coefficients of P are in $C^{l,\alpha}$, and that $P(v) = w$ for some $v \in C^{k,\alpha}(V)$ and $w \in C^{l,\alpha}(W)$. Then $v \in C^{k+l,\alpha}(V)$, and $\|v\|_{C^{k+l,\alpha}} \leqslant C\big(\|w\|_{C^{l,\alpha}} + \|v\|_{C^0}\big)$ for some constant C independent of v, w.*

1.4.1 How elliptic regularity results are proved

We shall now digress briefly to explain how the proofs of results like Theorems 1.4.1 and 1.4.2 work. For simplicity we will confine our attention to linear elliptic operators of order 2 on functions, but the proofs in the more general cases follow similar lines.

First, let $n > 2$ and consider $\mathbb{R}^n$ with coordinates $(x_1, \ldots, x_n)$, with the Euclidean metric $(\mathrm{d}x_1)^2 + \cdots + (\mathrm{d}x_n)^2$. The Laplacian Δ on $\mathbb{R}^n$ is given by

$$\Delta u = \sum_{j=1}^n \frac{\partial^2 u}{(\partial x_j)^2}.$$

Define a function $\Gamma : \mathbb{R}^n \setminus \{0\} \to \mathbb{R}$ by $\Gamma(x) = \frac{1}{(n-2)\Omega_{n-1}} |x|^{2-n}$, where Ω_{n-1} is the volume of the unit sphere $\mathcal{S}^{n-1}$ in $\mathbb{R}^n$. Then $\Delta\Gamma(x) = 0$ for $x \neq 0$ in $\mathbb{R}^n$. Now suppose that $\Delta u = f$, for u, f real functions on $\mathbb{R}^n$. It turns out that if $u(x)$ and $f(x)$ decay sufficiently fast as $x \to \infty$ in $\mathbb{R}^n$, we have

$$u(y) = \int_{x \in \mathbb{R}^n} \Gamma(x - y) f(x) \mathrm{d}x. \tag{1.8}$$

This is called *Green's representation* for u, and can be found in [126, §2.4].

Because (1.8) gives u in terms of f, if we know something about f or its derivatives, we can deduce something about u. For instance, differentiating (1.8) with respect to x_j, we see that

$$\frac{\partial u}{\partial x_j}(y) = -\int_{x \in \mathbb{R}^n} \frac{\partial \Gamma(x - y)}{\partial x_j} f(x) \mathrm{d}x = \int_{x \in \mathbb{R}^n} \Gamma(x - y) \frac{\partial f}{\partial x_j}(x) \mathrm{d}x$$

by integration by parts, provided $\partial f / \partial x_j$ exists, and using this equation one can deduce bounds on ∇u. Working directly from (1.8), one can deduce L^p estimates and Schauder estimates analogous to those in Theorem 1.4.1, for the operator Δ on $\mathbb{R}^n$.

Now Δ is an operator with *constant coefficients*, that is, the coefficients are constant in coordinates. The next stage in the proof is to extend the results to operators P with *variable coefficients*. The idea is to approximate P by an operator P' with constant coefficients in a small open set, and then use results about elliptic operators with constant coefficients proved using the Green's representation. For the approximation of P by P' to be a good approximation, it is necessary that the coefficients of P should not vary too quickly. This can be ensured, for instance, by supposing the coefficients of P to be Hölder continuous with some given bound on their Hölder norm.

As an example, here is a result on Schauder estimates for operators P with Hölder continuous coefficients, part of which will be needed in Chapter 6.

Theorem 1.4.3 *Let B_1, B_2 be the balls of radius $1, 2$ about 0 in $\mathbb{R}^n$. Suppose P is a linear elliptic operator of order 2 on functions on B_2, defined by*

$$Pu(x) = a^{ij}(x) \frac{\partial^2 u}{\partial x_i \partial x_j}(x) + b^i(x) \frac{\partial u}{\partial x_i}(x) + c(x) u(x).$$

Let $\alpha \in (0, 1)$. Suppose the coefficients a^{ij}, b^i and c lie in $C^{0,\alpha}(B_2)$ and there are constants $\lambda, \Lambda > 0$ such that $|a^{ij}(x)\xi_i\xi_j| \geqslant \lambda|\xi|^2$ for all $x \in B_2$ and $\xi \in \mathbb{R}^n$, and $\|a^{ij}\|_{C^{0,\alpha}} \leqslant \Lambda$, $\|b^i\|_{C^{0,\alpha}} \leqslant \Lambda$, and $\|c\|_{C^{0,\alpha}} \leqslant \Lambda$ on B_2 for all $i, j = 1, \dots, n$. Then there exist constants C, D depending on n, α, λ and Λ, such that whenever $u \in C^2(B_2)$ and $f \in C^{0,\alpha}(B_2)$ with $Pu = f$, we have $u|_{B_1} \in C^{2,\alpha}(B_1)$ and

$$\|u|_{B_1}\|_{C^{2,\alpha}} \leqslant C(\|f\|_{C^{0,\alpha}} + \|u\|_{C^0}), \tag{1.9}$$

and whenever $u \in C^2(B_2)$ and f is bounded, then $u|_{B_1} \in C^{1,\alpha}(B_1)$ and

$$\|u|_{B_1}\|_{C^{1,\alpha}} \leqslant D(\|f\|_{C^0} + \|u\|_{C^0}). \tag{1.10}$$

More generally, let $l \geqslant 0$ be an integer and $\alpha \in (0, 1)$. Suppose the coefficients a^{ij}, b^i and c lie in $C^{l,\alpha}(B_2)$ and there are constants $\lambda, \Lambda > 0$ such that $|a^{ij}(x)\xi_i\xi_j| \geqslant \lambda|\xi|^2$ for all $x \in B_2$ and $\xi \in \mathbb{R}^n$, and $\|a^{ij}\|_{C^{l,\alpha}} \leqslant \Lambda$, $\|b^i\|_{C^{l,\alpha}} \leqslant \Lambda$, and $\|c\|_{C^{l,\alpha}} \leqslant \Lambda$

on B_2 for all $i, j = 1, \ldots, n$. Then there exists a constant C depending on n, l, α, λ and Λ such that whenever $u \in C^2(B_2)$ and $f \in C^{l,\alpha}(B_2)$ with $Pu = f$, we have $u|_{B_1} \in C^{l+2,\alpha}(B_1)$ and

$$\left\| u|_{B_1} \right\|_{C^{l+2,\alpha}} \leqslant C \left(\|f\|_{C^{l,\alpha}} + \|u\|_{C^0} \right). \tag{1.11}$$

Here the estimates (1.9) and (1.11) follow from [126, Ths 6.2 and 6.17], and also from [267, Th. 5.6.2]. Estimate (1.10) follows from Morrey [267, Th. 5.5.5′(b)]. In fact, Morrey shows that the norm $\|f\|_{L^{n/(1-\alpha)}}$ rather than $\|f\|_{C^0}$ is sufficient in (1.10).

Theorem 1.4.3 specifies exactly what C and D in eqns (1.9)–(1.11) depend on, and this is worth looking at. The inequality $|a^{ij}(x)\xi_i\xi_j| \geqslant \lambda|\xi|^2$ implies that P is elliptic, by definition, so that the constant $\lambda > 0$ represents a sort of lower bound for the ellipticity of P. The constants C and D also depend on Λ, which is a bound for the coefficients of P in $C^{0,\alpha}$ or $C^{k,\alpha}$. Thus, Λ provides a measure of how close P is to being an operator with constant coefficients.

Notice that although u, f exist on B_2, the theorem only gives estimates of u on B_1, and these estimates depend on data on B_2. A result of this sort is called an *interior estimate*, because it estimates u only on the interior of the domain. Here is one reason why we must prove results of this structure. Consider the equation $\Delta u = 0$ on some domain Ω in $\mathbb{R}^n$. The *maximum principle* [126, §3] says that u cannot have a strict maximum at any point in the interior of Ω, roughly because $\Delta u > 0$ at that point. It follows that the maximum of u on Ω must occur at the *boundary* of Ω.

This illustrates the general principle that if P is a linear elliptic operator and $Pu = f$ on Ω, then u is likely to be most badly behaved, and most difficult to bound, near the boundary of Ω. Because of this, it is easier to prove an interior estimate like Theorem 1.4.3, than to estimate u on the whole of its domain.

Now Theorems 1.4.1 and 1.4.2 deal not with subsets of $\mathbb{R}^n$, but with compact manifolds. The final step in the proof of results like these goes as follows. Let M be a compact manifold. Using the compactness of M, we can find a finite set I and sets $\{X_i : i \in I\}$ and $\{Y_i : i \in I\}$, where each X_i, Y_i is an open set in M, the sets X_i form an *open cover* of M, and for each $i \in I$ we have $X_i \subset Y_i$ and the pair (X_i, Y_i) is diffeomorphic to the pair (B_1, B_2), where B_1, B_2 are the balls of radius 1, 2 in $\mathbb{R}^n$.

Suppose that we know an interior estimate for linear elliptic equations $Pv = w$ on the balls B_1, B_2 in $\mathbb{R}^n$, analogous to Theorem 1.4.3. Since (X_i, Y_i) is diffeomorphic to (B_1, B_2), we may apply this estimate to (X_i, Y_i), and the result is an estimate of $v|_{X_i}$, depending on norms of $v|_{Y_i}$ and $w|_{Y_i}$. Since the sets X_i form an open cover of M, in this way we estimate v on all of M.

Using this argument, we can use interior estimates for balls in $\mathbb{R}^n$ to prove results for compact manifolds M, that estimate the solution on the whole of M. Therefore, results such as Theorems 1.4.1 and 1.4.2 should be understood as purely *local* results, that do not encode any important global information about M and P.

1.5   Existence of solutions of linear elliptic equations

Now we will use the elliptic regularity results of §1.4 and the Kondrakov Theorem to prove some basic facts about linear elliptic operators. Our first result shows that the kernel of a linear elliptic operator on a compact manifold is very well-behaved.

Theorem 1.5.1 *Let V, W be vector bundles over a compact manifold M, and let P be a smooth linear elliptic operator of order k from V to W. Then P acts by $P :$ $C^\infty(V) \to C^\infty(W)$, $P : C^{k+l,\alpha}(V) \to C^{l,\alpha}(W)$ and $P : L^q_{k+l}(V) \to L^q_l(W)$. The kernel $\operatorname{Ker} P$ is the same for all of these actions, and is a finite-dimensional subspace of $C^\infty(V)$.*

Proof If $v \in \operatorname{Ker} P$ then $Pv = 0$. Since $0 \in C^\infty(W)$, Theorem 1.4.1 shows that $v \in C^\infty(V)$. Thus $\operatorname{Ker} P$ lies in $C^\infty(V)$, and is therefore the same for all three actions above. Let $\alpha \in (0, 1)$, and define $B = \{v \in \operatorname{Ker} P : \|v\|_{C^{k,\alpha}} \leqslant 1\}$, so that B is the unit ball in $\operatorname{Ker} P$ in the $C^{k,\alpha}$ norm. The Kondrakov Theorem, Theorem 1.2.3, shows that the inclusion $C^{k,\alpha}(V) \hookrightarrow C^k(V)$ is compact. Therefore B lies in a compact subset of $C^k(V)$, and the closure $\overline{B}$ of B in $C^k(V)$ is compact.

But $P : C^k(V) \to C^0(W)$ is continuous, and $P(b) = 0 \in C^0(W)$ if $b \in B$. Thus $P(b') = 0$ if $b' \in \overline{B}$, so $\overline{B} \subset \operatorname{Ker} P$. Since $\operatorname{Ker} P \subset C^{k,\alpha}(V)$, we see that $B = \overline{B}$, and B is a *compact* topological space. Now the only Banach spaces with compact unit balls are finite-dimensional, so $\operatorname{Ker} P$ is finite-dimensional, as we have to prove. $\square$

Now let M be a compact Riemannian manifold and V, W vector bundles over M equipped with metrics in the fibres. Let P be a smooth linear elliptic operator from V to W. Recall from §1.3.5 that $L^2(V)$ has an inner product $\langle\,,\,\rangle_V$. If A is a vector subspace of $L^2(V)$ and $v \in L^2(V)$, we say that $v \perp A$ if $\langle v, a \rangle_V = 0$ for all $a \in A$. Using this notation, we shall prove:

Proposition 1.5.2 *Let V, W be vector bundles over a compact Riemannian manifold M, equipped with metrics in the fibres, and let P be a smooth linear elliptic operator of order k from V to W. Let $l \geqslant 0$ be an integer, and let $\alpha \in (0, 1)$. Then there is a constant $D > 0$ such that if $v \in C^{k+l,\alpha}(V)$ and $v \perp \operatorname{Ker} P$, then $\|v\|_{C^{k+l,\alpha}} \leqslant D\|Pv\|_{C^{l,\alpha}}$.*

Similarly, if $p > 1$ and $l \geqslant 0$ is an integer, there is a constant $D > 0$ such that if $v \in L^p_{k+l}(V)$ and $v \perp \operatorname{Ker} P$, then $\|v\|_{L^p_{k+l}} \leqslant D\|Pv\|_{L^p_l}$.

Proof For simplicity, we will prove only the case $\|v\|_{C^{k,\alpha}} \leqslant D\|Pv\|_{C^{0,\alpha}}$. The proofs in the other cases work in exactly the same way, and are left to the reader. Define a subset S of $C^{k,\alpha}(V)$ by $S = \{v \in C^{k,\alpha}(V) : v \perp \operatorname{Ker} P \text{ and } \|v\|_{C^{k,\alpha}} = 1\}$, and let $\gamma = \inf\{\|Ps\|_{C^{0,\alpha}} : s \in S\}$. Suppose for a contradiction that $\gamma = 0$. Then we can choose a sequence $\{s_j\}_{j=1}^\infty$ in S such that $\|Ps_j\|_{C^{0,\alpha}} \to 0$ as $j \to \infty$. Now S is bounded in $C^{k,\alpha}(V)$ and the inclusion $C^{k,\alpha}(V) \hookrightarrow C^k(V)$ is compact, by the Kondrakov Theorem. Therefore there exists a subsequence $\{s_{i_j}\}_{j=1}^\infty$ that converges in $C^k(V)$ to some $s' \in C^k(V)$.

As $s_{i_j} \to s'$ in C^k we see that $Ps_{i_j} \to Ps'$ in C^0. But $\|Ps_{i_j}\|_{C^{0,\alpha}} \to 0$, and $\|Ps_{i_j}\|_{C^0} \leqslant \|Ps_{i_j}\|_{C^{0,\alpha}}$. Thus $Ps' = 0$ and $s' \in \operatorname{Ker} P$, so that $s' \in C^{k,\alpha}(V)$. Now by Theorem 1.4.1, there is a constant C such that $\|v\|_{C^{k,\alpha}} \leqslant C(\|Pv\|_{C^{0,\alpha}} + \|v\|_{C^0})$ for all $v \in C^{k,\alpha}(V)$. Therefore

$$\|s_{i_j} - s'\|_{C^{k,\alpha}} \leqslant C(\|Ps_{i_j}\|_{C^{0,\alpha}} + \|s_{i_j} - s'\|_{C^0})$$

for each j, since $Ps' = 0$. But $\|Ps_{i_j}\|_{C^{0,\alpha}} \to 0$ as $j \to \infty$, and $\|s_{i_j} - s'\|_{C^0} \to \infty$ as $j \to \infty$ because s_{i_j} converges to s' in C^k and so in C^0. Thus $\|s_{i_j} - s'\|_{C^{k,\alpha}} \to 0$ as $j \to \infty$. But S is closed in $C^{k,\alpha}(V)$, and therefore $s' \in S$.

As $s' \in S$, we have $s' \perp \operatorname{Ker} P$. But also $s' \in \operatorname{Ker} P$, from above. So $s' = 0$. However, $\|s'\|_{C^{k,\alpha}} = 1$ since $s' \in S$, a contradiction. Therefore $\gamma > 0$. Put $D = \gamma^{-1}$. Then for all $s \in S$ we have $\|s\|_{C^{k,\alpha}} = 1 \leqslant D\|Ps\|_{C^{0,\alpha}}$, by definition of γ. But any $v \in C^{k,\alpha}(V)$ with $v \perp \operatorname{Ker} P$ can be written $v = \lambda s$ for some $s \in S$, and so $\|v\|_{C^{k,\alpha}} \leqslant D\|Pv\|_{C^{0,\alpha}}$, as we have to prove. □

From §1.3.5, if V, W are vector bundles, with metrics on the fibres, over a compact Riemannian manifold M, and P is a smooth linear elliptic operator from V to W, then there is a smooth linear elliptic operator P^* from W to V. Our next result is an *existence* result for the equation $Pv = w$, as it gives a simple condition on w, that $w \perp \operatorname{Ker} P^*$, for there to exist a solution v. This is called the *Fredholm alternative*.

Theorem 1.5.3 *Suppose V, W are vector bundles over a compact Riemannian manifold M, equipped with metrics in the fibres, and P is a smooth linear elliptic operator of order k from V to W. Let $l \geqslant 0$ be an integer, let $p > 1$, and let $\alpha \in (0,1)$. Then the images of the maps*

$$P : C^{k+l,\alpha}(V) \to C^{l,\alpha}(W) \quad \text{and} \quad P : L_{k+l}^p(V) \to L_l^p(W)$$

are closed linear subspaces of $C^{l,\alpha}(W)$ and $L_l^p(W)$ respectively. If $w \in C^{l,\alpha}(W)$ then there exists $v \in C^{k+l,\alpha}(V)$ with $Pv = w$ if and only if $w \perp \operatorname{Ker} P^*$, and if $v \perp \operatorname{Ker} P$ then v is unique. Similarly, if $w \in L_l^p(W)$ then there exists $v \in L_{k+l}^p(V)$ with $Pv = w$ if and only if $w \perp \operatorname{Ker} P^*$, and if $v \perp \operatorname{Ker} P$ then v is unique.

Proof Let $\{w_j\}_{j=1}^\infty$ be a sequence in $P[C^{k+l,\alpha}(V)]$ that converges to some w' in $C^{l,\alpha}(W)$. Then for each w_j there exists a unique $v_j \in C^{k+l,\alpha}(V)$ such that $v_j \perp \operatorname{Ker} P$ and $Pv_j = w_j$. Applying Proposition 1.5.2 we see that for all i, j, $\|v_i - v_j\|_{C^{k+l,\alpha}} \leqslant D\|w_i - w_j\|_{C^{l,\alpha}}$, for D some constant. Since $\{w_j\}_{j=1}^\infty$ converges in $C^{l,\alpha}(W)$, $\|w_i - w_j\|_{C^{l,\alpha}} \to 0$ as $i, j \to \infty$, and therefore $\|v_i - v_j\|_{C^{k+l,\alpha}} \to 0$ as $i, j \to \infty$, and $\{v_j\}_{j=1}^\infty$ is a *Cauchy sequence* in $C^{k+l,\alpha}(V)$.

As $C^{k+l,\alpha}(V)$ is a Banach space and therefore complete, $\{v_j\}_{j=1}^\infty$ converges to some $v' \in C^{k+l,\alpha}(V)$. By continuity, $P(v') = w'$, so that $w' \in P[C^{k+l,\alpha}(V)]$. Therefore $P[C^{k+l,\alpha}(V)]$ contains its limit points, and is a *closed* linear subspace of $C^{l,\alpha}(W)$. Similarly, $P[L_{k+l}^p(V)]$ is closed in $L_l^p(W)$. This proves the first part.

By definition of P^*, if $v \in L_k^2(V)$ and $w \in L_k^2(W)$, then $\langle v, P^*w \rangle_V = \langle Pv, w \rangle_W$. It follows that if $w \in C^{l,\alpha}(W)$, then $w \subset \operatorname{Ker} P^*$ if and only if $\langle Pv, w \rangle_W = 0$ for all $v \in C^{k+l,\alpha}(V)$. So, $\operatorname{Ker} P^*$ is the orthogonal subspace to $P[C^{k+l,\alpha}(V)]$. But $P[C^{k+l,\alpha}(V)]$ is closed. Therefore, if $w \in C^{l,\alpha}(W)$, then $w \perp \operatorname{Ker} P^*$ if and only if $w \in P[C^{k+l,\alpha}(V)]$, that is, if and only if there exists $v \in C^{k+l,\alpha}(V)$ with $Pv = w$.

Clearly, we may add some element of $\operatorname{Ker} P$ to v to make $v \perp \operatorname{Ker} P$, and then v is unique. This proves the second part. The last part follows by the same method. $\square$

From elementary linear algebra, if A, B are finite-dimensional inner product spaces and $L : A \to B$ is a linear map, then $\operatorname{Ker} L$ and $\operatorname{Ker} L^*$ are finite-dimensional vector subspaces of A, B. For given $b \in B$, the equation $La = b$ has a solution $a \in A$ if and only if $b \perp \operatorname{Ker} L^*$, and two solutions differ by an element of $\operatorname{Ker} L$. Now by Theorems 1.5.1 and 1.5.3, these properties also hold for linear elliptic operators $P : C^{k+l,\alpha}(V) \to C^{l,\alpha}(W)$ or $P : L_{k+l}^p(V) \to L_k^p(W)$. Thus, linear elliptic operators behave very like linear operators on *finite-dimensional* vector spaces.

This gives us a way of thinking about linear elliptic operators. In the situation of theorem 1.5.3, define $E = \{v \in C^{k+l,\alpha}(V) : v \perp \operatorname{Ker} P\}$ and $F = \{w \in C^{l,\alpha}(W) : w \perp \operatorname{Ker} P^*\}$. Then $C^{k+l,\alpha}(V) = \operatorname{Ker} P \oplus E$ and $C^{l,\alpha}(W) = \operatorname{Ker} P^* \oplus F$, and the theorem implies that $P : E \to F$ is a *linear homeomorphism*, that is, it is both an invertible linear map and an isomorphism of E and F as topological spaces.

Now $\operatorname{Ker} P, \operatorname{Ker} P^*$ are finite-dimensional, and E, F infinite-dimensional. In some sense, P is close to being an *invertible* map between the infinite-dimensional spaces $C^{k+l,\alpha}(V)$ and $C^{l,\alpha}(W)$, as $P : E \to F$ is invertible, and it is only the finite-dimensional pieces $\operatorname{Ker} P$ and $\operatorname{Ker} P^*$ that cause the problem. Because of this, the existence and uniqueness of solutions of linear elliptic equations can be reduced, more-or-less, to finite-dimensional linear algebra. In contrast, non-elliptic linear differential equations are truly infinite-dimensional problems, and are more difficult to deal with.

Here is another example of the analogy between linear elliptic operators and finite-dimensional linear algebra. If $L : A \to B$ is a linear map of finite-dimensional inner product spaces A, B, then $\dim \operatorname{Ker} L - \dim \operatorname{Ker} L^* = \dim A - \dim B$. Thus the integer $\dim \operatorname{Ker} L - \dim \operatorname{Ker} L^*$ depends only on A and B, and is independent of L. Now let V, W be vector bundles over a compact Riemannian manifold M, with metrics in the fibres, and let P be a smooth linear elliptic operator of order k from V to W. Define the *index* $\operatorname{ind} P$ of P by $\operatorname{ind} P = \dim \operatorname{Ker} P - \dim \operatorname{Ker} P^*$ in $\mathbb{Z}$. The *Atiyah–Singer Index Theorem* [13] gives a formula for $\operatorname{ind} P$ in terms of topological invariants of the symbol $\sigma(P)$. That is, the index of P is actually a topological invariant. It is unchanged by deformations of P that preserve ellipticity, and in this sense is independent of P.

Finally, here is a version of the results of this section for operators with $C^{l,\alpha}$ coefficients. To prove it, follow the proofs above but apply Theorem 1.4.2 instead of Theorem 1.4.1 wherever it occurs. The reason for requiring $l \geqslant k$ is in order that P^* should exist.

Theorem 1.5.4 *Let $k > 0$ and $l \geqslant k$ be integers, and $\alpha \in (0,1)$. Suppose V, W are vector bundles over a compact Riemannian manifold M, equipped with metrics in the fibres, and P is a linear elliptic operator of order k from V to W with $C^{l,\alpha}$ coefficients. Then P^* is elliptic with $C^{l-k,\alpha}$ coefficients, and $\operatorname{Ker} P, \operatorname{Ker} P^*$ are finite-dimensional subspaces of $C^{k+l,\alpha}(V)$ and $C^{l,\alpha}(W)$ respectively. If $w \in C^{l,\alpha}(W)$ then there exists $v \in C^{k+l,\alpha}(V)$ with $Pv = w$ if and only if $w \perp \operatorname{Ker} P^*$, and if one requires that $v \perp \operatorname{Ker} P$ then v is unique.*

2

Introduction to connections, curvature and holonomy groups

In this chapter we will introduce the theory of connections, focussing in particular on two topics, the *curvature* and the *holonomy group* of a connection. Connections can be defined in two different sorts of bundle, that is, *vector bundles* and *principal bundles*. Both definitions will be given in §2.1.

Sections 2.2–2.4 define the holonomy group of a connection on a vector or principal bundle, and explain some of its basic properties, including its relationship with the curvature of the connection. The curvature is a *local* invariant of the connection, since it varies from point to point on the manifold, whereas the holonomy group is a *global* invariant, as it is independent of any base point in the manifold.

Section 2.5 considers connections on the tangent bundle TM of a manifold M, defines the *torsion* of a connection on TM, and discusses the holonomy groups of *torsion-free* connections. Finally, §2.6 defines *G-structures* on a manifold and considers the question of existence and uniqueness of torsion-free connections compatible with a G-structure. For a more detailed introduction to connections and holonomy groups, see Kobayashi and Nomizu [214, Ch. 2, App. 4,5,7].

2.1 Bundles, connections and curvature

We now discuss connections, and their curvature. Connections can be defined in two settings: vector bundles and principal bundles. These two concepts are different, but very closely related. We will define both kinds of connection, and explain the links between them.

2.1.1 Vector bundles and principal bundles

We begin by defining vector bundles and principal bundles.

Definition 2.1.1 Let M be a manifold. A *vector bundle* E over M is a fibre bundle whose fibres are (real or complex) vector spaces. That is, E is a manifold equipped with a smooth projection $\pi : E \rightarrow M$. For each $m \in M$ the fibre $E_m = \pi^{-1}(m)$ has the structure of a vector space, and there is an open neighbourhood U_m of m such that $\pi^{-1}(U_m) \cong U_m \times V$, where V is the fibre of E.

19

Now let M be a manifold, and G a Lie group. A *principal bundle* P over M with fibre G is a manifold P equipped with a smooth projection $\pi : P \to M$, and an action of G on P, which we will write as $p \overset{g}{\longmapsto} g \cdot p$, for $g \in G$ and $p \in P$. This G-action must be smooth and free, and the projection $\pi : P \to M$ must be a fibration, with fibres the orbits of the G-action, so that for each $m \in M$ the fibre $\pi^{-1}(m)$ is a copy of G.

Vector bundles and principal bundles are basic tools in differential geometry. Many geometric structures can be defined using either vector or principal bundles. Thus vector and principal bundles often provide two different but equivalent approaches to the same problem, and it is useful to understand both.

We shall explain the links between vector and principal bundle methods by showing how to translate from one to the other, and back. First, here is a way to go from vector to principal bundles.

Definition 2.1.2 Let M be a manifold, and $E \to M$ a vector bundle with fibre $\mathbb{R}^k$. Define a manifold F^E by

$$F^E = \big\{ (m, e_1, \ldots, e_k) : m \in M \text{ and } (e_1, \ldots, e_k) \text{ is a basis for } E_m \big\}.$$

Define $\pi : F^E \to M$ by $\pi : (m, e_1, \ldots, e_k) \mapsto m$. For each $A = (A_{ij})$ in $\mathrm{GL}(k, \mathbb{R})$ and $(m, e_1, \ldots, e_k)$ in F^E, define $A \cdot (m, e_1, \ldots, e_k) = (m, e_1', \ldots, e_k')$, where $e_i' = \sum_{j=1}^k A_{ij} e_j$. This gives an action of $\mathrm{GL}(k, \mathbb{R})$ on F^E, which makes F^E into a principal bundle over M, with fibre $\mathrm{GL}(k, \mathbb{R})$. We call F^E the *frame bundle* of E.

One frame bundle is of particular importance. When $E = TM$, the bundle F^{TM} will be written F, and called the *frame bundle of M*.

We can also pass from principal bundles to vector bundles.

Definition 2.1.3 Suppose M is a manifold, and P a principal bundle over M with fibre G, a Lie group. Let ρ be a representation of G on a vector space V. Then G acts on the product space $P \times V$ by the principal bundle action on the first factor, and ρ on the second. Define $\rho(P) = (P \times V)/G$, the quotient of $P \times V$ by this G-action. Now $P/G = M$, so the obvious map from $(P \times V)/G$ to P/G yields a projection from $\rho(P)$ to M. Since G acts freely on P, this projection has fibre V, and thus $\rho(P)$ is a *vector bundle* over M, with fibre V.

These two constructions are inverse, in the sense that if ρ is the canonical representation of $\mathrm{GL}(k, \mathbb{R})$ on $\mathbb{R}^k$ then $E \cong \rho(F^E)$. This gives a 1-1 correspondence between vector bundles over M with fibre $\mathbb{R}^k$, and principal bundles over M with fibre $\mathrm{GL}(k, \mathbb{R})$. But any Lie group G can be the fibre of a principal bundle, and not just $G = \mathrm{GL}(k, \mathbb{R})$, so principal bundles are more general than vector bundles.

Let P be a principal bundle over M with fibre G, let $\mathfrak{g}$ be the Lie algebra of G, and let $\mathrm{ad} : G \to \mathrm{GL}(\mathfrak{g})$ be the adjoint representation of G on $\mathfrak{g}$. Definition 2.1.3 gives a natural vector bundle $\mathrm{ad}(P)$ over M, with fibre $\mathfrak{g}$, called the *adjoint bundle*. This will be important later.

Let ρ be a representation of G on V, and $\pi : P \times V \to \rho(P)$ the natural projection. We may regard $P \times V$ as the trivial vector bundle over P with fibre V. Then if $e \in C^\infty\big(\rho(P)\big)$ is a smooth section of $\rho(P)$ over M, the pull-back $\pi^*(e)$ is a smooth section of $P \times V$ over P. Moreover, $\pi^*(e)$ is invariant under the action of G on $P \times V$, and this gives a 1-1 correspondence between sections of $\rho(P)$ over M and G-invariant sections of $P \times V$ over P.

2.1.2 Connections on vector bundles

Here is the definition of a connection on a vector bundle.

Definition 2.1.4 Let M be a manifold, and $E \to M$ a vector bundle. A *connection* ∇^E on E is a linear map $\nabla^E : C^\infty(E) \to C^\infty(E \otimes T^*M)$ satisfying the condition

$$\nabla^E(\alpha\, e) = \alpha \nabla^E e + e \otimes \mathrm{d}\alpha,$$

whenever $e \in C^\infty(E)$ is a smooth section of E and α is a smooth function on M. If ∇^E is such a connection, $e \in C^\infty(E)$, and $v \in C^\infty(TM)$ is a vector field, then we write $\nabla^E_v e = v \cdot \nabla^E e \in C^\infty(E)$, where '$\cdot$' contracts together the TM and T^*M factors in v and $\nabla^E e$. Then if $v \in C^\infty(TM)$ and $e \in C^\infty(E)$ and α, β are smooth functions on M, we have

$$\nabla^E_{\alpha v}(\beta e) = \alpha\beta\nabla^E_v e + \alpha(v \cdot \beta)e. \tag{2.1}$$

Here $v \cdot \beta$ is the Lie derivative of β by v. It is a smooth function on M, and could also be written $v \cdot \mathrm{d}\beta$.

Suppose E is a vector bundle with fibre $\mathbb{R}^k$ over M, and let $e_1, \ldots, e_k$ be smooth sections of E over some open set $U \subset M$, that form a basis of E at each point of U. Then every smooth section of E over U can be written uniquely as $\sum_{i=1}^k \alpha_i e_i$, where $\alpha_1, \ldots, \alpha_k$ are smooth functions on U. Let $f_1, \ldots, f_k$ be any smooth sections of $E \otimes T^*M$ over U, and define

$$\nabla^E\left[\sum_{i=1}^k \alpha_i e_i\right] = \sum_{i=1}^k (\alpha_i f_i + e_i \otimes \mathrm{d}\alpha_i)$$

for all smooth functions $\alpha_1, \ldots, \alpha_k$ on U. Then ∇^E is a connection on E over U, and moreover, every connection on E over U can be written uniquely in this way.

Next we explain how to define the *curvature* of a connection on a vector bundle. Curvature is a very important topic in geometry, and there are a number of ways to define it. The approach we take uses vector fields, and the Lie bracket of vector fields. Let M be a manifold, and E a vector bundle over M. Write $\mathrm{End}(E) = E \otimes E^*$, where E^* is the dual vector bundle to E. Let ∇^E be a connection on E. Then the curvature $R(\nabla^E)$ of the connection ∇^E is a smooth section of the vector bundle $\mathrm{End}(E) \otimes \Lambda^2 T^*M$, defined as follows.

Proposition 2.1.5 *Let M be a manifold, E a vector bundle over M, and ∇^E a connection on E. Suppose that $v, w \in C^\infty(TM)$ are vector fields and $e \in C^\infty(E)$, and that α, β, γ are smooth functions on M. Then*

$$\begin{aligned}\nabla^E_{\alpha v}\nabla^E_{\beta w}(\gamma e) &- \nabla^E_{\beta w}\nabla^E_{\alpha v}(\gamma e) - \nabla^E_{[\alpha v, \beta w]}(\gamma e) \\ &= \alpha\beta\gamma \cdot \left\{\nabla^E_v \nabla^E_w e - \nabla^E_w \nabla^E_v e - \nabla^E_{[v,w]} e\right\},\end{aligned} \tag{2.2}$$

where $[v, w]$ is the Lie bracket. Thus the expression $\nabla^E_v \nabla^E_w e - \nabla^E_w \nabla^E_v e - \nabla^E_{[v,w]} e$ is pointwise-linear in v, w and e. Also, it is clearly antisymmetric in v and w. Therefore *there exists a unique, smooth section $R(\nabla^E) \in C^\infty\big(\mathrm{End}(E) \otimes \Lambda^2 T^* M\big)$ called the curvature of ∇^E, that satisfies the equation*

$$R(\nabla^E) \cdot (e \otimes v \wedge w) = \nabla_v^E \nabla_w^E e - \nabla_w^E \nabla_v^E e - \nabla_{[v,w]}^E e \tag{2.3}$$

for all $v, w \in C^\infty(TM)$ and $e \in C^\infty(E)$.

Proof If $v, w \in C^\infty(TM)$ and α, β are smooth functions on M, then $[\alpha v, \beta w] = \alpha\beta[v, w] + \alpha(v \cdot \beta)w - \beta(w \cdot \alpha)v$. Using this and (2.1) to expand the terms on the left hand side of (2.2), we see that

$$\nabla_{\alpha v}^E \nabla_{\beta w}^E (\gamma e) = \alpha\beta\gamma \nabla_v^E \nabla_w^E e + \alpha\beta(w \cdot \gamma)\nabla_v^E e + \big\{\alpha\beta(v \cdot \gamma) + \alpha(v \cdot \beta)\gamma\big\}\nabla_w^E e$$
$$+ \big\{\alpha(v \cdot \beta)(w \cdot \gamma) + \alpha\beta(v \cdot (w \cdot \gamma))\big\}e,$$

$$\nabla_{\beta w}^E \nabla_{\alpha v}^E (\gamma e) = \alpha\beta\gamma \nabla_w^E \nabla_v^E e + \big\{\alpha\beta(w \cdot \gamma) + (w \cdot \alpha)\beta\gamma\big\}\nabla_v^E e + \alpha\beta(v \cdot \gamma)\nabla_w^E e$$
$$+ \big\{(w \cdot \alpha)\beta(v \cdot \gamma) + \alpha\beta(w \cdot (v \cdot \gamma))\big\}e,$$

$$\nabla_{[\alpha v, \beta w]}^E (\gamma e) = \alpha\beta\gamma \nabla_{[v,w]}^E e - (w \cdot \alpha)\beta\gamma \nabla_v^E e + \alpha(v \cdot \beta)\gamma \nabla_w^E e$$
$$+ \big\{\alpha\beta([v, w] \cdot \gamma) + \alpha(v \cdot \beta)(w \cdot \gamma) - (w \cdot \alpha)\beta(v \cdot \gamma)\big\}e.$$

Combining these equations with the identity $v \cdot (w \cdot \gamma) - w \cdot (v \cdot \gamma) = [v, w] \cdot \gamma$, after some cancellation we prove (2.2), and the proposition follows. $\square$

Here is one way to understand the curvature of ∇^E. Let $(x_1, \ldots, x_n)$ be local coordinates on M, and define $v_i = \partial/\partial x_i$ for $i = 1, \ldots, n$. Then v_i is a vector field on M, and $[v_i, v_j] = 0$. Let e be a smooth section of E. Then we may interpret $\nabla_{v_i}^E e$ as a kind of partial derivative $\partial e/\partial x_i$ of e. Using (or abusing) this partial derivative notation, eqn (2.3) implies that

$$R(\nabla^E) \cdot (e \otimes v_i \wedge v_j) = \frac{\partial^2 e}{\partial x_i \partial x_j} - \frac{\partial^2 e}{\partial x_j \partial x_i}. \tag{2.4}$$

Now, partial derivatives of functions commute, so $\partial^2 f/\partial x_i \partial x_j = \partial^2 f/\partial x_j \partial x_i$ if f is a smooth function on M. However, this does not hold for sections of E, as (2.4) shows that *the curvature $R(\nabla^E)$ measures how much partial derivatives in E fail to commute.*

2.1.3 Connections on principal bundles

Suppose P is a principal bundle over a manifold M, with fibre G and projection $\pi : P \to M$. Let $p \in P$, and set $m = \pi(p)$. Then the derivative of π gives a linear map $\mathrm{d}\pi_p : T_p P \to T_m M$. Define a subspace C_p of $T_p P$ by $C_p = \mathrm{Ker}(\mathrm{d}\pi_p)$. Then the subspaces C_p form a vector subbundle C of the tangent bundle TP, called the *vertical subbundle*. Note that C_p is $T_p(\pi^{-1}(m))$, the tangent space to the fibre of $\pi : P \to M$ over m. But the fibres of π are the orbits of the free G-action on P. It follows that there is a natural isomorphism $C_p \cong \mathfrak{g}$ between C_p and the Lie algebra $\mathfrak{g}$ of G.

Here is the definition of a connection on P.

Definition 2.1.6 Let M be a manifold, and P a principal bundle over M with fibre G, a Lie group. A *connection* on P is a vector subbundle D of TP called the *horizontal* *subbundle*, that is invariant under the G-action on P, and which satisfies $T_pP = C_p \oplus D_p$ for each $p \in P$. If $\pi(p) = m$, then $\mathrm{d}\pi_p$ maps $T_pP = C_p \oplus D_p$ onto T_mM, and as $C_p = \operatorname{Ker} \mathrm{d}\pi_p$, we see that $\mathrm{d}\pi_p$ induces an isomorphism between D_p and T_mM.

Thus the horizontal subbundle D is naturally isomorphic to $\pi^*(TM)$. So if $v \in C^\infty(TM)$ is a vector field on M, there is a unique section $\lambda(v)$ of the bundle $D \subset TP$ over P, such that $\mathrm{d}\pi_p\big(\lambda(v)|_p\big) = v|_{\pi(p)}$ for each $p \in P$. We call $\lambda(v)$ the *horizontal lift* of v. It is a vector field on P, and is invariant under the action of G on P.

We now define the *curvature* of a connection on a principal bundle. Let M be a manifold, P a principal bundle over M with fibre G, a Lie group with Lie algebra $\mathfrak{g}$, and D a connection on P. If $v, w \in C^\infty(TM)$ and α, β are smooth functions on M, then by a similar argument to the proof of (2.2) in Proposition 2.1.5, we can show that

$$\big[\lambda(\alpha v), \lambda(\beta w)\big] - \lambda\big([\alpha v, \beta w]\big) = \alpha\beta \cdot \big\{[\lambda(v), \lambda(w)] - \lambda([v, w])\big\},$$

where $[\,,\,]$ is the Lie bracket of vector fields. Thus the expression $\big[\lambda(v), \lambda(w)\big] - \lambda\big([v, w]\big)$ is pointwise-linear and antisymmetric in v, w. Also, as $\mathrm{d}\pi\big(\lambda(v)\big) = v$ for all vector fields v on M we see that

$$\mathrm{d}\pi\big([\lambda(v), \lambda(w)]\big) = \mathrm{d}\pi\big(\lambda([v, w])\big) = [v, w].$$

Therefore, $\big[\lambda(v), \lambda(w)\big] - \lambda\big([v, w]\big)$ lies in the kernel of $\mathrm{d}\pi$, which is the vertical subbundle C of TP. But there is a natural isomorphism $C_p \cong \mathfrak{g}$ for each $p \in P$, and thus we may regard $\big[\lambda(v), \lambda(w)\big] - \lambda\big([v, w]\big)$ as a section of the trivial vector bundle $P \times \mathfrak{g}$ over P.

As $\lambda(v)$, $\lambda(w)$ and $\lambda([v, w])$ are invariant under the action of G on P, this section of $P \times \mathfrak{g}$ is invariant under the natural action of G on $P \times \mathfrak{g}$. But from above there is a 1-1 correspondence between G-invariant sections of $P \times \mathfrak{g}$ over P, and sections of the adjoint bundle $\operatorname{ad}(P)$ over M. We use this to deduce the following result, which defines the *curvature* $R(P, D)$ of a connection D on P.

Proposition 2.1.7 *Let M be a manifold, G a Lie group with Lie algebra $\mathfrak{g}$, P a principal bundle over M with fibre G, and D a connection on P. Then there exists a unique, smooth section $R(P, D)$ of the vector bundle $\operatorname{ad}(P) \otimes \Lambda^2 T^*M$ called the curvature of D, that satisfies*

$$\pi^*\big(R(P, D) \cdot v \wedge w\big) = \big[\lambda(v), \lambda(w)\big] - \lambda\big([v, w]\big) \qquad (2.5)$$

for all $v, w \in C^\infty(TM)$. Here the left hand side is a $\mathfrak{g}$-valued function on P, the right hand side is a section of the subbundle $C \subset TP$, and the two sides are identified using the natural isomorphism $C_p \cong \mathfrak{g}$ for $p \in P$.

Next we relate connections on vector and principal bundles. Let M, P and G be as above. Let ρ be a representation of G on a vector space V, and define $E \to M$ to be the vector bundle $\rho(P)$ over M. Given a connection D on the principal bundle P, we will explain how to construct a unique connection ∇^E on E. Let $e \in C^\infty(E)$, so that $\pi^*(e)$ is a section of $P \times V$ over P. Then $\pi^*(e)$ is a function $\pi^*(e) : P \to V$, so its exterior derivative is a linear map $d\pi^*(e)|_p : T_pP \to V$ for each $p \in P$. Thus $d\pi^*(e)$ is a smooth section of the vector bundle $V \otimes T^*P$ over P.

Let D be a connection on P. Then for each $p \in P$ there are isomorphisms

$$T_pP \cong C_p \oplus D_p, \qquad C_p \cong \mathfrak{g} \qquad \text{and} \qquad D_p \cong \pi^*(T_{\pi(p)}M).$$

These give a natural splitting $V \otimes T^*P \cong V \otimes \mathfrak{g}^* \oplus V \otimes \pi^*(T^*M)$. Write $\pi_D(d\pi^*(e))$ for the component of $d\pi^*(e)$ in $C^\infty(V \otimes \pi^*(T^*M))$ in this splitting. Now both $\pi^*(e)$ and the vector bundle splitting are G-invariant, so $\pi_D(d\pi^*(e))$ must be G-invariant. But there is a 1-1 correspondence between G-invariant sections of $V \otimes \pi^*(T^*M)$ over P, and sections of the corresponding vector bundle $E \otimes T^*M$ over M. Therefore $\pi_D(d\pi^*(e))$ is the pull-back of a unique element of $C^\infty(E \otimes T^*M)$. We use this to define ∇^E.

Definition 2.1.8 Suppose M is a manifold, P a principal bundle over M with fibre G, and D a connection on P. Let ρ be a representation of G on a vector space V, and define E to be the vector bundle $\rho(P)$ over M. If $e \in C^\infty(E)$, then $\pi_D(d\pi^*(e))$ is a G-invariant section of $V \otimes \pi^*(T^*M)$ over P. Define $\nabla^E e \in C^\infty(E \otimes T^*M)$ to be the unique section of $E \otimes T^*M$ with pull-back $\pi_D(d\pi^*(e))$ under the natural projection $V \otimes \pi^*(T^*M) \to E$. This defines a connection ∇^E on the vector bundle E over M.

To each connection D on a principal bundle P, we have associated a unique connection ∇^E on the vector bundle $E = \rho(P)$. If $G = \mathrm{GL}(k, \mathbb{R})$ and ρ is the standard representation of G on $\mathbb{R}^k$, so that P is the frame bundle F^E of E, then this gives a 1-1 correspondence between connections on P and E. However, for general G and ρ the map $D \mapsto \nabla^E$ may be neither injective nor surjective.

Our final result, which follows quickly from the definitions, relates the ideas of curvature of connections in vector and principal bundles.

Proposition 2.1.9 *Suppose M is a manifold, G a Lie group with Lie algebra $\mathfrak{g}$, P a principal bundle over M with fibre G, and D a connection on P, with curvature $R(P, D)$. Let ρ be a representation of G on a vector space V, E the vector bundle $\rho(P)$ over M, and ∇^E the connection given in Definition 2.1.8, with curvature $R(\nabla^E)$.*

Now $\mathfrak{g}$ and $\mathrm{End}(V)$ are representations of G, and ρ gives a G-equivariant linear map $d\rho : \mathfrak{g} \to \mathrm{End}(V)$. This induces a map $d\rho : \mathrm{ad}(P) \to \mathrm{End}(E)$ of the vector bundles $\mathrm{ad}(P)$ and $\mathrm{End}(E)$ over M corresponding to $\mathfrak{g}$ and $\mathrm{End}(V)$. Let

$$d\rho \otimes \mathrm{id} : \mathrm{ad}(P) \otimes \Lambda^2 T^*M \to \mathrm{End}(E) \otimes \Lambda^2 T^*M$$

be the product with the identity on $\Lambda^2 T^*M$. Then $(d\rho \otimes \mathrm{id})(R(P, D)) = R(\nabla^E)$.

Thus, the definitions of curvature of connections in vector and principal bundles are essentially equivalent.

2.2 Vector bundles, connections and holonomy groups

We now define the holonomy group of a connection on a vector bundle, and prove some elementary facts about it. Let M be a manifold, $E \to M$ a vector bundle over M, and

∇^E a connection on E. Let $\gamma : [0, 1] \rightarrow M$ be a smooth curve in M. Then the pull-back $\gamma^*(E)$ of E to $[0, 1]$ is a vector bundle over $[0, 1]$ with fibre $E_{\gamma(t)}$ over $t \in [0, 1]$, where E_x is the fibre of E over $x \in M$.

Let s be a smooth section of $\gamma^*(E)$ over $[0, 1]$, so that $s(t) \in E_{\gamma(t)}$ for each $t \in [0, 1]$. The connection ∇^E pulls back under γ to give a connection on $\gamma^*(E)$ over $[0, 1]$. We say that s is *parallel* if its derivative under this pulled-back connection is zero, i.e. if $\nabla^E_{\dot{\gamma}(t)} s(t) = 0$ for all $t \in [0, 1]$, where $\dot{\gamma}(t)$ is $\frac{d}{dt}\gamma(t)$, regarded as a vector in $T_{\gamma(t)}M$.

Now this is a first-order ordinary differential equation in $s(t)$, and so for each possible initial value $e \in E_{\gamma(0)}$, there exists a unique, smooth solution s with $s(0) = e$. We shall use this to define the idea of *parallel transport* along γ.

Definition 2.2.1 Let M be a manifold, E a vector bundle over M, and ∇^E a connection on E. Suppose $\gamma : [0, 1] \rightarrow M$ is smooth, with $\gamma(0) = x$ and $\gamma(1) = y$, where $x, y \in M$. Then for each $e \in E_x$, there exists a unique smooth section s of $\gamma^*(E)$ satisfying $\nabla^E_{\dot{\gamma}(t)} s(t) = 0$ for $t \in [0, 1]$, with $s(0) = e$. Define $P_\gamma(e) = s(1)$. Then $P_\gamma : E_x \rightarrow E_y$ is a well-defined linear map, called the *parallel transport map*. This definition easily generalizes to the case when γ is continuous and piecewise-smooth, by requiring s to be continuous, and differentiable whenever γ is differentiable.

Here are some elementary properties of parallel transport. Let M, E and ∇^E be as above, let $x, y, z \in M$, and let α, β be piecewise-smooth paths in M with $\alpha(0) = x$, $\alpha(1) = y = \beta(0)$, and $\beta(1) = z$. Define paths α^{-1} and $\beta\alpha$ by

$$\alpha^{-1}(t) = \alpha(1 - t) \quad \text{and} \quad \beta\alpha(t) = \begin{cases} \alpha(2t) & \text{if } 0 \leqslant t \leqslant \frac{1}{2}, \\ \beta(2t - 1) & \text{if } \frac{1}{2} \leqslant t \leqslant 1. \end{cases}$$

Then α^{-1} and $\beta\alpha$ are piecewise-smooth paths in M with $\alpha^{-1}(0) = y$, $\alpha^{-1}(1) = x$, $\beta\alpha(0) = x$ and $\beta\alpha(1) = z$.

Suppose $e_x \in E_x$, and $P_\alpha(e_x) = e_y \in E_y$. Then there is a unique parallel section s of $\alpha^{-1}(E)$ with $s(0) = e_x$ and $s(1) = e_y$. Define $s'(t) = s(1 - t)$. Then s' is a parallel section of $(\alpha^{-1})^*(E)$. Since $s'(0) = e_y$ and $s'(1) = e_x$, it follows that $P_{\alpha^{-1}}(e_y) = e_x$. Thus, if $P_\alpha(e_x) = e_y$, then $P_{\alpha^{-1}}(e_y) = e_x$, and so P_α and $P_{\alpha^{-1}}$ are *inverse maps*. In particular, this implies that if γ is any piecewise-smooth path in M, then P_γ is *invertible*. By a similar argument, we can also show that $P_{\beta\alpha} = P_\beta \circ P_\alpha$.

Definition 2.2.2 Let M be a manifold, E a vector bundle over M, and ∇^E a connection on E. Fix a point $x \in M$. We say that γ is a *loop based at* x if $\gamma : [0, 1] \rightarrow M$ is a piecewise-smooth path with $\gamma(0) = \gamma(1) = x$. If γ is a loop based at x, then the parallel transport map $P_\gamma : E_x \rightarrow E_x$ is an invertible linear map, so that P_γ lies in $\mathrm{GL}(E_x)$, the group of invertible linear transformations of E_x. Define the *holonomy group* $\mathrm{Hol}_x(\nabla^E)$ of ∇^E based at x to be $\mathrm{Hol}_x(\nabla^E) = \{ P_\gamma : \gamma \text{ is a loop based at } x \} \subset \mathrm{GL}(E_x)$.

If α, β are loops based at x, then α^{-1} and $\beta\alpha$ are too, and from above we have $P_{\alpha^{-1}} = P_\alpha^{-1}$ and $P_{\beta\alpha} = P_\beta \circ P_\alpha$. Thus, if P_α and P_β lie in $\mathrm{Hol}_x(\nabla^E)$, then so do P_α^{-1} and $P_\beta \circ P_\alpha$. This shows $\mathrm{Hol}_x(\nabla^E)$ is closed under inverses and products in $\mathrm{GL}(E_x)$, and therefore $\mathrm{Hol}_x(\nabla^E)$ is a *subgroup* of $\mathrm{GL}(E_x)$, which justifies calling it a group.

Note that in this book we suppose all manifolds to be *connected*. Suppose $x, y \in M$. Since M is connected, we can find a piecewise-smooth path $\gamma : [0, 1] \to M$ with $\gamma(0) = x$ and $\gamma(1) = y$, so that $P_\gamma : E_x \to E_y$. If α is a loop based at x, then $\gamma\alpha\gamma^{-1}$ is a loop based at y, and $P_{\gamma\alpha\gamma^{-1}} = P_\gamma \circ P_\alpha \circ P_\gamma^{-1}$. Hence, if $P_\alpha \in \mathrm{Hol}_x(\nabla^E)$, then $P_\gamma \circ P_\alpha \circ P_\gamma^{-1} \in \mathrm{Hol}_y(\nabla^E)$. Thus

$$P_\gamma \, \mathrm{Hol}_x(\nabla^E) \, P_\gamma^{-1} = \mathrm{Hol}_y(\nabla^E). \tag{2.6}$$

Now this shows that the holonomy group $\mathrm{Hol}_x(\nabla^E)$ is *independent of the base point* x, in the following sense. Suppose E has fibre $\mathbb{R}^k$, say. Then any identification $E_x \cong \mathbb{R}^k$ induces an isomorphism $\mathrm{GL}(E_x) \cong \mathrm{GL}(k, \mathbb{R})$, and so we may regard $\mathrm{Hol}_x(\nabla^E)$ as a subgroup H of $\mathrm{GL}(k, \mathbb{R})$. If we choose a different identification $E_x \cong \mathbb{R}^k$, we instead get the subgroup aHa^{-1} of $\mathrm{GL}(k, \mathbb{R})$, for some $a \in \mathrm{GL}(k, \mathbb{R})$. Thus, the holonomy group is a subgroup of $\mathrm{GL}(k, \mathbb{R})$, defined up to conjugation. Moreover, (2.6) shows that if $x, y \in M$, then $\mathrm{Hol}_x(\nabla^E)$ and $\mathrm{Hol}_y(\nabla^E)$ yield the same subgroup of $\mathrm{GL}(k, \mathbb{R})$, up to conjugation. This proves:

Proposition 2.2.3 *Let M be a manifold, E a vector bundle over M with fibre $\mathbb{R}^k$, and ∇^E a connection on E. For each $x \in M$, the holonomy group $\mathrm{Hol}_x(\nabla^E)$ may be regarded as a subgroup of $\mathrm{GL}(k, \mathbb{R})$ defined up to conjugation in $\mathrm{GL}(k, \mathbb{R})$, and in this sense it is independent of the base point x.*

Because of this we may omit the subscript x and write the holonomy group of ∇^E as $\mathrm{Hol}(\nabla^E) \subset \mathrm{GL}(k, \mathbb{R})$, implicitly supposing that two subgroups of $\mathrm{GL}(k, \mathbb{R})$ are equivalent if they are conjugate in $\mathrm{GL}(k, \mathbb{R})$. In the same way, if E has fibre $\mathbb{C}^k$, then the holonomy group of ∇^E is a subgroup of $\mathrm{GL}(k, \mathbb{C})$, up to conjugation. The proposition shows that the holonomy group is a *global invariant* of the connection. Next we show that if M is simply-connected, then $\mathrm{Hol}(\nabla^E)$ is a *connected Lie group*.

Proposition 2.2.4 *Let M be a simply-connected manifold, E a vector bundle over M with fibre $\mathbb{R}^k$, and ∇^E a connection on E. Then $\mathrm{Hol}(\nabla^E)$ is a connected Lie subgroup of $\mathrm{GL}(k, \mathbb{R})$.*

Proof Choose a base point $x \in M$, and let γ be a loop in M based at x. Since M is simply-connected, the loop γ can be contracted to the constant loop at x, that is, there exists a family $\{\gamma_s : s \in [0, 1]\}$, where $\gamma_s : [0, 1] \to M$ satisfies $\gamma_s(0) = \gamma_s(1) = x$, $\gamma_0(t) = x$ for $t \in [0, 1]$ and $\gamma_1 = \gamma$, and $\gamma_s(t)$ depends continuously on s and t. In fact, as shown in [214, p. 73–75], one can also suppose that γ_s is piecewise-smooth, and depends on s in a piecewise-smooth way.

Therefore $s \mapsto P_{\gamma_s}$ is a piecewise-smooth map from $[0, 1]$ to $\mathrm{Hol}_x(\nabla^E)$. Since γ_0 is the constant loop at x we see that $P_{\gamma_0} = 1$, and $P_{\gamma_1} = P_\gamma$ as $\gamma_1 = \gamma$. Thus, each P_γ in $\mathrm{Hol}(\nabla^E)$ can be joined to the identity by a piecewise-smooth path in $\mathrm{Hol}(\nabla^E)$. Now by a theorem of Yamabe [343], every arcwise-connected subgroup of a Lie group is a connected Lie subgroup. So $\mathrm{Hol}(\nabla^E)$ is a connected Lie subgroup of $\mathrm{GL}(k, \mathbb{R})$. $\square$

When M is not simply-connected, it is convenient to consider the *restricted holonomy group* $\mathrm{Hol}^0(\nabla^E)$, which we now define.

Definition 2.2.5 Let M be a manifold, E a vector bundle over M with fibre $\mathbb{R}^k$, and ∇^E a connection on E. Fix $x \in M$. A loop γ based at x is called *null-homotopic* if it can be deformed to the constant loop at x. Define the *restricted holonomy group* $\mathrm{Hol}^0_x(\nabla^E)$ of ∇^E to be $\mathrm{Hol}^0_x(\nabla^E) = \{ P_\gamma : \gamma$ is a null-homotopic loop based at $x \}$. Then $\mathrm{Hol}^0_x(\nabla^E)$ is a subgroup of $\mathrm{GL}(E_x)$. As above we may regard $\mathrm{Hol}^0_x(\nabla^E)$ as a subgroup of $\mathrm{GL}(k, \mathbb{R})$ defined up to conjugation, and it is then independent of the base point x, and so is written $\mathrm{Hol}^0(\nabla^E) \subseteq \mathrm{GL}(k, \mathbb{R})$.

Here are some important properties of $\mathrm{Hol}^0(\nabla^E)$.

Proposition 2.2.6 *Let M be a manifold, E a vector bundle over M with fibre $\mathbb{R}^k$, and ∇^E a connection on E. Then $\mathrm{Hol}^0(\nabla^E)$ is a connected Lie subgroup of $\mathrm{GL}(k, \mathbb{R})$. It is the connected component of $\mathrm{Hol}(\nabla^E)$ containing the identity, and is a normal subgroup of $\mathrm{Hol}(\nabla^E)$. There is a natural, surjective group homomorphism $\phi : \pi_1(M) \to \mathrm{Hol}(\nabla^E)/\mathrm{Hol}^0(\nabla^E)$. Thus, if M is simply-connected, then $\mathrm{Hol}(\nabla^E) = \mathrm{Hol}^0(\nabla^E)$.*

Proof The argument used in Proposition 2.2.4 shows that the restricted holonomy group $\mathrm{Hol}^0(\nabla^E)$ is a connected Lie subgroup of $\mathrm{GL}(k, \mathbb{R})$. Fix $x \in M$. If α, β are loops based at x and β is null-homotopic, then $\alpha\beta\alpha^{-1}$ is null-homotopic. Thus, if $P_\alpha \in \mathrm{Hol}_x(\nabla^E)$ and $P_\beta \in \mathrm{Hol}^0_x(\nabla^E)$, then $P_{\alpha\beta\alpha^{-1}} = P_\alpha P_\beta P_\alpha^{-1}$ also lies in $\mathrm{Hol}^0_x(\nabla^E)$, and so $\mathrm{Hol}^0_x(\nabla^E)$ is a normal subgroup of $\mathrm{Hol}_x(\nabla^E)$.

The group homomorphism $\phi : \pi_1(M) \to \mathrm{Hol}_x(\nabla^E)/\mathrm{Hol}^0_x(\nabla^E)$ is given by $\phi([\gamma]) = P_\gamma \cdot \mathrm{Hol}^0_x(\nabla^E)$, where γ is a loop based at x and $[\gamma]$ the corresponding element of $\pi_1(M)$. It is easy to verify that ϕ is a surjective group homomorphism. Since $\pi_1(M)$ is countable, the quotient group $\mathrm{Hol}_x(\nabla^E)/\mathrm{Hol}^0_x(\nabla^E)$ is also countable. Therefore, $\mathrm{Hol}^0_x(\nabla^E)$ is the connected component of $\mathrm{Hol}_x(\nabla^E)$ containing the identity. $\square$

Now we can define the *Lie algebra* of $\mathrm{Hol}^0(\nabla^E)$.

Definition 2.2.7 Let M be a manifold, E a vector bundle over M with fibre $\mathbb{R}^k$, and ∇^E a connection on E. Then $\mathrm{Hol}^0(\nabla^E)$ is a Lie subgroup of $\mathrm{GL}(k, \mathbb{R})$, defined up to conjugation. Define the *holonomy algebra* $\mathfrak{hol}(\nabla^E)$ to be the Lie algebra of $\mathrm{Hol}^0(\nabla^E)$. It is a Lie subalgebra of $\mathfrak{gl}(k, \mathbb{R})$, defined up to the adjoint action of $\mathrm{GL}(k, \mathbb{R})$. Similarly, $\mathrm{Hol}^0_x(\nabla^E)$ is a Lie subgroup of $\mathrm{GL}(E_x)$ for all $x \in M$. Define $\mathfrak{hol}_x(\nabla^E)$ to be the Lie algebra of $\mathrm{Hol}^0_x(\nabla^E)$. It is a Lie subalgebra of $\mathrm{End}(E_x)$.

Note that because $\mathrm{Hol}^0(\nabla^E)$ is the identity component of $\mathrm{Hol}(\nabla^E)$, the Lie algebras of $\mathrm{Hol}^0(\nabla^E)$ and $\mathrm{Hol}(\nabla^E)$ coincide. Also, although $\mathrm{Hol}^0(\nabla^E)$ is a Lie subgroup of $\mathrm{GL}(k, \mathbb{R})$, it is not necessarily a *closed* subgroup, and so it may not be a submanifold of $\mathrm{GL}(k, \mathbb{R})$ in the strictest sense. (The inclusion of $\mathbb{R}$ in $T^2 = \mathbb{R}^2/\mathbb{Z}^2$ given by $t \mapsto (t + \mathbb{Z}, t\sqrt{2} + \mathbb{Z})$ for $t \in \mathbb{R}$ gives an example of a non-closed Lie subgroup of a Lie group, and this is the sort of behaviour we have in mind.) Even if $\mathrm{Hol}^0(\nabla^E)$ is closed, the full holonomy group $\mathrm{Hol}(\nabla^E)$ may not be closed in $\mathrm{GL}(k, \mathbb{R})$.

The term 'holonomy group' is in some ways misleading, as it suggests that the holonomy group is defined simply as an abstract Lie group. In fact, if ∇^E is a connection on a vector bundle E, then the holonomy group $\mathrm{Hol}(\nabla^E)$ comes equipped with a natural *representation* on the fibre $\mathbb{R}^k$ of E, or equivalently, $\mathrm{Hol}(\nabla^E)$ is embedded as a subgroup of $\mathrm{GL}(k, \mathbb{R})$. Thus, when we describe the holonomy group of a connection, we must specify not only a Lie group, but also a representation of this group. It is important to remember this. We will refer to the representation of $\mathrm{Hol}(\nabla^E)$ on the fibre of E as the *holonomy representation*.

2.3   Holonomy groups and principal bundles

Next we define holonomy groups of connections in principal bundles.

Definition 2.3.1 Let M be a manifold, P a principal bundle over M with fibre G, and D a connection on P. Let $\gamma : [0,1] \to P$ be a smooth curve in P. Then $\dot{\gamma}(t) \in T_{\gamma(t)}P$ is tangent to $\gamma([0,1])$ for each $t \in [0,1]$. We call γ a *horizontal* curve if its tangent vectors are horizontal, that is, $\dot{\gamma}(t) \in D_{\gamma(t)}$ for each $t \in [0,1]$. Similarly, if $\gamma : [0,1] \to P$ is piecewise-smooth, we say that γ is *horizontal* if $\dot{\gamma}(t) \in D_{\gamma(t)}$ for t in the open, dense subset of $[0,1]$ where $\dot{\gamma}(t)$ is well-defined.

Now, if $\gamma : [0,1] \to M$ is piecewise-smooth with $\gamma(0) = m$, and $p \in P$ with $\pi(p) = m$, then there exists a unique horizontal, piecewise-smooth map $\gamma' : [0,1] \to P$ such that $\gamma'(0) = p$ and $\pi \circ \gamma'$ is equal to γ, as maps $[0,1] \to M$. This follows from existence results for ordinary differential equations, and is analogous to the facts about existence and uniqueness of parallel sections of $\gamma^*(E)$ used in Definition 2.2.1. We call γ' a *horizontal lift* of γ.

Here is the definition of holonomy group.

Definition 2.3.2 Let M be a manifold, P a principal bundle over M with fibre G, and D a connection on P. For $p, q \in P$, write $p \sim q$ if there exists a piecewise-smooth horizontal curve in P joining p to q. Clearly, $\sim$ is an equivalence relation. Fix $p \in P$, and define the *holonomy group* of (P, D) based at p to be $\mathrm{Hol}_p(P, D) = \{g \in G : p \sim g \cdot p\}$. Similarly, define the *restricted holonomy group* $\mathrm{Hol}_p^0(P, D)$ to be the set of $g \in G$ for which there exists a piecewise-smooth, horizontal curve $\gamma : [0,1] \to P$ such that $\gamma(0) = p$, $\gamma(1) = g \cdot p$, and $\pi \circ \gamma$ is null-homotopic in M.

If $g \in G$ and $p, q \in P$ with $p \sim q$, then there is a horizontal curve γ in P joining p and q. Applying g to γ, we see that $g \cdot \gamma$ is a horizontal curve joining $g \cdot p$ and $g \cdot q$. Therefore if $g \in G$ and $p \sim q$, then $g \cdot p \sim g \cdot q$. If $g \in \mathrm{Hol}_p(P, D)$, then $p \sim g \cdot p$. Applying g^{-1} gives that $g^{-1} \cdot p \sim g^{-1} \cdot (g \cdot p) = p$. Thus $p \sim g^{-1} \cdot p$ and $g^{-1} \in \mathrm{Hol}_p(P, D)$, so that $\mathrm{Hol}_p(P, D)$ contains inverses of its elements.

Now suppose that $g, h \in \mathrm{Hol}_p(P, D)$. Applying g to $p \sim h \cdot p$ shows that $g \cdot p \sim (gh) \cdot p$. But $p \sim g \cdot p$, so $p \sim (gh) \cdot p$ as $\sim$ is an equivalence relation, and $gh \in \mathrm{Hol}_p(P, D)$. So $\mathrm{Hol}_p(P, D)$ is closed under products, and therefore it is a *subgroup* of G. A similar argument shows that $\mathrm{Hol}_p^0(P, D)$ is a subgroup of G.

Since $\sim$ is an equivalence relation, it is easy to see that if $p, q \in P$ and $p \sim q$, then $\mathrm{Hol}_p(P, D) = \mathrm{Hol}_q(P, D)$. Also, one can show that for all $g \in G$ and $p \in P$, we have $\mathrm{Hol}_{g \cdot p}(P, D) = g \, \mathrm{Hol}_p(P, D) g^{-1}$. Now if $p, q \in P$, then $\pi(p), \pi(q) \in M$. As M is connected, there exists a piecewise-smooth path γ in M with $\gamma(0) = \pi(p)$ and $\gamma(1) = \pi(q)$. There is a unique horizontal lift γ' of γ with $\gamma'(0) = p$ and $\gamma'(1) = q'$, for some $q' \in P$. As $\pi(q') = \pi(q)$, we see that $q' = g \cdot q$ for some $g \in G$, and as γ' is horizontal we have $p \sim q'$. Thus, whenever $p, q \in P$, there exists $g \in G$ with $q \sim g \cdot p$, and so from above $\operatorname{Hol}_q(P, D) = \operatorname{Hol}_{g \cdot p}(P, D) = g \operatorname{Hol}_p(P, D) g^{-1}$. This proves the following result, the analogue of Proposition 2.2.3.

Proposition 2.3.3 *Let M be a manifold, P a principal bundle over M with fibre G, and D a connection on P. Then the holonomy group $\operatorname{Hol}_p(P, D)$ depends on the base point $p \in P$ only up to conjugation in G. Thus we may regard the holonomy group as an equivalence class of subgroups of G under conjugation, and it is then independent of p and is written $\operatorname{Hol}(P, D)$. Similarly, we regard $\operatorname{Hol}^0_p(P, D)$ as an equivalence class of subgroups of G under conjugation, and write it $\operatorname{Hol}^0(P, D)$.*

By following the proofs of Propositions 2.2.4 and 2.2.6, we can show:

Proposition 2.3.4 *Let M be a manifold, P a principal bundle over M with fibre G, and D a connection on P. Then $\operatorname{Hol}^0(P, D)$ is a connected Lie subgroup of G. It is the connected component of $\operatorname{Hol}(P, D)$ containing the identity, and is normal in $\operatorname{Hol}(P, D)$. There is a natural, surjective homomorphism $\phi : \pi_1(M) \to \operatorname{Hol}(P, D)/\operatorname{Hol}^0(P, D)$. If M is simply-connected, then $\operatorname{Hol}(P, D) = \operatorname{Hol}^0(P, D)$.*

We define the Lie algebra of $\operatorname{Hol}^0(P, D)$.

Definition 2.3.5 Suppose M is a manifold, P a principal bundle over M with fibre G, and D a connection on P. Then $\operatorname{Hol}^0(P, D)$ is a connected Lie subgroup of G, defined up to conjugation. Define the *holonomy algebra* $\mathfrak{hol}(P, D)$ to be the Lie algebra of $\operatorname{Hol}^0(P, D)$. Then $\mathfrak{hol}(P, D)$ is a Lie subalgebra of the Lie algebra $\mathfrak{g}$ of G, and is defined up to the adjoint action of G on $\mathfrak{g}$.

Similarly, $\operatorname{Hol}^0_p(P, D)$ is a Lie subgroup of G for all $p \in P$. Let $\mathfrak{hol}_p(P, D) \subseteq \mathfrak{g}$ be the Lie algebra of $\operatorname{Hol}^0_p(P, D)$. Let $\pi(p) = m \in M$, and define $\mathfrak{hol}_m(P, D) = \pi(\mathfrak{hol}_p(P, D))$, where $\pi : P \times \mathfrak{g} \to \operatorname{ad}(P)$ is as in Definition 2.1.3. Then $\mathfrak{hol}_m(P, D)$ is a vector subspace of $\operatorname{ad}(P)_m$. As $\mathfrak{hol}_{g \cdot p}(P, D) = \operatorname{Ad}(g)\big[\mathfrak{hol}_p(P, D)\big]$ for g in G, we see that $\mathfrak{hol}_m(P, D)$ is independent of the choice of $p \in \pi^{-1}(m)$. Thus $\mathfrak{hol}_m(P, D)$ is well-defined.

Now let M, P, G and D be as above, and fix $p \in P$. Write $H = \operatorname{Hol}_p(P, D)$, and suppose H is a closed Lie subgroup of G. Define $Q = \{q \in P : p \sim q\}$. Clearly, Q is preserved by the action of H on P, and thus H acts freely on Q. Also, π restricts to Q giving a projection $\pi : Q \to M$, and it is easy to see that the fibres of $\pi : Q \to M$ are actually the orbits of H. As H is a closed subgroup of G, it is a Lie group, and one can also show that Q is a submanifold of P, and thus a manifold. If H is not closed in G, then Q is not a submanifold of P in the strict sense.

All this shows that Q is a *principal subbundle* of P, with fibre H. A subbundle of this sort is called a *reduction* of P. Let C' be the vertical subbundle of Q. A point q lies in Q if it can be joined to p by a horizontal curve. Therefore, any horizontal curve starting in Q must remain in Q, and so T_qQ must contain all horizontal vectors at q, giving that $D_q \subset T_qQ$. Now $T_qP = C_q \oplus D_q$, and $D_q \subset T_qQ$, and clearly $C'_q = C_q \cap T_qQ$. But these equations imply that $T_qQ = C'_q \oplus D_q$. Therefore, the restriction D' of the distribution D to Q is in fact a *connection* on Q. Thus we have proved:

Theorem 2.3.6. (Reduction Theorem) *Let M be a manifold, P a principal bundle over M with fibre G, and D a connection on P. Fix $p \in P$, let $H = \text{Hol}_p(P, D)$, and suppose that H is a closed Lie subgroup of G. Define $Q = \{q \in P : p \sim q\}$. Then Q is a principal subbundle of P with fibre H, and the connection D on P restricts to a connection D' on Q. In other words, P reduces to Q, and the connection D on P reduces to D' on Q.*

The hypothesis that H is closed in G here may be dropped, but then Q may not be closed in P. An example of such a subgroup $H \subset G$ was given in the previous section. Using this theorem, we can interpret holonomy groups in the following way. Suppose P is a principal bundle over M, with fibre G, and a connection D. Then $\text{Hol}(P, D)$ is the smallest subgroup $H \subseteq G$, up to conjugation, for which it is possible to find a reduction Q of P with fibre H, such that the connection D reduces to Q.

Finally, we shall compare the holonomy groups of connections in vector bundles and in principal bundles, using the ideas of §2.1. The relation between the two is given by the following proposition, which is easy to prove.

Proposition 2.3.7 *Let M be a manifold, and P a principal bundle over M with fibre G. Suppose $\rho : G \to GL(V)$ is a representation of G on a vector space V, and set $E = \rho(V)$. Let D be a connection on P, and ∇^E the connection on E given in Definition 2.1.8. Then $\text{Hol}(P, D)$ and $\text{Hol}(\nabla^E)$ are subgroups of G and $GL(V)$ respectively, each defined up to conjugation, and $\rho\big(\text{Hol}(P, D)\big) = \text{Hol}(\nabla^E)$.*

Similarly, suppose M is a manifold, E a vector bundle over M with fibre $\mathbb{R}^k$, and F^E the frame bundle of E. Then F^E is a principal bundle with fibre $GL(k, \mathbb{R})$. Let ∇^E be a connection on E, and D^E the corresponding connection on F^E. Then $\text{Hol}(\nabla^E)$ and $\text{Hol}(F^E, D^E)$ are both subgroups of $GL(k, \mathbb{R})$ defined up to conjugation, and $\text{Hol}(\nabla^E) = \text{Hol}(F^E, D^E)$.

Thus the two definitions of holonomy group are essentially equivalent.

2.4 Holonomy groups and curvature

Given a connection on a vector bundle or a principal bundle, there is a fundamental relationship between the holonomy group (or its Lie algebra) and the curvature of the connection. The holonomy algebra both constrains the curvature, and is determined by it. Here are two results showing that the curvature of a connection lies in a vector bundle derived from the holonomy algebra.

Proposition 2.4.1 *Let M be a manifold, E a vector bundle over M, and ∇^E a connection on E. Then for each $m \in M$ the curvature $R(\nabla^E)_m$ of ∇^E at m lies in $\mathfrak{hol}_m(\nabla^E) \otimes \Lambda^2 T_m^* M$, where $\mathfrak{hol}_m(\nabla^E)$ is the vector subspace of $\text{End}(E_m)$ given in Definition 2.2.7.*

Proposition 2.4.2 *Let M be a manifold, P a principal bundle over M with fibre G, and D a connection on P. Then for each $m \in M$ the curvature $R(P, D)_m$ of D at m lies in $\mathfrak{hol}_m(P, D) \otimes \Lambda^2 T_m^* M$, where $\mathfrak{hol}_m(P, D)$ is the vector subspace of $\text{ad}(P)_m$ given in Definition 2.3.5.*

We will only prove the second proposition, as the first follows from it.

Proof of Proposition 2.4.2 It is enough to show that if v, w are vector fields on M then $(R(P, D) \cdot v \wedge w)_m$ lies in $\mathfrak{hol}_m(P, D)$. Choose $p \in M$ with $m = \pi(p)$. Then $(R(P, D) \cdot v \wedge w)_m$ lies in $\mathfrak{hol}_m(P, D)$ if and only if $\pi^*(R(P, D) \cdot v \wedge w)_p$ lies in $\mathfrak{hol}_p(P, D)$.

So by (2.5), we must show that for all $v, w \in C^\infty(TM)$ and $p \in P$, we have

$$\big[\lambda(v), \lambda(w)\big]\big|_p - \lambda\big([v, w]\big)\big|_p \in \mathfrak{hol}_p(P, D). \tag{2.7}$$

Here $[\lambda(v), \lambda(w)]|_p - \lambda([v, w])|_p \in C_p$, which is identified with $\mathfrak{g}$, and $\mathfrak{hol}_p(P, D) \subseteq \mathfrak{g}$. Let Q be the subset $\{q \in P : p \sim q\} \subseteq P$ considered in §2.3. Then by Theorem 2.3.6, Q is a principal subbundle of P with fibre $\mathrm{Hol}_p(P, D)$, and the connection D reduces to Q. This means that at $q \in Q$, we have $D_q \subset T_qQ$.

Consider the restriction of $\lambda(v)$ to Q. Since it is horizontal, it lies in D_q and hence in T_qQ at each $q \in Q$. Thus, $\lambda(v)|_Q$ is a vector field on Q. Similarly, $\lambda(w)|_Q$ and $\lambda([v, w])|_Q$ are vector fields on Q, so $[\lambda(v), \lambda(w)]|_Q$ is a vector field on Q. Since $p \in Q$, we see that $[\lambda(v), \lambda(w)]|_p - \lambda([v, w])|_p \in T_pQ$. But we already know this lies in C_p, so it lies in $C_p \cap T_pQ$. However, $C_p \cap T_pQ$ is identified with $\mathfrak{hol}_p(P, D)$ under the isomorphism $C_p \cong \mathfrak{g}$. This verifies eqn (2.7), and the proof is complete. □

There is a kind of converse to Propositions 2.4.1 and 2.4.2, known as the *Ambrose–Singer Holonomy Theorem* [10], [214, p. 89]. We state it here in two forms, for connections in vector bundles and principal bundles.

Theorem 2.4.3 (a) *Let M be a manifold, E a vector bundle over M, and ∇^E a connection on E. Fix $x \in M$, so that $\mathfrak{hol}_x(\nabla^E)$ is a Lie subalgebra of $\mathrm{End}(E_x)$. Then $\mathfrak{hol}_x(\nabla^E)$ is the vector subspace of $\mathrm{End}(E_x)$ spanned by all elements of $\mathrm{End}(E_x)$ of the form $P_\gamma^{-1}\big[R(\nabla^E)_y \cdot (v \wedge w)\big]P_\gamma$, where $x \in M$ is a point, $\gamma : [0, 1] \to M$ is piecewise smooth with $\gamma(0) = x$ and $\gamma(1) = y$, $P_\gamma : E_x \to E_y$ is the parallel translation map, and $v, w \in T_yM$.*

(b) *Let M be a manifold, P a principal bundle over M with fibre G, and D a connection on P. Fix $p \in P$, and define $Q = \{q \in P : p \sim q\}$, as in §2.3. Then $\mathfrak{hol}_p(P, D)$ is the vector subspace of the Lie algebra $\mathfrak{g}$ of G spanned by the elements of the form $\pi^*(R(P, D) \cdot v \wedge w)_q$ for all $q \in Q$ and $v, w \in C^\infty(TM)$, where π maps $P \times \mathfrak{g}$ to $\mathrm{ad}(P)$.*

This shows that $R(\nabla^E)$ determines $\mathfrak{hol}(\nabla^E)$, and hence $\mathrm{Hol}^0(\nabla^E)$. For instance, if ∇^E is flat, so that $R(\nabla^E) = 0$, then $\mathfrak{hol}(\nabla^E) = 0$, and therefore $\mathrm{Hol}^0(\nabla^E) = \{1\}$. The theorem is used by Kobayashi and Nomizu [214, Th. 8.2, p. 90] to prove the next proposition.

Proposition 2.4.4 *Let M be a manifold and P a principal fibre bundle over M with fibre G. If $\dim M \geqslant 2$ and G is connected, then there exists a connection D on P with $\mathrm{Hol}(P, D) = G$.*

As a corollary we have the following result, which can be seen as a sort of converse to Theorem 2.3.6.

Theorem 2.4.5 *Let M be a manifold, and P a principal bundle over M with fibre G. Suppose $\dim M \geqslant 2$. Then for each connected Lie subgroup $H \subset G$, there exists a* connection D on P with holonomy group $\mathrm{Hol}(P, D) = H$ if and only if P reduces to a principal bundle Q with fibre H.

This shows that the question of which groups can appear as the holonomy group of a connection on a general vector or principal bundle is determined entirely by global topological issues: it comes down to asking when the principal bundle admits a reduction to a subgroup, which can be answered using algebraic topology. Therefore, the question of which groups can be the holonomy groups of a connection on a general bundle is not very interesting from the geometrical point of view. To make the question interesting we must impose additional conditions on the connection, as we will see in the next section.

2.5 Connections on the tangent bundle, and torsion

We now consider connections ∇ on the tangent bundle TM of a manifold M. We shall show that ∇ also acts on the tensors on M, and the *constant tensors* on M are determined by the holonomy group $\mathrm{Hol}(\nabla)$. We also define an invariant $T(\nabla)$ called the *torsion*, and discuss the holonomy groups of torsion-free connections.

Suppose M is a manifold of dimension n, and let F be the *frame bundle* of M, as in Definition 2.1.2. Then TM is a vector bundle over M with fibre $\mathbb{R}^n$, and F a principal bundle over M with fibre $\mathrm{GL}(n, \mathbb{R})$. As in §2.1, there is a 1-1 correspondence between connections ∇ on TM, and connections D on F.

Now §2.1–§2.4 developed the theories of connections in vector bundles and principal bundles in parallel, comparing them, but keeping the two theories distinct. From now on we will not make this distinction. Instead, we will *identify* a connection ∇ on TM with the corresponding connection D on F. We will refer to both as *connections on M*, and we will make use of both vector and principal bundle methods, according to which picture is most helpful.

2.5.1 Holonomy groups and constant tensors

Let M be a manifold of dimension n, and ∇ a connection on M. Then ∇ is identified with a connection D on the frame bundle F of M, a principal bundle with fibre $\mathrm{GL}(n, \mathbb{R})$. Now if ρ is a representation of $\mathrm{GL}(n, \mathbb{R})$ on a vector space V then Definitions 2.1.3 and 2.1.8 define a vector bundle $\rho(F)$ on M and a connection, ∇^ρ say, on $\rho(F)$.

Let ρ be the usual representation of $\mathrm{GL}(n, \mathbb{R})$ on $V = \mathbb{R}^n$. Then $\rho(F)$ is just TM, and $\nabla^\rho = \nabla$. However, starting from V we can construct many other representations of $\mathrm{GL}(n, \mathbb{R})$ by taking duals, tensor products, exterior products and so on. From each of these representations, we get a vector bundle with a connection. For instance, the representation of $\mathrm{GL}(n, \mathbb{R})$ on V^* gives the cotangent bundle T^*M, and the representation $\bigotimes^k V \otimes \bigotimes^l V^*$ yields the bundle $\bigotimes^k TM \otimes \bigotimes^l T^*M$.

In fact, all of the vector bundles of tensors $\bigotimes^k TM \otimes \bigotimes^l T^*M$, and subbundles of these such as the symmetric tensors $S^k TM$ or the exterior forms $\Lambda^l T^*M$, arise through the construction of Definition 2.1.3. Thus, Definition 2.1.8 yields a connection on each of these bundles. This gives:

Lemma 2.5.1 *Let M be a manifold. Then a connection ∇ on TM induces connections on all the vector bundles of tensors on M, such as $\bigotimes^k TM \otimes \bigotimes^l T^*M$. All of these induced connections on tensors will also be written ∇.*

Let M be a manifold, ∇ a connection on M, and S a tensor on M, so that $S \in C^\infty(\bigotimes^k TM \otimes \bigotimes^l T^*M)$ for some k, l. We say that S is a *constant tensor* if $\nabla S = 0$. Our next result shows that the constant tensors on M are determined entirely by the holonomy group $\mathrm{Hol}(\nabla)$.

Proposition 2.5.2 *Let M be a manifold, and ∇ a connection on TM. Fix $x \in M$, and let $H = \mathrm{Hol}_x(\nabla)$. Then H is a subgroup of $\mathrm{GL}(T_xM)$. Let E be the vector bundle $\bigotimes^k TM \otimes \bigotimes^l T^*M$ over M. Then the connection ∇ on TM induces a connection ∇^E on E, and H has a natural representation on the fibre E_x of E at x.*

Suppose $S \in C^\infty(E)$ is a constant tensor, so that $\nabla^E S = 0$. Then $S|_x$ is fixed by the action of H on E_x. Conversely, if $S_x \in E_x$ is fixed by the action of H, then there exists a unique tensor $S \in C^\infty(E)$ such that $\nabla^E S = 0$ and $S|_x = S_x$.

Proof Let $\rho : H \to \mathrm{GL}(E_x)$ be the natural representation. Then Proposition 2.3.7 shows that $\mathrm{Hol}_x(\nabla^E) = \rho(H)$. Let γ be a loop in M based at x, and $P_\gamma \in \mathrm{GL}(E_x)$ the parallel translation map using ∇^E in E. Then $P_\gamma \in \mathrm{Hol}_x(\nabla^E)$, so $P_\gamma \in \rho(H)$, and $P_\gamma = \rho(h)$ for some $h \in H$. Moreover, for every $h \in H$ we have $P_\gamma = \rho(h)$ for some loop γ in M based at x.

Now $\nabla^E S = 0$, and therefore the pull-back $\gamma^*(S)$ is a *parallel* section of $\gamma^*(E)$ over $[0, 1]$. Therefore $P_\gamma\big(S|_{\gamma(0)}\big) = S|_{\gamma(1)}$. But $\gamma(0) = \gamma(1) = x$, so $P_\gamma\big(S|_x\big) = S|_x$. Thus $\rho(h)\big(S|_x\big) = S|_x$ for all $h \in H$, and $S|_x$ is fixed by the action of H on E_x.

For the second part, suppose $S_x \in E_x$ is fixed by $\rho(H)$. We will define $S \in C^\infty(E)$ with the required properties. Let $y \in M$ be any point. As M is connected, there is a piecewise-smooth path $\alpha : [0, 1] \to M$ with $\alpha(0) = x$ and $\alpha(1) = y$. Let α and β be two such paths, and let $P_\alpha, P_\beta : E_x \to E_y$ be the parallel transport maps, so that $P_{\alpha^{-1}\beta} = P_\alpha^{-1} P_\beta$. But $\alpha^{-1}\beta$ is a loop based at x, and thus $P_{\alpha^{-1}\beta} = P_\alpha^{-1} P_\beta = \rho(h)$ for some $h \in H$.

Now $\rho(h)(S_x) = S_x$ by assumption. Hence $P_\alpha^{-1} P_\beta(S_x) = S_x$, giving $P_\alpha(S_x) = P_\beta(S_x)$. Therefore, if $\alpha, \beta : [0, 1] \to M$ are any two piecewise-smooth paths from x to y then $P_\alpha(S_x) = P_\beta(S_x)$, and the element $P_\alpha(S_x) \in E_y$ depends only on y, and not on α. Define a section S of E by $S|_y = P_\alpha(S_x)$, where α is any piecewise-smooth path from x to y. Then S is well-defined. If γ is any path in M then $\gamma^*(S)$ is parallel, and thus S is differentiable with $\nabla^E S = 0$. Also $S|_x = S_x$ by definition, and clearly $S \in C^\infty(E)$, which finishes the proof. $\square$

In the proposition we wrote ∇^E for the connection on E, in order to distinguish it from the connection ∇ on TM. Usually we will not make this distinction, but will write ∇ for the connections on all the tensor bundles of M. As a corollary we have:

Corollary 2.5.3 *Let M be a manifold and ∇ a connection on TM, and fix $x \in M$. Define $G \subset \mathrm{GL}(T_xM)$ to be the subgroup of $\mathrm{GL}(T_xM)$ that fixes $S|_x$ for all constant tensors S on M. Then $\mathrm{Hol}_x(\nabla)$ is a subgroup of G.*

Now, in nearly all of the geometrical situations that interest us—if, for instance, $\mathrm{Hol}_x(\nabla)$ is compact and connected—we actually have $\mathrm{Hol}_x(\nabla) = G$ in this corollary. This is not true in every case, as for instance G is closed in $\mathrm{GL}(T_xM)$ but $\mathrm{Hol}_x(\nabla)$ is not always closed, but it is a good general rule. The point is that $\mathrm{Hol}(\nabla) = G$ if $\mathrm{Hol}(\nabla)$ can be defined as the subgroup of $\mathrm{GL}(n, \mathbb{R})$ fixing a collection of elements in the finite-dimensional representations of $\mathrm{GL}(n, \mathbb{R})$, and this is true for most of the subgroups of $\mathrm{GL}(n, \mathbb{R})$ of any geometrical interest.

Thus, given a manifold M and a connection ∇ on TM, the holonomy group $\mathrm{Hol}(\nabla)$ determines the constant tensors on M, and the constant tensors on M usually determine the holonomy group $\mathrm{Hol}(\nabla)$. Therefore, studying the holonomy of a connection, and studying its constant tensors, come down to the same thing.

2.5.2 The torsion of a connection on M

Let M be a manifold, and ∇ a connection on M. Then the *torsion* $T(\nabla)$ of ∇ is a tensor on M defined in the following proposition. We leave the proof as an easy exercise, as it is similar to that of Proposition 2.1.5.

Proposition 2.5.4 *Let M be a manifold, and ∇ a connection on TM. Suppose $v, w \in C^\infty(TM)$ are vector fields and α, β smooth functions on M. Then*

$$\nabla_{\alpha v}(\beta w) - \nabla_{\beta w}(\alpha v) - [\alpha v, \beta w] = \alpha \beta \cdot \{\nabla_v w - \nabla_w v - [v, w]\},$$

where $[v, w]$ is the Lie bracket. Thus the expression $\nabla_v w - \nabla_w v - [v, w]$ is pointwise-linear in v and w, and it is clearly antisymmetric in v and w. Therefore there exists a unique, smooth section $T(\nabla) \in C^\infty(TM \otimes \Lambda^2 T^*M)$ called the torsion of ∇, that satisfies the equation

$$T(\nabla) \cdot (v \wedge w) = \nabla_v w - \nabla_w v - [v, w] \quad \text{for all } v, w \in C^\infty(TM). \tag{2.8}$$

The torsion $T(\nabla)$ of a connection ∇ is a tensor invariant, similar to the curvature $R(\nabla)$. The definition of $T(\nabla)$ uses ∇ once, but that of $R(\nabla)$ uses ∇ twice. In fact, the torsion is a much simpler invariant than the curvature. Note also that we can only define the torsion of a connection on TM, as the definition makes no sense for an arbitrary vector bundle E over M. A connection ∇ on TM with $T(\nabla) = 0$ is called *torsion-free*, or of *zero torsion*. Torsion-free connections are an important class of connections.

Let M be a manifold and ∇ a connection on M. For simplicity we will write T for the torsion $T(\nabla)$ and R for the curvature $R(\nabla)$ of ∇. Then T and R are tensors on M. Using the index notation, we have

$$T = T^a_{bc} \quad \text{with} \quad T^a_{bc} = -T^a_{cb}, \quad \text{and} \quad R = R^a{}_{bcd} \quad \text{with} \quad R^a{}_{bcd} = -R^a{}_{bdc}.$$

For a *torsion-free* connection ∇, the curvature R and its derivative ∇R have certain extra symmetries, known as the *Bianchi identities*.

Proposition 2.5.5 *Let M be a manifold, and ∇ a torsion-free connection on TM. Then the curvature $R^a{}_{bcd}$ of ∇ satisfies the tensor equations*

$$R^a{}_{bcd} + R^a{}_{cdb} + R^a{}_{dbc} = 0, \quad \text{and} \tag{2.9}$$

$$\nabla_e R^a{}_{bcd} + \nabla_c R^a{}_{bde} + \nabla_d R^a{}_{bec} = 0. \tag{2.10}$$

These are known as the first and second Bianchi identities, respectively.

Proof Equation (2.9) is equivalent to the condition that

$$R \cdot (u \otimes v \wedge w) + R \cdot (v \otimes w \wedge u) + R \cdot (w \otimes u \wedge v) = 0 \tag{2.11}$$

for all vector fields $u, v, w \in C^\infty(TM)$. By definition of R we have

$$R \cdot (u \otimes v \wedge w) = \nabla_v \nabla_w u - \nabla_w \nabla_v u - \nabla_{[v,w]} u. \tag{2.12}$$

But ∇ is torsion-free, so $T = 0$, and therefore

$$T \cdot \big(u \wedge [v, w] \big) = \nabla_u [v, w] - \nabla_{[v,w]} u - [u, [v, w]] = 0.$$

Substituting this into (2.12) gives

$$R \cdot (u \otimes v \wedge w) = \nabla_v \nabla_w u - \nabla_w \nabla_v u - \nabla_u [v, w] + [u, [v, w]].$$

Applying the three cyclic permutations of u, v, w to this equation and adding the results together, we find that the left hand side of (2.11) is equal to

$$\nabla_u \big(\nabla_v w - \nabla_w v - [v, w] \big) + \nabla_v \big(\nabla_w u - \nabla_u w - [w, u] \big)$$
$$+ \nabla_w \big(\nabla_u v - \nabla_v u - [u, v] \big) + [u, [v, w]] + [v, [w, u]] + [w, [u, v]].$$

The first term is $\nabla_u \big(T \cdot (v \wedge w) \big)$, which is zero as $T = 0$, and the two similar terms vanish in the same way. But the remaining three terms sum to zero, by the Jacobi identity for vector fields. Thus (2.11) holds, and this proves (2.9). Equation (2.10) can be proved by similar methods. $\qquad\square$

2.5.3 The holonomy of torsion-free connections

It was explained in §2.4 that Theorem 2.4.5 describes exactly the possible holonomy groups of a connection on a bundle, and so the problem of which groups can be the holonomy groups of a general connection on a bundle, is not a very fruitful one. However, the problem can be made much more interesting by restricting attention to *torsion-free* connections. We state this as the following question, which is one of the main motivating problems in the field of holonomy groups.

Question 1: *What are the possible holonomy groups* $\mathrm{Hol}(\nabla)$ *of torsion-free connections* ∇ *on a given manifold M?*

In general, the problem of determining which holonomy groups are realized by torsion-free connections on a given manifold M, a compact manifold for instance, is very difficult and depends strongly on the topology of M. So, let us consider instead the corresponding *local* problem, that is:

Question 2: *What are the possible holonomy groups* $\mathrm{Hol}(\nabla)$ *of torsion-free connections* ∇ *on an open ball in* $\mathbb{R}^n$ *?*

This question is still rather difficult, but some powerful algebraic techniques can be applied to the problem, and a fairly complete answer to the question is known.

The classification of Lie groups, and representations of Lie groups, is well understood. Therefore one could, in principle, write down a list of all possible connected Lie subgroups of $\mathrm{GL}(n, \mathbb{R})$ up to conjugation, for each n. This list is of course infinite, and rather complicated. The idea is to test every subgroup on this list, to see whether or not it can be a holonomy group.

Naturally, because of the complexity of the classification of Lie groups and representations, this is a huge task, and so one looks for short cuts. There is an algebraic method that excludes many groups H from being holonomy groups. This is to study the space $\mathfrak{R}^H$ of possible curvature tensors for H. Proposition 2.4.1 and the first Bianchi identity restrict $\mathfrak{R}^H$, making it small. But the Ambrose–Singer Holonomy Theorem shows that $\mathfrak{R}^H$ must be large enough to generate the Lie algebra $\mathfrak{h}$ of H. If these requirements are not consistent, then H cannot be a holonomy group.

The list of groups H that pass this test is shorter and more manageable. Berger [27, Th. 3–Th. 5, p. 318–320] published, without proof, a list of these groups which is substantially complete, but with some omissions. Later, a number of holonomy groups of torsion-free connections that did not appear on Berger's list were discovered by Bryant, Chi, Merkulov and Schwachhöfer, who called them *exotic holonomy groups*. A (hopefully) complete classification of holonomy groups of torsion-free connections, with proof, has been published by Merkulov and Schwachhöfer [260], to which the reader is referred for further details and references.

These algebraic methods eventually yield a list of candidates H for possible holonomy groups, but they do not prove that every such H actually occurs as a holonomy group. There is another approach to the classification problem using the machinery of *Cartan–Kähler theory*, which is a way of describing how many solutions there are to a given partial differential equation. The advantage of this approach is that it systematically determines whether each H occurs as a holonomy group or not. Bryant [57] uses Cartan–Kähler theory to give a unified treatment of the classification of holonomy groups. We shall discuss the problem of classification of *Riemannian holonomy groups* at greater length in §3.4.

2.6 G-structures and intrinsic torsion

We will now discuss G-structures on manifolds, and their torsion. The theory of G-structures gives a different way of looking at connections on M and their holonomy groups, and is a useful framework for studying geometrical structures.

Definition 2.6.1 Let M be a manifold of dimension n, and F the frame bundle of M, as in §2.5. Then F is a principal bundle over M with fibre $\mathrm{GL}(n, \mathbb{R})$. Let G be a Lie subgroup of $\mathrm{GL}(n, \mathbb{R})$. Then a *$G$-structure* on M is a principal subbundle P of F, with fibre G.

The G-structures, for the many possible Lie subgroups $G \subseteq \mathrm{GL}(n, \mathbb{R})$, provide a large family of interesting geometrical structures on manifolds. Other geometrical objects such as Riemannian metrics and complex structures can also be interpreted as G-structures, as the following example shows.

Example 2.6.2 Let (M, g) be a Riemannian n-manifold, and F the frame bundle of M. Each point of F is $(x, e_1, \ldots, e_n)$, where $x \in M$ and $(e_1, \ldots, e_n)$ is a basis for $T_x M$. Define P to be the subset of F for which $(e_1, \ldots, e_n)$ is *orthonormal* with respect to g. Then P is a principal subbundle of F with fibre $O(n)$, so P is an $O(n)$-*structure* on M. In fact, this gives a 1-1 correspondence between $O(n)$-structures and Riemannian metrics on M.

Now let M be a manifold of dimension n with frame bundle F, let G be a Lie subgroup of $GL(n, \mathbb{R})$, and P a G-structure on M. Suppose D is a connection on P. Then there is a unique connection D' on F that reduces to D on P. Conversely, a connection D' on F reduces to a connection D on P if and only if for each $p \in P$, the subspace $D'|_p$ of $T_p F$ lies in $T_p P$.

As we explained in §2.5, connections D' on the principal bundle F are equivalent to connections ∇ on the vector bundle TM. We call a connection ∇ on TM *compatible with the G-structure P*, if the corresponding connection on F reduces to P. Thus we see that every connection D on P induces a unique connection ∇ on TM, and conversely, a connection ∇ on TM arises from a connection D on P if and only if ∇ is compatible with P. Our next result shows that if ∇ is a fixed connection on TM, then there is a compatible G-structure P if and only if $\mathrm{Hol}(\nabla) \subseteq G$.

Proposition 2.6.3 *Let M be a connected manifold of dimension n, with frame bundle F, and fix $f \in F$. Let ∇ be a connection on TM. Then for each Lie subgroup $G \subset GL(n, \mathbb{R})$, there exists a G-structure P on M compatible with ∇ and containing f if and only if $\mathrm{Hol}_f(\nabla) \subseteq G \subseteq GL(n, \mathbb{R})$. If P exists then it is unique. More generally, there is a 1-1 correspondence between the set of G-structures on M compatible with ∇, but not necessarily containing f, and the homogeneous space $G \backslash \{a \in GL(n, \mathbb{R}) : a \, \mathrm{Hol}_f(\nabla) a^{-1} \subseteq G\}$.*

Proof The proof is similar to that of Theorem 2.3.6, so we will be brief. If P exists then it contains f and is closed under G, so it contains $g \cdot f$ for each $g \in G$. As P is compatible with ∇, any horizontal curve starting in P remains in P. Thus, if $p \in P$ and $q \in F$ with $p \sim q$ then $q \in P$, where $\sim$ is the equivalence relation defined in §2.3.

Combining these two facts shows that if $p \in F$ and $p \sim g \cdot f$ for any $g \in G$, then $p \in P$. But as M is connected, every $p \in P$ must satisfy $p \sim g \cdot f$ for some $g \in G$. So P must be $\{p \in F : p \sim g \cdot f$ for some $g \in G\}$ if it exists. It is easy to show that this set is a principal bundle over M, with fibre the subgroup of $GL(n, \mathbb{R})$ generated by G and $\mathrm{Hol}_f(\nabla)$. Hence, P exists if and only if $\mathrm{Hol}_f(\nabla) \subseteq G$, and if it exists it is unique.

Now if $a \in GL(n, \mathbb{R})$, then $\mathrm{Hol}_{a \cdot f}(\nabla) = a \, \mathrm{Hol}_f(\nabla) a^{-1}$. Thus from above, there is a unique G-structure P containing $a \cdot f$ if and only if $a \, \mathrm{Hol}_f(\nabla) a^{-1} \subseteq G$. But any G-structure containing $a \cdot f$ also contains $(ga) \cdot f$ for all $g \in G$. So the set of G-structures on M compatible with ∇ is in 1-1 correspondence with the given set. □

This proposition gives a good picture of the set of G-structures compatible with a fixed connection ∇ on TM. So let us turn the problem around, and ask about the set of connections ∇ on TM compatible with a fixed G-structure P on M. We have seen above that these are in 1-1 correspondence with connections on P, of which there are many. So we shall restrict our attention to *torsion-free* connections ∇, and ask the question: given a G-structure P on a manifold M, how many torsion-free connections ∇ are there on TM compatible with P?

If ∇ and ∇' are two connections on P, then the difference $\alpha = \nabla' - \nabla$ is a smooth section of $\mathrm{ad}(P) \otimes T^*M$. But $\mathrm{ad}(P)$ is a vector subbundle of $TM \otimes T^*M$. So α is a tensor, written α^a_{bc} in index notation, and if v, w are vector fields then $(\nabla'_v w - \nabla_v w)^a = \alpha^a_{bc} w^b v^c$. Substituting this into (2.8) we see that

$$T(\nabla')^a_{bc} = T(\nabla)^a_{bc} - \alpha^a_{bc} + \alpha^a_{cb}.$$

Let ∇ be an arbitrary, fixed connection on P. Clearly, there exists a *torsion-free* connection ∇' on P if and only if there is an $\alpha \in C^\infty\big(\mathrm{ad}(P) \otimes T^*M\big)$ with $T(\nabla)^a_{bc} = \alpha^a_{bc} - \alpha^a_{cb}$. Moreover, if some such ∇' does exist, then the set of all torsion-free connections ∇' on P is in 1-1 correspondence with the vector space of $\alpha \in C^\infty\big(\mathrm{ad}(P) \otimes T^*M\big)$ for which $\alpha^a_{bc} = \alpha^a_{cb}$.

Here is an alternative way to explain this.

Definition 2.6.4 Let G be a Lie subgroup of $\mathrm{GL}(n, \mathbb{R})$, and let V be $\mathbb{R}^n$. Then G acts faithfully on V, and $\mathfrak{g} \subset V \otimes V^*$. Define $\sigma : \mathfrak{g} \otimes V^* \to V \otimes \Lambda^2 V^*$ by $\sigma(\alpha^a_{bc}) = \alpha^a_{bc} - \alpha^a_{cb}$, in index notation. Define vector spaces $W_1, \ldots, W_4$ by

$$W_1 = V \otimes \Lambda^2 V^*, \quad W_2 = \mathrm{Im}\,\sigma, \quad W_3 = V \otimes \Lambda^2 V^* / \mathrm{Im}\,\sigma \quad \text{and} \quad W_4 = \mathrm{Ker}\,\sigma,$$

and let $\rho_j : G \to \mathrm{GL}(W_j)$ be the natural representations of G on $W_1, \ldots, W_4$. Now suppose M is a manifold of dimension n, and P a G-structure on M. Then we can associate a vector bundle $\rho(P)$ over M to each representation ρ of G, as in §2.1.1. Thus $\rho_1(P), \ldots, \rho_4(P)$ are vector bundles over M. Clearly, $\rho_2(P)$ is a vector subbundle of $\rho_1(P)$, and the quotient bundle $\rho_1(P)/\rho_2(P)$ is $\rho_3(P)$.

If ∇ is any connection on P, then its torsion $T(\nabla)$ lies in $C^\infty(\rho_1(P))$, and if ∇, ∇' are two connections on P, then $T(\nabla') - T(\nabla)$ lies in the subspace $C^\infty(\rho_2(P))$ of $C^\infty(\rho_1(P))$. Therefore, the projections of $T(\nabla)$ and $T(\nabla')$ to the quotient bundle $\rho_3(P) = \rho_1(P)/\rho_2(P)$ are equal. Define the *intrinsic torsion* $T^i(P)$ of P to be the projection to $\rho_3(P)$ of the torsion $T(\nabla)$ of any connection ∇ on P. Then $T^i(P)$ lies in $C^\infty(\rho_3(P))$, and depends only on the G-structure P and not on the choice of ∇.

We call the G-structure P *torsion-free* if $T^i(P) = 0$. Clearly, there exists a torsion-free connection ∇ on P if and only if P is torsion-free, and so the intrinsic torsion $T^i(P)$ is the obstruction to finding a torsion-free connection on P. Any two torsion-free connections differ by an element of $C^\infty(\rho_4(P))$. Thus, if $T^i(P) = 0$ then the torsion-free connections ∇ on P are in 1-1 correspondence with $C^\infty(\rho_4(P))$. If $\mathrm{Ker}\,\sigma = 0$, this set is a single point, so ∇ is unique.

The proof of the next result is similar to that of Proposition 2.6.3.

Proposition 2.6.5 *Let M be a manifold of dimension n, and G a Lie subgroup of $\mathrm{GL}(n, \mathbb{R})$. Then M admits a torsion-free G-structure P if and only if there exists a torsion-free connection ∇ on TM with $\mathrm{Hol}(\nabla) = H$, for some subgroup H of G.*

This shows that torsion-free G-structures on a manifold M are intimately related to torsion-free connections ∇ on TM with $\mathrm{Hol}(\nabla) = G$. However, torsion-free G-structures are simpler, and often easier to work with, than torsion-free connections with prescribed holonomy. This is because a torsion-free G-structure P is defined by a differential equation $T^i(P) = 0$, whereas the condition $\mathrm{Hol}(\nabla) = G$ involves both differentiation and integration, and is rather more complicated.

A number of familiar geometric structures are in fact torsion-free G-structures in disguise. Here are some examples. We saw in Example 2.6.2 that a Riemannian metric g is equivalent to an $\mathrm{O}(n)$-structure P. But when $G = \mathrm{O}(n)$ in Definition 2.6.4 it turns out that σ is both injective and surjective. Therefore, every $\mathrm{O}(n)$-structure P is torsion-free, and there is a unique torsion-free connection ∇ on P. This is the *Levi-Civita connection*, and will be discussed at greater length in §3.1.1.

Set $n = 2m$, and let G be the subgroup $\mathrm{GL}(m, \mathbb{C}) \subset \mathrm{GL}(2m, \mathbb{R})$. Then an *almost complex structure* J on a manifold M is equivalent to a $\mathrm{GL}(m, \mathbb{C})$-structure on M, and J is a *complex structure* if and only if this $\mathrm{GL}(m, \mathbb{C})$-structure is torsion-free. (Complex structures and almost complex structures will be defined in §5.1.) Thus, a complex structure is equivalent to a torsion-free $\mathrm{GL}(m, \mathbb{C})$-structure.

Note that in this case, because $\mathrm{Ker}\,\sigma$ is nonzero, a complex manifold admits infinitely many torsion-free connections preserving the complex structure. In a similar way, a *symplectic structure* on a manifold M of dimension $2m$ is the same thing as a torsion-free $\mathrm{Sp}(m, \mathbb{R})$-structure, where $\mathrm{Sp}(m, \mathbb{R}) \subset \mathrm{GL}(2m, \mathbb{R})$ is the symplectic group, and a *Kähler structure* on M is the same as a torsion-free $\mathrm{U}(m)$-structure.

3

Riemannian holonomy groups

Let M be a manifold, and g a Riemannian metric on M. Then there is a unique, preferred connection ∇ on TM called the *Levi-Civita connection*, which is torsion-free and satisfies $\nabla g = 0$. The curvature $R(\nabla)$ of the Levi-Civita connection is called the *Riemann curvature*, and its holonomy group $\mathrm{Hol}(\nabla)$ the *Riemannian holonomy group* $\mathrm{Hol}(g)$ of g.

In 1955, Marcel Berger proved that if (M, g) is a Riemannian manifold with M simply-connected and g irreducible and nonsymmetric, then $\mathrm{Hol}(g)$ must be one of $\mathrm{SO}(n)$, $\mathrm{U}(m)$, $\mathrm{SU}(m)$, $\mathrm{Sp}(m)$, $\mathrm{Sp}(m)\,\mathrm{Sp}(1)$, G_2 or $\mathrm{Spin}(7)$. The goal of §3.1–§3.4 is to explain what this result means and how it is proved. We start with the Levi-Civita connection, Riemann curvature, and Riemannian holonomy groups. After sections on reducible Riemannian manifolds and symmetric spaces we move onto Berger's classification, describing the proof and the groups on Berger's list. Sections 3.5 and 3.6 explore the relationship between the holonomy group $\mathrm{Hol}(g)$ and the topology of the underlying manifold M, in particular when M is compact.

For more information on the material of §3.1–§3.4, see Kobayashi and Nomizu [214, §III, §IV]. The treatments by Besse [30, §10] and Salamon [294, §2, §10] are also helpful, and Bryant [57] approaches the classification of holonomy groups from a different point of view, that of Cartan–Kähler theory.

3.1 Introduction to Riemannian holonomy groups

We define the Levi-Civita connection ∇ and Riemann curvature tensor R of a Riemannian metric g, and prove some symmetries of R and ∇R. Then we discuss the elementary properties of Riemannian holonomy groups and their relation to torsion-free G-structures.

3.1.1 The Levi-Civita connection

Each Riemannian manifold (M, g) has a natural *Levi-Civita connection* ∇ on TM, which is torsion-free with $\nabla g = 0$. This result is called the *Fundamental Theorem of Riemannian Geometry*, and is very important. Here is a proof.

Theorem 3.1.1 *Let M be a manifold and g a Riemannian metric on M. Then there exists a unique, torsion-free connection ∇ on TM with $\nabla g = 0$, called the Levi-Civita connection.*

Proof Suppose first that ∇ is a torsion-free connection on TM with $\nabla g = 0$. Let $u, v, w \in C^\infty(TM)$ be vector fields on M. Then $g(v, w)$ is a smooth function on M, and so u acts on $g(v, w)$ to give another smooth function $u \cdot g(v, w)$ on M. Since $\nabla g = 0$, using the properties of connections we find that

$$u \cdot g(v, w) = g(\nabla_u v, w) + g(v, \nabla_u w).$$

Combining this with similar expressions for $v \cdot g(u, w)$ and $w \cdot g(u, v)$ gives

$$
\begin{aligned}
u \cdot g&(v, w) + v \cdot g(u, w) - w \cdot g(u, v) \\
&= g(\nabla_u v, w) + g(v, \nabla_u w) + g(\nabla_v u, w) + g(u, \nabla_v w) \\
&\quad - g(\nabla_w u, v) - g(u, \nabla_w v) \\
&= g(\nabla_u v + \nabla_v u, w) + g(\nabla_v w - \nabla_w v, u) + g(\nabla_u w - \nabla_w u, v) \\
&= g(2\nabla_u v - [u, v], w) + g([v, w], u) + g([u, w], v).
\end{aligned}
$$

Here we have used $\nabla_u v - \nabla_v u = [u, v]$, and two similar equations, which hold because ∇ is torsion-free. Rearranging this equation shows that

$$
\begin{aligned}
2g(\nabla_u v, w) = \ &u \cdot g(v, w) + v \cdot g(u, w) - w \cdot g(u, v) \\
&+ g([u, v], w) - g([v, w], u) - g([u, w], v).
\end{aligned}
\tag{3.1}
$$

It is easy to show that for fixed u, v, there is a unique vector field $\nabla_u v$ which satisfies (3.1) for all $w \in C^\infty(TM)$. This defines ∇ uniquely, and it turns out that ∇ is indeed a torsion-free connection with $\nabla g = 0$. $\qquad\square$

In §2.5 we saw that a connection on the tangent bundle TM of a manifold M induces connections on vector bundles of tensors on M. Thus, the Levi-Civita connection ∇ of a Riemannian metric g on M induces connections on all the tensors on M. These connections will also be written ∇.

3.1.2   The Riemann curvature

Suppose M is a Riemannian manifold, with metric g and Levi-Civita connection ∇. Then the curvature $R(\nabla)$ of ∇ is a tensor $R^a{}_{bcd}$ on M. Define $R_{abcd} = g_{ae} R^e{}_{bcd}$. We shall refer to both $R^a{}_{bcd}$ and R_{abcd} as the *Riemann curvature* of g. The following theorem gives a number of symmetries of R_{abcd}. Equations (3.3) and (3.4) are known as the *first* and *second Bianchi identities*, respectively.

Theorem 3.1.2 *Let (M, g) be a Riemannian manifold, ∇ the Levi-Civita connection of g, and R_{abcd} the Riemann curvature of g. Then R_{abcd} and $\nabla_e R_{abcd}$ satisfy*

$$R_{abcd} = -R_{abdc} = -R_{bacd} = R_{cdab}, \tag{3.2}$$

$$R_{abcd} + R_{adbc} + R_{acdb} = 0, \tag{3.3}$$

$$\text{and} \qquad \nabla_e R_{abcd} + \nabla_c R_{abde} + \nabla_d R_{abec} = 0. \tag{3.4}$$

Proof Since ∇ is torsion-free, by Proposition 2.5.5 the Bianchi identities (2.9) and (2.10) hold for $R^a{}_{bcd}$. Contracting these with g and substituting in $\nabla g = 0$, we get (3.3) and (3.4). Also $R(\nabla) \in C^\infty \big(\mathrm{End}(TM) \otimes \Lambda^2 T^* M\big)$, and thus $R^a{}_{bcd} = -R^a{}_{bdc}$, which gives $R_{abcd} = -R_{abdc}$, the first part of (3.2).

Now ∇ also acts on tensors such as g, and by properties of curvature we deduce that

$$\nabla_c \nabla_d g_{ab} - \nabla_d \nabla_c g_{ab} = -R^e{}_{acd} g_{eb} - R^e{}_{bcd} g_{ae}.$$

But the left hand side is zero as $\nabla g = 0$, and the right hand side is $-R_{bacd} - R_{abcd}$ by definition. Therefore $R_{abcd} = -R_{bacd}$, the second part of (3.2).

To prove the third part of (3.2), by permuting a, b, c, d in (3.3) we get

$$R_{abcd} + R_{adbc} + R_{acdb} = 0, \qquad R_{dabc} + R_{dcab} + R_{dbca} = 0, \qquad (3.5)$$
$$R_{bcda} + R_{bacd} + R_{bdac} = 0, \qquad R_{cdab} + R_{cbda} + R_{cabd} = 0. \qquad (3.6)$$

Adding together eqns (3.5), subtracting eqns (3.6), and applying the first two parts of (3.2) we get $2R_{abcd} - 2R_{cdab} = 0$, and thus $R_{abcd} = R_{cdab}$, as we want. $\qquad\square$

Next we define two components of the Riemann curvature tensor, the Ricci curvature and the scalar curvature.

Definition 3.1.3 Let (M, g) be a Riemannian manifold, with Riemann curvature $R^a{}_{bcd}$. Then g is called *flat* if $R^a{}_{bcd} = 0$. The *Ricci curvature* of g is $R_{ab} = R^c{}_{acb}$, and the *scalar curvature* of g is $s = g^{ab} R_{ab} = g^{ab} R^c{}_{acb}$. By (3.2), the Ricci curvature satisfies $R_{ab} = R_{ba}$. We say that g is *Einstein* if $R_{ab} = \lambda g_{ab}$ for some constant $\lambda \in \mathbb{R}$, and that g is *Ricci-flat* if $R_{ab} = 0$.

Einstein and Ricci-flat metrics are interesting for a number of reasons. There are of course a huge number of Riemannian metrics on any manifold of dimension at least two. The Einstein and Ricci-flat metrics provide a natural way of picking out a much smaller set of special, 'best' metrics on the manifold. Also, Einstein and Ricci-flat metrics are of great importance to physicists, because in general relativity, empty space is described by a Ricci-flat Lorentzian metric.

3.1.3 Riemannian holonomy groups

Let (M, g) be a Riemannian manifold of dimension n with Levi-Civita connection ∇. Then $\nabla g = 0$, and so g is a *constant tensor* in the sense of §2.5.1. Therefore, by Proposition 2.5.2, if $x \in M$ then the action of $\mathrm{Hol}_x(\nabla)$ on $T_x M$ preserves the metric $g|_x$ on $T_x M$. But the group of transformations of $T_x M$ preserving $g|_x$ is the *orthogonal group* $\mathrm{O}(n)$. Therefore the holonomy group $\mathrm{Hol}(\nabla)$ is a subgroup of $\mathrm{O}(n)$.

Here is another way to see this, using the ideas of Theorem 2.3.6. Let F be the frame bundle of M. Then each point of F is a basis $(e_1, \dots, e_n)$ for one of the tangent spaces $T_x M$ of M. Define P to be the subset of points $(e_1, \dots, e_n)$ in F such that $e_1, \dots, e_n$ are orthonormal with respect to the metric g. Then P is a principal subbundle of F with fibre $\mathrm{O}(n)$, that is, a reduction of F. Moreover, because the connection ∇ in F satisfies $\nabla g = 0$, the connection ∇ reduces to P. Again, we see that $\mathrm{Hol}(\nabla)$ is a subgroup of $\mathrm{O}(n)$, defined up to conjugation.

Definition 3.1.4 Let (M, g) be a Riemannian manifold with Levi-Civita connection ∇. Define the *holonomy group* $\mathrm{Hol}(g)$ of g to be $\mathrm{Hol}(\nabla)$. Then $\mathrm{Hol}(g)$ is a subgroup of $\mathrm{O}(n)$, defined up to conjugation in $\mathrm{O}(n)$. We shall refer to the holonomy group of a Riemannian metric as a *Riemannian holonomy group*. Similarly, define the *restricted holonomy group* $\mathrm{Hol}^0(g)$ of g to be $\mathrm{Hol}^0(\nabla)$. Then $\mathrm{Hol}^0(g)$ is a connected Lie subgroup of $\mathrm{SO}(n)$ defined up to conjugation in $\mathrm{O}(n)$.

Using the results of §2.5 it is easy to prove the following proposition.

Proposition 3.1.5 *Let M be an n-manifold, and ∇ a torsion-free connection on TM. Then ∇ is the Levi-Civita connection of a Riemannian metric g on M if and only if $\mathrm{Hol}(\nabla)$ is conjugate in $\mathrm{GL}(n, \mathbb{R})$ to a subgroup of $\mathrm{O}(n)$.*

Thus, Riemannian holonomy is really part of the wider subject of holonomy groups of torsion-free connections.

Definition 3.1.6 Let (M, g) be a Riemannian manifold with Levi-Civita connection ∇. Define the *holonomy algebra* $\mathfrak{hol}(g)$ of g to be $\mathfrak{hol}(\nabla)$. Then $\mathfrak{hol}(g)$ is a Lie subalgebra of $\mathfrak{so}(n)$, defined up to the adjoint action of $\mathrm{O}(n)$. Let $x \in M$. Then $\mathfrak{hol}_x(\nabla)$ is a vector subspace of $T_x M \otimes T_x^* M$. We may use the metric g to identify $T_x M \otimes T_x^* M$ and $\bigotimes^2 T_x^* M$, by equating $T^a_{\ b}$ with $T_{ab} = g_{ac} T^c_{\ b}$. This identifies $\mathfrak{hol}_x(\nabla)$ with a vector subspace of $\bigotimes^2 T_x^* M$ that we will write as $\mathfrak{hol}_x(g)$. It is easy to see that $\mathfrak{hol}_x(g)$ actually lies in $\Lambda^2 T_x^* M$.

Now Proposition 2.4.1 shows that $R^a_{\ bcd}$ lies in $\mathfrak{hol}_x(\nabla) \otimes \Lambda^2 T_x^* M$ at x. Lowering the index a to get R_{abcd} as above, we see that R_{abcd} lies in $\mathfrak{hol}_x(g) \otimes \Lambda^2 T_x^* M$ at x. Using this and eqn (3.2), we have:

Theorem 3.1.7 *Let (M, g) be a Riemannian manifold with Riemann curvature R_{abcd}. Then R_{abcd} lies in the subspace $S^2 \mathfrak{hol}_x(g)$ of $\Lambda^2 T_x^* M \otimes \Lambda^2 T_x^* M$ at each $x \in M$.*

Combining this theorem with the Bianchi identities, (3.3) and (3.4), gives quite strong restrictions on the curvature tensor R_{abcd} of a Riemannian metric g with a prescribed holonomy group $\mathrm{Hol}(g)$. These restrictions are the basis of the classification of Riemannian holonomy groups, which will be explained in §3.4.

3.1.4 Riemannian holonomy groups and torsion-free G-structures

We now apply the ideas of §2.6 to Riemannian holonomy groups. Suppose M is an n-manifold, and g a Riemannian metric on M. Then Example 2.6.2 defines a unique $\mathrm{O}(n)$-structure P on M. If G is a Lie subgroup of $\mathrm{O}(n)$ and Q a G-structure on M, we say that Q is *compatible with* g if Q is a subbundle of P. Equivalently, Q is compatible with g if each point of Q, which is a basis of some tangent space $T_x M$, is orthonormal with respect to g.

If Q is compatible with g, then $P = \mathrm{O}(n) \cdot Q$ as a subset of the frame bundle F of M, and so one can reconstruct P, and hence g, from Q. Thus, if G is a Lie subgroup of $\mathrm{O}(n)$ then a G-structure Q on M gives us a Riemannian metric g on M, and some additional geometric data as well. For instance, an $\mathrm{SO}(n)$-structure on M is equivalent to a metric g, together with a choice of orientation on M.

Putting $G = \mathrm{O}(n)$ in Definition 2.6.4, the map $\sigma : \mathfrak{o}(n) \otimes (\mathbb{R}^n)^* \to \mathbb{R}^n \otimes \Lambda^2(\mathbb{R}^n)^*$ turns out to be an isomorphism. As σ is surjective, every $\mathrm{O}(n)$-structure P on M is torsion-free, and so there exists a torsion-free connection ∇ on TM compatible with P. And since σ is injective, ∇ is unique. Thus, given a Riemannian metric g on M, there is a unique torsion-free connection ∇ on TM compatible with the $\mathrm{O}(n)$-structure P corresponding to g. This is the Levi-Civita connection of g, and we have found an alternative proof of Theorem 3.1.1.

When G is a Lie subgroup of $\mathrm{O}(n)$, then the map σ of Definition 2.6.4 is injective, but not in general surjective. Hence, if Q is a G-structure on M, then the condition $T^i(Q) = 0$ for Q to be torsion-free is in general nontrivial. If Q is torsion-free, then there is a unique torsion-free connection ∇ on TM compatible with Q, which is the Levi-Civita connection of the unique metric g compatible with Q.

If Q is a torsion-free G-structure compatible with g then $\mathrm{Hol}(g) \subseteq G$. Because of this, torsion-free G-structures are a useful tool for studying Riemannian holonomy groups. The following result is easily deduced from Proposition 2.6.3, and shows the relationship between torsion-free G-structures and metrics with prescribed holonomy.

Proposition 3.1.8 *Let (M, g) be a connected Riemannian n-manifold. Then $\mathrm{Hol}(g)$ is a subgroup of $\mathrm{O}(n)$, defined up to conjugation. Let G be a Lie subgroup of $\mathrm{O}(n)$. Then M admits a torsion-free G-structure Q compatible with g if and only if $\mathrm{Hol}(g)$ is conjugate to a subgroup of G. Moreover, there is a 1-1 correspondence between the set of such G-structures Q, and the homogeneous space*

$$ G \backslash \{ a \in \mathrm{O}(n) : a \, \mathrm{Hol}(g) a^{-1} \subseteq G \}. $$

G-structures are often used in constructions of Riemannian metrics g with holonomy G. Here is a sketch of an argument used in Chapter 11. One writes down an explicit G-structure Q on a manifold M with intrinsic torsion $T^i(Q)$ *small*, in some suitable sense. Then one proves that Q can be deformed to a nearby G-structure $\tilde{Q}$ with $T^i(\tilde{Q}) = 0$. The metric $\tilde{g}$ associated to $\tilde{Q}$ has $\mathrm{Hol}(\tilde{g}) \subseteq G$. Finally, if M satisfies certain topological conditions, it can be shown that $\mathrm{Hol}(\tilde{g}) = G$.

3.2 Reducible Riemannian manifolds

Let M_1, M_2 be manifolds, and $M_1 \times M_2$ the product manifold. Then at each point (p_1, p_2) of $M_1 \times M_2$, we have $T_{(p_1,p_2)}(M_1 \times M_2) \cong T_{p_1}M_1 \oplus T_{p_2}M_2$. Let g_1, g_2 be Riemannian metrics on M_1, M_2. Then $g_1|_{p_1} + g_2|_{p_2}$ is a metric on $T_{p_1}M_1 \oplus T_{p_2}M_2$. Define the *product metric* $g_1 \times g_2$ on $M_1 \times M_2$ by $g_1 \times g_2|_{(p_1,p_2)} = g_1|_{p_1} + g_2|_{p_2}$ for all $p_1 \in M_1$ and $p_2 \in M_2$. Then $g_1 \times g_2$ is a Riemannian metric on $M_1 \times M_2$, and $M_1 \times M_2$ is a Riemannian manifold. We call $(M_1 \times M_2, g_1 \times g_2)$ a *Riemannian product*.

A Riemannian manifold (M, g) is said to be *reducible* if it is isometric to a Riemannian product $(M_1 \times M_2, g_1 \times g_2)$, with $\dim M_i > 0$. Also, (M, g) is said to be *locally reducible* if every point has a reducible open neighbourhood. We shall call (M, g) *irreducible* if it is not locally reducible. The following proposition, which is easy to prove, gives the holonomy group of a product metric $g_1 \times g_2$.

Proposition 3.2.1 *Let* (M_1, g_1), (M_2, g_2) *be Riemannian manifolds. Then the product metric* $g_1 \times g_2$ *has holonomy* $\mathrm{Hol}(g_1 \times g_2) = \mathrm{Hol}(g_1) \times \mathrm{Hol}(g_2)$.

In our next three propositions we shall show that if g is a Riemannian metric and the holonomy representation (see the end of §2.2) of g is reducible, then the metric itself is at least locally reducible, and its holonomy group is a product. The first proposition is left as an exercise for the reader.

Proposition 3.2.2 *Let* (M, g) *be a Riemannian manifold with Levi-Civita connection* ∇, *and fix* $p \in M$, *so that* $\mathrm{Hol}_p(g)$ *acts on* T_pM. *Suppose that* $T_pM = V_p \oplus W_p$, *where* V_p, W_p *are proper vector subspaces of* T_pM *preserved by* $\mathrm{Hol}_p(g)$, *and orthogonal with respect to* g. *Then there are natural vector subbundles* V, W *of* TM *with fibres* V_p, W_p *at* p. *These subbundles* V, W *are orthogonal with respect to* g *and closed under parallel translation, and satisfy* $TM = V \oplus W$ *and* $T^*M = V^* \oplus W^*$.

Proposition 3.2.3 *In the situation of Proposition 3.2.2, let* R_{abcd} *be the Riemann curvature of* g. *Then* R_{abcd} *is a section of the subbundle* $S^2(\Lambda^2 V^*) \oplus S^2(\Lambda^2 W^*)$ *of* $S^2(\Lambda^2 T^* M)$. *Also, the reduced holonomy group* $\mathrm{Hol}_p^0(g)$ *is a product group* $H_V \times H_W$, *where* H_V *is a subgroup of* $\mathrm{SO}(V_p)$ *and acts trivially on* W_p, *and* H_W *is a subgroup of* $\mathrm{SO}(W_p)$ *and acts trivially on* V_p.

Proof As $T^*M \cong V^* \oplus W^*$, we see that $\Lambda^2 T^* M \cong \Lambda^2 V^* \oplus \Lambda^2 W^* \oplus V^* \otimes W^*$. By Theorem 3.1.7 we know that R_{abcd} lies in $S^2 \mathfrak{hol}_p(g)$ at $p \in M$, where $\mathfrak{hol}_p(g)$ is a vector subspace of $\Lambda^2 T^* M$ identified with the holonomy algebra $\mathfrak{hol}(g)$. But, because the holonomy algebra preserves the splitting $TM = V \oplus W$, we see that $\mathfrak{hol}_p(g)$ lies in the subspace $\Lambda^2 V_p^* \oplus \Lambda^2 W_p^*$ of $\Lambda^2 T_p^* M$. Therefore, R_{abcd} is a section of

$$\Lambda^2 V^* \otimes \Lambda^2 V^* \oplus \Lambda^2 W^* \otimes \Lambda^2 W^* \oplus \Lambda^2 V^* \otimes \Lambda^2 W^* \oplus \Lambda^2 W^* \otimes \Lambda^2 V^*. \qquad (3.7)$$

Now, R_{abcd} satisfies the first Bianchi identity (3.3). Using this, we find that the components of R_{abcd} in the last two components of (3.7) are zero. Therefore, since R_{abcd} is symmetric in the two $\Lambda^2 T^* M$ factors we see that it is a section of $S^2(\Lambda^2 V^*) \oplus S^2(\Lambda^2 W^*)$, as we have to prove. We deduce that $R^a{}_{bcd}$ is a section of the bundle

$$V \otimes V^* \otimes \Lambda^2 V^* \quad \oplus \quad W \otimes W^* \otimes \Lambda^2 W^*. \qquad (3.8)$$

Let $q \in M$, let $\gamma : [0, 1] \to M$ be piecewise-smooth with $\gamma(0) = p$ and $\gamma(1) = q$, and let $P_\gamma : T_pM \to T_qM$ be the parallel translation map. Because R lies in the subbundle (3.8), it follows that $\langle R_q \cdot (u \wedge v) : u, v \in T_qM \rangle = A_q \oplus B_q$, where A_q is a subspace of $V_q \otimes V_q^*$ and B_q a subspace of $W_q \otimes W_q^*$. Now V and W are closed under parallel translation, so P_γ takes V_p to V_q and W_p to W_q, and thus $P_\gamma^{-1} A_q P_\gamma$ lies in $V_p \otimes V_p^*$ and $P_\gamma^{-1} B_q P_\gamma$ lies in $W_p \otimes W_p^*$.

But Theorem 2.4.3 says that $\mathfrak{hol}_p(\nabla)$ is spanned by the elements of $\mathrm{End}(T_pM)$ of the form $P_\gamma^{-1}[R_q \cdot (u \wedge v)]P_\gamma$, for all $q \in M$ and $u, v \in T_qM$, and we have just shown that for each fixed $q \in M$, the subspace generated by these elements splits into a direct sum of a piece in $V_p \otimes V_p^*$, and a piece in $W_p \otimes W_p^*$. Therefore, the span of these elements for all $q \in M$ splits in the same way, so by Theorem 2.4.3 we see that $\mathfrak{hol}_p(\nabla) = \mathfrak{h}_V \oplus \mathfrak{h}_W$, where $\mathfrak{h}_V$ is a subspace of $V_p \otimes V_p^*$ and $\mathfrak{h}_W$ a subspace of $W_p \otimes W_p^*$.

As $\mathfrak{hol}_p(\nabla)$ is the Lie algebra of $\mathrm{Hol}_p^0(g)$, which is a connected Lie group, we see that $\mathfrak{h}_v$, $\mathfrak{h}_w$ are the Lie algebras of subgroups H_v, H_w of $\mathrm{SO}(V_p)$ and $\mathrm{SO}(W_p)$ respectively, and $\mathrm{Hol}_p^0(g) = H_v \times H_w$. This completes the proof. $\square$

Proposition 3.2.4 *In the situation of Proposition 3.2.3, there is a connected open neighbourhood N of p in M and a diffeomorphism $N \cong X \times Y$ for manifolds X, Y, such that under the isomorphism $T(X \times Y) \cong TX \oplus TY$, we have $V|_N = TX$ and $W|_N = TY$. There are Riemannian metrics g_X on X and g_Y on Y such that $g|_N$ is isometric to the product metric $g_X \times g_Y$. Thus g is locally reducible, and moreover $\mathrm{Hol}_q^0(g_X) \subseteq H_V$ and $\mathrm{Hol}_r^0(g_Y) \subseteq H_W$, where $p \in N$ is identified with $(q, r) \in X \times Y$.*

Proof Since V is closed under parallel translation, we deduce that if $u \in C^\infty(TM)$ and $v \in C^\infty(V)$, then $\nabla_u v \in C^\infty(V)$. Suppose that $v, v' \in C^\infty(V)$. Then $\nabla_v v'$ and $\nabla_{v'} v \in C^\infty(V)$. But ∇ is torsion-free, so that $[v, v'] = \nabla_v v' - \nabla_{v'} v$. Thus, if $v, v' \in C^\infty(V)$, then $[v, v'] \in C^\infty(V)$. This proves that V is an *integrable distribution*, so that by the Frobenius Theorem [214, p. 10] we see that locally M is fibred by a family of submanifolds of M, with tangent spaces V.

Similarly, we deduce that W is an integrable distribution. But $TM = V \oplus W$, so these two integrable distributions define a *local product structure* on M. This means that we can identify a connected open neighbourhood N of $p \in M$ with a product manifold $X \times Y$, such that the isomorphism $T(X \times Y) \cong TX \oplus TY$ identifies V with TX and W with TY, as we want.

As V and W are orthogonal, we may write $g = g_V + g_W$, where $g_V \in C^\infty(S^2 V^*)$ and $g_W \in C^\infty(S^2 W^*)$. Since ∇ is torsion-free and N is connected, it is not difficult to show that the restriction of g_V to $N = X \times Y$ is independent of the Y directions, and is therefore the pull-back to $X \times Y$ of a metric g_X on X. Similarly, $g_W|_N$ is the pull-back of a metric g_Y on Y. Therefore $g|_N$ is isometric to $g_X \times g_Y$, as we have to prove. The rest of the proposition follows from Proposition 3.2.1 and the definition of local reducibility. $\square$

From Propositions 3.2.2–3.2.4 we immediately deduce:

Corollary 3.2.5 *Let M be an n-manifold, and g an irreducible Riemannian metric on M. Then the representations of $\mathrm{Hol}(g)$ and $\mathrm{Hol}^0(g)$ on $\mathbb{R}^n$ are irreducible.*

More generally, if (M, g) is a Riemannian manifold of dimension n, then $\mathrm{Hol}^0(g)$ is a subgroup of $\mathrm{SO}(n)$, and has a natural representation on $\mathbb{R}^n$. By the representation theory of Lie groups, we may decompose $\mathbb{R}^n$ into a finite direct sum of irreducible representations of $\mathrm{Hol}^0(g)$. By applying Propositions 3.2.2–3.2.4 and using induction on k, we easily prove the following theorem.

Theorem 3.2.6 *Let (M, g) be a Riemannian n-manifold, so that $\mathrm{Hol}^0(g)$ is a subgroup of $\mathrm{SO}(n)$ acting on $\mathbb{R}^n$. Then there is a splitting $\mathbb{R}^n = \mathbb{R}^{n_1} \oplus \cdots \oplus \mathbb{R}^{n_k}$, where $n_j > 0$, and a corresponding isomorphism $\mathrm{Hol}^0(g) = H_1 \times \cdots \times H_k$, where H_j is a connected Lie subgroup of $\mathrm{SO}(n_j)$ acting irreducibly on $\mathbb{R}^{n_j}$.*

The theorem shows that if the holonomy representation of $\mathrm{Hol}^0(g)$ is reducible, then $\mathrm{Hol}^0(g)$ is in fact a product group, and the holonomy representation a direct sum of irreducible representations of each factor. Therefore, if G is a connected Lie group and V a representation of G, such that G and V cannot be written $G = G_1 \times \cdots \times G_k$ and $V = V_1 \oplus \cdots \oplus V_k$ where V_j is an irreducible representation of G_j, then G and V cannot be the reduced holonomy group and holonomy representation of any Riemannian metric. This is a strong statement, because there are very many such pairs G, V, and all are excluded as possible holonomy groups.

Notice that Theorem 3.2.6 does *not* claim that the groups H_j are the holonomy groups of metrics on manifolds of dimension n_j. In fact, a careful examination of the proof shows that the group H_j is generated by subgroups which are the holonomy groups of metrics on small open subsets of $\mathbb{R}^{n_j}$, but there are topological difficulties in assembling these patches into a single Riemannian manifold: there is a natural way to do it, but the resulting topological space may not be Hausdorff.

The reason this is important to us is in the classification of Riemannian holonomy groups. Theorem 3.2.6 nearly says that if the holonomy representation of $\mathrm{Hol}^0(g)$ is reducible, then $\mathrm{Hol}^0(g)$ is a product of holonomy groups of metrics in lower dimensions. If we knew this was true, then a classification of groups $\mathrm{Hol}^0(g)$ with irreducible holonomy representations would imply a classification of all groups $\mathrm{Hol}^0(g)$, and thus we could restrict our attention to holonomy groups with irreducible representations.

In fact, this problem turns out not to matter. The classification theory for irreducible holonomy groups that we will summarize in §3.4 also works to classify the groups H_j arising in Theorem 3.2.6, and the classification is the same. This is because the failure of the Hausdorff condition we referred to above does not affect the proof. For further discussion of this point, see Besse [30, §10.42, §10.107].

Next, suppose that the manifold M in Propositions 3.2.2–3.2.4 is *simply-connected*, and the metric g is *complete*. In this case, by a result of de Rham [93], the local product structure constructed in Proposition 3.2.4 is actually a global product structure, so that $M \cong X \times Y$, and g is globally isometric to a Riemannian product metric $g_X \times g_Y$. Therefore, using similar arguments to the previous theorem we may prove the following result, which is a sort of converse to Proposition 3.2.1.

Theorem 3.2.7 *Let* (M, g) *be a complete, simply-connected Riemannian manifold. Then there exist complete, simply-connected Riemannian manifolds* (M_j, g_j) *for* $j = 1, \ldots, k$, *such that the holonomy representation of* $\mathrm{Hol}(g_j)$ *is irreducible,* (M, g) *is isometric to* $(M_1 \times \cdots \times M_k, g_1 \times \cdots \times g_k)$, *and* $\mathrm{Hol}(g) = \mathrm{Hol}(g_1) \times \cdots \times \mathrm{Hol}(g_k)$.

It is shown in [214, App. 5], using the theory of Lie groups, that every connected Lie subgroup of $\mathrm{SO}(n)$ that acts irreducibly on $\mathbb{R}^n$ is *closed* in $\mathrm{SO}(n)$. From this result and Theorem 3.2.6 we deduce:

Theorem 3.2.8 *Let* (M, g) *be a Riemannian* n-*manifold. Then* $\mathrm{Hol}^0(g)$ *is a closed, connected Lie subgroup of* $\mathrm{SO}(n)$.

Since $\mathrm{SO}(n)$ is compact, this implies that $\mathrm{Hol}^0(g)$ is also compact. By a study of the fundamental group of a compact, irreducible Riemannian manifold, Cheeger and Gromoll [75, Th. 6] prove the following result.

Theorem 3.2.9 *Let* (M, g) *be a compact, irreducible Riemannian* n-*manifold. Then* $\mathrm{Hol}(g)$ *is a compact Lie subgroup of* $\mathrm{O}(n)$.

3.3 Riemannian symmetric spaces

We will now briefly describe the theory of Riemannian symmetric spaces. These were introduced in 1925 by Élie Cartan, who classified them completely, by applying his own classification of irreducible representations of Lie groups. For a very thorough treatment of symmetric spaces, see Helgason's book [157]. Also, the treatments by Kobayashi and Nomizu [215, Chap. XI] and Besse [30, §§7.F, 10.G, 10.K] are helpful.

Definition 3.3.1 A Riemannian manifold (M, g) is said to be a *Riemannian symmetric space* if for every point $p \in M$ there exists an isometry $s_p : M \to M$ that is an involution (that is, s_p^2 is the identity), such that p is an isolated fixed point of s_p.

Let (M, g) be a *complete* Riemannian manifold, and $p \in M$. Then for each unit vector $u \in T_p M$, there is a unique geodesic $\gamma_u : \mathbb{R} \to M$ parametrized by arc length, such that $\gamma_u(0) = p$ and $\dot{\gamma}_u(0) = u$. Define the *exponential map* $\exp_p : T_p M \to M$ by $\exp_p(tu) = \gamma_u(t)$ for all $t \in \mathbb{R}$ and unit vectors $u \in T_p M$. Then $\exp_p$ induces a diffeomorphism between neighbourhoods of 0 in $T_p M$ and p in M.

Identifying $T_p M \cong \mathbb{R}^n$ yields a coordinate system on M near p. These are called *normal coordinates* or *geodesic normal coordinates* at p. If we start with a Riemannian manifold (M, g) that is not complete, then $\exp_p$ can still be defined in a neighbourhood of 0 in $T_p M$. The following lemma shows that the isometries in Definition 3.3.1 assume a particularly simple form in normal coordinates.

Lemma 3.3.2 *Let (M, g) be a complete Riemannian manifold, let $p \in M$, and suppose $s : M \to M$ is an involutive isometry with isolated fixed point p. Then $s\big(\exp_p(v)\big) = \exp_p(-v)$ for all $v \in T_p M$.*

Proof The derivative $\mathrm{d}s$ of s maps $T_p M$ to itself. Since s^2 is the identity, $(\mathrm{d}s)^2$ is the identity on $T_p M$, and as p is an isolated fixed point, 0 is the sole fixed point of $\mathrm{d}s$ on $T_p M$. Clearly, this implies that $\mathrm{d}s(v) = -v$ for all $v \in T_p M$. Now s preserves g, as it is an isometry, and the map $\exp_p$ depends solely on g. Therefore, $\exp_p$ must commute with s, in the sense that $s\big(\exp_p(v)\big) = \exp_p\big(\mathrm{d}s(v)\big)$ for $v \in T_p M$. Substituting $\mathrm{d}s(v) = -v$ gives the result. $\qquad\square$

The next three propositions explore the geometry of Riemannian symmetric spaces.

Proposition 3.3.3 *Let (M, g) be a connected, simply-connected Riemannian symmetric space. Then g is complete. Let G be the group of isometries of (M, g) generated by elements of the form $s_q \circ s_r$ for $q, r \in M$. Then G is a connected Lie group acting transitively on M. Choose $p \in M$, and let H be the subgroup of G fixing p. Then H is a closed, connected Lie subgroup of G, and M is the homogeneous space G/H.*

Proof Let $\gamma : (-\epsilon, \epsilon) \to M$ be a geodesic segment in M, parametrized by arc length. Then the lemma shows that $s_{\gamma(0)}(\gamma(y)) = \gamma(-y)$ for $y \in (-\epsilon, \epsilon)$. More generally we see that $s_{\gamma(x/2)}(\gamma(y)) = \gamma(x - y)$ whenever $\frac{1}{2}x, y$ and $x - y$ lie in $(-\epsilon, \epsilon)$. This gives

$$s_{\gamma(x/2)} \circ s_{\gamma(0)}\big(\gamma(y)\big) = \gamma(x + y),$$

provided $\frac{1}{2}x, y$ and $x + y$ are in $(-\epsilon, \epsilon)$. Thus, the map $\alpha_x = s_{\gamma(x/2)} \circ s_{\gamma(0)}$ moves points a distance x along γ. But α_x is defined on the whole of M. By applying α_x and its inverse α_{-x} many times, one can define the geodesic γ not just on $(-\epsilon, \epsilon)$, but on $\mathbb{R}$. Therefore, every geodesic in M can be extended indefinitely, and g is complete.

Now α_x lies in G by definition. Hence, if p, q are two points in M joined by a geodesic segment of length x, there exists an element α_x in G such that $\alpha_x(p) = q$. But M is connected, and so every two points p, q in M can be joined by a finite number of geodesic segments put end to end. Composing the corresponding elements of G, we get an element α of G with $\alpha(p) = q$. Therefore G acts transitively on M.

Let $q, r \in M$. Then, as M is connected, there exists a smooth path $\gamma : [0, 1] \rightarrow M$ with $\gamma(0) = q$ and $\gamma(1) = r$. Consider the family of isometries $s_q \circ s_{\gamma(t)}$ for $t \in [0, 1]$. This is a smooth path in G, joining the identity at $t = 0$ with $s_q \circ s_r$ at $t = 1$. Thus, the generating elements $s_q \circ s_r$ of G can be joined to the identity by smooth paths, and so G is arcwise-connected.

By the Myers–Steenrod Theorem [30, p. 39], the isometry group of (M, g) is a Lie group acting smoothly on M. Thus G is an arcwise-connected subgroup of a Lie group, so by the theorem of Yamabe quoted in §2.2, G is a connected Lie group. Also G acts smoothly on M, and so H is a closed Lie subgroup of G. Since G acts transitively on M, we have $M \cong G/H$. Because M is simply-connected, G/H is simply-connected, and this implies that H is connected. □

Note that the group G in the proposition may not be the full isometry group of (M, g), or even the identity component of the isometry group. For example, if M is $\mathbb{R}^n$ with the Euclidean metric, then $G = \mathbb{R}^n$ acting by translations and $H = 0$, but the full isometry group also includes the rotations $O(n)$ acting on $\mathbb{R}^n$.

Proposition 3.3.4 *Let* M, g, G, p *and* H *be as above, and let* $\mathfrak{g}$ *be the Lie algebra of* G. *Then there is an involutive Lie group isomorphism* $\sigma : G \rightarrow G$, *and a splitting* $\mathfrak{g} = \mathfrak{h} \oplus \mathfrak{m}$, *where* $\mathfrak{h}$ *is the Lie algebra of* H, *and* $\mathfrak{h}$ *and* $\mathfrak{m}$ *are the eigenspaces of the involution* $d\sigma : \mathfrak{g} \rightarrow \mathfrak{g}$ *with eigenvalues* 1 *and* -1 *respectively. These subspaces satisfy*

$$[\mathfrak{h}, \mathfrak{h}] \subseteq \mathfrak{h}, \qquad [\mathfrak{h}, \mathfrak{m}] \subseteq \mathfrak{m} \quad and \quad [\mathfrak{m}, \mathfrak{m}] \subseteq \mathfrak{h}. \tag{3.9}$$

There is a natural isomorphism $\mathfrak{m} \cong T_pM$. *The adjoint action of* H *on* $\mathfrak{g}$ *induces a representation of* H *on* $\mathfrak{m}$, *or equivalently* T_pM, *and this representation is faithful. Also,* H *is the identity component of the fixed point set of* σ.

Proof Define $\sigma : G \rightarrow G$ by $\sigma(\alpha) = s_p \circ \alpha \circ s_p$. Clearly σ does map G to G, and is a group isomorphism, so $d\sigma : \mathfrak{g} \rightarrow \mathfrak{g}$ is a Lie algebra isomorphism. Also, σ is an involution, as $s_p^2 = 1$, so $(d\sigma)^2$ is the identity. Therefore $d\sigma$ has eigenvalues ± 1, and $\mathfrak{g}$ is the direct sum of the corresponding eigenspaces. As M is connected, the isometries s_p in Definition 3.3.1 are unique. Hence, for $p \in M$ and $\alpha \in G$ we have $\alpha \circ s_p \circ \alpha^{-1} = s_{\alpha(p)}$. Therefore $h \circ s_p \circ h^{-1} = s_p$ for $h \in H$, so that $\sigma(h) = s_p \circ h \circ s_p = h$. Thus, H is fixed by σ, and $d\sigma$ is the identity on $\mathfrak{h}$.

The identification of G/H with M identifies p with the coset H and T_pM with $\mathfrak{g}/\mathfrak{h}$. Under this identification, the maps $ds_p : T_pM \rightarrow T_pM$ and $d\sigma : \mathfrak{g}/\mathfrak{h} \rightarrow \mathfrak{g}/\mathfrak{h}$ coincide. But ds_p multiplies by -1, as in Lemma 3.3.2. Thus, $d\sigma$ is the identity on $\mathfrak{h} \subset \mathfrak{g}$, and acts as -1 on $\mathfrak{g}/\mathfrak{h}$. Therefore there exists a unique splitting $\mathfrak{g} = \mathfrak{h} \oplus \mathfrak{m}$ where $\mathfrak{h}, \mathfrak{m}$ are the 1 and -1 eigenspaces of $d\sigma$, as we want. The relations in (3.9) then follow easily from the fact that $d\sigma$ is a Lie algebra isomorphism.

The splitting $\mathfrak{g} = \mathfrak{h} \oplus \mathfrak{m}$ and the isomorphism $T_pM \cong \mathfrak{g}/\mathfrak{h}$ give an isomorphism $\mathfrak{m} \cong T_pM$, as required. Clearly, $\mathfrak{m}$ is preserved by the adjoint action of H on $\mathfrak{g}$, and this gives a representation of H on $\mathfrak{m}$, and so on T_pM. This action of H on T_pM can also be described as follows: each element $h \in H$ fixes p, and so $\mathrm{d}h$ maps T_pM to itself. Now g is complete, so the exponential map $\exp_p : T_pM \to M$ is well-defined, and clearly satisfies $\exp_p(\mathrm{d}h(v)) = h \cdot \exp_p(v)$ for all $h \in H$ and $v \in T_pM$.

Thus, $\mathrm{d}h : T_pM \to T_pM$ determines the action of h on the subset $\exp_p(T_pM)$ of M, and so on all of M, as M is connected. It follows that if $\mathrm{d}h$ is the identity then h is the identity. Therefore, the representation of H on T_pM, or equivalently on $\mathfrak{m}$, is faithful. Finally, we know that H is part of the fixed point set of σ, that H is connected, and that the subspace of $\mathfrak{g}$ fixed by $\mathrm{d}\sigma$ is $\mathfrak{h}$. Together these show that H is the identity component of the fixed points of σ. $\square$

Proposition 3.3.5 *In the situation of Proposition 3.3.4, the Riemann curvature R_p of g at p lies in $T_pM \otimes T_p^*M \otimes \Lambda^2 T_p^*M$. Identifying $\mathfrak{m}$ and T_pM in the natural way, R_p is given by the equation*

$$R_p \cdot (u \otimes v \wedge w) = [u, [v, w]], \tag{3.10}$$

for all $u, v, w \in \mathfrak{m}$. Moreover, the holonomy group $\mathrm{Hol}_p(g)$ is H, with the above representation on T_pM, and the Riemann curvature R of g satisfies $\nabla R = 0$.

Proof Using the splitting $\mathfrak{g} = \mathfrak{h} \oplus \mathfrak{m}$, one can construct a unique torsion-free, G-invariant connection ∇ on TM. This satisfies $\nabla g = 0$, and so ∇ is the Levi-Civita connection of g. Explicit computation then yields the formula (3.10) for R_p. For more details of this argument, see [215, §X.2, §XI.3]. The formula shows that for $v, w \in \mathfrak{m}$, we have $R_p \cdot (v \wedge w) = -\mathrm{ad}([v, w])$ in $\mathrm{End}(\mathfrak{m})$. But $[\mathfrak{m}, \mathfrak{m}] \subset \mathfrak{h}$ by (3.9), so that $R_p \cdot (v \wedge w) \in \mathrm{ad}(\mathfrak{h})$.

By definition, G is generated by $s_q \circ s_r$ for $q, r \in M$. By fixing $r = p$ and letting q approach p we can prove an infinitesimal version of this statement, which says that $\mathfrak{g}$ is generated by the elements $\frac{\mathrm{d}}{\mathrm{d}t}\big(s_{\exp_p(tv)} \circ s_p\big)\big|_{t=0}$ of $\mathfrak{g}$, for all $v \in T_pM$. But these are exactly the elements of $\mathfrak{m}$, and thus $\mathfrak{g}$ is generated as a Lie algebra by $\mathfrak{m}$. Therefore (3.9) implies that $[\mathfrak{m}, \mathfrak{m}] = \mathfrak{h}$, and so the equation $R_p \cdot (v \wedge w) = -\mathrm{ad}([v, w])$ gives

$$\big\langle R_p \cdot (v \wedge w) : v, w \in \mathfrak{m} \big\rangle = \mathrm{ad}(\mathfrak{h}). \tag{3.11}$$

Proposition 2.4.1 then shows $\mathrm{ad}(\mathfrak{h})$ is a subset of $\mathfrak{hol}_p(g)$, the Lie algebra of $\mathrm{Hol}_p(g)$.

Let $q \in M$. Then $s_p : M \to M$ is an isometry, and so preserves R and ∇R. But $\mathrm{d}s_q$ acts as -1 on T_qM, and thus $\mathrm{d}s_q(\nabla R|_q) = -\nabla R|_q$. Therefore $\nabla R|_q = 0$ for all $q \in M$, giving and $\nabla R = 0$ as we want. Let $\gamma : [0, 1] \to M$ be piecewise-smooth with $\gamma(0) = p$ and $\gamma(1) = q$, and let $P_\gamma : T_pM \to T_qM$ be the parallel translation map. Since $\nabla R = 0$, it follows that $P_\gamma^{-1}[R_q \cdot (v \wedge w)]P_\gamma = R_p \cdot (P_\gamma^{-1}v \wedge P_\gamma^{-1}w)$, for all $v, w \in T_qM$.

Now Theorem 2.4.3 shows that $\mathfrak{hol}_p(\nabla)$ is spanned by elements of $\mathrm{End}(T_pM)$ of the form $P_\gamma^{-1}[R_q \cdot (v \wedge w)]P_\gamma$. Therefore (3.11) implies that $\mathfrak{hol}_p(\nabla) = \mathrm{ad}(\mathfrak{h})$. But H is connected by Proposition 3.3.3, and $\mathrm{Hol}_p(g)$ is connected as M is simply-connected, and so $\mathrm{Hol}_p(g) = \mathrm{Ad}(H)$, where Ad is the adjoint representation of H on $\mathfrak{m}$, and we identify $\mathfrak{m}$ and $T_p M$. Since the representation is faithful, we have $\mathrm{Hol}_p(g) \cong H$, and the proof is complete. $\qquad\qquad\qquad\qquad\qquad\qquad\qquad\qquad\qquad\qquad\qquad\qquad\quad$ □

Propositions 3.3.3–3.3.5 reduce the problem of classifying simply-connected Riemannian symmetric spaces to a problem in the theory of Lie groups. This was solved completely by E. Cartan in 1926–7, who was able to write down a complete list of all simply-connected Riemannian symmetric spaces. Helgason [157, Chap. IX] discusses Cartan's proof, and Besse [30, §7.H, §10.K] gives tables of all the possibilities.

Using the results above, the holonomy group of a Riemannian symmetric space is easily found. Therefore, Cartan's classification implies the classification of the holonomy groups of Riemannian symmetric spaces. A considerable number of Riemannian holonomy groups arise in this way: for example, every connected, compact, simple Lie group is (up to a finite cover) the holonomy group of an irreducible Riemannian symmetric space, with holonomy representation the adjoint representation.

Some well-known examples of Riemannian symmetric spaces are $\mathbb{R}^n$ with the Euclidean metric, $\mathcal{S}^n$ with the round metric, $\mathcal{H}^n$ with the hyperbolic metric, and $\mathbb{CP}^n$ with the Fubini–Study metric. The corresponding groups are $G = \mathbb{R}^n$ and $H = \{1\}$ for $\mathbb{R}^n$, so that $\mathbb{R}^n$ has holonomy group $\{1\}$, $G = \mathrm{SO}(n+1)$ and $H = \mathrm{SO}(n)$ for $\mathcal{S}^n$ and $G = \mathrm{SO}(n,1)$ and $H = \mathrm{SO}(n)$ for $\mathcal{H}^n$, so that $\mathcal{S}^n$ and $\mathcal{H}^n$ both have holonomy group $\mathrm{SO}(n)$, and $G = \mathrm{U}(n+1)/\mathrm{U}(1)$ and $H = \mathrm{U}(n)$ for $\mathbb{CP}^n$, so that $\mathbb{CP}^n$ has holonomy $\mathrm{U}(n)$.

Next we shall discuss *locally symmetric* Riemannian manifolds, which satisfy a local version of the symmetric space condition.

Definition 3.3.6 We call a Riemannian manifold (M, g) *locally symmetric* if every point $p \in M$ admits an open neighbourhood U_p in M, and an involutive isometry $s_p : U_p \to U_p$, with unique fixed point p. We call (M, g) *nonsymmetric* if it is not locally symmetric.

Clearly, every Riemannian symmetric space is locally symmetric. Conversely, one can show that every locally symmetric Riemannian manifold is locally isometric to a Riemannian symmetric space.

Theorem 3.3.7 *Suppose (M, g) is a locally symmetric Riemannian manifold. Then there is a unique simply-connected Riemannian symmetric space (N, h) with (M, g) locally isometric to (N, h). In other words, given any points $p \in M$ and $q \in N$, there exist isometric open neighbourhoods U of p in M and V of q in N.*

A proof of this theorem can be found in [157, p. 183], but we will not give it. Here is one way the result can be proved. The problem is that because the isometries s_p are defined only locally, they cannot be put together to form Lie groups of isometries G, H, as in Proposition 3.3.3. However, the Lie algebras $\mathfrak{g}, \mathfrak{h}$ of G, H can be defined in the locally symmetric case, as Lie algebras of Killing vector fields, defined locally. Let G be the unique connected, simply-connected Lie group with Lie algebra $\mathfrak{g}$, and let H be the unique connected Lie subgroup of G with Lie algebra $\mathfrak{h}$. Then $N = G/H$ is a Riemannian symmetric space, with the desired properties.

Let (M, g) be a Riemannian manifold. If (M, g) is locally symmetric, then by Theorem 3.3.7 it is locally isometric to a Riemannian symmetric space, and so Proposition

3.3.5 shows that $\nabla R = 0$, where ∇ is the Levi-Civita connection, and R the Riemann curvature. Surprisingly, the converse is also true: if $\nabla R = 0$, then (M, g) is locally symmetric.

Theorem 3.3.8 *Let (M, g) be a Riemannian manifold, with Levi-Civita connection ∇ and Riemann curvature R. Then (M, g) is locally symmetric if and only if $\nabla R = 0$.*

See [215, p. 244] for a proof of this. Here is how it works. Let (M, g) be a Riemannian manifold, and let $p \in M$. Then $\exp_p : T_p M \to M$ is well-defined and injective in a small ball $B_\epsilon(0)$ about the origin in $T_p M$. Define $U_p = \exp_p\big(B_\epsilon(0)\big)$, and define $s_p : U_p \to U_p$ by $s_p\big(\exp_p(v)\big) = \exp_p(-v)$. This map s_p is called the *geodesic symmetry* about p, and is clearly an involution.

Now, if $\nabla R = 0$ it can be shown that s_p is an *isometry* on U_p. This is because the Jacobi fields along a geodesic are the solutions of a differential equation with constant coefficients, and therefore the metric g on U_p assumes a simple form in normal coordinates at p, determined entirely by the metric and curvature at p. A similar argument is given in detail in [214, §VI, Th. 7.2, Th. 7.4].

Theorem 3.3.8 is important in the classification of Riemannian holonomy groups. For suppose (M, g) is a Riemannian manifold with $\nabla R = 0$. Then (M, g) is locally isometric to a simply-connected Riemannian symmetric space (N, h) by Theorems 3.3.7 and 3.3.8, and therefore $\mathrm{Hol}^0(g) = \mathrm{Hol}(h)$. But, as we have seen above, the classification of holonomy groups of Riemannian symmetric spaces comes out of Cartan's classification of Riemannian symmetric spaces, and is already well understood.

Therefore, we may restrict our attention to holonomy groups of Riemannian metrics that are nonsymmetric, for which $\nabla R \neq 0$. Now, this condition $\nabla R \neq 0$ can be used to exclude many candidate holonomy groups, in the following way. Theorems 3.1.2 and 3.1.7 show that if a metric g has a prescribed holonomy group $H \subset \mathrm{O}(n)$, then the Riemann curvature R and its derivative ∇R have certain symmetries, and also lie in vector subspaces determined by the Lie algebra $\mathfrak{h}$ of H. For some groups H, these conditions force $\nabla R = 0$, so that H cannot be a nonsymmetric holonomy group.

3.4 The classification of Riemannian holonomy groups

In this section we shall discuss the question: which subgroups of $\mathrm{O}(n)$ can be the holonomy group of a Riemannian n-manifold (M, g)? To simplify the answer, it is convenient to restrict the question in three ways. Firstly, we suppose that M is simply-connected, or equivalently, we study the restricted holonomy group $\mathrm{Hol}^0(g)$ instead of the holonomy group $\mathrm{Hol}(g)$. This eliminates issues to do with the fundamental group and global topology of M.

Secondly, we know from §3.2 that if g is locally reducible then $\mathrm{Hol}^0(g)$ is a product of holonomy groups in lower dimensions. Therefore, we suppose that g is irreducible. And thirdly, from §3.3, if g is locally symmetric then $\mathrm{Hol}^0(g)$ lies on the list of holonomy groups of Riemannian symmetric spaces, which are already known. So we suppose that g is not locally symmetric, and ask another question: which subgroups of $\mathrm{SO}(n)$ can be the holonomy group of an irreducible, nonsymmetric Riemannian metric g on a simply-connected n-manifold M? In 1955, Berger [27, Th. 3, p. 318] proved the following result, which is the first part of the answer to this question.

Theorem 3.4.1. (Berger) *Suppose M is a simply-connected manifold of dimension n, and g is an irreducible, nonsymmetric Riemannian metric on M. Then exactly one of the following seven cases holds:*

(i) $\mathrm{Hol}(g) = \mathrm{SO}(n)$,

(ii) $n = 2m$ *with* $m \geqslant 2$, *and* $\mathrm{Hol}(g) = \mathrm{U}(m)$ *in* $\mathrm{SO}(2m)$,

(iii) $n = 2m$ *with* $m \geqslant 2$, *and* $\mathrm{Hol}(g) = \mathrm{SU}(m)$ *in* $\mathrm{SO}(2m)$,

(iv) $n = 4m$ *with* $m \geqslant 2$, *and* $\mathrm{Hol}(g) = \mathrm{Sp}(m)$ *in* $\mathrm{SO}(4m)$,

(v) $n = 4m$ *with* $m \geqslant 2$, *and* $\mathrm{Hol}(g) = \mathrm{Sp}(m)\,\mathrm{Sp}(1)$ *in* $\mathrm{SO}(4m)$,

(vi) $n = 7$ *and* $\mathrm{Hol}(g) = G_2$ *in* $\mathrm{SO}(7)$, *or*

(vii) $n = 8$ *and* $\mathrm{Hol}(g) = \mathrm{Spin}(7)$ *in* $\mathrm{SO}(8)$.

In fact Berger also included the eighth case $n = 16$ and $\mathrm{Hol}(g) = \mathrm{Spin}(9)$ in $\mathrm{SO}(16)$, but it was shown by Alekseevskii [7] and also by Brown and Gray [53] that any Riemannian metric with holonomy group $\mathrm{Spin}(9)$ is symmetric. We shall refer to Theorem 3.4.1 as *Berger's Theorem*, and to the groups in parts (i)–(vii) as *Berger's list*. Berger's proof will be discussed in §3.4.3. It is rather algebraic, and uses the classification of Lie groups and their representations, and the symmetry properties of the curvature tensor.

Berger proved that the groups on his list were the only possibilities, but he did not show whether the groups actually do occur as holonomy groups. It is now known (but this took another thirty years to find out) that all of the groups on Berger's list do occur as the holonomy groups of irreducible, nonsymmetric metrics.

Here are a few remarks about Berger's Theorem.

- Theorem 3.4.1 gives the holonomy group $\mathrm{Hol}(g)$ not just as an abstract group, but as a particular subgroup of $\mathrm{SO}(n)$. In other words, the holonomy representation of $\mathrm{Hol}(g)$ on $\mathbb{R}^n$ is completely specified.

- Combining Theorem 3.4.1 with the results of §3.2 and §3.3, we see that the restricted holonomy group $\mathrm{Hol}^0(g)$ of any Riemannian manifold (M, g) is a product of groups from Berger's list and the holonomy groups of Riemannian symmetric spaces, which are known from Cartan's classification.

- In cases (ii)–(v) of Theorem 3.4.1, we require $m \geqslant 2$ to avoid repeating holonomy groups. In case (ii), $\mathrm{U}(1) = \mathrm{SO}(2)$, coinciding with $n = 2$ in case (i). In (iii) $\mathrm{SU}(1) = \{1\}$, which acts reducibly, and can be regarded as $\mathrm{SO}(1) \times \mathrm{SO}(1)$ acting on $\mathbb{R}^2 = \mathbb{R} \oplus \mathbb{R}$. In (iv) $\mathrm{Sp}(1) = \mathrm{SU}(2)$ in $\mathrm{SO}(4)$, and in (v) $\mathrm{Sp}(1)\,\mathrm{Sp}(1) = \mathrm{SO}(4)$.

In §3.4.1, we will say a little bit about each of the groups on Berger's list. Chapters 5, 7, 10 and 11 discuss cases (ii)–(vii) in much more detail. Then §3.4.2 will discuss Berger's list as a whole, bringing out various common features and themes. Finally, §3.4.3 explains the principles behind the proofs by Berger and Simons of Theorem 3.4.1.

3.4.1 The groups on Berger's list

We make some brief remarks, with references, about each group on Berger's list.

(i) $\mathrm{SO}(n)$ is the holonomy group of generic Riemannian metrics.

(ii) Riemannian metrics g with $\mathrm{Hol}(g) \subseteq \mathrm{U}(m)$ are called *Kähler metrics*. Kähler metrics are a natural class of metrics on complex manifolds, and generic Kähler metrics on a given complex manifold have holonomy $U(m)$. Kähler geometry is covered by Griffiths and Harris [132], and Kobayashi and Nomizu [215, §IX].

(iii) Metrics g with $\mathrm{Hol}(g) \subseteq SU(m)$ are called *Calabi–Yau metrics*. Since $SU(m)$ is a subgroup of $U(m)$, all Calabi–Yau metrics are Kähler. If g is Kähler, then $\mathrm{Hol}^0(g) \subseteq SU(m)$ if and only if g is Ricci-flat. Thus Calabi–Yau metrics are locally the same as Ricci-flat Kähler metrics.

Explicit examples of *complete* metrics with holonomy $SU(m)$ were given by Calabi [69]. The existence of metrics with holonomy $SU(m)$ on *compact* manifolds follows from Yau's solution of the Calabi Conjecture, [345]. The most well-known example is the $K3$ surface, which admits a family of metrics with holonomy $SU(2)$.

(iv) Metrics g with $\mathrm{Hol}(g) \subseteq Sp(m)$ are called *hyperkähler metrics*. As $Sp(m) \subseteq SU(2m) \subset U(2m)$, hyperkähler metrics are Ricci-flat and Kähler. Explicit examples of *complete* metrics with holonomy $Sp(m)$ were found by Calabi [69]. Yau's solution of the Calabi Conjecture can be used to construct metrics with holonomy $Sp(m)$ on *compact* manifolds; examples were given by Fujiki [114] in the case $Sp(2)$, and Beauville [25] in the case $Sp(m)$.

(v) Metrics g with holonomy group $Sp(m)\,Sp(1)$ for $m \geqslant 2$ are called *quaternionic Kähler metrics*. (Note that quaternionic Kähler metrics are not in fact Kähler.) They are Einstein, but not Ricci-flat. For the theory of quaternionic Kähler manifolds, see Salamon [294], and for explicit examples, see Galicki and Lawson [120, 121]. It is an important open question whether there exist compact, non-symmetric quaternionic Kähler manifolds with positive scalar curvature.

(vi) and (vii) The holonomy groups G_2 and $\mathrm{Spin}(7)$ are called the *exceptional holonomy groups*. The existence of metrics with holonomy G_2 and $\mathrm{Spin}(7)$ was first established in 1985 by Bryant [56], using the theory of exterior differential systems. Explicit examples of complete metrics with holonomy G_2 and $\mathrm{Spin}(7)$ were found by Bryant and Salamon [64]. Metrics with holonomy G_2 and $\mathrm{Spin}(7)$ on compact manifolds were constructed by the author in [183, 184] for the case of G_2, and [185] for the case of $\mathrm{Spin}(7)$.

3.4.2 A discussion of Berger's list

Attempts to generalize the concept of number from real numbers to complex numbers and beyond led to the discovery of the four *division algebras*: the *real numbers* $\mathbb{R}$, the *complex numbers* $\mathbb{C}$, the *quaternions* $\mathbb{H}$, and the *octonions* or *Cayley numbers* $\mathbb{O}$. At each step in this sequence, the dimension doubles, and one algebraic property is lost. So, the complex numbers have dimension 2 (over $\mathbb{R}$) and are not ordered; the quaternions have dimension 4 and are not commutative; and the octonions have dimension 8 and are not associative. The sequence stops here, possibly because there are no algebraic properties left to lose.

The groups on Berger's list correspond to the division algebras. First consider cases (i)–(v). The group $SO(n)$ is a group of automorphisms of $\mathbb{R}^n$. Both $U(m)$ and $SU(m)$ are groups of automorphisms of $\mathbb{C}^m$, and $Sp(m)$ and $Sp(m)\,Sp(1)$ are automorphism groups of $\mathbb{H}^m$. To make the analogy between $\mathbb{R}$, $\mathbb{C}$ and $\mathbb{H}$ more complete, we add the holonomy group $O(n)$. Then $O(n)$, $U(m)$ and $Sp(m) Sp(1)$ are automorphism groups of $\mathbb{R}^n$, $\mathbb{C}^m$ and $\mathbb{H}^m$ respectively, preserving a metric. And $SO(n)$, $SU(m)$ and $Sp(m)$ are the subgroups of $O(n)$, $U(m)$ and $Sp(m) Sp(1)$ with 'determinant 1' in an appropriate sense.

The exceptional cases (vi) and (vii) also fit into this pattern, although not as neatly. One can regard G_2 and $Spin(7)$ as automorphism groups of $\mathbb{O}$. The octonions split as $\mathbb{O} \cong \mathbb{R} \oplus \operatorname{Im} \mathbb{O}$, where $\operatorname{Im} \mathbb{O} \cong \mathbb{R}^7$ is the *imaginary octonions*. The automorphism group of $\operatorname{Im} \mathbb{O}$ is G_2.

In some sense, G_2 is the group of 'determinant 1' automorphisms of $\mathbb{O}$, so it fits into the sequence $SO(n)$, $SU(m)$, $Sp(m)$, G_2. Similarly, $Spin(7)$ is the group of automorphisms of $\mathbb{O} \cong \mathbb{R}^8$ which preserve a certain part of the multiplicative structure of $\mathbb{O}$, and it fits into the sequence $O(n)$, $U(m)$, $Sp(m) Sp(1)$, $Spin(7)$.

Here are three ways in which we can gather together the holonomy groups on Berger's list into subsets with common features.

- The *Kähler holonomy groups* are $U(m)$, $SU(m)$ and $Sp(m)$. Any Riemannian manifold with one of these holonomy groups is a Kähler manifold, and thus a complex manifold. Therefore one can use complex geometry to study the Kähler holonomy groups, and this is a tremendous advantage.

  As complex manifolds are locally trivial, complex geometry has a very different character to Riemannian geometry, and a great deal is known about the *global* geometry of complex manifolds, particularly through complex algebraic geometry, which has no real parallel in Riemannian geometry.

  Although metrics with holonomy $Sp(m) Sp(1)$ for $m > 1$ are not Kähler, they should be considered along with the Kähler holonomy groups. A Riemannian manifold M with holonomy $Sp(m) Sp(1)$ has a *twistor space* [294], a complex manifold Z of real dimension $4m + 2$, which fibres over M with fibre $\mathbb{CP}^1$. If M has positive scalar curvature, then Z is Kähler. Thus, metrics with this holonomy group can also be studied using complex and Kähler geometry.

- The *Ricci-flat holonomy groups* are $SU(m)$, $Sp(m)$, G_2 and $Spin(7)$. Any metric with one of these holonomy groups is Ricci-flat. As irreducible symmetric spaces (other than $\mathbb{R}$) are Einstein with nonzero scalar curvature, none of the Ricci-flat holonomy groups can be the holonomy group of a symmetric space, or more generally of a homogeneous space. Because of this, simple examples of metrics with the Ricci-flat holonomy groups are difficult to find, and one has to work harder to get a feel for what the geometry of these metrics is like.

- The *exceptional holonomy groups* are G_2 and $Spin(7)$. They are the exceptional cases in Berger's classification, and they are rather different from the other holonomy groups. The holonomy groups $U(m)$, $SU(m)$, $Sp(m)$ and $Sp(m) Sp(1)$ can all be approached through complex geometry, and $SO(n)$ is uninteresting for obvious reasons.

  This leaves G_2 and $Spin(7)$, which are similar to one another but stand out from the rest. Since we cannot use complex manifold theory to tell us about the *global* geometry of manifolds with holonomy G_2 and $Spin(7)$, at present our understanding of them is essentially *local* in nature.

3.4.3   A sketch of the proof of Berger's Theorem

Here is a sketch of Berger's proof of Theorem 3.4.1. As M is simply-connected, Theorem 3.2.8 shows that $\mathrm{Hol}(g)$ is a closed, connected Lie subgroup of $\mathrm{SO}(n)$, and since g is irreducible, Corollary 3.2.5 shows that the representation of $\mathrm{Hol}(g)$ on $\mathbb{R}^n$ is irreducible. So, suppose that H is a closed, connected subgroup of $\mathrm{SO}(n)$ with irreducible representation on $\mathbb{R}^n$, and Lie algebra $\mathfrak{h}$. The classification of all such groups H follows from the classification of Lie groups (and is of considerable complexity).

Berger's method was to take the list of all such groups H, and to apply two tests to each possibility to find out if it could be a holonomy group. The only groups H which passed both tests are those in the theorem. Berger's tests are algebraic and involve the curvature tensor. Suppose that R_{abcd} is the Riemann curvature of a metric g with $\mathrm{Hol}(g) = H$. Then Theorem 3.1.2 shows that $R_{abcd} \in S^2\mathfrak{h}$, and the first Bianchi identity (3.3) applies.

If $\mathfrak{h}$ has large codimension in $\mathfrak{so}(n)$, then the vector space $\mathfrak{R}^H$ of elements of $S^2\mathfrak{h}$ satisfying (3.3) will be small, or even zero. But Theorem 2.4.3 shows that $\mathfrak{R}^H$ must be big enough to generate $\mathfrak{h}$. For many of the candidate groups H this does not hold, and so H cannot be a holonomy group. This is the first test. Now $\nabla_e R_{abcd}$ lies in $(\mathbb{R}^n)^* \otimes \mathfrak{R}^H$, and also satisfies the second Bianchi identity (3.4). Frequently these requirements imply that $\nabla R = 0$, so that g is locally symmetric. Therefore we may exclude such H, and this is Berger's second test.

Later, with the benefit of hindsight, Simons [313] found a shorter (but still difficult) proof of Theorem 3.4.1 based on showing that if g is irreducible and nonsymmetric, then $\mathrm{Hol}(g)$ must act transitively on the unit sphere in $\mathbb{R}^n$. But the list of compact, connected Lie groups acting transitively and effectively on spheres had already been found by Montgomery and Samelson [262], [38].

They turn out to be the groups on Berger's list, plus two others, $\mathrm{Sp}(m)\,\mathrm{U}(1)$ acting on S^{4m-1} for $m > 1$, and $\mathrm{Spin}(9)$ acting on S^{15}. Thus, to complete the second proof one must show that these two cannot occur as holonomy groups. Short accounts of Simons' proof are given by Besse [30, p. 303–305] and Salamon [294, p. 149–151].

3.5   Holonomy groups, exterior forms and cohomology

Let (M, g) be a compact Riemannian manifold. In this section we explore the links between $\mathrm{Hol}(g)$ and the de Rham cohomology $H^*(M, \mathbb{R})$. We explain how a G-structure on M divides the bundle of k-forms on M into a sum of vector subbundles, corresponding to irreducible representations of G. If the G-structure is torsion-free for $G \subseteq \mathrm{O}(n)$ then $H^k(M, \mathbb{R})$ has an analogous decomposition into vector subspaces by Hodge theory. Finally we show that if $\mathrm{Hol}(g)$ is one of the Ricci-flat holonomy groups then $H^1(M, \mathbb{R}) = 0$, and $\pi_1(M)$ is finite.

3.5.1   Decomposition of exterior forms

Let M be an n-manifold, G a Lie subgroup of $\mathrm{GL}(n, \mathbb{R})$, and Q a G-structure on M. Then from Definition 2.1.3, to each representation ρ of G on a vector space V we can associate a vector bundle $\rho(Q)$ over M, with fibre V. In particular, if ρ is the restriction to G of the natural representation of $\mathrm{GL}(n, \mathbb{R})$ on $V = \mathbb{R}^n$, then $\rho(Q) = TM$.

The representation of G on V induces representations of G on the dual vector space V^* and its exterior powers $\Lambda^k V^*$. Write ρ^k for the representation of G on $\Lambda^k V^*$. Then $\rho^k(Q) = \Lambda^k T^* M$, the bundle of k-forms on M. Now $\Lambda^k V^*$ is an *irreducible* representation of $\mathrm{GL}(n, \mathbb{R})$ for every k. However, if G is a proper subgroup of $\mathrm{GL}(n, \mathbb{R})$, it can happen that the representation ρ^k of G on $\Lambda^k V^*$ is *reducible*. Then we can write $\Lambda^k V^* = \bigoplus_{i \in I^k} W_i^k$ and $\rho^k = \bigoplus_{i \in I^k} \rho_i^k$, where each (ρ_i^k, W_i^k) is an irreducible representation of G, and I^k is a finite indexing set.

But then $\Lambda^k T^* M = \rho^k(Q) = \bigoplus_{i \in I^k} \rho_i^k(Q)$. This means that a G-structure on M induces a splitting of the vector bundle $\Lambda^k T^* M$ of k-forms on M into a direct sum of vector subbundles $\rho_i^k(Q)$ corresponding to irreducible representations of G. We shall use the notation Λ_i^k for $\rho_i^k(Q)$, and $\pi_i : \Lambda^k T^* M \to \Lambda_i^k$ for the projection to Λ_i^k in the decomposition $\Lambda^k T^* M = \bigoplus_{i \in I^k} \Lambda_i^k$.

Note that analogous decompositions hold for tensor bundles $\bigotimes^k TM \otimes \bigotimes^l T^* M$ on a manifold with a G-structure Q, and also, if G is a subgroup of $\mathrm{O}(n)$ and M is spin, for the spin bundles of M with respect to the Riemannian metric induced by Q. We will not make much use of these, though. The following proposition, which is trivial to prove, summarizes the material above.

Proposition 3.5.1 *Let G be a Lie subgroup of $\mathrm{GL}(n, \mathbb{R})$. Write (ρ, V) for the natural representation of G on $\mathbb{R}^n$, and let ρ^k be the induced representation of G on $\Lambda^k V^*$. Then $(\rho^k, \Lambda^k V^*)$ is a direct sum of irreducible representations (ρ_i^k, W_i^k) of G, for $i \in I^k$, a finite indexing set. Suppose M is an n-manifold, and Q a G-structure on M. Then there is a natural decomposition*

$$\Lambda^k T^* M = \bigoplus_{i \in I^k} \Lambda_i^k, \tag{3.12}$$

where Λ_i^k is a vector subbundle of $\Lambda^k T^* M$ with fibre W_i^k. If two representations (ρ_i^k, W_i^k) and (ρ_j^l, W_j^l) are isomorphic, then Λ_i^k and Λ_j^l are isomorphic. If $\phi : \Lambda^k V^* \to \Lambda^l V^*$ is a G-equivariant linear map, there is a corresponding map $\Phi : \Lambda^k T^* M \to \Lambda^l T^* M$ of vector bundles.

We can also do the same thing, but working with *complex* forms $\Lambda^k V^* \otimes_{\mathbb{R}} \mathbb{C}$, $\Lambda^k T^* M \otimes_{\mathbb{R}} \mathbb{C}$, and their decomposition into irreducible representations *over* $\mathbb{C}$. The advantage of doing this is that G-representations which are irreducible over $\mathbb{R}$ may be reducible over $\mathbb{C}$ when complexified, so working over $\mathbb{C}$ can give a finer decomposition into smaller pieces. This occurs when $G = \mathrm{U}(m)$, as will be explained in §5.2.2.

As an example of these ideas, we explain the Hodge star of §1.1.2.

Example 3.5.2 Let G be the subgroup $\mathrm{SO}(n)$ of $\mathrm{GL}(n, \mathbb{R})$. Then a G-structure on an n-manifold M is equivalent to a Riemannian metric g and an orientation on M. There is an isomorphism $* : \Lambda^k V^* \to \Lambda^{n-k} V^*$ between the representations of $\mathrm{SO}(n)$. By Proposition 3.5.1, this induces an isomorphism $* : \Lambda^k T^* M \to \Lambda^{n-k} T^* M$ called the *Hodge star*.

In the case $n = 4m$, the map $* : \Lambda^{2m} V^* \to \Lambda^{2m} V^*$ satisfies $*^2 = 1$, and so $\Lambda^{2m} V^*$ splits as $\Lambda^{2m} V^* = W_+^{2m} \oplus W_-^{2m}$, where $W_\pm^{2m}$ are the eigenspaces of $*$ with eigenvalues ± 1. Here W_+^{2m} and W_-^{2m} are in fact irreducible representations of $\mathrm{SO}(4m)$

of equal dimension, and we choose the indexing set $I^{2m} = \{+, -\}$. Thus, there is a corresponding splitting $\Lambda^{2m}T^*M \cong \Lambda^{2m}_+ \oplus \Lambda^{2m}_-$.

We are principally interested in the case in which G is a Lie subgroup of $\mathrm{O}(n)$ in $\mathrm{GL}(n, \mathbb{R})$, and Q a torsion-free G-structure. The condition that Q be torsion-free implies that the exterior derivative d on k-forms, and its formal adjoint d*, behave in a special way with regard to the splitting $\Lambda^k T^*M = \bigoplus_{i \in I^k} \Lambda^k_i$, and this has important consequences for the de Rham cohomology of M.

3.5.2 Hodge theory and the splitting of de Rham cohomology

Suppose (M, g) is a compact Riemannian n-manifold. Let G be a Lie subgroup of $\mathrm{O}(n)$, and suppose that $\mathrm{Hol}(g) \subseteq G$. Then from §3.1.4, there is a unique, torsion-free $\mathrm{O}(n)$-structure P on M induced by g, and a torsion-free G-structure Q on M contained in P. Moreover, the Levi-Civita connection ∇ of g reduces to Q.

Our goal is to study the *Hodge theory* of M. Hodge theory is described in §1.1.3, and concerns the *Laplacian* $\Delta = \mathrm{dd}^* + \mathrm{d}^*\mathrm{d}$ acting on k-forms on M, and its kernel $\mathcal{H}^k$, which is a finite-dimensional vector space of k-forms. The *Weitzenbock formula* for k-forms [30, §1.I], [296, Prop. 4.10] is

$$(\mathrm{dd}^* + \mathrm{d}^*\mathrm{d})\xi = \nabla^*\nabla\xi - 2\tilde{R}(\xi), \tag{3.13}$$

where, using the index notation and writing R_{ab} for the Ricci curvature and $R^a{}_{bcd}$ for the Riemann curvature of g, we have

$$
\begin{aligned}
\tilde{R}(\xi)_{r_1 \dots r_k} = &\sum_{1 \leqslant i < j \leqslant k} g^{bc} R^a{}_{r_i c r_j} \, \xi_{r_1 \dots r_{i-1} a r_{i+1} \dots r_{j-1} b r_{j+1} \dots r_k} \\
&- \tfrac{1}{2} \sum_{j=1}^{k} g^{ab} R_{r_j b} \, \xi_{r_1 \dots r_{j-1} a r_{j+1} \dots r_k}.
\end{aligned} \tag{3.14}
$$

Proposition 3.5.1 gives an isomorphism

$$C^\infty(\Lambda^k T^*M) = \bigoplus_{i \in I^k} C^\infty(\Lambda^k_i). \tag{3.15}$$

Suppose that $\xi \in C^\infty(\Lambda^k_i)$ for some i. As ∇ preserves the G-structure Q, it preserves the decomposition (3.12). Thus $\nabla\xi \in C^\infty(T^*M \otimes \Lambda^k_i)$, and so $\nabla^*\nabla\xi \in C^\infty(\Lambda^k_i)$.

Now Theorem 3.1.7 shows that the Riemann curvature R_{abcd} lies in $S^2\mathfrak{hol}_x(g)$ at each $x \in M$, where $\mathfrak{hol}_x(g)$ is a vector subspace of $\Lambda^2 T^*_x M$ isomorphic to the holonomy algebra $\mathfrak{hol}(g)$. As $\mathrm{Hol}(g) \subseteq G$, we have $\mathfrak{hol}(g) \subset \mathfrak{g}$, where $\mathfrak{g}$ is the Lie algebra of G. Using this, one can show that the linear map $\tilde{R} : \Lambda^k T^*M \to \Lambda^k T^*M$ defined by (3.14) preserves the splitting (3.12), and thus $\tilde{R}(\xi) \in C^\infty(\Lambda^k_i)$. Therefore we see that if $\xi \in C^\infty(\Lambda^k_i)$, then both $\nabla^*\nabla\xi$ and $\tilde{R}(\xi)$ lie in $C^\infty(\Lambda^k_i)$, and so $(\mathrm{dd}^* + \mathrm{d}^*\mathrm{d})\xi$ lies in $C^\infty(\Lambda^k_i)$ by (3.13). This proves that the Laplacian $\Delta = \mathrm{dd}^* + \mathrm{d}^*\mathrm{d}$ maps $C^\infty(\Lambda^k_i)$ into itself for each $i \in I^k$.

Thus in (3.15), the Laplacian Δ takes each factor $C^\infty(\Lambda^k_i)$ to itself. Therefore

$$\mathcal{H}^k = \bigoplus_{i \in I^k} \mathcal{H}^k_i, \quad \text{where} \quad \mathcal{H}^k = \mathrm{Ker}\,\Delta \quad \text{and} \quad \mathcal{H}^k_i = \mathrm{Ker}\big(\Delta|_{\Lambda^k_i}\big). \tag{3.16}$$

In Proposition 3.5.1 we saw that if W^k_i and W^l_j are isomorphic representations of G, then the vector bundles Λ^k_i and Λ^l_j are isomorphic. Now it turns out that $\nabla^*\nabla$ and $\tilde{R}$

depend only on the representation of G, and not on its particular embedding in $\Lambda^k V^*$. By (3.13) this is true also of $\mathrm{dd}^* + \mathrm{d}^*\mathrm{d}$, and so if W_i^k and W_j^l are isomorphic representations of G, it follows that $\mathscr{H}_i^k$ and $\mathscr{H}_j^l$ are isomorphic.

But $H^k(M,\mathbb{R})$ is isomorphic to $\mathscr{H}^k$ by Theorem 1.1.3. Thus, (3.16) gives a decomposition of $H^k(M,\mathbb{R})$ corresponding to the splitting (3.12). This yields:

Theorem 3.5.3 *Suppose M is a compact n-manifold and Q a torsion-free G-structure on M, where G is a Lie subgroup of $O(n)$, and let g be the metric associated to Q. Then Proposition 3.5.1 gives a splitting $\Lambda^k T^* M = \bigoplus_{i \in I^k} \Lambda_i^k$, corresponding to the decomposition of $\Lambda^k(\mathbb{R}^n)^*$ into irreducible representations of G.*

The Laplacian $\Delta = \mathrm{dd}^* + \mathrm{d}^*\mathrm{d}$ of g maps $C^\infty(\Lambda_i^k)$ to itself. Define $\mathscr{H}_i^k = \mathrm{Ker}\,\Delta|_{\Lambda_i^k}$, and let $H_i^k(M,\mathbb{R})$ be the subspace of the de Rham cohomology group $H^k(M,\mathbb{R})$ with representatives in $\mathscr{H}_i^k$. Then $H_i^k(M,\mathbb{R}) \cong \mathscr{H}_i^k$, and we have the direct sum

$$H^k(M,\mathbb{R}) = \bigoplus_{i \in I^k} H_i^k(M,\mathbb{R}). \tag{3.17}$$

If W_i^k and W_j^l are isomorphic as representations of G, then $H_i^k(M,\mathbb{R}) \cong H_j^l(M,\mathbb{R})$.

The theorem shows that if a Riemannian metric g on a compact manifold M has $\mathrm{Hol}(g) = G$, then the de Rham cohomology $H^*(M,\mathbb{R})$ has a natural decomposition into smaller pieces, which depend on G and its representations. The Betti numbers of M are $b^k(M) = \dim H^k(M,\mathbb{R})$. Define the *refined Betti numbers* $b_i^k(M)$ to be $b_i^k(M) = \dim H_i^k(M,\mathbb{R})$, for $i \in I^k$. Then (3.17) shows that $b^k(M) = \sum_{i \in I^k} b_i^k(M)$. The refined Betti numbers carry both *topological* information about M, and *geometrical* information about the G-structure Q.

If a compact manifold M admits a metric g with $\mathrm{Hol}(g) = G$, then the theorem forces its cohomology $H^*(M,\mathbb{R})$ to assume a certain form. Conversely, if one can show that $H^*(M,\mathbb{R})$ cannot be written in this way, then M cannot admit any metric g with $\mathrm{Hol}(g) = G$. Thus, one can prove that some compact manifolds do not admit metrics with a given holonomy group, for purely topological reasons. For example, when $G = U(m)$ one finds that $b^k(M)$ must be even when k is odd; roughly speaking, this is because each irreducible representation occurs twice.

Suppose that for some k and $i \in I^k$, we have $W_i^k \cong \mathbb{R}$, the trivial representation of G. Now $\Lambda^0(\mathbb{R}^n)^* = W_1^0$ is also a copy of $\mathbb{R}$, the trivial representation. So W_i^k and W_1^0 are isomorphic as representations of G, and $H_i^k(M,\mathbb{R}) \cong H_1^0(M,\mathbb{R}) = H^0(M,\mathbb{R})$ by Theorem 3.5.3. But $H^0(M,\mathbb{R}) = \mathbb{R}$, as M is connected. Thus, if W_i^k is the trivial representation $\mathbb{R}$, then there is a natural isomorphism $H_i^k(M,\mathbb{R}) \cong \mathbb{R}$.

The explanation for this is simple. Proposition 2.5.2 shows that there is a 1-1 correspondence between *constant tensors* on M, and invariant elements of the corresponding representation of $\mathrm{Hol}(g)$. Since $\mathrm{Hol}(g) \subseteq G$, if W_i^k is a trivial representation its elements are invariant under $\mathrm{Hol}(g)$, and so correspond to *constant k-forms*. Now if ξ is a constant k-form, then $\nabla \xi = 0$. But $\mathrm{d}\xi$ and $\mathrm{d}^*\xi$ are components of $\nabla \xi$, and so $\mathrm{d}\xi = \mathrm{d}^*\xi = 0$. Hence $(\mathrm{dd}^* + \mathrm{d}^*\mathrm{d})\xi = 0$, and ζ lies in $\mathscr{H}^k$. Thus, if W_i^k is a trivial representation, then $\mathscr{H}_i^k$ is a vector space of constant k-forms isomorphic to W_i^k.

We have shown that if Q is a torsion-free G-structure on a compact manifold M, then to each G-invariant element of $\Lambda^k V^*$ there corresponds a constant k-form ξ, and this defines a cohomology class $[\xi] \in H^k(M, \mathbb{R})$. So, to each torsion-free G-structure Q we associate a collection of cohomology classes in $H^k(M, \mathbb{R})$, corresponding to the G-invariant part of $\Lambda^k V^*$. Now $H^k(M, \mathbb{R})$ is a topological invariant of M, and does not depend on Q. Thus we can compare the cohomology classes associated to *different* torsion-free G-structures Q, Q' on M.

Often, the family of torsion-free G-structures on M, when divided by the group of diffeomorphisms of M isotopic to the identity, forms a finite-dimensional manifold $\mathscr{M}_G$ called a *moduli space*. The cohomology classes associated to a torsion-free G-structure give maps $\mathscr{M}_G \to H^k(M, \mathbb{R})$, and these provide a natural coordinate system on $\mathscr{M}_G$. More generally, the splitting (3.17) may also be regarded in this light. This topological information can be exploited to give a local (and, in good cases, global) description of the moduli spaces $\mathscr{M}_G$, when G is one of the Ricci-flat holonomy groups.

3.5.3   One-forms and the Ricci-flat holonomy groups

Let (M, g) be a compact Riemannian n-manifold. From eqns (3.13) and (3.14) we see that if ξ is a 1-form, then

$$(\mathrm{dd}^* + \mathrm{d}^*\mathrm{d})\xi_a = \nabla^*\nabla\xi_a + R_{ab}g^{bc}\xi_c,$$

where R_{ab} is the Ricci curvature of g. Suppose now that $\xi \in \mathscr{H}^1$, the kernel of $\mathrm{dd}^* + \mathrm{d}^*\mathrm{d}$ on 1-forms. Then we have $\nabla^*\nabla\xi_a + R_{ab}g^{bc}\xi_c = 0$. Taking the inner product of this equation with ξ and integrating by parts yields

$$\|\nabla\xi\|_{L^2}^2 + \int_M R_{ab}g^{bc}g^{ad}\xi_c\xi_d\mathrm{d}V = 0. \tag{3.18}$$

If the Ricci curvature R_{ab} is zero, this shows that $\|\nabla\xi\|_{L^2} = 0$, so that $\nabla\xi = 0$.

More generally, if R_{ab} is *nonnegative*, then the second term in (3.18) is nonnegative. But the first term is also nonnegative, so both must be zero. Thus, if R_{ab} is nonnegative, we have $\nabla\xi = 0$. If R_{ab} is *positive definite*, then either $\xi \equiv 0$, or the second term in (3.18) is positive, which is a contradiction. This shows that if R_{ab} is zero or nonnegative, then all 1-forms ξ in $\mathscr{H}^1$ are constant, and if R_{ab} is positive definite, then all 1-forms ξ in $\mathscr{H}^1$ are zero.

Suppose that R_{ab} is nonnegative, and that $\dim \mathscr{H}^1 = k > 0$. Choose a basis $\xi_1, \ldots, \xi_k$ for $\mathscr{H}^1$. Then $\xi_1, \ldots, \xi_k$ are constant 1-forms, and so by Proposition 2.5.2 they correspond to elements of $(\mathbb{R}^n)^*$ fixed by $\mathrm{Hol}(g)$. Clearly, this implies that $k \leqslant n$ and $\mathbb{R}^n$ splits orthogonally as $\mathbb{R}^n = \mathbb{R}^k \oplus \mathbb{R}^{n-k}$, where $\mathrm{Hol}(g)$ preserves the splitting and acts trivially on $\mathbb{R}^k$.

Thus, the action of $\mathrm{Hol}(g)$ on $\mathbb{R}^n$ is *reducible*, and so by Corollary 3.2.5 the metric g is locally reducible. Moreover, let $(\tilde{M}, \tilde{g})$ be the universal cover of (M, g). Then $\mathrm{Hol}(\tilde{g}) = \mathrm{Hol}^0(g) \subseteq \mathrm{Hol}(g)$, and $\tilde{g}$ is *complete* as M is compact. So Theorem 3.2.7 applies to show that $(\tilde{M}, \tilde{g})$ is globally reducible. In fact $(\tilde{M}, \tilde{g})$ is isometric to a product $\mathbb{R}^k \times N$, where $\mathbb{R}^k$ carries the Euclidean metric and N is a Riemannian $(n-k)$-manifold.

But $\mathscr{H}^1 \cong H^1(M, \mathbb{R})$ by Theorem 1.1.3. Thus we have proved the following result, known as the *Bochner Theorem*.

Theorem 3.5.4. (Bochner [32]) *Let* (M, g) *be a compact Riemannian manifold. If the Ricci curvature of* g *is nonnegative, then* $\dim H^1(M, \mathbb{R}) = k \leqslant \dim M$, *and the universal cover* $(\tilde{M}, \tilde{g})$ *of* (M, g) *is isometric to a product* $\mathbb{R}^k \times N$, *where* $\mathbb{R}^k$ *has the Euclidean metric. If the Ricci curvature of* g *is positive definite, then* $H^1(M, \mathbb{R}) = 0$.

One can also prove strong results on the fundamental group of a compact manifold with nonnegative Ricci curvature. As a consequence of the *Cheeger–Gromoll Splitting Theorem* [75], [30, §6.G] we get the following result, from Besse [30, Cor. 6.67].

Theorem 3.5.5 *Suppose* (M, g) *is a compact Riemannian manifold. If* g *is Ricci-flat then* M *admits a finite cover isometric to* $T^k \times N$, *where* T^k *carries a flat metric and* N *is a compact, simply-connected Riemannian manifold. If the Ricci curvature of* M *is positive definite then* $\pi_1(M)$ *is finite.*

Now suppose that $\mathrm{Hol}(g)$ is one of the Ricci-flat holonomy groups $\mathrm{SU}(m)$, $\mathrm{Sp}(m)$, G_2 and $\mathrm{Spin}(7)$. Then $R_{ab} = 0$, so that $\mathscr{H}^1$ consists of constant 1-forms, from above. But by Proposition 2.5.2, the constant tensors on M are entirely determined by $\mathrm{Hol}(g)$. Since $\mathrm{SU}(m)$, $\mathrm{Sp}(m)$, G_2 and $\mathrm{Spin}(7)$ all fix no nonzero elements in $(\mathbb{R}^n)^*$, there are no nonzero constant 1-forms. Thus $\mathscr{H}^1 = 0$, and so $H^1(M, \mathbb{R}) = 0$. Moreover, it follows from Theorem 3.5.5 that $\pi_1(M)$ is finite. Therefore we have:

Corollary 3.5.6 *Suppose* (M, g) *is a compact Riemannian manifold and* $\mathrm{Hol}(g)$ *is one of the Ricci-flat holonomy groups* $\mathrm{SU}(m)$, $\mathrm{Sp}(m)$, G_2 *and* $\mathrm{Spin}(7)$. *Then* $H^1(M, \mathbb{R}) = 0$, *so that* $b^1(M) = 0$, *and the fundamental group* $\pi_1(M)$ *is finite.*

This is an example of how the topology of a compact manifold M can impose constraints on the possible holonomy groups $\mathrm{Hol}(g)$ of Riemannian metrics g on M.

3.6 Spinors and holonomy groups

If (M, g) is a Riemannian spin manifold, then there is a natural vector bundle S over M called the *spin bundle*, and sections of S are called *spinors*. Spinors are closely related to tensors, and have similar properties. In particular, just as the constant tensors on M are determined by $\mathrm{Hol}(g)$, so the constant spinors on M are also determined by $\mathrm{Hol}(g)$, with the right choice of spin structure. In this section we explore the link between spin geometry and holonomy groups, and deduce some topological information about compact $4m$-manifolds M with the Ricci-flat holonomy groups.

3.6.1 Introduction to spin geometry

Here is a brief explanation of some ideas from spin geometry that we will need later. Some general references are Lawson and Michelson [233] and Harvey [150].

For each $n \geqslant 3$, the Lie group $\mathrm{SO}(n)$ is connected and has fundamental group $\pi_1(\mathrm{SO}(n)) = \mathbb{Z}_2$. Therefore it has a double cover, the Lie group $\mathrm{Spin}(n)$, which is a compact, connected, simply-connected Lie group. The covering map $\pi : \mathrm{Spin}(n) \rightarrow \mathrm{SO}(n)$ is a Lie group homomorphism. There is a natural representation Δ^n of $\mathrm{Spin}(n)$, called the *spin representation*. It has the following properties:

- Δ^{2m} is a complex representation of $\mathrm{Spin}(2m)$, with complex dimension 2^m. It splits into a direct sum $\Delta^{2m} = \Delta^{2m}_+ \oplus \Delta^{2m}_-$, where $\Delta^{2m}_{\pm}$ are irreducible representations of $\mathrm{Spin}(2m)$ with complex dimension 2^{m-1}.

- Δ^{2m+1} is a complex representation of $\mathrm{Spin}(2m+1)$, with complex dimension 2^m. It is irreducible, and does not split into positive and negative parts.
- When $n = 8k-1, 8k$ or $8k+1$, $\Delta^n = \Delta^n_{\mathbb{R}} \otimes_{\mathbb{R}} \mathbb{C}$, where $\Delta^n_{\mathbb{R}}$ is a *real* representation of $\mathrm{Spin}(n)$.

Now let (M, g) be an oriented Riemannian n-manifold. The metric and orientation on M induce a unique $\mathrm{SO}(n)$-structure P on M. A *spin structure* $(\tilde{P}, \pi)$ on M is a principal bundle $\tilde{P}$ over M with fibre $\mathrm{Spin}(n)$, together with a map of bundles $\pi : \tilde{P} \to P$, that is locally modelled on the projection $\pi : \mathrm{Spin}(n) \to \mathrm{SO}(n)$. We may regard $\tilde{P}$ as a double cover of P, and π as the covering map.

Spin structures do not exist on every manifold. In fact, an oriented Riemannian manifold M admits a spin structure if and only if $w_2(M) = 0$, where $w_2(M) \in H^2(M, \mathbb{Z}_2)$ is the second Stiefel–Whitney class of M. Also, if a spin structure exists it may not be unique: when $w_2(M) = 0$, the family of spin structures on M is parametrized by $H^1(M, \mathbb{Z}_2)$. This is finite if M is compact, and zero if M is simply-connected. We call M a *spin manifold* if $w_2(M) = 0$, that is, if M admits a spin structure.

Let (M, g) be an oriented, spin Riemannian n-manifold, and choose a spin structure $(\tilde{P}, \pi)$ on M. Define the *(complex) spin bundle* $S \to M$ to be $S = \tilde{P} \times_{\mathrm{Spin}(n)} \Delta^n$. Then S is a complex vector bundle over M, with fibre Δ^n. Sections of S are called *spinors*. If $n = 2m$, then Δ^n splits as $\Delta^n = \Delta^n_+ \oplus \Delta^n_-$, and so S also splits as $S = S_+ \oplus S_-$, where $S_\pm$ are vector subbundles of S with fibre $\Delta^n_\pm$. Sections of S_+, S_- are called *positive* and *negative spinors* respectively. In dimensions $8k-1, 8k$ and $8k+1$ there is a real spin representation $\Delta^n_{\mathbb{R}}$ as well as a complex one Δ^n. In this case one defines the *real spin bundle* $S_{\mathbb{R}} = \tilde{P} \times_{\mathrm{Spin}(n)} \Delta^n_{\mathbb{R}}$. We shall always work with complex spinors, unless we explicitly say otherwise.

The $\mathrm{SO}(n)$-bundle P over M has a natural connection, the Levi-Civita connection ∇ of g. Because $\pi : \tilde{P} \to P$ is locally an isomorphism, we may lift ∇ to $\tilde{P}$. Thus, $\tilde{P}$ also carries a natural connection, and as in §2.1, this induces a connection $\nabla^S : C^\infty(S) \to C^\infty(T^*M \otimes S)$ on S, called the *spin connection*. Now, there is a natural linear map from $T^*M \otimes S$ to S, defined by *Clifford multiplication*. Composing this map with ∇^S gives a first-order, linear partial differential operator $D : C^\infty(S) \to C^\infty(S)$ called the *Dirac operator*.

The Dirac operator is self-adjoint and elliptic. In even dimensions, it splits as a sum $D = D_+ \oplus D_-$, where D_+ maps $C^\infty(S_+) \to C^\infty(S_-)$ and D_- maps $C^\infty(S_-) \to C^\infty(S_+)$. Here $D_\pm$ are both first-order linear elliptic operators, and D_- is the formal adjoint of D_+, and vice versa. The result of changing the orientation of M is, in even dimensions, to exchange S_+ and S_-, and D_+ and D_-.

3.6.2 Parallel spinors and holonomy groups

Let (M, g) be an oriented Riemannian n-manifold with a spin structure. Then the holonomy group $\mathrm{Hol}(\nabla^s)$ is a subgroup of $\mathrm{Spin}(n)$. Moreover, under the projection $\pi : \mathrm{Spin}(n) \to \mathrm{SO}(n)$, the image of $\mathrm{Hol}(\nabla^s)$ is exactly $\mathrm{Hol}(g)$. The projection $\pi : \mathrm{Hol}(\nabla^s) \to \mathrm{Hol}(g)$ may be an isomorphism, or it may be a double cover; in general, this depends on the choice of spin structure. However, if M is simply-connected, then both $\mathrm{Hol}(g)$ and $\mathrm{Hol}(\nabla^s)$ are connected, which forces $\mathrm{Hol}(\nabla^s)$ to be the identity component of $\pi^{-1}(\mathrm{Hol}(g))$ in $\mathrm{Spin}(n)$. Thus, for simply-connected spin manifolds, the classification of holonomy groups of spin connections ∇^s follows from that of Riemannian holonomy groups.

Suppose that $\sigma \in C^\infty(S)$ satisfies $\nabla^s\sigma = 0$, so that σ is a *parallel spinor*, or *constant spinor*. Just as in §2.5.1 we found a 1-1 correspondence between constant tensors and elements of the appropriate representation invariant under $\mathrm{Hol}(\nabla)$, so there is a 1-1 correspondence between parallel spinors and elements of Δ^n invariant under $\mathrm{Hol}(\nabla^s)$. Therefore, one can apply Berger's classification of Riemannian holonomy groups to classify the holonomy groups of metrics with parallel spinors. This has been done by Wang [336, p. 59], in the following result.

Theorem 3.6.1 *Let M be an orientable, connected, simply-connected spin n-manifold for $n \geqslant 3$, and g an irreducible Riemannian metric on M. Define N to be the dimension of the space of parallel spinors on M. If n is even, define $N_\pm$ to be the dimensions of the spaces of parallel spinors in $C^\infty(S_\pm)$, so that $N = N_+ + N_-$.*

Suppose $N \geqslant 1$. Then, after making an appropriate choice of orientation for M, exactly one of the following holds:

 (i) *$n = 4m$ for $m \geqslant 1$ and $\mathrm{Hol}(g) = \mathrm{SU}(2m)$, with $N_+ = 2$ and $N_- = 0$,*
 (ii) *$n = 4m$ for $m \geqslant 2$ and $\mathrm{Hol}(g) = \mathrm{Sp}(m)$, with $N_+ = m + 1$ and $N_- = 0$,*
 (iii) *$n = 4m + 2$ for $m \geqslant 1$ and $\mathrm{Hol}(g) = \mathrm{SU}(2m + 1)$, with $N_+ = 1$ and $N_- = 1$,*
 (iv) *$n = 7$ and $\mathrm{Hol}(g) = G_2$, with $N = 1$, and*
 (v) *$n = 8$ and $\mathrm{Hol}(g) = \mathrm{Spin}(7)$, with $N_+ = 1$ and $N_- = 0$.*

With the opposite orientation, the values of $N_\pm$ are exchanged.

Notice that the holonomy groups appearing here are exactly the Ricci-flat holonomy groups. Hence, every Riemannian spin manifold that admits a nonzero parallel spinor is Ricci-flat. (In fact, this can be proved directly.) Conversely, it is natural to ask whether every Riemannian manifold with one of the Ricci-flat holonomy groups is in fact a spin manifold, and possesses constant spinors. The answer to this is yes, and it follows from the next proposition.

Proposition 3.6.2 *Suppose M is an n-manifold that admits a G-structure Q, where $n \geqslant 3$ and G is a connected, simply-connected subgroup of $\mathrm{SO}(n)$. Then M is spin, and has a natural spin structure $\tilde{P}$ induced by Q.*

Proof Since G and $\mathrm{SO}(n)$ are connected, the embedding $\iota : G \hookrightarrow \mathrm{SO}(n)$ lifts to a homomorphism $\tilde{\iota}$ between the universal covers of G and $\mathrm{SO}(n)$. Because G is simply-connected and the universal cover of $\mathrm{SO}(n)$ is $\mathrm{Spin}(n)$, as $n \geqslant 3$, this gives an injective Lie group homomorphism $\tilde{\iota} : G \hookrightarrow \mathrm{Spin}(n)$ such that $\pi \circ \tilde{\iota} = \iota$, where $\pi : \mathrm{Spin}(n) \to \mathrm{SO}(n)$ is the covering map.

Now the G-structure Q on M induces an $\mathrm{SO}(n)$-structure $P = \mathrm{SO}(n) \cdot Q$ on M. By definition, M is spin if and only if P admits a double cover $\tilde{P}$, which fibres over M with fibre $\mathrm{Spin}(n)$. But using the embedding $\tilde{\iota} : G \hookrightarrow \mathrm{Spin}(n)$ we may define $\tilde{P} = Q \times_G \mathrm{Spin}(n)$, and this is indeed a double cover of P that fibres over M with fibre $\mathrm{Spin}(n)$. Thus M is spin. $\qquad\square$

Since all of the Ricci-flat holonomy groups $\mathrm{SU}(m)$, $\mathrm{Sp}(m)$, G_2 and $\mathrm{Spin}(7)$ are connected and simply-connected, the next corollary quickly follows.

Corollary 3.6.3 *Let (M, g) be a Riemannian manifold, and suppose that $\mathrm{Hol}(g)$ is one of the Ricci-flat holonomy groups $\mathrm{SU}(m)$, $\mathrm{Sp}(m)$, G_2 and $\mathrm{Spin}(7)$. Then M is spin, with a preferred spin structure. With this spin structure, the spaces of parallel spinors on M are nonzero and have the dimensions prescribed by Theorem 3.6.1.*

Thus, an irreducible metric has one of the Ricci-flat holonomy groups if and only if it admits a nonzero constant spinor.

3.6.3   Harmonic spinors and the $\hat{A}$-genus

Let (M, g) be a compact Riemannian spin manifold, with spin bundle S and Dirac operator D. The Weitzenbock formula of Lichnerowicz [244, p. 8, eqn (7)] states that if $\sigma \in C^\infty(S)$ then $D^2\sigma = (\nabla^s)^*\nabla^s\sigma + \frac{1}{4}s\,\sigma$, where s is the *scalar curvature* of g. We call σ a *harmonic spinor* if $D\sigma = 0$, that is, if $\sigma \in \mathrm{Ker}\,D$. By the 'Bochner argument' used to prove Theorem 3.5.4, we find:

Proposition 3.6.4 *Let (M, g) be a compact Riemannian spin manifold. If the scalar curvature s of g is zero, then every harmonic spinor on M is parallel. If the scalar curvature s of g is positive, then there are no nonzero harmonic spinors.*

Suppose (M, g) is a compact Riemannian spin manifold of dimension $4m$, for $m \geqslant 1$. Then D_+ is a linear elliptic operator, with adjoint D_-, which has *index* $\mathrm{ind}\,D_+ = \dim \mathrm{Ker}\,D_+ - \dim \mathrm{Ker}\,D_-$. Now the *Atiyah–Singer Index Theorem* [13] gives a topological formula for $\mathrm{ind}\,D_+$, and by [13, Th. 5.3], $\mathrm{ind}\,D_+$ is equal to $\hat{A}(M)$, a characteristic class of M called the $\hat{A}$-*genus*. From Proposition 3.6.4, if M is a compact Riemannian spin manifold of dimension $4m$ with positive scalar curvature, then $\mathrm{Ker}\,D_\pm = 0$, and so $\hat{A}(M) = 0$. Conversely, if M is a compact spin manifold of dimension $4m$ and $\hat{A}(M) \neq 0$, then there are no Riemannian metrics with positive scalar curvature on M.

If M is a compact Riemannian spin manifold with zero scalar curvature, Proposition 3.6.4 shows that $\mathrm{Ker}\,D_\pm$ are the spaces of parallel positive and negative spinors, which are determined by $\mathrm{Hol}(\nabla^s)$. Thus, $\mathrm{ind}\,D_+$ is determined by $\mathrm{Hol}(\nabla^s)$, and in fact by $\mathrm{Hol}(g)$. When $\mathrm{Hol}(g)$ is $\mathrm{SU}(2m)$, $\mathrm{Sp}(m)$ or $\mathrm{Spin}(7)$, by Corollary 3.6.3 the dimensions of $\mathrm{Ker}\,D_\pm$ are those given in Theorem 3.6.1, and this gives $\hat{A}(M)$ explicitly. Thus we have proved the following result.

Theorem 3.6.5 *Let (M, g) be a compact Riemannian spin manifold of dimension $4m$, for $m \geqslant 1$. If the scalar curvature of g is positive, then $\hat{A}(M) = 0$. If the scalar curvature of g is zero, then $\hat{A}(M)$ is an integer determined by the holonomy group $\mathrm{Hol}(g)$ of g. In particular, if $\mathrm{Hol}(g) = \mathrm{SU}(2m)$ then $\hat{A}(M) = 2$. If $\mathrm{Hol}(g) = \mathrm{SU}(2k) \times \mathrm{SU}(2m-2k)$ for $0 < k < m$ then $\hat{A}(M) = 4$. If $\mathrm{Hol}(g) = \mathrm{Sp}(m)$ then $\hat{A}(M) = m + 1$, and if $m = 2$ and $\mathrm{Hol}(g) = \mathrm{Spin}(7)$ then $\hat{A}(M) = 1$.*

Here is one way this theorem is applied. Let G be $\mathrm{SU}(2m)$, $\mathrm{Sp}(m)$ or $\mathrm{Spin}(7)$. Then we can use analytic methods to construct a torsion-free G-structure Q on a compact manifold M. The Riemannian metric g associated to Q must then have $\mathrm{Hol}(g) \subseteq G$, but it does not immediately follow that $\mathrm{Hol}(g) = G$. However, by studying the topology of M we may compute $\hat{A}(M)$, and so use Theorem 3.6.5 to distinguish between the different possibilities for $\mathrm{Hol}(g)$.

4

Calibrated geometry

The theory of *calibrated geometry* was invented by Harvey and Lawson in their seminal paper [151], which is still an excellent reference. It concerns calibrated submanifolds, a special kind of minimal submanifold of a Riemannian manifold M, which are defined using a closed form on M called a calibration. It is closely connected with the theory of Riemannian holonomy groups because Riemannian manifolds with reduced holonomy usually come equipped with one or more natural calibrations.

We begin in §4.1 by discussing minimal and calibrated submanifolds. Section 4.2 explains the relation between calibrated geometry and holonomy groups, central to this book, and §4.3 considers the problem of classifying constant calibrations on $\mathbb{R}^n$. Finally, §4.4 describes *geometric measure theory*, which studies a class of measure-theoretic generalizations of submanifolds called *integral currents* that have good compactness properties and are natural in calibrated geometry problems.

4.1 Minimal submanifolds and calibrated submanifolds

For clarity we first define *submanifolds*, following Kobayashi and Nomizu [214, §1.1].

Definition 4.1.1 Let M and N be smooth manifolds, and $\iota : N \rightarrow M$ a smooth map. We call ι an *immersion* if for each $x \in N$, the linear map $d\iota|_x : T_x N \rightarrow T_{\iota(x)} M$ is injective. We then say that N (or its image $\iota(N)$ in M) is an *immersed submanifold* in M. We call ι an *embedding* if it is an injective immersion. We then say that N (or its image $\iota(N)$ in M) is an *embedded submanifold* in M. Two submanifolds $\iota : N \rightarrow M$ and $\iota' : N' \rightarrow M$ are *isomorphic* if there exists a diffeomorphism $\delta : N \rightarrow N'$ with $\iota = \iota' \circ \delta$. We consider isomorphic submanifolds to be the same. In this book we do not require submanifolds N to be *connected* (though the ambient manifold M is assumed connected), nor the image $\iota(N)$ to be *closed* in M.

As usual, we generally think of a submanifold $\iota : N \rightarrow M$ as a special kind of subset of M, that is, we implicitly identify N with its image $\iota(N)$ in M, and suppress all mention of the immersion ι. For embedded submanifolds this is reasonable, as the set $\iota(N)$ can be given the structure of a smooth manifold uniquely such that the inclusion $\iota(N) \hookrightarrow N$ is an embedding, and then $\iota(N) \hookrightarrow N$ is isomorphic to $\iota : N \rightarrow M$.

But for immersed submanifolds there can be distinct points $x, y \in N$ with $\iota(x) = \iota(y)$ in M, so that $\iota(x)$ is a 'self-intersection point' of $\iota(N)$. Then the image $\iota(N)$ may be *singular* at $\iota(x)$, that is, we cannot give $\iota(N)$ the structure of a manifold such that the inclusion $\iota(N) \hookrightarrow N$ is an embedding. Alternatively, $\iota(N)$ could be a nonsingular embedded submanifold with $\iota : N \to \iota(N)$ a nontrivial multiple cover. In these cases one cannot reconstruct N, ι up to isomorphism from $\iota(N)$, so an immersed submanifold is more than just a subset of M. Next we discuss *minimal submanifolds*. Two introductory books on minimal submanifolds are Lawson [232] and Xin [342].

Definition 4.1.2 Let (M, g) be a Riemannian manifold, and N a compact submanifold of M with immersion $\iota : N \to M$. Then $\iota^*(g)$ is a Riemannian metric on N, so we can define the *volume* $\mathrm{vol}(N)$ of N by integrating the volume form of $\iota^*(g)$ over N. We call N a *minimal submanifold* if its volume is stationary under small variations of the immersion $\iota : N \to M$.

For *noncompact* submanifolds we modify this definition as follows, as $\mathrm{vol}(N)$ may be infinite (undefined), or N might have a boundary. Let $S \subset N$ be an open subset whose closure $\bar{S}$ in N is compact. Then the volume $\mathrm{vol}(S)$ is well-defined and finite. We call N a *minimal submanifold* if for all such subsets S, the volume $\mathrm{vol}(S)$ is stationary under small variations of $\iota : N \to M$ which are supported in S, that is, we consider only variations $\tilde{\iota} : N \to M$ with $\tilde{\iota}|_{N \setminus S} \equiv \iota|_{N \setminus S}$.

One-dimensional minimal submanifolds are *geodesics*. It is natural to think of minimal submanifolds as submanifolds with minimal volume, but in fact we only require the volume to be stationary. For example, the equator in $\mathcal{S}^2$ is minimal, but does not minimize length amongst lines of latitude.

We can also define minimal submanifolds by a partial differential equation. Suppose N is a submanifold in a Riemannian manifold (M, g). Let $\nu \to N$ be the normal bundle of N in M, so that $TM|_N = TN \oplus \nu$ is an orthogonal direct sum. The *second fundamental form* is a section B of $S^2 T^* N \otimes \nu$ such that whenever v, w are vector fields on M with $v|_N, w|_N$ sections of TN over N, then $B \cdot (v|_N \otimes w|_N) = \pi_\nu(\nabla_v w|_N)$, where '$\cdot$' contracts $S^2 T^* N$ with $TN \otimes TN$, ∇ is the Levi-Civita connection of g, and π_ν is the projection to ν in the splitting $TM|_N = TN \oplus \nu$.

The *mean curvature vector* κ of N is the trace of the second fundamental form B taken using the metric g on N. It is a section of the normal bundle ν. It can be shown by the Euler–Lagrange method that a submanifold N is minimal if and only if its mean curvature vector κ is zero. Thus, an equivalent definition of minimal submanifold is a submanifold with zero mean curvature.

If $\iota : N \to M$ is an immersed submanifold, then the mean curvature κ of N depends on ι and its first and second derivatives, so the condition that N be minimal is a *nonlinear second-order p.d.e.* on ι. In a certain sense (after factoring out by diffeomorphisms of N) this p.d.e. is *elliptic*, as in §1.3. Thus elliptic regularity results apply, as in §1.4. Using results of Morrey [266], one can show that if $\iota : N \to M$ is a C^2 immersion with zero mean curvature then $\iota : N \to M$ is isomorphic to a smooth immersion $\iota' : N' \to M$, and if (M, g) is real analytic then we can take N', ι' to be real analytic.

Now we can define *calibrated submanifolds*, following Harvey and Lawson [151].

Definition 4.1.3 Let (M, g) be a Riemannian manifold. An *oriented tangent k-plane* V on M is a vector subspace V of some tangent space $T_x M$ to M with $\dim V = k$, equipped with an orientation. If V is an oriented tangent k-plane on M then $g|_V$ is a

Euclidean metric on V, so combining $g|_V$ with the orientation on V gives a natural *volume form* vol_V on V, which is a k-form on V.

Now let φ be a closed k-form on M. We say that φ is a *calibration* on M if for every oriented k-plane V on M we have $\varphi|_V \leqslant \mathrm{vol}_V$. Here $\varphi|_V = \alpha \cdot \mathrm{vol}_V$ for some $\alpha \in \mathbb{R}$, and $\varphi|_V \leqslant \mathrm{vol}_V$ if $\alpha \leqslant 1$. Let N be an oriented submanifold of M with dimension k. Then each tangent space $T_x N$ for $x \in N$ is an oriented tangent k-plane. We say that N is a *calibrated submanifold* or *φ-submanifold* if $\varphi|_{T_x N} = \mathrm{vol}_{T_x N}$ for all $x \in N$.

It is easy to show that calibrated submanifolds are automatically *minimal submanifolds* [151, Th. II.4.2]. We prove this in the compact case, but noncompact calibrated submanifolds are locally volume-minimizing as well.

Proposition 4.1.4 *Let (M, g) be a Riemannian manifold, φ a calibration on M, and N a compact φ-submanifold in M. Then N is volume-minimizing in its homology class.*

Proof Let $\dim N = k$, and let $[N] \in H_k(M, \mathbb{R})$ and $[\varphi] \in H^k(M, \mathbb{R})$ be the homology and cohomology classes of N and φ. Then

$$[\varphi] \cdot [N] = \int_{x \in N} \varphi|_{T_x N} = \int_{x \in N} \mathrm{vol}_{T_x N} = \mathrm{vol}(N),$$

since $\varphi|_{T_x N} = \mathrm{vol}_{T_x N}$ for each $x \in N$, as N is a calibrated submanifold. If N' is any other compact k-submanifold of M with $[N'] = [N]$ in $H_k(M, \mathbb{R})$, then

$$[\varphi] \cdot [N] = [\varphi] \cdot [N'] = \int_{x \in N'} \varphi|_{T_x N'} \leqslant \int_{x \in N'} \mathrm{vol}_{T_x N'} = \mathrm{vol}(N'),$$

since $\varphi|_{T_x N'} \leqslant \mathrm{vol}_{T_x N'}$ as φ is a calibration. The last two equations give $\mathrm{vol}(N) \leqslant \mathrm{vol}(N')$. Thus N is volume-minimizing in its homology class. $\qquad\square$

Now let (M, g) be a Riemannian manifold with a calibration φ, and let $\iota : N \to M$ be an immersed submanifold. Whether N is a φ-submanifold depends upon the tangent spaces of N. That is, it depends on ι and its first derivative. So, to be calibrated with respect to φ is a *first-order* p.d.e. on ι. But if N is calibrated then N is minimal, so it has zero mean curvature. As above this is a *second-order* p.d.e. on ι, which is implied by the calibration first-order p.d.e. One moral is that the calibrated equations, being first-order, are often easier to solve than the minimal submanifold equations, which are second-order. So calibrated geometry is a fertile source of examples of minimal submanifolds.

4.2 Calibrated geometry and Riemannian holonomy groups

A calibration φ on (M, g) can only have nontrivial calibrated submanifolds if there exist oriented tangent k-planes V on M with $\varphi|_V = \mathrm{vol}_V$. For instance, $\varphi = 0$ is a calibration on M, but has no calibrated submanifolds. This means that a calibration φ is only interesting if the set of oriented tangent k-planes V on M with $\varphi|_V = \mathrm{vol}_V$ has reasonably large dimension. We now explain a natural method of constructing interesting calibrations φ on Riemannian manifolds (M, g) with special holonomy, which automatically have families of calibrated tangent k-planes with reasonably large dimension.

Let $G \subset O(n)$ be a possible holonomy group of a Riemannian metric. In particular, we can take G to be one of Berger's list $U(m), SU(m), Sp(m), Sp(m) Sp(1), G_2$ or

Spin(7). Then G acts on the k-forms $\Lambda^k(\mathbb{R}^n)^*$ on $\mathbb{R}^n$, so we can look for G-invariant k-forms on $\mathbb{R}^n$. Suppose φ_0 is a nonzero, G-invariant k-form on $\mathbb{R}^n$. By rescaling φ_0 we can arrange that for each oriented k-plane $U \subset \mathbb{R}^n$ we have $\varphi_0|_U \leqslant \mathrm{vol}_U$, and that $\varphi_0|_U = \mathrm{vol}_U$ for at least one such U. Let $\mathcal{F}$ be the family of oriented k-planes U in $\mathbb{R}^n$ with $\varphi_0|_U = \mathrm{vol}_U$, and let $l = \dim \mathcal{F}$. Then $\mathcal{F}$ is nonempty. Since φ_0 is G-invariant, if $U \in \mathcal{F}$ then $\gamma \cdot U \in \mathcal{F}$ for all $\gamma \in G$. This usually means that $l > 0$.

Let M be a manifold of dimension n, and g a metric on M with Levi-Civita connection ∇ and holonomy group G. Then by Proposition 2.5.2 there is a k-form φ on M with $\nabla\varphi = 0$, corresponding to φ_0. Hence $\mathrm{d}\varphi = 0$, and φ is closed. Also, the condition $\varphi_0|_U \leqslant \mathrm{vol}_U$ for all oriented k-planes U in $\mathbb{R}^n$ implies that $\varphi|_V \leqslant \mathrm{vol}_V$ for all oriented tangent k-planes in M. Thus φ is a *calibration* on M.

At each point $x \in M$ there is an l-dimensional family $\mathcal{F}_x$ of oriented tangent k-planes V with $\varphi|_V = \mathrm{vol}_V$, isomorphic to $\mathcal{F}$. Hence, the set of oriented tangent k-planes V in M with $\varphi|_V = \mathrm{vol}_V$ has dimension $l + n$, which is reasonably large. This suggests that locally there should exist many φ-submanifolds N in M, so the calibrated geometry of φ on (M, g) is nontrivial.

This gives us a general method for finding interesting calibrations on manifolds with reduced holonomy. Here are some examples of this, taken from [151].

- Let $G = \mathrm{U}(m) \subset \mathrm{O}(2m)$. Then G preserves a 2-form ω_0 on $\mathbb{R}^{2m}$. If g is a metric on M with holonomy $\mathrm{U}(m)$ then g is *Kähler* with complex structure J, and the 2-form ω on M associated to ω_0 is the *Kähler form* of g, as in Chapter 5.

  One can show that ω is a calibration on (M, g), and the calibrated submanifolds are exactly the *holomorphic curves* in (M, J). More generally $\omega^k/k!$ is a calibration on M for $1 \leqslant k \leqslant m$, and the corresponding calibrated submanifolds are the complex k-dimensional submanifolds of (M, J).

- Let $G = \mathrm{SU}(m) \subset \mathrm{O}(2m)$. Riemannian manifolds (M, g) with holonomy $\mathrm{SU}(m)$ are called *Calabi–Yau manifolds*, and are the subject of Chapter 7. A Calabi–Yau manifold comes equipped with a complex m-form θ called a *holomorphic volume form*. The real part $\mathrm{Re}\,\theta$ is a calibration on M. Its calibrated submanifolds are called *special Lagrangian submanifolds*, and are the subject of Chapter 8.

- The group $G_2 \subset \mathrm{O}(7)$ preserves a 3-form φ_0 and a 4-form $*\varphi_0$ on $\mathbb{R}^7$, which will be given explicitly in §11.1. Thus a Riemannian 7-manifold (M, g) with holonomy G_2 comes with a 3-form φ and 4-form $*\varphi$, which are both calibrations. We call φ-submanifolds *associative 3-folds*, and $*\varphi$-submanifolds *coassociative 4-folds*. They will be studied in Chapter 12.

- The group $\mathrm{Spin}(7) \subset \mathrm{O}(8)$ preserves a 4-form Ω_0 on $\mathbb{R}^8$, which will be given explicitly in §11.4. Thus a Riemannian 8-manifold (M, g) with holonomy $\mathrm{Spin}(7)$ has a 4-form Ω, which is a calibration. We call Ω-submanifolds *Cayley 4-folds*. They will be discussed in Chapter 12.

It is an important general principle that to each calibration φ on an n-manifold (M, g) with special holonomy we construct in this way, there corresponds a constant calibration φ_0 on $\mathbb{R}^n$. Locally, φ-submanifolds in M look very like φ_0-submanifolds in $\mathbb{R}^n$, and have many of the same properties. Thus, to understand the calibrated submanifolds in a manifold with special holonomy, it is often a good idea to start by studying the corresponding calibrated submanifolds of $\mathbb{R}^n$.

In particular, *singularities* of φ-submanifolds in M will be locally modelled on singularities of φ_0-submanifolds in $\mathbb{R}^n$. So by studying singular φ_0-submanifolds in $\mathbb{R}^n$, we may understand the singular behaviour of φ-submanifolds in M.

4.3 Classification of calibrations on $\mathbb{R}^n$

Just as we classified Riemannian holonomy groups in Chapter 3, it is an interesting problem to look for some form of *classification* of calibrated geometries. In particular, we will see in Chapters 8 and 12 that the special Lagrangian, associative, coassociative and Cayley calibrations mentioned in §4.2 have a beautiful, rich theory with many calibrated submanifolds, and a classification should reveal if there are other interesting calibrated geometries that we have missed.

We now explain some ideas from Harvey and Lawson [151, §II.7] and Morgan [264] which go some way towards answering this question. We restrict to calibrations φ on Euclidean space $(\mathbb{R}^n, g)$ with constant coefficients. As in §4.2, if such a φ is invariant under a possible holonomy group $G \subset O(n)$ then Riemannian n-manifolds with holonomy G also carry calibrations modelled on φ, so classifying constant calibrations φ on $\mathbb{R}^n$ implies a classification of constant calibrations on manifolds with special holonomy.

Let $\varphi \in \Lambda^k(\mathbb{R}^n)^*$ be a calibration on $\mathbb{R}^n$. Write $\mathrm{Gr}_+(k, \mathbb{R}^n)$ for the Grassmannian of oriented k-planes V in $\mathbb{R}^n$, so that $\varphi|_V \leqslant \mathrm{vol}_V$ for all $V \in \mathrm{Gr}_+(k, \mathbb{R}^n)$ as φ is a calibration. Define $\mathcal{F}_\varphi \subseteq \mathrm{Gr}_+(k, \mathbb{R}^n)$ to be the subset of V which are calibrated with respect to φ, that is, those V with $\varphi|_V = \mathrm{vol}_V$. Then an oriented k-submanifold $N \subseteq \mathbb{R}^n$ is a φ-submanifold if and only if $T_x N \in \mathcal{F}_\varphi$ for all $x \in N$. So $\mathcal{F}_\varphi$ determines the family of φ-submanifolds in $\mathbb{R}^n$.

The first principle in our classification is that two calibrations $\varphi, \psi \in \Lambda^k(\mathbb{R}^n)^*$ are equivalent if $\mathcal{F}_\varphi = \mathcal{F}_\psi$, that is, if they determine the same calibrated submanifolds. Also we will consider two calibrations φ, ψ equivalent if they are conjugate under $O(n)$. Thus what we want to determine is the family of possible subsets $\mathcal{F}_\varphi$ in $\mathrm{Gr}_+(k, \mathbb{R}^n)$ realized by calibrations $\varphi \in \Lambda^k(\mathbb{R}^n)^*$, up to the action of $O(n)$ on $\mathrm{Gr}_+(k, \mathbb{R}^n)$. Our next definition explains a useful point of view on these sets $\mathcal{F}_\varphi$.

Definition 4.3.1 For integers $0 < k < n$, define a map $\mathrm{Gr}_+(k, \mathbb{R}^n) \to \Lambda^k \mathbb{R}^n$ as follows: map each V in $\mathrm{Gr}_+(k, \mathbb{R}^n)$, which is an oriented k-dimensional subspace of $\mathbb{R}^n$, to $e_1 \wedge e_2 \wedge \cdots \wedge e_k$ in $\Lambda^k \mathbb{R}^n$, where $(e_1, \dots, e_k)$ is an oriented orthonormal basis of V. The point $e_1 \wedge e_2 \wedge \cdots \wedge e_k$ is independent of choice of $(e_1, \dots, e_k)$, so this defines a map $\mathrm{Gr}_+(k, \mathbb{R}^n) \to \Lambda^k \mathbb{R}^n$, which is in fact an embedding. We use this to regard $\mathrm{Gr}_+(k, \mathbb{R}^n)$ as a compact submanifold of $\Lambda^k \mathbb{R}^n$.

Now $\Lambda^k(\mathbb{R}^n)^* \cong (\Lambda^k \mathbb{R}^n)^*$, so each k-form φ on $\mathbb{R}^n$ defines a linear map $L_\phi : \Lambda^k \mathbb{R}^n \to \mathbb{R}$. It is easy to see that for $V \in \mathrm{Gr}_+(k, \mathbb{R}^n) \subset \Lambda^k \mathbb{R}^n$ we have $\varphi|_V = L_\phi(V) \, \mathrm{vol}_V$. Therefore φ is a *calibration* on $\mathbb{R}^n$ if and only if $\big| L_\phi|_{\mathrm{Gr}_+(k, \mathbb{R}^n)} \big| \leqslant 1$, and if this holds then the set $\mathcal{F}_\varphi$ is the set of $V \in \mathrm{Gr}_+(k, \mathbb{R}^n)$ with $L_\varphi(V) = 1$. That is, $\mathcal{F}_\varphi$ is the intersection of $\mathrm{Gr}_+(k, \mathbb{R}^n)$ with the real hyperplane $L_\varphi = 1$ in $\Lambda^k \mathbb{R}^n$.

Thus we have the following picture. The oriented Grassmannian $\mathrm{Gr}_+(k, \mathbb{R}^n)$ is a compact submanifold of $\Lambda^k \mathbb{R}^n$. If we bring a real hyperplane in from infinity in the vector space $\Lambda^k \mathbb{R}^n \cong \mathbb{R}^{n!/k!(n-k)!}$ until it first touches the submanifold $\mathrm{Gr}_+(k, \mathbb{R}^n)$

then the contact set is $\mathcal{F}_\varphi$ for some calibration φ, and all nonempty $\mathcal{F}_\varphi$ arise this way. Therefore these sets $\mathcal{F}_\varphi$ are called *faces* of the Grassmannian $\mathrm{Gr}_+(k, \mathbb{R}^n)$, by analogy with faces of polyhedra in Euclidean geometry.

The *Hodge star* $*$ of §1.1.2 gives linear isometries $* : \Lambda^k(\mathbb{R}^n)^* \to \Lambda^{n-k}(\mathbb{R}^n)^*$ and $* : \Lambda^k \mathbb{R}^n \to \Lambda^{n-k} \mathbb{R}^n$, which satisfy $* \, \mathrm{Gr}_+(k, \mathbb{R}^n) = \mathrm{Gr}_+(n-k, \mathbb{R}^n)$ and $*\mathcal{F}_\varphi = \mathcal{F}_{*\varphi}$. To see this on the level of k-planes, note that each oriented k-plane V has a unique orthogonal oriented $(n-k)$-plane $V^\perp$, such that if $\varphi \in \Lambda^k(\mathbb{R}^n)^*$ with $\varphi|_V = \alpha \, \mathrm{vol}_V$ for $\alpha \in \mathbb{R}$ then $*\varphi|_{V^\perp} = \alpha \, \mathrm{vol}_{V^\perp}$. Thus, φ is a calibration if and only if $*\varphi$ is a calibration, and then V is calibrated w.r.t. φ if and only if $V^\perp$ is calibrated w.r.t. $*\varphi$. Therefore, classifying faces of $\mathrm{Gr}_+(k, \mathbb{R}^n)$ is equivalent to classifying faces of $\mathrm{Gr}_+(n-k, \mathbb{R}^n)$.

The case $k = 1$ is trivial as $\mathrm{Gr}_+(1, \mathbb{R}^n)$ is the unit sphere $\mathcal{S}^{n-1}$ in $\mathbb{R}^n$, and nonempty faces are single points. Faces of $\mathrm{Gr}_+(2, \mathbb{R}^n)$ are classified by Harvey and Lawson [151, Th. II.7.16], and come from symplectic calibrations on subspaces $\mathbb{R}^{2m} \subseteq \mathbb{R}^n$. Applying the Hodge star gives classifications of faces of $\mathrm{Gr}_+(n-1, \mathbb{R}^n)$ and $\mathrm{Gr}_+(n-2, \mathbb{R}^n)$. This gives a complete description of calibrations of degree $1, 2, n-2$ and $n-1$ in $\mathbb{R}^n$. We express it in terms of choices of orthonormal coordinates in $\mathbb{R}^n$, which is equivalent to working up to the action of $\mathrm{O}(n)$.

Theorem 4.3.2 *Let $\varphi \in \Lambda^k(\mathbb{R}^n)^*$ be a calibration with $\mathcal{F}_\varphi \neq \emptyset$ for $k = 1, 2, n-2$ or $n-1$. Then there exist orthonormal coordinates $(x_1, \ldots, x_n)$ on $\mathbb{R}^n$ such that either:*

(i) $k = 1$, $\varphi = \mathrm{d}x_1$ *and* $\mathcal{F}_\varphi = \{(1, 0, \ldots, 0)\}$, *and φ-submanifolds in $\mathbb{R}^n$ are real affine lines parallel to $(1, 0, \ldots, 0)$;*

(ii) $k = 2$ *and* $\mathcal{F}_\varphi = \mathcal{F}_{\psi_m} \cong \mathbb{CP}^{m-1}$ *for some $1 \leqslant m \leqslant n/2$ where $\psi_m = \sum_{j=1}^m \mathrm{d}x_{2j-1} \wedge \mathrm{d}x_{2j}$, and φ-submanifolds are of the form $\Sigma \times \{v\}$ for Σ a holomorphic curve in $\mathbb{R}^{2m} = \mathbb{C}^{2m}$ with complex coordinates $z_j = x_{2j-1} + ix_{2j}$ for $j = 1, \ldots, m$, and $v \in \mathbb{R}^{n-2m}$ with coordinates $(x_{2m+1}, \ldots, x_n)$;*

(iii) $k = n - 2$ *and* $\mathcal{F}_\varphi = \mathcal{F}_{*\psi_m} \cong \mathbb{CP}^{m-1}$ *for some $1 \leqslant m \leqslant n/2$ and ψ_m as in (ii), and φ-submanifolds are of the form $\Sigma \times \mathbb{R}^{n-2m}$ for Σ a complex hypersurface in $\mathbb{R}^{2m} = \mathbb{C}^{2m}$ with complex coordinates $z_j = x_{2j-1} + ix_{2j}$ for $j = 1, \ldots, m$, where $\mathbb{R}^{n-2m}$ has coordinates $(x_{2m+1}, \ldots, x_n)$; or*

(iv) $k = n - 1$, $\varphi = \mathrm{d}x_2 \wedge \cdots \wedge \mathrm{d}x_n$ *and* $\mathcal{F}_\varphi = \{\frac{\partial}{\partial x_2} \wedge \cdots \wedge \frac{\partial}{\partial x_n}\}$, *and φ-submanifolds in $\mathbb{R}^n$ are real hyperplanes $x_1 = c$ for $c \in \mathbb{R}$.*

Theorem 4.3.2 classifies all calibrations on $\mathbb{R}^n$ for $n \leqslant 5$, so the first new case is 3-forms on $\mathbb{R}^6$. These were classified by Dadok and Harvey [90] and Morgan [263, §4].

Theorem 4.3.3 *Let $\varphi \in \Lambda^3(\mathbb{R}^6)^*$ be a calibration with $\mathcal{F}_\varphi \neq \emptyset$. Then there exist orthonormal coordinates $(x_1, \ldots, x_6)$ on $\mathbb{R}^6$ such that either:*

(i) $\mathcal{F}_\varphi = \mathcal{F}_{\mathrm{d}x_1 \wedge \mathrm{d}x_2 \wedge \mathrm{d}x_3} = \{\frac{\partial}{\partial x_1} \wedge \frac{\partial}{\partial x_2} \wedge \frac{\partial}{\partial x_3}\}$, *and φ-submanifolds in $\mathbb{R}^6$ are affine 3-planes $(x_4, x_5, x_6) = \mathbf{c}$ for $\mathbf{c} \in \mathbb{R}^3$;*

(ii) $\mathcal{F}_\varphi = \{V_1, V_2\}$, *where V_1, V_2 have oriented orthonormal bases $\frac{\partial}{\partial x_1}, \frac{\partial}{\partial x_2}, \frac{\partial}{\partial x_3}$ and $\cos\theta_1 \frac{\partial}{\partial x_1} + \sin\theta_1 \frac{\partial}{\partial x_4}, \cos\theta_2 \frac{\partial}{\partial x_2} + \sin\theta_2 \frac{\partial}{\partial x_5}, \cos\theta_3 \frac{\partial}{\partial x_3} + \sin\theta_3 \frac{\partial}{\partial x_6}$ for some $0 < \theta_1 \leqslant \theta_2 \leqslant \theta_3 < \pi$ with $\theta_3 < \theta_1 + \theta_2$, and φ-submanifolds in $\mathbb{R}^6$ are affine 3-planes parallel to V_1 or V_2. Furthermore, for each choice of $\theta_1, \theta_2, \theta_3$ as above there is a unique calibration $\varphi \in \Lambda^3(\mathbb{R}^6)^*$ with $\mathcal{F}_\varphi = \{V_1, V_2\}$;*

(iii) $\mathcal{F}_\varphi = \mathcal{F}_{d x_1 \wedge d x_2 \wedge d x_5 + d x_3 \wedge d x_4 \wedge d x_5} \cong \mathbb{CP}^1$, and φ-submanifolds are of the form $\Sigma \times \mathbb{R} \times \{c\}$ for Σ a holomorphic curve in $\mathbb{R}^4 = \mathbb{C}^2$ with complex coordinates $(z_1, z_2) = (x_1 + i x_2, x_3 + i x_4)$, and $\mathbb{R}$ has coordinate x_5, and $x_6 = c \in \mathbb{R}$; or

(iv) $\varphi = \mathrm{Re}\big((d x_1 + i d x_2) \wedge (d x_3 + i d x_4) \wedge (d x_5 + i d x_6)\big)$ and $\mathcal{F}_\varphi \cong \mathrm{SU}(3)/\mathrm{SO}(3)$, and φ-submanifolds are special Lagrangian 3-folds in $\mathbb{R}^6 = \mathbb{C}^3$, as in Chapter 8.

The faces of $\mathrm{Gr}_+(3, \mathbb{R}^7)$ were classified by Harvey and Morgan [152, Th. 6.2]. There are five discrete and five infinite types of families. Here is a partial statement.

Theorem 4.3.4 Let $\varphi \in \Lambda^3(\mathbb{R}^7)^*$ be a calibration with $\mathcal{F}_\varphi \neq \emptyset$. Then there exist orthonormal coordinates $(x_1, \ldots, x_7)$ on $\mathbb{R}^7$ such that either:

(i) $\varphi = d x_{123} + d x_{145} + d x_{167} + d x_{246} - d x_{257} - d x_{347} - d x_{356}$ as in (11.1), where $d x_{abc} = d x_a \wedge d x_b \wedge d x_c$, and $\mathcal{F}_\varphi \cong G_2/\mathrm{SO}(4)$, and φ-submanifolds are associative 3-folds in $\mathbb{R}^7$, as in Chapter 12;

(ii) $\varphi = \mathrm{Re}\big((d x_1 + i d x_2) \wedge (d x_3 + i d x_4) \wedge (d x_5 + i d x_6)\big)$ and $\mathcal{F}_\varphi \cong \mathrm{SU}(3)/\mathrm{SO}(3)$, and φ-submanifolds are of the form $\Sigma \times \{c\}$ for Σ a special Lagrangian 3-fold in $\mathbb{R}^6 \cong \mathbb{C}^3$, as in Chapter 8, and $x_7 = c \in \mathbb{R}$;

(iii) $\varphi = (d x_1 \wedge d x_2 + d x_3 \wedge d x_4 + d x_5 \wedge d x_6) \wedge d x_7$ and $\mathcal{F}_\varphi \cong \mathbb{CP}^2$, and φ-submanifolds are of the form $\Sigma \times \mathbb{R}$ for Σ a holomorphic curve in $\mathbb{R}^6 = \mathbb{C}^3$ with complex coordinates $(z_1, z_2, z_3) = (x_1 + i x_2, x_3 + i x_4, x_5 + i x_6)$, and $\mathbb{R}$ has coordinate x_7;

(iv) $\mathcal{F}_\varphi = \mathcal{F}_{d x_1 \wedge d x_2 \wedge d x_5 + d x_3 \wedge d x_4 \wedge d x_5} \cong \mathbb{CP}^1$, and φ-submanifolds are of the form $\Sigma \times \mathbb{R} \times \{\mathbf{c}\}$ for Σ a holomorphic curve in $\mathbb{R}^4 = \mathbb{C}^2$ with complex coordinates $(z_1, z_2) = (x_1 + i x_2, x_3 + i x_4)$, and $\mathbb{R}$ has coordinate x_5, and $(x_6, x_7) = \mathbf{c} \in \mathbb{R}^2$; or

(v) in the remaining six cases $\mathcal{F}_\varphi$ is one point, or two points, or two $\mathbb{CP}^1$'s intersecting in a point, or diffeomorphic to $\mathcal{S}^1, \mathcal{S}^2$ or $\mathcal{S}^3$.

Applying the Hodge star classifies faces of $\mathrm{Gr}_+(4, \mathbb{R}^7)$. In particular, from case (i) we get coassociative 4-folds in $\mathbb{R}^7$, which will be studied in Chapter 12. One conclusion we can draw from Theorems 4.3.2–4.3.4 is that for all constant calibrations φ on $\mathbb{R}^n$ for $n \leqslant 7$ with $\dim \mathcal{F}_\varphi$ reasonably large (that is, with $\dim \mathcal{F}_\varphi > 0$ for $n \leqslant 6$ and $\dim \mathcal{F}_\varphi > 3$ for $n = 7$), the φ-submanifolds are derived from one of: (a) complex curves in $\mathbb{C}^2$ or $\mathbb{C}^3$, (b) complex surfaces in $\mathbb{C}^3$, (c) special Lagrangian 3-folds in $\mathbb{C}^3$, (d) associative 3-folds in $\mathbb{R}^7$, or (e) coassociative 4-folds in $\mathbb{R}^7$. So we have not missed any interesting calibrated geometries in 7 dimensions or less.

No complete classification of faces of $\mathrm{Gr}_+(k, \mathbb{R}^n)$ is known for any $n \geqslant 8$, and it seems certain the full answer will be very complex and messy. Dadok, Harvey and Morgan [91] classify $\mathcal{F}_\varphi$ for φ in several large subspaces of $\Lambda^4(\mathbb{R}^8)^*$, including $\Lambda_+^4(\mathbb{R}^8)^*$. Calibrations in $\Lambda_+^4(\mathbb{R}^8)^*$ include the following interesting examples:

- The Cayley 4-form Ω_0 of (11.12), which has $\mathcal{F}_{\Omega_0} \cong \mathrm{Spin}(7)/(\mathrm{SU}(2)^3/\mathbb{Z}_2)$, with $\dim \mathcal{F}_{\Omega_0} - 12$. Its calibrated submanifolds are Cayley 4-folds, as in Chapter 12.
- The special Lagrangian calibration $\mathrm{Re}(dz_1 \wedge \cdots \wedge dz_4)$ on $\mathbb{C}^4 = \mathbb{R}^8$, which has $\mathcal{F}_{\mathrm{Re}(\cdots)} \cong \mathrm{SU}(4)/\mathrm{SO}(4)$, with $\dim \mathcal{F}_{\mathrm{Re}(\cdots)} = 9$. Its calibrated submanifolds are special Lagrangian 4-folds, as in Chapter 8.

- The Kähler calibration $\frac{1}{2}\omega \wedge \omega$ for ω the Kähler form on $\mathbb{C}^4 = \mathbb{R}^8$, which has $\mathcal{F}_{\omega \wedge \omega / 2} \cong U(4)/U(2) \times U(2)$, with $\dim \mathcal{F}_{\omega \wedge \omega / 2} = 8$. Its calibrated submanifolds are *complex surfaces* in $\mathbb{C}^4$.
- A calibration φ with $\mathcal{F}_\varphi \cong Sp(2)/U(2)$ and $\dim \mathcal{F}_\varphi = 6$, whose calibrated submanifolds are *complex Lagrangian surfaces* in $\mathbb{C}^4 = \mathbb{R}^8$ with respect to the complex symplectic form $dz_1 \wedge dz_2 + dz_3 \wedge dz_4$.
- A calibration φ with $\mathcal{F}_\varphi \cong Sp(2)/Sp(1) \times Sp(1)$ and $\dim \mathcal{F}_\varphi = 4$, whose calibrated submanifolds are *affine quaternionic lines* $\mathbb{H}$ in $\mathbb{H}^2 = \mathbb{R}^8$.

For all five examples we have $\mathcal{F}_\varphi \subseteq \mathcal{F}_{\Omega_0}$, so that all these different kinds of calibrated 4-folds in $\mathbb{R}^8$ are examples of Cayley 4-folds. Also, if A^3 is an associative 3-fold and C^4 a coassociative 4-fold in $\mathbb{R}^7$ then $\mathbb{R} \times A^3$ and $\{c\} \times C^4$ for $c \in \mathbb{R}$ are Cayley 4-folds in $\mathbb{R} \times \mathbb{R}^7 = \mathbb{R}^8$. Thus, Cayley 4-folds in $\mathbb{R}^8$ include as special cases all the other interesting classes of calibrated 4-submanifolds we have found.

4.4   Geometric measure theory and tangent cones

We now review some *geometric measure theory*, and its application to calibrated geometry. An introduction to the subject is provided by Morgan [265] and an in-depth (but dated) treatment by Federer [102], and Harvey and Lawson [151, §II] relate geometric measure theory to calibrated geometry. Geometric measure theory studies measure-theoretic generalizations of submanifolds called *rectifiable* and *integral currents*, which may be very singular, and is particularly powerful for *minimal* submanifolds.

Let (M, g) be a complete Riemannian manifold. For $k = 0, \dots, \dim M$ write $\mathscr{D}^k$ for the vector space of smooth k-forms on M with compact support, and $\mathscr{D}_k$ for its dual vector space. Elements of $\mathscr{D}_k$ are called *currents*. We equip $\mathscr{D}_k$ with the weak topology, that is, $T_j \to T$ in $\mathscr{D}_k$ as $j \to \infty$ if and only if $T_j(\varphi) \to T(\varphi)$ in $\mathbb{R}$ for all $\varphi \in \mathscr{D}^k$.

Let N be a compact, oriented k-dimensional submanifold of M, possibly with boundary. Then N defines a current $\overline{N} \in \mathscr{D}_k$ by $\overline{N}(\varphi) = \int_N \varphi$ for all $\varphi \in \mathscr{D}^k$. There is a natural boundary operator $\partial : \mathscr{D}_k \to \mathscr{D}_{k-1}$ given by $(\partial T)(\varphi) = T(d\varphi)$ for all $\varphi \in \mathscr{D}^{k-1}$. This is compatible with the boundary operator on submanifolds, for if N is an oriented k-submanifold with boundary and $\varphi \in \mathscr{D}^{k-1}$ then $\int_{\partial N} \varphi = \int_N d\varphi$ by Stokes' Theorem, so $\partial \overline{N} = \overline{(\partial N)}$.

Thus currents can be regarded as generalizations of oriented submanifolds. The space $\mathscr{D}_k$ is too huge to be useful, so we introduce two subspaces, rectifiable and integral currents. Roughly speaking, *rectifiable currents* are compactly-supported, finite area, countable $\mathbb{Z}$-linear combinations of currents of the following form: let $A \subset \mathbb{R}^k$ have finite Hausdorff k-measure and $F : \mathbb{R}^k \to M$ be a Lipschitz map, then we define a current by $\varphi \mapsto \int_A F^*(\varphi)$. As F is Lipschitz it is differentiable almost everywhere, so $F^*(\varphi)$ exists almost everywhere as a bounded k-form on $\mathbb{R}^k$, and $\int_A F^*(\varphi)$ is well-defined. Rectifiable k-currents T have a *volume* $\mathrm{vol}(T)$ defined using Hausdorff k-measure. We call $T \in \mathscr{D}_k$ an *integral current* if T and ∂T are rectifiable.

Here is a very important property of integral currents [265, 5.5], [102, 4.2.17].

Theorem 4.4.1. (The Compactness Theorem) *Let (M, g) be a complete Riemannian manifold, $U \subseteq M$ be compact, $0 \leqslant k \leqslant \dim M$ and $C \geqslant 0$. Then the subset of k-*

dimensional integral currents T in M supported in U with $\mathrm{vol}(T) \leqslant C$ *and* $\mathrm{vol}(\partial T) \leqslant$
C *is compact in the weak topology on* $\mathscr{D}_k$.

The following argument is useful in the study of minimal submanifolds. Let (M, g) be a compact Riemannian manifold, and α a nonzero homology class in $H_k(M, \mathbb{Z})$. We would like to find a compact minimal k-submanifold N in M with homology class $[N] = \alpha$. To do this, we choose a minimizing sequence $(N_i)_{i=1}^{\infty}$ of compact submanifolds N_i with $[N_i] = \alpha$, such that $\mathrm{vol}(N_i)$ approaches the infimum of volumes of submanifolds with homology class α as $i \to \infty$.

Pretend for the moment that the set of all closed k-dimensional submanifolds N without boundary with $\mathrm{vol}(N) \leqslant C$ is a *compact* topological space. Then there exists a subsequence $(N_{i_j})_{j=1}^{\infty}$ which converges to some submanifold N, which is the minimal submanifold we want. In fact this does not work, because the set of submanifolds N does not have the compactness properties we need. However, Theorem 4.4.1 implies that integral currents do have these properties, and so every integral homology class α in $H_k(M, \mathbb{Z})$ is represented by a volume-minimizing integral current.

The question remains: how close are these volume-minimizing integral currents to being submanifolds? Here is a major result of Almgren [9], [265, Th. 8.3]. The *interior* T° of an integral current T is $\mathrm{supp}\, T \setminus \mathrm{supp}\, \partial T$. It is not known whether the singular set of T° has finite Hausdorff $(k{-}2)$-measure. When $k = 2$ or $k = \dim M - 1$ one can go further; for a survey, see Morgan [265, §8].

Theorem 4.4.2. (Almgren [9]) *Let (M, g) be a complete Riemannian manifold and $T \in \mathscr{D}_k$ be a volume-minimizing rectifiable current in M. Then the interior T° of T is a smooth, embedded minimal submanifold of M except for a singular set of Hausdorff dimension at most $k - 2$.*

Harvey and Lawson [151, §II] discuss calibrated geometry and geometric measure theory. They show that on a Riemannian manifold (M, g) with calibration k-form φ one can define *integral φ-currents*, that is, integral currents which are calibrated w.r.t. φ, and that they are *volume-minimizing* in their homology class. An integral current T in M is a φ-current if $\int_T \varphi = \mathrm{vol}(T)$, or equivalently if the tangent k-planes to T are calibrated by φ almost everywhere in Hausdorff k-measure. If N is a compact φ-submanifold with boundary then $\overline{N}$ is an integral φ-current.

Next we discuss *tangent cones* of volume-minimizing integral currents, a generalization of tangent spaces of submanifolds, as in [265, 9.7]. Since cones in $\mathbb{R}^n$ except $\{0\}$ are never compactly-supported, tangent cones are not rectifiable currents, but they are *locally rectifiable*, that is, their restriction to compact sets in $\mathbb{R}^n$ is rectifiable.

Definition 4.4.3 A locally rectifiable current C in $\mathbb{R}^n$ is called a *cone* if $C = tC$ for all $t > 0$, where $t : \mathbb{R}^n \to \mathbb{R}^n$ acts by dilations in the obvious way. Let T be a locally rectifiable current in $\mathbb{R}^n$, and let $\mathbf{x} \in T^\circ$. We say that C is a *tangent cone* to T at $\mathbf{x}$ if there exists a decreasing sequence $r_1 > r_2 > \cdots$ tending to zero such that $r_j^{-1}(T - \mathbf{x})$ converges to C as a locally rectifiable current as $j \to \infty$.

More generally, if (M, g) is a complete Riemannian n-manifold, T is a locally rectifiable current in M, and $x \in T^\circ$, then one can define a *tangent cone* C to T at x, which is a locally rectifiable current cone in the Euclidean vector space $T_x M$. Identifying M with $\mathbb{R}^n$ near x using a coordinate system, the two notions of tangent cone coincide.

The next result follows from Morgan [265, p. 94-95], Federer [102, 5.4.3] and Harvey and Lawson [151, Th. II.5.15].

Theorem 4.4.4 *Let* (M, g) *be a complete Riemannian manifold, and* T *a volume-minimizing integral current in* M. *Then for all* $x \in T^\circ$, *there exists a tangent cone* C *to* T *at* x. *Moreover* C *is itself a volume-minimizing locally rectifiable current in* $T_x M$ *with* $\partial C = \emptyset$, *and if* T *is calibrated with respect to a calibration* φ *on* (M, g), *then* C *is calibrated with respect to the constant calibration* $\varphi|_x$ *on* $T_x M$.

Note that the theorem does *not* claim that the tangent cone C is unique, and in fact it is an important open question whether a volume-minimizing integral current has a unique tangent cone at each point of T°. However, Simon [312] shows that if some tangent cone C is nonsingular and multiplicity 1 away from 0, then C is the unique tangent cone, and T converges to C in a C^1 sense. Simon claims only that ϕ is C^2 rather than smooth, but smoothness follows by elliptic regularity as in §4.1.

Theorem 4.4.5 *Let* C *be an* m-*dimensional oriented minimal cone in* $\mathbb{R}^n$ *with* $C' = C \setminus \{0\}$ *nonsingular, so that* $\Sigma = C \cap \mathcal{S}^{n-1}$ *is a compact, oriented, nonsingular, embedded, minimal* $(m-1)$-*submanifold of* $\mathcal{S}^{n-1}$. *Define* $\iota : \Sigma \times (0, \infty) \to C' \subset \mathbb{R}^n$ *by* $\iota(\sigma, r) = r\sigma$. *Let* (M, g) *be a complete Riemannian* n-*manifold and* $x \in M$. *Fix an isometry* $\upsilon : \mathbb{R}^n \to T_x M$, *and choose an embedding* $\Upsilon : B_R \to M$ *with* $\Upsilon(0) = x$ *and* $d\Upsilon|_0 = \upsilon$, *where* B_R *is the ball of radius* $R > 0$ *about* $0 \in \mathbb{R}^n$.

Suppose T *is a minimal integral current in* M *with* $x \in T^\circ$, *and* $\upsilon_*(C)$ *is a tangent cone to* T *at* x *with multiplicity* 1. *Then* $\upsilon_*(C)$ *is the unique tangent cone to* T *at* x. *Furthermore there exists* $R' \in (0, R]$ *and an embedding* $\phi : \Sigma \times (0, R') \to B_{R'} \subseteq B_R$ *with*

$$|\phi(\sigma, r)| \equiv r, \quad |\phi - \iota| = o(r) \quad \text{and} \quad |\nabla(\phi - \iota)| = o(1) \quad \text{as } r \to 0, \qquad (4.1)$$

such that $T \cap \big(\Upsilon(B_{R'}) \setminus \{x\}\big)$ *is the embedded submanifold* $\Upsilon \circ \phi(\Sigma \times (0, R'))$, *with multiplicity* 1.

In §8.5 we will discuss special Lagrangian m-folds with *isolated conical singularities*, which have this kind of behaviour near their singular points. One moral we can draw from these ideas is that the tangent cone at a singular point of a calibrated submanifold captures the leading order behaviour of the submanifold near the singular point, and so to understand the possible singularities of calibrated submanifolds we should start by constructing and studying calibrated cones in $\mathbb{R}^n$ with respect to the corresponding constant calibration.

5

Kähler manifolds

In this chapter we shall introduce the very rich geometry of *complex* and *Kähler manifolds*. Complex manifolds are manifolds with a geometric structure called a *complex structure*, which gives every tangent space the structure of a complex vector space. They are defined in §5.1, together with complex submanifolds and holomorphic maps. Section 5.2 discusses tensors on complex manifolds, and their decomposition into components using the complex structure, and §5.3 defines holomorphic vector bundles over a complex manifold.

Sections 5.4–5.7 deal with *Kähler metrics* on complex manifolds. A Kähler metric is a Riemannian metric on a complex manifold, that is compatible with the complex structure in a natural way. Also, Kähler metrics have special holonomy groups: if g is a Kähler metric on a complex manifold of dimension m, then the holonomy group $\mathrm{Hol}(g)$ is a subgroup of $\mathrm{U}(m)$. Quite a lot of the geometry of Kähler metrics that we will describe has parallels in the geometry of other holonomy groups. We discuss Kähler potentials, the curvature of Kähler metrics, and exterior forms on Kähler manifolds.

In §5.1–§5.7 we treat complex and Kähler manifolds using differential and Riemannian geometry. But one can also study complex and Kähler manifolds using *complex algebraic geometry*, and we give an introduction to this in §5.8–§5.10. Algebraic geometry is a very large subject and we cannot do it justice in a few pages, so we aim only to provide enough background for the reader to understand the algebraic parts of the rest of the book, which occur mostly in Chapters 7, 10 and 11.

Section 5.8 introduces *complex algebraic varieties*, the objects studied in complex algebraic geometry, and briefly describes some of the fundamental ideas—morphisms, rational maps, sheaves and so on. We then cover two areas in more detail: singularities, resolutions and deformations in §5.9, and holomorphic line bundles and divisors in §5.10.

And now a word about notation. Unfortunately, the literature on Kähler geometry is rather inconsistent about the notation it uses. For example, while writing this chapter I found four different definitions of the Kähler form ω of §5.4 in various books, that differ from the definition we shall give by constant factors. In the same way, the operator d^c of §5.2, the Ricci form ρ of §5.6 and the Laplacian Δ on a Kähler manifold all have several definitions, differing by constant factors. I have done my best to make the formulae in this book consistent with each other, but readers are warned that other books and papers have other conventions.

5.1   Introduction to complex manifolds

A complex manifold is a real, even-dimensional manifold equipped with a geometric structure called a *complex structure*. Here are three ways to define this.

First Definition. Let M be a real manifold of dimension $2m$. A *complex chart* on M is a pair (U, ψ), where U is open in M and $\psi : U \to \mathbb{C}^m$ is a diffeomorphism between U and some open set in $\mathbb{C}^m$. Equivalently, ψ gives a set of complex coordinates $z_1, \ldots, z_m$ on U. If (U_1, ψ_1) and (U_2, ψ_2) are two complex charts, then the *transition function* is $\psi_{12} : \psi_1(U_1 \cap U_2) \to \psi_2(U_1 \cap U_2)$, given by $\psi_{12} = \psi_2 \circ \psi_1^{-1}$. We say M is a *complex manifold* if it has an atlas of complex charts (U, ψ), such that all the transition functions are *holomorphic*, as maps from $\mathbb{C}^m$ to itself.

This is the traditional definition of complex manifold, using holomorphic coordinates. However, in this book we prefer to take a more differential geometric point of view, and to define geometric structures using tensors. So, here are the preliminaries to our second definition of complex manifold.

Let M be a real manifold of dimension $2m$. We define an *almost complex structure* J on M to be a smooth tensor J_a^b on M satisfying $J_a^b J_b^c = -\delta_a^c$. Let v be a smooth vector field on M, written v^a in index notation, and define a new vector field Jv by $(Jv)^b = J_a^b v^a$. Thus J acts linearly on vector fields. The equation $J_a^b J_b^c = -\delta_a^c$ implies that $J(Jv) = -v$, so that $J^2 = -1$. Observe that J gives each tangent space $T_p M$ the structure of a *complex vector space*.

For all smooth vector fields v, w on M, define a vector field $N_J(v, w)$ by

$$N_J(v, w) = [v, w] + J\big([Jv, w] + [v, Jw]\big) - [Jv, Jw],$$

where $[\,,\,]$ is the Lie bracket of vector fields. It turns out that N_J is a *tensor*, meaning that $N_J(v, w)$ is pointwise bilinear in v and w. We call N_J the *Nijenhuis tensor* of J.

Second Definition. Let M be a real manifold of dimension $2m$, and J an almost complex structure on M. We call J a *complex structure* if $N_J \equiv 0$ on M. A *complex manifold* is a manifold M equipped with a complex structure J. We shall often use the notation (M, J) to refer to a manifold and its complex structure.

Here is why the first two definitions are equivalent. Let $f : M \to \mathbb{C}$ be a smooth complex function on M. We say that f is *holomorphic* if $J_a^b(\mathrm{d}f)_b \equiv i(\mathrm{d}f)_a$ on M. These are called the Cauchy–Riemann equations. It turns out that if $m > 1$, the equations are overdetermined, and the Nijenhuis tensor N_J is an obstruction to the existence of holomorphic functions. Simply put, if $N_J \equiv 0$ there are many holomorphic functions locally, but if $N_J \not\equiv 0$ there are few holomorphic functions.

Let (U, ψ) be a complex chart on M. Then ψ is a set of complex coordinates $(z_1, \ldots, z_m)$ on U, where $z_j : U \to \mathbb{C}$ is a smooth function. We call (U, ψ) a *holomorphic chart* if each of the functions $z_1, \ldots, z_m$ is holomorphic in the above sense.

The *Newlander–Nirenberg Theorem* shows that a necessary and sufficient condition for there to exist a holomorphic chart around each point of M, is the vanishing of the Nijenhuis tensor N_J of J. Therefore, if (M, J) is a complex manifold in the sense of the second definition, then M has an atlas of holomorphic charts. This atlas makes M into a complex manifold in the sense of the first definition.

Our third way to define the idea of complex structure uses the language of G-structures and intrinsic torsion, that was defined in §2.6. Let M be a real $2m$-manifold with frame bundle F, and let J be an almost complex structure on M. Define P to be the subbundle of frames in F in which the components of J assume the standard form

$$
J_b^a = \begin{cases} 1 & \text{if } a = b+m, \\ -1 & \text{if } a = b-m, \\ 0 & \text{otherwise,} \end{cases}
$$

for $a, b = 1, \ldots, 2m$. Then P is a principal subbundle of F with fibre $\mathrm{GL}(m, \mathbb{C})$, which is a Lie subgroup of $\mathrm{GL}(2m, \mathbb{R})$ in the obvious way, and so P is a $\mathrm{GL}(m, \mathbb{C})$-structure on M. Clearly, this defines a 1-1 correspondence between almost complex structures J and $\mathrm{GL}(m, \mathbb{C})$-structures P on M.

It turns out that the Nijenhuis tensor N_J of J is equivalent to the *intrinsic torsion* $T^i(P)$ of P, in the sense of Definition 2.6.4. Thus, $N_J \equiv 0$ if and only if P is *torsion-free*. So, the second definition of complex structure is equivalent to the following:

Third Definition. Let M be a real manifold of dimension $2m$. Then a *complex structure* on M is a torsion-free $\mathrm{GL}(m, \mathbb{C})$-structure on M.

All three definitions are useful for different purposes. The second definition is convenient for differential geometric calculations, and we will use it most often. But the first definition is best for defining complex manifolds explicitly. Here is a simple example.

Example 5.1.1 A very important family of complex manifolds are the *complex projective spaces* $\mathbb{CP}^m$. Define $\mathbb{CP}^m$ to be the set of one-dimensional vector subspaces of $\mathbb{C}^{m+1}$. Let $(z_0, \ldots, z_m)$ be a point in $\mathbb{C}^{m+1} \setminus \{0\}$. Then we write

$$
[z_0, \ldots, z_m] = \big\{ (\alpha z_0, \ldots, \alpha z_m) : \alpha \in \mathbb{C} \big\} \in \mathbb{CP}^m.
$$

Every point in $\mathbb{CP}^m$ is of the form $[z_0, \ldots, z_m]$ for some $(z_0, \ldots, z_m) \in \mathbb{C}^{m+1} \setminus \{0\}$. This notation is called *homogeneous coordinates* for $\mathbb{CP}^m$. The homogeneous coordinates of a point are not unique, since if $\lambda \in \mathbb{C}$ is nonzero then $[z_0, \ldots, z_m]$ and $[\lambda z_0, \ldots, \lambda z_m]$ represent the same point in $\mathbb{CP}^m$.

We will define a set of complex charts on $\mathbb{CP}^m$, that make it into a compact complex manifold of dimension m, in the sense of the first definition above. For each $j = 0, 1, \ldots, m$, define $U_j = \big\{ [z_0, \ldots, z_m] \in \mathbb{CP}^m : z_j \neq 0 \big\}$. Then U_j is an open set in $\mathbb{CP}^m$, and every point in U_j can be written uniquely as $[z_0, \ldots, z_{j-1}, 1, z_{j+1}, \ldots, z_m]$, that is, with $z_j = 1$. Define a map $\psi_j : U_j \to \mathbb{C}^m$ by

$$
\psi_j \big([z_0, \ldots, z_{j-1}, 1, z_{j+1}, \ldots, z_m] \big) = (z_0, \ldots, z_{j-1}, z_{j+1}, \ldots, z_m).
$$

Then ψ_j is a diffeomorphism, and (U_j, ψ_j) is a *complex chart* on $\mathbb{CP}^m$. It is easy to show that the charts (U_j, ψ_j) for $j = 0, 1, \ldots, m$ form a *holomorphic atlas* for $\mathbb{CP}^m$, and thus $\mathbb{CP}^m$ is a *complex manifold*.

5.1.1   Holomorphic maps and complex submanifolds

Many concepts in real differential geometry have natural analogues in the world of complex manifolds. We will now examine the ideas of holomorphic maps and complex submanifolds, which are the complex analogues of smooth maps and submanifolds.

Let M, N be complex manifolds with complex structures J_M, J_N, and let $f : M \to N$ be a smooth map. At each point $p \in M$, the derivative of f is a linear map $\mathrm{d}_p f : T_p M \to T_{f(p)} N$. We say that f is a *holomorphic map* if, for each point $p \in M$ and each $v \in T_p M$, the equation $J_N\big(\mathrm{d}_p f(v)\big) = \mathrm{d}_p f\big(J_M(v)\big)$ holds. In other words, we must have $J_N \circ \mathrm{d}f = \mathrm{d}f \circ J_M$ as maps between the vector bundles TM and $f^*(TN)$ over M.

A map $f : M \to N$ between complex manifolds is called *biholomorphic*, or a *biholomorphism*, if an inverse map $f^{-1} : N \to M$ exists, and both f and f^{-1} are holomorphic maps. Biholomorphic maps are the natural notion of isomorphism of complex manifolds, just as diffeomorphisms are isomorphisms of smooth manifolds.

Now let M be a complex manifold with complex structure J, and let N be a submanifold of M. Then for each $p \in N$, the tangent space $T_p N$ is a vector subspace of the tangent space $T_p M$. We say that N is a *complex submanifold* of M if $J(T_p N) = T_p N$ for each $p \in N$, that is, if the tangent spaces of N are closed under J.

If N is a complex submanifold, then the restriction of J to TN is a complex structure on N, so that N is a complex manifold, and the inclusion map $i : N \to M$ is holomorphic. Also, a submanifold $N \subset M$ is a complex submanifold if and only if it can locally be written as the zeros of a finite number of holomorphic functions.

The complex projective spaces $\mathbb{CP}^m$ have many complex submanifolds, which are defined as the set of zeros of a collection of polynomials. Such submanifolds are called *complex algebraic varieties*, and will be the subject of §5.8. Here are two examples, to illustrate the ideas of holomorphic map and complex submanifold.

Example 5.1.2 Define a map $f : \mathbb{CP}^1 \to \mathbb{CP}^2$ by

$$f\big([x, y]\big) = [x^2, xy, y^2].$$

As above, if $\lambda \in \mathbb{C}$ is nonzero then the points $[\lambda x, \lambda y]$ and $[x, y]$ are the same. To see that f is well-defined, we must show that the definition is independent of the choice of homogeneous coordinates. But this is obvious because

$$\big[(\lambda x)^2, (\lambda x)(\lambda y), (\lambda y)^2\big] = \big[\lambda^2 x^2, \lambda^2 xy, \lambda^2 y^2\big] = \big[x^2, xy, y^2\big],$$

and so f is well-defined. The reason that this works is that the polynomials x^2, xy and y^2 are all homogeneous of the same degree.

This map f is a holomorphic map between the complex manifolds $\mathbb{CP}^1$ and $\mathbb{CP}^2$. Its image $N = \mathrm{Im}\, f$ is a complex submanifold of $\mathbb{CP}^2$, which is isomorphic as a complex manifold to $\mathbb{CP}^1$. We can also define N as a subset of $\mathbb{CP}^2$ by

$$N = \big\{ [z_0, z_1, z_2] \in \mathbb{CP}^2 : z_0 z_2 - z_1^2 = 0 \big\}.$$

Thus N is an example of a conic in $\mathbb{CP}^2$.

Example 5.1.3 Let C be the cubic in $\mathbb{CP}^2$ given by

$$C = \left\{ [z_0, z_1, z_2] \in \mathbb{CP}^2 : z_0^3 + z_1^3 + z_2^3 = 0 \right\}.$$

This is a compact, complex submanifold of $\mathbb{CP}^2$. As a real manifold, C is diffeomorphic to the torus T^2. More generally, let $P(z_0, z_1, z_2)$ be a homogeneous complex polynomial of degree $d \geqslant 1$, and define a set $C_P \subset \mathbb{CP}^2$ by

$$C_P = \left\{ [z_0, z_1, z_2] \in \mathbb{CP}^2 : P(z_0, z_1, z_2) = 0 \right\}.$$

For generic polynomials P the curve C_P is a compact complex submanifold of $\mathbb{CP}^2$, which is diffeomorphic as a real manifold to a surface of genus $g = \frac{1}{2}(d-1)(d-2)$. For more details, see [132, §2.1].

5.2   Tensors on complex manifolds

We showed in §3.5.1 that if G is a Lie subgroup of $\mathrm{GL}(n, \mathbb{R})$ and Q is a G-structure on an n-manifold M, then the bundles of tensors and exterior forms on M decompose into a direct sum of subbundles corresponding to irreducible representations of G. Now §5.1 defined a complex structure to be a torsion-free $\mathrm{GL}(m, \mathbb{C})$-structure on a $2m$-manifold. Thus, the bundles of tensors and exterior forms on a complex manifold split into subbundles corresponding to irreducible representations of $\mathrm{GL}(m, \mathbb{C})$. In this section we will explain these splittings, and some of their consequences.

Let M be a manifold of dimension $2m$, and J a complex structure on M. Then J acts linearly on vector fields v by $v \mapsto Jv$, such that $J(Jv) = -v$. As vector fields are sections of the tangent bundle TM of M, we may regard J as a bundle-linear map $J : TM \to TM$. At a point p in M, this gives a linear map $J_p : T_pM \to T_pM$. Now T_pM is a real vector space isomorphic to $\mathbb{R}^{2m}$. It is convenient to *complexify* T_pM to get $T_pM \otimes_{\mathbb{R}} \mathbb{C}$, which is a complex vector space isomorphic to $\mathbb{C}^{2m}$. (Note that this operation of complexification is independent of J.) The map J_p extends naturally to a map $J_p : T_pM \otimes_{\mathbb{R}} \mathbb{C} \to T_pM \otimes_{\mathbb{R}} \mathbb{C}$, which is linear over $\mathbb{C}$.

Consider the eigenvalues and eigenvectors of J_p in $T_pM \otimes_{\mathbb{R}} \mathbb{C}$. Since $J_p^2 = -\,\mathrm{id}$, where id is the identity, any eigenvalue λ of J_p must satisfy $\lambda^2 = -1$. Hence $\lambda = \pm i$. Define $T_p^{(1,0)}M$ to be the eigenspace of J_p in $T_pM \otimes_{\mathbb{R}} \mathbb{C}$ with eigenvalue i, and $T_p^{(0,1)}M$ to be the eigenspace with eigenvalue $-i$. It is easy to show that $T_p^{(1,0)}M \cong \mathbb{C}^m \cong T_p^{(0,1)}M$, that $T_pM \otimes_{\mathbb{R}} \mathbb{C} = T_p^{(1,0)}M \oplus T_p^{(0,1)}M$, and that $T_p^{(1,0)}M$ and $T_p^{(0,1)}M$ are complex conjugate subspaces under the natural complex conjugation on $T_pM \otimes_{\mathbb{R}} \mathbb{C}$.

As this works at every point p in M, we have defined two subbundles $T^{(1,0)}M$ and $T^{(0,1)}M$ of $TM \otimes_{\mathbb{R}} \mathbb{C}$, with $TM \otimes_{\mathbb{R}} \mathbb{C} = T^{(1,0)}M \oplus T^{(0,1)}M$. What we have shown is that a complex structure on M splits the complexified tangent bundle into two subbundles. In a similar way, the complexified cotangent bundle, and in fact complexified tensors of all kinds on M, are split into subbundles by the complex structure.

This is an important idea in complex geometry, and to make use of it we will usually work with *complex-valued tensors* on complex manifolds, that is, all vector bundles will be complexified, as above with the tangent bundle. We will now develop the idea in two different ways. Firstly, a notation for tensors on complex manifolds will be defined, and secondly, the decomposition of exterior forms by the complex structure is explained, and the ∂ and $\bar{\partial}$ operators are defined.

5.2.1 The decomposition of complex tensors

Let M be a complex manifold with complex structure J, which will be written with indices as J_j^k. Let $S = S^{a\cdots}_{\cdots}$ be a tensor on M, taking values in $\mathbb{C}$. Here a is a contravariant index of S, and any other indices of S are represented by dots. The Greek characters $\alpha, \beta, \gamma, \delta, \epsilon$ and their conjugates $\bar{\alpha}, \bar{\beta}, \bar{\gamma}, \bar{\delta}, \bar{\epsilon}$, will be used in place of the Roman indices a, b, c, d, e respectively. They are tensor indices in the normal sense, and their use is actually a shorthand indicating a modification to the tensor itself.

Define $S^{\alpha\cdots}_{\cdots} = \frac{1}{2}(S^{a\cdots}_{\cdots} - iJ_j^a S^{j\cdots}_{\cdots})$ and $S^{\bar{\alpha}\cdots}_{\cdots} = \frac{1}{2}(S^{a\cdots}_{\cdots} + iJ_j^a S^{j\cdots}_{\cdots})$. In the same way, if b is a covariant index on a complex-valued tensor $T_{b\cdots}^{\cdots}$, define $T_{\beta\cdots}^{\cdots} = \frac{1}{2}(T_{b\cdots}^{\cdots} - iJ_b^j T_{j\cdots}^{\cdots})$ and $T_{\bar{\beta}\cdots}^{\cdots} = \frac{1}{2}(T_{b\cdots}^{\cdots} + iJ_b^j T_{j\cdots}^{\cdots})$. These operations on tensors are projections, and satisfy $S^{a\cdots}_{\cdots} = S^{\alpha\cdots}_{\cdots} + S^{\bar{\alpha}\cdots}_{\cdots}$ and $T_{b\cdots}^{\cdots} = T_{\beta\cdots}^{\cdots} + T_{\bar{\beta}\cdots}^{\cdots}$.

Let δ_a^b be the Kronecker delta, regarded as a tensor on M. Then $\delta_a^b = \delta_\alpha^\beta + \delta_{\bar{\alpha}}^{\bar{\beta}}$ in this notation. It is also easy to show that $J_a^b = i\delta_\alpha^\beta - i\delta_{\bar{\alpha}}^{\bar{\beta}}$. Thus J acts on tensor indices of the form $\alpha, \beta, \ldots$ by multiplication by i, so we may think of these as *complex linear* components with respect to J. Similarly, J acts on indices of the form $\bar{\alpha}, \bar{\beta}, \ldots$ by multiplication by $-i$, and we may think of these as *complex antilinear* components.

5.2.2 Exterior forms on complex manifolds

By the argument used above, the complexified cotangent bundle $T^*M \otimes_{\mathbb{R}} \mathbb{C}$ splits into pieces: $T^*M \otimes_{\mathbb{R}} \mathbb{C} = T^{*(1,0)}M \oplus T^{*(0,1)}M$. Now if U, V, W are vector spaces with $U = V \oplus W$, then the exterior powers of U, V and W are related by $\Lambda^k U = \bigoplus_{j=0}^k \Lambda^j V \otimes \Lambda^{k-j}W$. Using the splitting of $T^*M \otimes_{\mathbb{R}} C$, it follows that

$$\Lambda^k T^*M \otimes_{\mathbb{R}} \mathbb{C} = \bigoplus_{j=0}^k \Lambda^j T^{*(1,0)}M \otimes_{\mathbb{C}} \Lambda^{k-j} T^{*(0,1)}M. \tag{5.1}$$

Define $\Lambda^{p,q}M$ to be the bundle $\Lambda^p T^{*(1,0)}M \otimes_{\mathbb{C}} \Lambda^q T^{*(0,1)}M$. Then (5.1) gives

$$\Lambda^k T^*M \otimes_{\mathbb{R}} \mathbb{C} = \bigoplus_{j=0}^k \Lambda^{j,k-j}M. \tag{5.2}$$

This is the decomposition of the exterior k-forms on M induced by the complex structure J. A section of $\Lambda^{p,q}M$ is called a *(p,q)-form*.

We may use the splittings of $\Lambda^k T^*M \otimes_{\mathbb{R}} \mathbb{C}$ and $\Lambda^{k+1}T^*M \otimes_{\mathbb{R}} \mathbb{C}$ to divide the exterior derivative d on complex k-forms into components, each component mapping sections of $\Lambda^{p,q}M$ to sections of $\Lambda^{r,s}M$, where $p + q = k$ and $r + s = k + 1$. Provided J is a complex structure (not just an almost complex structure), the only nonzero components are those that map $\Lambda^{p,q}M$ to $\Lambda^{p+1,q}M$ and to $\Lambda^{p,q+1}M$.

Define ∂ to be the component of d mapping $C^\infty(\Lambda^{p,q}M)$ to $C^\infty(\Lambda^{p+1,q}M)$, and $\bar{\partial}$ to be the component of d mapping $C^\infty(\Lambda^{p,q}M)$ to $C^\infty(\Lambda^{p,q+1}M)$. Then $\partial, \bar{\partial}$ are first-order partial differential operators on complex k-forms which satisfy $d = \partial + \bar{\partial}$. The identity $d^2 = 0$ implies that $\partial^2 = \bar{\partial}^2 = 0$ and $\partial\bar{\partial} + \bar{\partial}\partial = 0$. As $\bar{\partial}^2 = 0$, we may define the *Dolbeault cohomology groups* $H_{\bar{\partial}}^{p,q}(M)$ of a complex manifold, by

$$H^{p,q}_{\bar\partial}(M) = \frac{\mathrm{Ker}\big(\bar\partial : C^\infty(\Lambda^{p,q}M) \to C^\infty(\Lambda^{p,q+1}M)\big)}{\mathrm{Im}\big(\bar\partial : C^\infty(\Lambda^{p-1,q}M) \to C^\infty(\Lambda^{p,q}M)\big)}. \qquad (5.3)$$

The Dolbeault cohomology groups depend on the complex structure of M.

Now define an operator $\mathrm{d}^c : C^\infty(\Lambda^k T^*M \otimes_\mathbb{R} \mathbb{C}) \to C^\infty(\Lambda^{k+1}T^*M \otimes_\mathbb{R} \mathbb{C})$ by $\mathrm{d}^c = i(\bar\partial - \partial)$. It is easy to show that

$$\mathrm{d}\mathrm{d}^c + \mathrm{d}^c\mathrm{d} = 0, \quad (\mathrm{d}^c)^2 = 0, \quad \partial = \tfrac{1}{2}(\mathrm{d} + i\mathrm{d}^c), \quad \bar\partial = \tfrac{1}{2}(\mathrm{d} - i\mathrm{d}^c) \text{ and } \mathrm{d}\mathrm{d}^c = 2i\partial\bar\partial.$$

Also d^c is a *real* operator, that is, if α is a real k-form then $\mathrm{d}^c\alpha$ is a real $(k+1)$-form.

5.3 Holomorphic vector bundles

Next we define holomorphic vector bundles over a complex manifold, which are the analogues in complex geometry of smooth vector bundles over real manifolds. A good reference for the material in this section is [132, §0.5 & §1.1].

Definition 5.3.1 Let M be a complex manifold. Let $\{E_p : p \in M\}$ be a family of complex vector spaces of dimension k, parametrized by M. Let E be the total space of this family, and $\pi : E \to M$ be the natural projection. Suppose also that E has the structure of a complex manifold. This collection of data (the family of complex vector spaces, with a complex structure on its total space) is called a *holomorphic vector bundle* with fibre $\mathbb{C}^k$, if the following conditions hold.

 (i) The map $\pi : E \to M$ is a holomorphic map of complex manifolds.
 (ii) For each $p \in M$ there exists an open neighbourhood $U \subset M$, and a biholomorphic map $\varphi_U : \pi^{-1}(U) \to U \times \mathbb{C}^k$.
 (iii) In part (ii), for each $u \in U$ the map φ_U takes E_u to $\{u\} \times \mathbb{C}^k$, and this is an isomorphism between E_u and $\mathbb{C}^k$ as complex vector spaces.

The vector space E_p is called the *fibre* of E over p. Usually we will refer to E as the holomorphic vector bundle, implicitly assuming that the rest of the structure is given.

Let E and F be holomorphic vector bundles over M. Then E^* and $E \otimes F$ are also holomorphic bundles in a natural way, where E^* is the *dual vector bundle* to E with fibre E^*_p at $p \in M$, and $E \otimes F$ is the *tensor product bundle*, with fibre $E_p \otimes F_p$.

Suppose E is a holomorphic vector bundle over M, with projection $\pi : E \to M$. A *holomorphic section* s of E is a holomorphic map $s : M \to E$, such that $\pi \circ s$ is the identity map on M. Because the fibres of E are complex vector spaces, holomorphic sections of E can be added together and multiplied by complex constants. Thus the holomorphic sections of E form a complex vector space, which is finite-dimensional if M is compact.

Now, every complex manifold M comes equipped with a number of natural holomorphic vector bundles. For example, the product $M \times \mathbb{C}^k$ is a holomorphic vector bundle over M, called the *trivial* vector bundle with fibre $\mathbb{C}^k$. Also, the tangent bundle TM and the cotangent bundle T^*M are both real vector bundles over M, but we may make them into complex vector bundles by identifying J with multiplication by $i \in \mathbb{C}$. The total spaces of TM and T^*M both have natural complex structures, which make them into *holomorphic vector bundles*.

From TM and T^*M we can make other holomorphic vector bundles by tensor products. We will consider the bundles of exterior forms. The vector bundles $\Lambda^{p,q}M$ of §5.2.2 are *complex vector bundles*, smooth vector bundles with complex vector spaces as fibres. But which of these are holomorphic vector bundles? It turns out that $\Lambda^{p,q}M$ is a holomorphic vector bundle in a natural way if and only if $q = 0$, so that $\Lambda^{p,0}M$ is a holomorphic vector bundle for $p = 0, 1, \ldots, m$. There are natural isomorphisms

$$\Lambda^{0,0}M \cong M \times \mathbb{C}, \quad \Lambda^{1,0}M \cong T^*M, \quad \text{and} \quad \Lambda^{p,0}M \cong \Lambda_{\mathbb{C}}^p T^*M,$$

as holomorphic vector bundles.

Now let $s \in C^\infty(\Lambda^{p,0}M)$, so that s is a smooth section of $\Lambda^{p,0}M$. Then s is a *holomorphic section* of $\Lambda^{p,0}M$ if and only if $\bar{\partial}s = 0$ in $C^\infty(\Lambda^{p,1}M)$. A holomorphic section of $\Lambda^{p,0}M$ is called a *holomorphic p-form*. From eqn (5.3) we see that the Dolbeault group $H_{\bar{\partial}}^{p,0}(M)$ is actually the vector space of holomorphic p-forms on M.

5.4 Introduction to Kähler manifolds

Let (M, J) be a complex manifold, and let g be a Riemannian metric on M. We call g a *Hermitian metric* if three equivalent conditions hold:

(i) $g(v, w) = g(Jv, Jw)$ for all vector fields v, w on M,

(ii) in index notation, $g_{ab} \equiv J_a^c J_b^d g_{cd}$,

(iii) in the notation of §5.2, $g_{ab} \equiv g_{\alpha\bar{\beta}} + g_{\bar{\alpha}\beta}$. That is, $g_{\alpha\beta} \equiv g_{\bar{\alpha}\bar{\beta}} \equiv 0$.

This is a natural compatibility condition between a complex structure and a Riemannian metric. If g is a Hermitian metric, we define a 2-form ω on M called the *Hermitian form* of g in three equivalent ways:

(i) $\omega(v, w) = g(Jv, w)$ for all vector fields v, w on M,

(ii) in index notation, $\omega_{ac} = J_a^b g_{bc}$,

(iii) in the notation of §5.2, $\omega_{ab} = ig_{\alpha\bar{\beta}} - ig_{\bar{\alpha}\beta}$.

Then ω is a (1,1)-form, and we may reconstruct g from ω using the equation $g(v, w) = \omega(v, Jw)$. Define a (1,1)-form ω on a complex manifold to be *positive* if $\omega(v, Jv) > 0$ for all nonzero vectors v. It is easy to see that if ω is a (1,1)-form on a complex manifold, then ω is the Hermitian form of a Hermitian metric if and only if ω is positive. The idea of Hermitian metric also makes sense for J an almost complex structure.

Definition 5.4.1 Let (M, J) be a complex manifold, and g a Hermitian metric on M, with Hermitian form ω. We say g is a *Kähler metric* if $d\omega = 0$. In this case we call ω the *Kähler form*, and the triple (M, J, g) a *Kähler manifold*.

Here are some important facts about Kähler metrics.

Proposition 5.4.2 *Let M be a manifold of dimension $2m$, J an almost complex structure on M, and g a Hermitian metric, with Hermitian form ω. Let ∇ be the Levi-Civita connection of g. Then the following conditions are equivalent:*

(i) *J is a complex structure and g is Kähler.*

(ii) *$\nabla J = 0$.*

(iii) *$\nabla\omega = 0$.*

(iv) *The holonomy group of g is contained in* $U(m)$, *and* J *is associated to the corresponding* $U(m)$-*structure.*

In §5.1 we saw that one way to define a complex structure is as a torsion-free $GL(m, \mathbb{C})$-structure on a $2m$-manifold M. In the same way, a Kähler structure can be defined to be a *torsion-free* $U(m)$-*structure* on M, as in §2.6.

Notice that if g is a Hermitian metric on a complex manifold, then the rather weak condition $d\omega = 0$ implies the much stronger conditions that $\nabla\omega = \nabla J = 0$. One moral is that Kähler metrics are easy to construct, as closed 2-forms are easy to find, but they have many interesting properties following from $\nabla\omega = \nabla J = 0$. Also, note that a Kähler metric is just a Riemannian metric with holonomy contained in $U(m)$. The highest exterior power ω^m of ω is proportional to the volume form dV_g of g, and with the conventions used in this book, the relationship is

$$\omega^m = m! \cdot dV_g. \tag{5.4}$$

Let M be a Kähler manifold with Kähler metric g and Kähler form ω. If N is a *complex submanifold* of M in the sense of §5.1.1, then the restriction of g to N is also Kähler. (One way to see this is that the restriction of ω to N is clearly a closed, positive (1,1)-form.) Thus, any complex submanifold of a Kähler manifold is a Kähler manifold in its own right.

Example 5.4.3 The complex manifold $\mathbb{CP}^m$, described in Example 5.1.1, carries a natural Kähler metric. Here is one way to define it. There is a natural projection

$$\pi : \mathbb{C}^{m+1} \setminus \{0\} \to \mathbb{CP}^m, \quad \text{defined by} \quad \pi : (z_0, \ldots, z_m) \mapsto [z_0, \ldots, z_m].$$

Define a real function $u : \mathbb{C}^{m+1} \setminus \{0\} \to (0, \infty)$ by $u(z_0, \ldots, z_m) = |z_0|^2 + \cdots + |z_m|^2$. Define a closed (1,1)-form α on $\mathbb{C}^{m+1} \setminus \{0\}$ by $\alpha = dd^c(\log u)$. Now α is not the Kähler form of any Kähler metric on $\mathbb{C}^{m+1} \setminus \{0\}$, because it is not positive. However, there does exist a unique positive (1,1)-form ω on $\mathbb{CP}^m$, such that $\alpha = \pi^*(\omega)$. The Kähler metric g on $\mathbb{CP}^m$ with Kähler form ω is the *Fubini–Study metric*.

The idea here of using dd^c to make Kähler forms will be explored in §5.5. Since $\mathbb{CP}^m$ is Kähler, it follows that any complex submanifold of $\mathbb{CP}^m$ is also a Kähler manifold. Now there are lots of complex submanifolds in the complex projective spaces $\mathbb{CP}^m$. They are studied in the subject of *complex algebraic geometry*, which will be introduced in §5.8. This gives a huge number of examples of Kähler manifolds.

5.5 Kähler potentials

Let (M, J) be a complex manifold. We have seen that to each Kähler metric g on M there is associated a closed real (1,1)-form ω, called the Kähler form. Conversely, if ω is a closed real (1,1) form on M, then ω is the Kähler form of a Kähler metric if and only if ω is *positive* (that is, $\omega(v, Jv) > 0$ for all nonzero vectors v). Positivity is an *open condition* on closed real (1,1)-forms, meaning that it holds on an open set in the space of closed real (1,1)-forms.

Let ϕ be a smooth real function on M. Then $\mathrm{dd}^c\phi$ is clearly a closed (and in fact exact) real 2-form, as both d and d^c are real operators. But since $\mathrm{dd}^c = 2i\partial\bar{\partial}$, it follows that $\mathrm{dd}^c\phi$ is also a (1,1)-form. Thus if ϕ is a real function then $\mathrm{dd}^c\phi$ is a closed real (1,1)-form. The following are converses to this.

The Local dd^c-Lemma. *Let η be a smooth, closed, real $(1,1)$-form on the unit disc in $\mathbb{C}^m$. Then there is a smooth real function ϕ on the unit disc with $\eta = \mathrm{dd}^c\phi$.*

The Global dd^c-Lemma. *Let M be a compact Kähler manifold, and η a smooth, exact, real $(1,1)$-form on M. Then there is a smooth real function ϕ on M with $\eta = \mathrm{dd}^c\phi$.*

In the second it is necessary that M should be Kähler, rather than just complex, for the result to hold. From the Local dd^c-Lemma it follows that if g is a Kähler metric on M with Kähler form ω, then locally in M we may write $\omega = \mathrm{dd}^c\phi$ for some real function ϕ. Such a function ϕ is called a *Kähler potential* for the metric g. However, in general we cannot find a global Kähler potential for g, for the following reason.

Suppose M is a compact Kähler manifold of dimension $2m$, with Kähler form ω. As ω is closed, it defines a de Rham cohomology class $[\omega] \in H^2(M, \mathbb{R})$, called the *Kähler class*. Now $[\omega]^m = \int_M \omega^m = m!\,\mathrm{vol}(M) > 0$ by (5.4), so $[\omega] \neq 0$. However, $\mathrm{dd}^c\phi$ is exact, so that $[\mathrm{dd}^c\phi] = 0$ in $H^2(M, \mathbb{R})$. Therefore, on a compact Kähler manifold it is impossible to find a global Kähler potential; but we do have the following useful result.

Lemma 5.5.1 *Let M be a compact complex manifold and g, g' Kähler metrics on M with Kähler forms ω, ω'. Suppose $[\omega] = [\omega']$ in $H^2(M, \mathbb{R})$. Then there is a smooth real function ϕ on M with $\omega' = \omega + \mathrm{dd}^c\phi$, which is unique up to addition of a constant.*

Proof Since $[\omega] = [\omega']$, $\omega' - \omega$ is an exact, real (1,1)-form. So, by the Global dd^c-Lemma, a function ϕ exists with $\omega' - \omega = \mathrm{dd}^c\phi$, and $\omega' = \omega + \mathrm{dd}^c\phi$ as we want. If ϕ_1 and ϕ_2 are both solutions, then by subtraction $\mathrm{dd}^c(\phi_1 - \phi_2) = 0$ on M, which implies that $\phi_1 - \phi_2$ is constant, as M is compact. Therefore ϕ is unique up to a constant. $\qquad\square$

The lemma gives a parametrization of the Kähler metrics with a fixed Kähler class, by smooth functions on the manifold. We may also express the metric g' in terms of g and ϕ. As $\omega' = \omega + \mathrm{dd}^c\phi = \omega + 2i\partial\bar{\partial}\phi$, we have $\omega'_{\alpha\bar{\beta}} = \omega_{\alpha\bar{\beta}} + i\partial_\alpha\bar{\partial}_{\bar{\beta}}\phi$ and $\omega'_{\bar{\alpha}\beta} = \omega_{\bar{\alpha}\beta} - i\bar{\partial}_{\bar{\alpha}}\partial_\beta\phi$. But $g_{\alpha\bar{\beta}} = -i\omega_{\alpha\bar{\beta}}, g_{\bar{\alpha}\beta} = i\omega_{\bar{\alpha}\beta}, g'_{\alpha\bar{\beta}} = -i\omega'_{\alpha\bar{\beta}}$ and $g'_{\bar{\alpha}\beta} = i\omega'_{\bar{\alpha}\beta}$, and therefore $g'_{\alpha\bar{\beta}} = g_{\alpha\bar{\beta}} + \partial_\alpha\bar{\partial}_{\bar{\beta}}\phi$ and $g'_{\bar{\alpha}\beta} = g_{\bar{\alpha}\beta} + \bar{\partial}_{\bar{\alpha}}\partial_\beta\phi$.

5.6 Curvature of Kähler manifolds

Let M be a $2m$-manifold, and g a Kähler metric on M. Then $\mathrm{Hol}(g) \subseteq \mathrm{U}(m)$, by Proposition 5.4.2. Applying Theorem 3.1.7, one can show that in the notation of §5.2, the Riemann curvature tensor of g satisfies

$$R^a{}_{bcd} = R^\alpha{}_{\beta\gamma\bar{\delta}} + R^\alpha{}_{\beta\bar{\gamma}\delta} + R^{\bar{\alpha}}{}_{\bar{\beta}\gamma\bar{\delta}} + R^{\bar{\alpha}}{}_{\bar{\beta}\bar{\gamma}\delta}. \tag{5.5}$$

Now a general tensor $T^a{}_{bcd}$ has 16 components in its complex decomposition. Equation (5.5) says that 12 of these components vanish for the curvature tensor of a Kähler manifold, leaving only 4 components. However, using symmetries of Riemann curvature, and complex conjugation, we may identify $R^\alpha{}_{\beta\bar{\gamma}\delta}$ with $R^\alpha{}_{\beta\gamma\bar{\delta}}$, and identify both

$R^{\bar{\alpha}}{}_{\beta\gamma\bar{\delta}}$ and $R^{\bar{\alpha}}{}_{\bar{\beta}\bar{\gamma}\delta}$ with the complex conjugate of $R^{\alpha}{}_{\beta\gamma\bar{\delta}}$. Thus the Kähler curvature is determined by the single component $R^{\alpha}{}_{\beta\gamma\bar{\delta}}$.

The *Ricci curvature* is $R_{bd} = R^{a}{}_{bad}$. From (5.5) we see that $R_{bd} = R^{\alpha}{}_{\beta\alpha\bar{\delta}} + R^{\bar{\alpha}}{}_{\bar{\beta}\bar{\alpha}\delta}$. Hence $R_{ab} = R_{\alpha\bar{\beta}} + R_{\bar{\alpha}\beta}$, and $R_{\alpha\beta} = R_{\bar{\alpha}\bar{\beta}} = 0$. Also, $R_{ab} = R_{ba}$ by symmetries of curvature. Therefore, the Ricci curvature satisfies the same conditions as a Hermitian metric. From a Hermitian metric we can make a Hermitian form, so we will try the same trick with the Ricci curvature. Define the *Ricci form* ρ by $\rho_{ab} = iR_{\alpha\bar{\beta}} - iR_{\bar{\alpha}\beta}$, or equivalently $\rho_{ac} = J_a^b R_{bc}$. Then ρ is a real (1,1)-form, and we may recover the Ricci curvature from ρ using the equation $R_{ab} = \rho_{ac} J_b^c$. It is a remarkable fact that ρ is a *closed* 2-form. The cohomology class $[\rho] \in H^2(M, \mathbb{R})$ depends only on the complex structure of M, and is equal to $2\pi c_1(M)$, where $c_1(M)$ is the first Chern class of M.

To see why this is so, we will give an explicit expression for the Ricci curvature in coordinates. Let $(z_1, \ldots, z_m)$ be holomorphic coordinates on an open set in M. Let $g_{ab} = g_{\alpha\bar{\beta}} + g_{\bar{\alpha}\beta}$ be the Kähler metric. We may regard α as an index for $dz_1, \ldots, dz_m$, and $\bar{\beta}$ as an index for $d\bar{z}_1, \ldots, d\bar{z}_m$. Hence, α and $\bar{\beta}$ are both indices running from 1 to m, and $g_{\alpha\bar{\beta}}$ is an $m \times m$ complex matrix.

It is easy to see $g_{\alpha\bar{\beta}}$ is a *Hermitian* matrix (that is, $g_{\alpha\bar{\beta}} = \overline{g_{\beta\bar{\alpha}}}$), so it has real eigenvalues, and $\det(g_{\alpha\bar{\beta}})$ is a *real* function. This determinant is also given by

$$\omega^m = i^m m! \det(g_{\alpha\bar{\beta}}) \, dz_1 \wedge d\bar{z}_1 \wedge dz_2 \wedge d\bar{z}_2 \wedge \cdots \wedge dz_m \wedge d\bar{z}_m. \tag{5.6}$$

Here ω^m is the m-fold wedge product of ω. We see from (5.6) that $\det(g_{\alpha\bar{\beta}})$ is *positive*, as ω^m is a positive $2m$-form by (5.4).

It can be shown that the Ricci curvature is given by $R_{\alpha\bar{\beta}} = -\partial_\alpha \bar{\partial}_{\bar{\beta}} [\log \det(g_{\gamma\bar{\delta}})]$, and therefore the Ricci form is

$$\rho = -i\partial\bar{\partial} [\log \det(g_{\gamma\bar{\delta}})] = -\tfrac{1}{2} dd^c [\log \det(g_{\gamma\bar{\delta}})]. \tag{5.7}$$

Thus locally we may write $\rho = -\tfrac{1}{2} dd^c f$ for a smooth real function f, so ρ is closed. As the determinant only makes sense in a holomorphic coordinate system, and we cannot find holomorphic coordinates on the whole of M, this is only a local expression for ρ.

As we remarked in §3.4.1, a Kähler metric g on M has $\mathrm{Hol}^0(g) \subseteq \mathrm{SU}(m)$ if and only if it is *Ricci-flat*, and thus if and only if it has Ricci form $\rho = 0$. Such metrics are called *Calabi–Yau metrics*, because they can be constructed using Yau's solution of the Calabi Conjecture, as in Chapter 6.

5.7 Exterior forms on Kähler manifolds

Section 1.1.2 defined the Hodge star $*$ and the operators d^* and Δ_d on an oriented Riemannian manifold. We begin by defining analogues of these on a Kähler manifold. Let M be a Kähler manifold of real dimension $2m$, with Kähler metric g. The complex structure induces a natural orientation on M, and the metric and the orientation combine to give a volume form dV_g on M, which is a real $2m$ form.

Let α, β be *complex* k-forms on M. Define a pointwise inner product (α, β) by

$$(\alpha, \beta) = \alpha_{a_1 \ldots a_k} \overline{\beta_{b_1 \ldots b_k}} g^{a_1 b_1} \ldots g^{a_k b_k}$$

in index notation. Here (α, β) is a *complex* function on M, which is linear in α and antilinear in β, that is, linear in the complex conjugate $\bar{\beta}$ of β. Note that $(\beta, \alpha) = \overline{(\alpha, \beta)}$, and that (α, α) is a nonnegative real function on M.

When M is compact, define the L^2 inner product of complex k-forms α, β by $\langle \alpha, \beta \rangle = \int_M (\alpha, \beta) \mathrm{d}V_g$. Then $\langle \, , \, \rangle$ is a *Hermitian inner product* on the space of complex k-forms. That is, $\langle \alpha, \beta \rangle$ is a complex number, bilinear in α and $\bar{\beta}$, such that $\langle \beta, \alpha \rangle = \overline{\langle \alpha, \beta \rangle}$, and $\langle \alpha, \alpha \rangle = \|\alpha\|^2_{L^2}$ is real and nonnegative.

Now let the *Hodge star on Kähler manifolds* be the unique map $* : \Lambda^k T^* M \otimes_{\mathbb{R}} \mathbb{C} \to \Lambda^{2m-k} T^* M \otimes_{\mathbb{R}} \mathbb{C}$ satisfying $\alpha \wedge (*\beta) = (\alpha, \beta) \mathrm{d}V_g$ for all complex k-forms α, β. Then $*\beta$ is antilinear in β. The relation to the Hodge star on real forms defined in §1.1.2, is that if $\beta = \beta_1 + i\beta_2$ for β_1, β_2 real k-forms, then $*\beta = *\beta_1 - i*\beta_2$. It satisfies $*1 = \mathrm{d}V_g$ and $*(*\beta) = (-1)^k \beta$, for β a complex k-form, so that $*^{-1} = (-1)^k *$.

Since M is complex, we have operators d, ∂ and $\bar{\partial}$ taking complex k-forms to complex $(k+1)$-forms. Define operators d^*, ∂^* and $\bar{\partial}^*$ by

$$\mathrm{d}^*\alpha = -*\mathrm{d}(*\alpha), \quad \partial^*\alpha = -*\partial(*\alpha) \quad \text{and} \quad \bar{\partial}^*\alpha = -*\bar{\partial}(*\alpha). \tag{5.8}$$

Then d^*, ∂^* and $\bar{\partial}^*$ all take complex k-forms to complex $(k-1)$-forms. Moreover, the argument used in §1.1.2 to show that $\langle \alpha, \mathrm{d}^*\beta \rangle = \langle \mathrm{d}\alpha, \beta \rangle$ for α a $(k-1)$-form and β a k-form on a compact oriented Riemannian manifold, also shows that

$$\langle \alpha, \mathrm{d}^*\beta \rangle = \langle \mathrm{d}\alpha, \beta \rangle, \quad \langle \alpha, \partial^*\beta \rangle = \langle \partial\alpha, \beta \rangle \quad \text{and} \quad \langle \alpha, \bar{\partial}^*\beta \rangle = \langle \bar{\partial}\alpha, \beta \rangle,$$

where α is a complex $(k-1)$-form and β a complex k-form on M.

In §1.1 we defined the Laplacian $\Delta = \mathrm{d}\mathrm{d}^* + \mathrm{d}^*\mathrm{d}$ on Riemannian manifolds. By analogy, from $\mathrm{d}, \partial, \bar{\partial}, \mathrm{d}^*, \partial^*, \bar{\partial}^*$ we can make three Laplacians on complex k-forms:

$$\Delta_{\mathrm{d}} = \mathrm{d}\mathrm{d}^* + \mathrm{d}^*\mathrm{d}, \quad \Delta_{\partial} = \partial\partial^* + \partial^*\partial \quad \text{and} \quad \Delta_{\bar{\partial}} = \bar{\partial}\bar{\partial}^* + \bar{\partial}^*\bar{\partial}.$$

We call Δ_{d} the d-*Laplacian*, Δ_{∂} the ∂-*Laplacian* and $\Delta_{\bar{\partial}}$ the $\bar{\partial}$-*Laplacian*. It can be shown (see [132, p. 115]) that these satisfy

$$\Delta_{\partial} = \Delta_{\bar{\partial}} = \tfrac{1}{2}\Delta_{\mathrm{d}}. \tag{5.9}$$

Now $*$ takes $\Lambda^{p,q} M$ to $\Lambda^{m-p,m-q} M$. Since $\partial : C^\infty(\Lambda^{p,q} M) \to C^\infty(\Lambda^{p+1,q} M)$ and $\bar{\partial} : C^\infty(\Lambda^{p,q} M) \to C^\infty(\Lambda^{p,q+1} M)$, we see from (5.8) that $\partial^* : C^\infty(\Lambda^{p,q} M) \to C^\infty(\Lambda^{p-1,q} M)$ and $\bar{\partial}^* : C^\infty(\Lambda^{p,q} M) \to C^\infty(\Lambda^{p,q-1} M)$. Thus $\Delta_{\partial}, \Delta_{\bar{\partial}}$ and Δ_{d} map $C^\infty(\Lambda^{p,q} M) \to C^\infty(\Lambda^{p,q} M)$.

It is conventional to call the $\bar{\partial}$-Laplacian on a Kähler manifold the Laplacian, and to write it Δ rather than $\Delta_{\bar{\partial}}$. This can lead to confusion, because on a Riemannian manifold we call the d-Laplacian Δ_{d} the Laplacian, so that the Laplacian on a Kähler manifold is half the Laplacian on a Riemannian manifold.

5.7.1 Hodge theory on Kähler manifolds

In §1.1.3 we summarized the ideas of Hodge theory for a compact Riemannian manifold (M, g). Then in §3.5.2 we showed that the space $\mathscr{H}^k$ of Hodge k-forms is a direct sum of subspaces $\mathscr{H}^k_i$ corresponding to irreducible representations of the holonomy group

Hol(g) of g, and deduced that the de Rham cohomology group $H^k(M, \mathbb{R})$ decomposes in the same way.

Since a Kähler metric g has $\mathrm{Hol}(g) \subseteq \mathrm{U}(m)$, these ideas apply to compact Kähler manifolds, and we will now work them out in more detail. Our notation differs slightly from §1.1.3 and §3.5.2, in that we deal with complex rather than real k-forms, and write the summands $\mathscr{H}^{p,q}$ rather than $\mathscr{H}_i^k$. Let M be a compact Kähler manifold, and define

$$\mathscr{H}^{p,q} = \mathrm{Ker}\big(\Delta : C^\infty(\Lambda^{p,q}M) \to C^\infty(\Lambda^{p,q}M)\big)$$

so that $\mathscr{H}^{p,q}$ is the vector space of harmonic (p,q)-forms on M. It is easy to show by (5.9) and integration by parts that $\alpha \in \mathscr{H}^{p,q}$ if and only if $\partial \alpha = \bar{\partial} \alpha = \partial^* \alpha = \bar{\partial}^* \alpha = 0$. Here is a version of the Hodge decomposition theorem for the $\bar{\partial}$ operator, proved in [132, p. 84].

Theorem 5.7.1 *Let M be a compact Kähler manifold. Then*

$$C^\infty(\Lambda^{p,q}M) = \mathscr{H}^{p,q} \oplus \bar{\partial}\big[C^\infty(\Lambda^{p,q-1}M)\big] \oplus \bar{\partial}^*\big[C^\infty(\Lambda^{p,q+1}M)\big], \text{ where}$$

$$\mathrm{Ker}\,\bar{\partial} = \mathscr{H}^{p,q} \oplus \bar{\partial}\big[C^\infty(\Lambda^{p,q-1}M)\big] \quad \text{and} \quad \mathrm{Ker}\,\bar{\partial}^* = \mathscr{H}^{p,q} \oplus \bar{\partial}^*\big[C^\infty(\Lambda^{p,q+1}M)\big].$$

Comparing Theorem 5.7.1 with (5.3) defining the Dolbeault groups $H_{\bar{\partial}}^{p,q}(M)$ of M we see that $H_{\bar{\partial}}^{p,q}(M) \cong \mathscr{H}^{p,q}$. Now define

$$\mathscr{H}^k = \mathrm{Ker}\big(\Delta : C^\infty(\Lambda^k T^*M \otimes_{\mathbb{R}} \mathbb{C}) \to C^\infty(\Lambda^k T^*M \otimes_{\mathbb{R}} \mathbb{C})\big).$$

As $\Delta = \frac{1}{2}\Delta_{\mathrm{d}}$ by (5.9), Theorem 1.1.4 implies that there is a natural isomorphism between $\mathscr{H}^k$ and the complex cohomology $H^k(M, \mathbb{C})$ of M. But

$$\mathscr{H}^k = \bigoplus_{j=0}^k \mathscr{H}^{j,k-j}.$$

Define $H^{p,q}(M)$ to be the vector subspace of $H^{p+q}(M, \mathbb{C})$ with representatives in $\mathscr{H}^{p,q}$. Then we have:

Theorem 5.7.2 *Let M be a compact Kähler manifold of real dimension $2m$. Then $H^k(M, \mathbb{C}) = \bigoplus_{j=0}^k H^{j,k-j}(M)$. Every element of $H^{p,q}(M)$ is represented by a unique harmonic (p,q)-form. Moreover for all p, q we have $H^{p,q}(M) \cong H_{\bar{\partial}}^{p,q}(M)$,*

$$H^{p,q}(M) \cong \overline{H^{q,p}(M)} \quad \text{and} \quad H^{p,q}(M) \cong \big(H^{m-p,m-q}(M)\big)^*.$$

Note that if M is a compact complex manifold admitting Kähler metrics, then the decomposition $H^k(M, \mathbb{C}) = \bigoplus_{j=0}^k H^{j,k-j}(M)$ above depends only on the complex structure of M, and not on the choice of a particular Kähler structure. Define the *Hodge numbers* $h^{p,q}$ or $h^{p,q}(M)$ by $h^{p,q} = \dim H^{p,q}(M)$. Theorem 5.7.2 implies that

$$b^k = \sum_{j=0}^k h^{j,k-j} \quad \text{and} \quad h^{p,q} = h^{q,p} = h^{m-p,m-q} = h^{m-q,m-p}. \tag{5.10}$$

From these equations one can deduce that some compact manifolds, even complex manifolds, cannot admit a Kähler metric for topological reasons. Here are two ways this can happen. First, if $k = 2l + 1$ then $b^k = 2\sum_{j=0}^{l} h^{j,k-j}$, so that if M is a compact Kähler manifold, then b^k is even when k is odd. Thus, any compact manifold that admits a Kähler metric must satisfy this topological condition. For instance, the 4-manifold $\mathcal{S}^3 \times \mathcal{S}^1$ has a complex structure, but as $b^1 = 1$ it can have no Kähler metric.

Secondly, a compact complex manifold M has Dolbeault groups $H_{\bar{\partial}}^{p,q}(M)$ depending on its complex structure. Theorem 5.7.2 shows that a necessary condition for M to admit a Kähler metric with this complex structure is that $\dim H_{\bar{\partial}}^{p,q}(M) = \dim H_{\bar{\partial}}^{q,p}(M)$ for all p, q, which is not always the case.

Now in the splitting (5.2) of complex k-forms on a complex manifold into (p, q)-forms, the summands $\Lambda^{p,q}M$ correspond to irreducible representations of $\mathrm{GL}(m, \mathbb{C})$. However, when $1 \leqslant p, q \leqslant m - 1$, the corresponding representation of $\mathrm{U}(m)$ is *not* irreducible, but is the sum of several irreducible subrepresentations. This means that $\mathscr{H}^{p,q}$ and $H^{p,q}(M)$ can be split into smaller pieces, using the $\mathrm{U}(m)$-structure. The simplest example of this is that $\Lambda^{1,1}M = \langle \omega \rangle \oplus \Lambda_0^{1,1}M$, where ω is the Kähler form and $\Lambda_0^{1,1}M$ is the bundle of (1,1)-forms orthogonal to ω. However, we prefer to work with the splitting into (p, q)-forms, as it is simpler and loses little information.

Finally we consider *Kähler classes* and the *Kähler cone*.

Definition 5.7.3 Let (M, J) be a compact complex manifold admitting Kähler metrics. If g is a Kähler metric on M, then the Kähler form ω of g is a closed real 2-form, and so defines a de Rham cohomology class $[\omega] \in H^2(M, \mathbb{R})$, called the *Kähler class* of g. It is a *topological invariant* of g. Since ω is also a (1,1)-form, $[\omega]$ lies in the intersection $H^{1,1}(M) \cap H^2(M, \mathbb{R})$, regarding $H^{1,1}(M)$ and $H^2(M, \mathbb{R})$ as vector subspaces of $H^2(M, \mathbb{C})$. Define the *Kähler cone* $\mathcal{K}_M$ of M to be the set of Kähler classes $[\omega] \in H^{1,1}(M) \cap H^2(M, \mathbb{R})$ of Kähler metrics on M.

If g_1, g_2 are Kähler metrics on M and $t_1, t_2 > 0$, then $t_1 g_1 + t_2 g_2$ is also Kähler. Thus, if $\alpha_1, \alpha_2 \in \mathcal{K}_M$ and $t_1, t_2 > 0$ then $t_1 \alpha_1 + t_2 \alpha_2 \in \mathcal{K}_M$, so that $\mathcal{K}_M$ is a *convex cone*. Furthermore, if ω is the Kähler form of a Kähler metric and η is a smooth, closed real (1,1)-form on M with $|\eta| < 1$ on M, then $\omega + \eta$ is also the Kähler form of a Kähler metric on M. As M is compact, this implies that $\mathcal{K}_M$ is *open* in $H^{1,1}(M) \cap H^2(M, \mathbb{R})$.

Suppose that Σ is a compact complex curve in a complex manifold (M, J), and that ω is the Kähler form of a Kähler metric g on M. Then Σ defines a homology class $[\Sigma] \in H_2(M, \mathbb{R})$, and the area of Σ with respect to g is $[\omega] \cdot [\Sigma] \in \mathbb{R}$. But this area must be positive. Therefore, each α in the Kähler cone $\mathcal{K}_M$ of M must satisfy $\alpha \cdot [\Sigma] > 0$, for each compact complex curve $\Sigma \subset M$. In simple cases $\mathcal{K}_M$ is exactly the subset of $H^{1,1}(M) \cap H^2(M, \mathbb{R})$ satisfying these inequalities, and is a polyhedral cone bounded by a finite number of hyperplanes, but this is not always true.

5.8 Complex algebraic varieties

This section is designed as a rather brief introduction to complex algebraic geometry. We shall define complex algebraic varieties, and discuss related ideas such as the Zariski topology, sheaves, and schemes. Here are some introductory books on algebraic geometry. Griffiths and Harris [132] cover complex algebraic geometry, taking quite a differential geometric point of view, and discussing manifolds, Kähler metrics, Hodge theory, and so on. Hartshorne [149] has a more algebraic approach, and this section is largely based on [149, §1 & §2]. Two other books, both rather algebraic, are Harris [148] and Iitaka [171]. Harris' book is more elementary and contains lots of examples.

Let $\mathbb{C}^m$ have complex coordinates $(z_1, \ldots, z_m)$, and $\mathbb{C}[z_1, \ldots, z_m]$ be the ring of polynomials in the variables $z_1, \ldots, z_m$, with complex coefficients. Then $\mathbb{C}[z_1, \ldots, z_m]$ is a *ring of complex functions* on $\mathbb{C}^m$. Now $\mathbb{C}^m$ is a topological space, with the usual manifold topology. However, there is another natural topology on $\mathbb{C}^m$ called the *Zariski topology*, which is more useful for the purposes of algebraic geometry.

Definition 5.8.1 An *algebraic set* in $\mathbb{C}^m$ is the set of common zeros of a finite number of polynomials in $\mathbb{C}[z_1, \ldots, z_m]$. It is easy to show that if X, Y are algebraic sets in $\mathbb{C}^m$, then $X \cap Y$ and $X \cup Y$ are algebraic sets. Also, $\emptyset$ and $\mathbb{C}^m$ are algebraic sets.

Define the *Zariski topology* on $\mathbb{C}^m$ by taking the open subsets to be $\mathbb{C}^m \setminus X$, for all algebraic sets X. This gives a topology on $\mathbb{C}^m$, in which a subset $X \subset \mathbb{C}^m$ is closed if and only if it is algebraic.

In this section, we shall regard $\mathbb{C}^m$ as a topological space with the Zariski topology, rather than the usual topology. The simplest sort of complex algebraic varieties are affine varieties, which we define now.

Definition 5.8.2 An algebraic set X in $\mathbb{C}^m$ is said to be *irreducible* if it is not the union $X_1 \cup X_2$ of two proper subsets, which are also algebraic sets in $\mathbb{C}^m$.

An *affine algebraic variety*, or simply *affine variety*, is an irreducible algebraic set in $\mathbb{C}^m$. It is considered to be a topological space, with the induced (Zariski) topology. A *quasi-affine variety* is an open set in an affine variety.

Let X be an affine variety in $\mathbb{C}^m$. Let $I(X)$ be the set of $f \in \mathbb{C}[z_1, \ldots, z_m]$ that vanish on X, and $A(X)$ the set $\{f|_X : f \in \mathbb{C}[z_1, \ldots, z_m]\}$ of functions on X. Then $A(X)$ is a ring of functions on X, and $I(X)$ is an ideal in the ring $\mathbb{C}[z_1, \ldots, z_m]$, with $A(X) \cong \mathbb{C}[z_1, \ldots, z_m]/I(X)$. We call $A(X)$ the *affine coordinate ring* of X.

Choose $x \in X$, and define I_x to be the set of functions in $A(X)$ that are zero at x. Clearly, I_x is an ideal in $A(X)$. In fact, I_x is a *maximal ideal* in $A(X)$. Moreover, every maximal ideal in $A(X)$ is of the form I_x for some $x \in X$, and if $x_1, x_2 \in X$, then $I_{x_1} = I_{x_2}$ if and only if $x_1 = x_2$. It follows that there is a 1-1 correspondence between the points of X, and the set of maximal ideals in $A(X)$. This is the beginning of the subject of *affine algebraic geometry*. The idea is that the ring $A(X)$ is regarded as the primary object, and then X is derived from $A(X)$. The philosophy is to investigate affine varieties by using the algebraic properties of rings of functions on them.

Next, we will discuss *projective varieties*. Projective varieties are subsets of $\mathbb{CP}^m$, just as affine varieties are subsets of $\mathbb{C}^m$. Since we cannot define polynomials on $\mathbb{CP}^m$, instead we use *homogeneous polynomials* on $\mathbb{C}^{m+1}$.

Definition 5.8.3 Let $\mathbb{CP}^m$ be the complex projective space, and let $[z_0, \ldots, z_m]$ be homogeneous coordinates on $\mathbb{CP}^m$, as described in Example 5.1.1. Let d be a non-negative integer. A polynomial $f \in \mathbb{C}[z_0, \ldots, z_m]$ is called *homogeneous of degree* d if $f(\lambda z_0, \ldots, \lambda z_m) = \lambda^d f(z_0, \ldots, z_m)$ for all λ and $z_0, \ldots, z_m \in \mathbb{C}$. Let f be a homogeneous polynomial in $\mathbb{C}[z_0, \ldots, z_m]$, and let $[z_0, \ldots, z_m] \in \mathbb{CP}^m$. We say that $[z_0, \ldots, z_m]$ is a *zero* of f if $f(z_0, \ldots, z_m) = 0$. As f is homogeneous, this definition does not depend on the choice of homogeneous coordinates for $[z_0, \ldots, z_m]$.

Here are the analogous definitions of the Zariski topology on $\mathbb{CP}^m$, and projective and quasi-projective varieties.

Definition 5.8.4 Define an *algebraic set* in $\mathbb{CP}^m$ to be the set of common zeros of a finite number of homogeneous polynomials in $\mathbb{C}[z_0, \ldots, z_m]$. Define the *Zariski topology* on $\mathbb{CP}^m$ by taking the open subsets to be $\mathbb{CP}^m \setminus X$, for all algebraic sets $X \subset \mathbb{CP}^m$. An algebraic set X in $\mathbb{CP}^m$ is said to be *irreducible* if it is not the union of two proper algebraic subsets. A *projective algebraic variety*, or simply *projective variety*, is defined to be an irreducible algebraic set in $\mathbb{CP}^m$. A *quasi-projective variety* is defined to be an open subset of a projective variety, in the Zariski topology.

Now $\mathbb{C}^m$ can be identified with the open set $\{[z_0, \ldots, z_m] \in \mathbb{CP}^m : z_0 \neq 0\}$ in $\mathbb{CP}^m$. Making this identification, we see that affine, quasi-affine and projective varieties are all examples of quasi-projective varieties. Because of this, quasi-projective varieties are often referred to as *algebraic varieties*, or simply *varieties*. In this section we consider varieties to be topological spaces with the Zariski topology, unless we specify otherwise.

An affine variety is studied using the ring of polynomials on it. On projective varieties we cannot consider polynomials, so we consider two other classes of functions, the *rational functions* and *regular functions*.

Definition 5.8.5 Let X be a quasi-projective variety in $\mathbb{CP}^m$. Let $g, h \in \mathbb{C}[z_0, \ldots, z_m]$ be homogeneous polynomials of the same degree d. Define a subset U_h in X by $U_h = \{[z_0, \ldots, z_m] \in X : h(z_0, \ldots, z_m) \neq 0\}$. Then U_h is a Zariski open set. Define the *rational function* $f : U_h \to \mathbb{C}$ by

$$f([z_0, \ldots, z_m]) = \frac{g(z_0, \ldots, z_m)}{h(z_0, \ldots, z_m)}.$$

As g and h are both homogeneous of the same degree, f is independent of the choice of homogeneous coordinates for each point, and so is well-defined.

Let U be open in X and let $f : U \to \mathbb{C}$ be a function. If $p \in U$, we say that f is *regular* at p if there is an open set $U' \subset U$ containing p, and f is equal to a rational function on U'. If f is regular at every point $p \in U$, we say f is *regular*. A regular function is one that is *locally* equal to a rational function.

Next we define two ideas of map between varieties, *morphisms* and *rational maps*.

Definition 5.8.6 Let X and Y be varieties. A *morphism* $\phi : X \to Y$ is a continuous map (with the Zariski topologies) such that whenever V is open in Y and $f : V \to \mathbb{C}$ is regular, then $f \circ \phi : \phi^{-1}(V) \to \mathbb{C}$ is also regular. Clearly, if $\phi : X \to Y$ and $\psi : Y \to Z$ are morphisms of varieties, then $\psi \circ \phi : X \to Z$ is also a morphism. A map $\phi : X \to Y$ is called an *isomorphism* if ϕ is bijective, so that it has an inverse $\phi^{-1} : Y \to X$, and both ϕ and ϕ^{-1} are morphisms.

Definition 5.8.7 Let X and Y be varieties. A *rational map* $\phi : X \dashrightarrow Y$ is an equivalence class of morphisms $\phi_U : U \to Y$, where U is a dense open set in X, and morphisms $\phi_U : U \to Y$ and $\phi_V : V \to Y$ are equivalent if $\phi_U|_{U \cap V} = \phi_V|_{U \cap V}$. Note that a rational map is *not* in general a map of the set X to the set Y.

A *birational map* $\phi : X \dashrightarrow Y$ is a rational map which admits a rational inverse. That is, ϕ is an equivalence class of maps $\phi_U : U \to V$, where U, V are dense open sets in X, Y respectively, and ϕ_U is an isomorphism of varieties. If there is a birational map between X and Y, we say X and Y are *birationally equivalent*, or simply *birational*. This is an equivalence relation.

A *birational morphism* $\phi : X \to Y$ is a morphism of varieties which is also a birational map. That is, there should exist dense open subsets $U \subset X$ and $V \subset Y$ such that $\phi(U) = V$ and $\phi|_U : U \to V$ is an isomorphism. A birational morphism $\phi : X \to Y$ is a genuine map from the set X to Y.

If X and Y are isomorphic varieties, then they are birational. However, if X and Y are birational, they need not be isomorphic. Thus, birationality is a cruder equivalence relation on varieties than isomorphism.

Let X be a variety, and U be open in X. Define A_U to be the set of regular functions on U. Then A_U is a *ring of functions on U*. If U, V are open sets with $U \subseteq V$, then restriction from V to U gives a natural map $r_{V,U} : A_V \to A_U$, which is a ring homomorphism. All this information is packaged together in a composite mathematical object called a *sheaf of rings* on X (see [132, p. 35] or [171, p. 27]).

Definition 5.8.8 Let X be a topological space with topology $\mathcal{T}$. A *sheaf of rings* $\mathscr{F}$ on X associates to each open set $U \in \mathcal{T}$ a ring $\mathscr{F}(U)$, called the *sections* of $\mathscr{F}$ over U, and to each pair $U \subset V$ in $\mathcal{T}$ a ring homomorphism $r_{V,U} : \mathscr{F}(V) \to \mathscr{F}(U)$, such that conditions (i)–(v) below are satisfied. The map $r_{V,U}$ is called the *restriction map*, and for $\sigma \in \mathscr{F}(V)$ we write $r_{V,U}(\sigma) = \sigma|_U$. Here are the necessary axioms.

(i) $\mathscr{F}(\emptyset) = \{0\}$.
(ii) $r_{U,U} : \mathscr{F}(U) \to \mathscr{F}(U)$ is the identity for all $U \in \mathcal{T}$.
(iii) If $U, V, W \in \mathcal{T}$ with $U \subset V \subset W$, then $r_{W,U} = r_{W,V} \circ r_{V,U}$.
(iv) If $U, V \in \mathcal{T}$ and $\sigma \in \mathscr{F}(U), \tau \in \mathscr{F}(V)$ satisfy $\sigma|_{U \cap V} = \tau|_{U \cap V}$, then there exists $\rho \in \mathscr{F}(U \cup V)$ such that $\rho|_U = \sigma$ and $\rho|_V = \tau$.
(v) If $U, V \in \mathcal{T}$ and $\sigma \in \mathscr{F}(U \cup V)$ satisfies $\sigma|_U = 0$ and $\sigma|_V = 0$, then $\sigma = 0$.

A *sheaf of groups* on X is defined in exactly the same way, except that $\mathscr{F}(U)$ should be a group rather than a ring, and the restrictions $r_{V,U}$ should be group homomorphisms. A *ringed space* is defined to be a pair $(X, \mathcal{O})$, where X is a topological space and $\mathcal{O}$ a sheaf of rings on X. We call X the *base space*, and $\mathcal{O}$ the *structure sheaf*.

If $\mathscr{F}$ is a sheaf of groups or rings over a topological space X, then one can define the *sheaf cohomology groups* $H^j(X, \mathscr{F})$ for $j = 0, 1, 2, \ldots$. They are an important tool in algebraic geometry. The group $H^0(X, \mathscr{F})$ is $\mathscr{F}(X)$, the group of global sections of $\mathscr{F}$ over X, but the groups $H^k(X, \mathscr{F})$ for $k \geqslant 1$ are more difficult to interpret. For more details, see [132, §0.3], [149, §3] or [171, §4].

Let X be a projective variety. Then X is a topological space, with the Zariski topology, and the regular functions on X form a sheaf $\mathcal{O}$ of rings on X, called the *sheaf of regular functions on X*. Thus $(X, \mathcal{O})$ is a *ringed space*. Moreover, for each open set U, the ring $\mathcal{O}(U)$ is actually a ring of complex functions on U.

Complex algebraic geometry can be described as *the study of complex algebraic varieties up to isomorphisms*. Let X and Y be varieties, and $\phi : X \to Y$ an isomorphism of varieties. Consider the question: if X and Y are isomorphic, what features of X and Y have to be 'the same'? Well, from the definition we see that X and Y have to be isomorphic as topological spaces, with the Zariski topologies, and the sheaves of regular functions must also agree. However, there is no need for the embeddings $X \hookrightarrow \mathbb{CP}^m$ and $Y \hookrightarrow \mathbb{CP}^n$ to be related at all.

Because of this, it is useful to think of a variety not as a particular subset of $\mathbb{CP}^m$, but as a topological space X equipped with a sheaf of rings $\mathcal{O}$. Following this idea, one can define the concept of an *abstract variety*, which is a variety without a given embedding in $\mathbb{CP}^m$. In affine algebraic geometry, the primary object is a ring A. The topological space X is derived from A as the set of maximal ideals, and is studied using algebraic tools and a lot of ring theory.

In more general algebraic geometry, the primary object is often an abstract variety, regarded as a topological space X equipped with a sheaf of rings $\mathcal{O}$. It is also studied from a very algebraic point of view. In fact, a lot of algebraic geometry is written in terms of *schemes* [149, §2], which are closely related to varieties. Recall that in an affine variety, the points of the topological space are the maximal ideals of a ring. In an *affine scheme* the points of the topological space are instead the *prime ideals* of a ring. A *scheme* is a ringed space $(X, \mathcal{O})$ that is locally isomorphic to an affine scheme.

5.9   Singular varieties, resolutions, and deformations

Let X be a variety in $\mathbb{CP}^m$, and let $x \in X$. We say that x is a *nonsingular point* if X is a complex submanifold of $\mathbb{CP}^m$ in a neighbourhood of x. We call x *singular* if it is not nonsingular. The variety X is called *singular* if it has singular points, and *nonsingular* otherwise. In general, the nonsingular points form a dense open subset of X, and the singular points are a finite union of subvarieties of X. There is also an equivalent, algebraic way to define singular points, using the idea of *local ring*.

For example, let $p(z_1, \ldots, z_m)$ be a complex polynomial that is not constant with no repeated factors, and X be the hypersurface $\{(z_1, \ldots, z_m) \in \mathbb{C}^m : p(z_1, \ldots, z_m) = 0\}$ in $\mathbb{C}^m$. Then a point $x \in X$ is singular if and only if $\partial p / \partial z_j = 0$ at x for $j = 1, \ldots, m$. So, for instance, the quadric $z_1^2 + z_2^2 + z_3^2 = 0$ in $\mathbb{C}^3$ has just one singular point at $(0, 0, 0)$.

Clearly, a complex algebraic variety X is a *complex manifold* if and only if it is nonsingular. The converse, however, is not true: not every complex manifold is an algebraic variety. Let X be a compact complex manifold. A *meromorphic function* f on X is a singular holomorphic function, that can be written locally as the quotient of two holomorphic functions. On an algebraic variety, all the regular functions are meromorphic. Therefore a variety must have a lot of meromorphic functions—enough to form a holomorphic coordinate system near each point, for instance.

So, if a compact complex manifold has only a few meromorphic functions, then it cannot be an algebraic variety. There are many compact complex manifolds that admit no nonconstant meromorphic functions at all, and these are not algebraic varieties. However, *Chow's Theorem* [132, p. 167] states that any compact complex submanifold of $\mathbb{CP}^m$ is algebraic. The study of nonalgebraic complex manifolds is sometimes called *transcendental complex geometry*.

There is a natural generalization of the idea of complex manifold to include singularities, called a *complex analytic variety*.

Definition 5.9.1 Let U be an open set in $\mathbb{C}^m$, in the usual topology, rather than the Zariski topology. An *analytic subset of* U is a subset $S \subseteq U$ defined by the vanishing of a finite number of holomorphic functions on U. The restriction to S of the sheaf $\mathscr{O}_U$ of holomorphic functions on U is a sheaf of rings $\mathscr{O}_S$ on S. We define a *(complex) analytic variety* to be a ringed space $(X, \mathscr{O}_X)$ such that X is Hausdorff, and $(X, \mathscr{O}_X)$ is locally isomorphic to $(S, \mathscr{O}_S)$ for analytic subsets $S \subseteq U \subseteq \mathbb{C}^m$. Here sheaves and ringed spaces are defined in Definition 5.8.8.

We call a point $x \in X$ *nonsingular* if $(X, \mathscr{O}_X)$ is locally isomorphic to $(\mathbb{C}^k, \mathscr{O}_{\mathbb{C}^k})$ near x, where k is the dimension of X near x. We call x *singular* if it is not nonsingular. If X contains no singular points then it is a *complex manifold*. Otherwise it is a *singular complex manifold*. For more information about analytic varieties, see [132, p. 12–14].

Complex algebraic varieties are examples of complex analytic varieties. Conversely, complex analytic varieties are locally isomorphic to complex algebraic varieties, but not necessarily globally isomorphic. All the ideas in the rest of this section work equally well in the setting of algebraic varieties or analytic varieties, but we will only give definitions for one of the two.

Now a singular point in a variety X is a point where the manifold structure of X breaks down in some way. Given a singular variety X, it is an important problem in algebraic geometry to understand how to repair the singularities of X, and make a new, nonsingular variety $\tilde{X}$ closely related to X. There are two main strategies used to do this, called *resolution* and *deformation*.

5.9.1 Resolutions of singular varieties

Definition 5.9.2 Let X be a singular variety. A *resolution* $(\tilde{X}, \pi)$ of X is a normal, nonsingular variety $\tilde{X}$ with a proper birational morphism $\pi : \tilde{X} \to X$. Here *normal varieties* are defined by Griffiths and Harris [132, p. 177] and Iitaka [171, §2], and *proper morphisms* by Hartshorne [149, p. 95–105]. From [149, p. 95], a morphism $f : X \to Y$ of complex algebraic varieties is proper if it pulls back compact sets in Y to compact sets in X, using the manifold topologies on X, Y, not the Zariski topologies.

This means that $\tilde{X}$ is a complex manifold, and the map $\pi : \tilde{X} \to X$ is surjective. There are dense open sets of X and $\tilde{X}$ on which π is also injective, and in fact biholomorphic. But if x is a singular point of X, then $\pi^{-1}(x)$ is in general a compact subvariety of $\tilde{X}$, rather than a single point. Often $\pi^{-1}(x)$ is a submanifold of $\tilde{X}$, or a finite union of submanifolds. Thus, in a resolution we repair the singularities by replacing each singular point by a submanifold, or more general subvariety.

One way to construct resolutions is to use a technique called *blowing up*, which we define first for affine varieties.

Definition 5.9.3 Let $X \subset \mathbb{C}^m$ be an affine variety with affine coordinate ring $A(X)$, let Y be a closed subvariety of X, and let $I_Y \subset A(X)$ be the ideal of functions in $A(X)$ that are zero on Y. Then I_Y is finitely generated, and we can choose a set of generators $f_0, \ldots, f_n \in I_Y$ for I_Y. Define a map $\phi : X \setminus Y \to \mathbb{CP}^n$ by $\phi(x) = [f_0(x), \ldots, f_n(x)]$.

The *(algebraic) blow-up* $\tilde{X}$ *of* X *along* Y is the closure in $X \times \mathbb{CP}^n$ of the graph of ϕ, that is, $\tilde{X} = \overline{\{(x, \phi(x)) : x \in X \setminus Y\}} \subset X \times \mathbb{CP}^n$. The *projection* $\pi : \tilde{X} \to X$ is the map $\pi : (x, z) \mapsto x$. Then $\tilde{X}$ is a variety, and $\pi : \tilde{X} \to X$ is a birational morphism. In a similar way, one can define the blow-up $(\tilde{X}, \pi)$ of a general algebraic or analytic variety X along a closed subvariety Y. For more information, see [132, p. 182, p. 602] and [148, p. 82].

Here $\pi : \tilde{X} \to X$ is surjective, and $\pi : \tilde{X} \setminus \pi^{-1}(Y) \to X \setminus Y$ is an isomorphism. The pull-back $\pi^{-1}(Y)$ is a finite union of closed subvarieties of $\tilde{X}$ of codimension one, called the *exceptional divisor*. If X is nonsingular and Y is a submanifold of X, then $\pi^{-1}(Y)$ is the projectivized normal bundle of Y in X. In particular, if Y is the single point y, then $\pi^{-1}(y)$ is the complex projective space $P(T_y X)$.

Now suppose X is a singular variety, and let $Y \subset X$ be the set of singular points in X. Then Y is a finite union of subvarieties of X, and we can consider the blow-up $\tilde{X}$ of X along Y. Although $\tilde{X}$ may not be nonsingular, it is a general principle that the singularities of $\tilde{X}$ are usually of a less severe kind, and easier to resolve, than those of X. Our next result, sometimes called the *Resolution of Singularities Theorem*, shows that the singularities of any variety can be resolved by a finite number of blow-ups.

Theorem 5.9.4. (Hironaka [159]) *Let X be a complex algebraic variety. Then there exists a resolution $\pi : \tilde{X} \to X$, which is the result of a finite sequence of blow-ups of X. That is, there are varieties $X = X_0, X_1, \ldots, X_n = \tilde{X}$, such that X_j is a blow-up of X_{j-1} along some subvariety, with projection $\pi_j : X_j \to X_{j-1}$, and the map $\pi : \tilde{X} \to X$ is $\pi = \pi_1 \circ \cdots \circ \pi_n$.*

5.9.2   Deformations of singular and nonsingular varieties

Definition 5.9.5 Let X be a complex analytic variety of dimension m. A *1-parameter family of deformations* of X is a complex analytic variety $\mathscr{X}$ of dimension $m + 1$, together with a proper holomorphic map $f : \mathscr{X} \to \Delta$ where Δ is the unit disc in $\mathbb{C}$, such that $X_0 = f^{-1}(0)$ is isomorphic to X. The other fibres $X_t = f^{-1}(t)$ for $t \neq 0$ are called *deformations* of X.

If the deformations X_t are nonsingular for $t \neq 0$, they are called *smoothings* of X. By a *small deformation* of X we mean a deformation X_t where t is small. That is, when we say something is true for all small deformations of X, we mean that in any 1-parameter family of deformations $\{X_t : t \in \Delta\}$ of X, the statement holds for all sufficiently small t. We say that X is *rigid* if all small deformations X_t of X are biholomorphic to X.

We shall be interested in deformations of complex analytic varieties for two reasons. Firstly, a singular variety X may admit a family of nonsingular deformations X_t. Thus, as with resolutions, deformation gives a way of repairing the singularities of X to get a nonsingular variety.

From the point of view of algebraic geometry, there is a big difference between resolution and deformation. If X is a singular variety and $\tilde{X}$ a resolution of X, then X and $\tilde{X}$ are birationally equivalent, and share the same field of meromorphic functions. So to algebraic geometers, who often try to classify varieties up to birational equivalence, X and $\tilde{X}$ are nearly the same thing. But a variety X and its deformations X_t can be algebraically very different.

The second reason we will be interested in deformations is when we wish to describe the family of all integrable complex structures upon a particular compact manifold, up to isomorphism. Suppose that (X, J) is a compact complex manifold. Then all small deformations X_t of X are nonsingular, and are diffeomorphic to X as real manifolds. Thus small deformations of X are equivalent to complex structures J_t on X that are close to the complex structure J, in a suitable sense. So, to understand the local geometry of the moduli space of complex structures on X, we need a way to study the collection of *all* small deformations of a complex analytic variety.

Definition 5.9.6 Let X be a complex analytic variety. A *family of deformations of* X consists of a (possibly singular) complex analytic variety T called the *base space* containing a *base point* t_0, and a complex analytic variety $\mathscr{X}$ with a flat holomorphic map $f : \mathscr{X} \to T$, such that $X_{t_0} = f^{-1}(t_0)$ is isomorphic to X. The other fibres $X_t = f^{-1}(t_0)$ are then deformations of X. Here *flatness* is a technical condition upon morphisms of algebraic or analytic varieties defined in [149, §III.9]. It implies, in particular, that $\dim X_t = \dim X$ for all $t \in T$.

If (S, s_0) is another complex analytic variety with base point and $F : S \to T$ is a holomorphic map with $F(s_0) = t_0$, then we get an induced family of deformations $F^*(\mathscr{X})$ of X over S. A family of deformations of X is called *versal* or *semi-universal* if any other family of small deformations of X can be induced from it by a suitable map F. It is called *universal* if this map F is unique.

Note that some authors (e.g. Slodowy [315, p. 7]) define semi-universality differently, and distinguish between versal and semi-universal deformations. If $\{X_t : t \in T\}$ is a *universal* family of deformations of X, then every small deformation of X appears exactly once in the family. Thus the collection of all deformations of X is locally isomorphic to the base space T of the universal family, and has the structure of a complex analytic variety. In particular, if (X, J) is a compact complex manifold and $\{X_t : t \in T\}$ is a universal family of deformations of (X, J), then the moduli space of all complex structures J_t on X is locally isomorphic to T.

However, there exist compact complex manifolds (X, J) which have no universal family of deformations. The moduli space of complex structures on X up to isomorphism has a natural topology. If this topology is not Hausdorff near J then (X, J) cannot have a universal family of deformations, because the base space T would be non-Hausdorff, contradicting its definition as a complex analytic variety.

Rather than working with moduli spaces of pathological topology, it is helpful in this case to consider a *versal* family of deformations of X. In a versal family every small deformation of X is represented at least once, but some may appear many times. Now the theory of deformations of compact complex manifolds was developed by Kodaira, Spencer and Kuranishi, and is described in Kodaira [217]. The main result in this theory is that a versal family of deformations exists for any compact complex manifold, and can be constructed using sheaf cohomology.

Let X be a compact complex manifold, and let Θ_X be the sheaf of holomorphic vector fields of X. Then the *sheaf cohomology groups* $H^*(X, \Theta_X)$ are the cohomology of the complex

$$0 \to C^\infty(T^{1,0}X) \xrightarrow{\bar\partial} C^\infty(T^{1,0}X \otimes \Lambda^{0,1}X) \xrightarrow{\bar\partial} C^\infty(T^{1,0}X \otimes \Lambda^{0,2}X) \xrightarrow{\bar\partial} \cdots .$$

Here we interpret $H^1(X, \Theta_X)$ as the space of infinitesimal deformations of the complex structure of X, and $H^2(X, \Theta_X)$ as the space of obstructions to lifting an infinitesimal deformation to an actual deformation of the complex structure of X. Kodaira, Spencer and Kuranishi prove that there is an open neighbourhood U of 0 in $H^1(X, \Theta_X)$ and a holomorphic map $\Phi : U \to H^2(X, \Theta_X)$ with $\Phi(0) = d\Phi(0) = 0$, such that $T = \Phi^{-1}(0)$ is the base of a versal family of deformations of X, with base point 0, called the *Kuranishi family* of X. The group $H^0(X, \Theta_X)$ also has an interpretation as the Lie algebra of the group of holomorphic automorphisms of X, and if $H^0(X, \Theta_X) = 0$ then the Kuranishi family is universal.

5.10 Line bundles and divisors

Let M be a complex manifold. A *holomorphic line bundle* over M is a holomorphic vector bundle with fibre $\mathbb{C}$, the complex line. Holomorphic lines bundles are important in algebraic geometry. If L, L' and L'' are holomorphic line bundles over M, then the dual bundle L^* and the tensor product $L \otimes L'$ are also holomorphic line bundles. These operations satisfy the equations $L \otimes L' \cong L' \otimes L$, $(L \otimes L') \otimes L'' \cong L \otimes (L' \otimes L'')$, and $L \otimes L^* \cong \bar{\mathbb{C}}$, where $\bar{\mathbb{C}}$ is the trivial line bundle $M \times \mathbb{C}$.

Define $\mathcal{P}_M$ to be the set of isomorphism classes of holomorphic line bundles over M. From the equations above we see that $\mathcal{P}_M$ is an abelian group, where multiplication is given by the tensor product, inverses are dual bundles, and the identity is the trivial bundle $\bar{\mathbb{C}}$. This group is called the *Picard group* of M. It can be identified with the *sheaf cohomology group* $H^1(M, \mathcal{O}^*)$ [132, p. 133], but we will not explain this.

Because of the group structure on $\mathcal{P}_M$, it is convenient to use a multiplicative notation for line bundles. Let M be a complex manifold, L a holomorphic line bundle over M, and $k \in \mathbb{Z}$. Then we write $L^k = \bigotimes^k L$ if $k > 0$, $L^k = \bigotimes^{-k} L^*$ if $k < 0$, and $L^0 = \bar{\mathbb{C}}$. In particular, the dual L^* is written L^{-1}.

If M is a complex manifold of dimension m, then $\Lambda^{p,0}M$ is a holomorphic vector bundle with fibre dimension $\binom{m}{p}$. Thus, when $p = m$, the fibre of $\Lambda^{m,0}M$ is $\mathbb{C}$, and $\Lambda^{m,0}M$ is a holomorphic line bundle. This is called the *canonical bundle* of M, and is written K_M. It is the bundle of *complex volume forms* on M, and is an important tool in algebraic geometry.

Let L be a holomorphic line bundle over a complex manifold M. The *first Chern class* $c_1(L)$ of L is a topological invariant of L called a *characteristic class*, which lies in the cohomology group $H^2(M, \mathbb{Z})$. Characteristic classes are described in [261]. The first Chern class classifies line bundles as smooth vector bundles. It satisfies

$$c_1(L^*) = -c_1(L) \quad \text{and} \quad c_1(L \otimes L') = c_1(L) + c_1(L').$$

Thus $c_1 : \mathcal{P}_M \to H^2(M, \mathbb{Z})$ is a homomorphism of abelian groups.

Let M be a complex manifold, and L a holomorphic line bundle over M. For each open set $U \subset M$, define $\mathcal{O}_L(U)$ to be the vector space of holomorphic sections of L over U, and if U, V are open in M with $U \subset V$, let $r_{V,U} : \mathcal{O}_L(V) \to \mathcal{O}_L(U)$ be the restriction map. Then $\mathcal{O}_L$ is a sheaf of groups over M, the *sheaf of holomorphic sections* of L. Now we describe the line bundles over the projective space $\mathbb{CP}^m$.

Example 5.10.1 Recall that in Example 5.1.1, $\mathbb{CP}^m$ was defined to be the set of one-dimensional vector spaces of $\mathbb{C}^{m+1}$. Define the *tautological line bundle* L^{-1} over $\mathbb{CP}^m$ to be the subbundle of the trivial bundle $\mathbb{CP}^m \times \mathbb{C}^{m+1}$, whose fibre at $x \in \mathbb{CP}^m$ is the line in $\mathbb{C}^{m+1}$ represented by x. Then L^{-1} is a vector bundle over $\mathbb{CP}^m$ with fibre $\mathbb{C}$. The total space of L^{-1} is a complex submanifold of $\mathbb{CP}^m \times \mathbb{C}^{m+1}$, and has the structure of a complex manifold. Thus, L^{-1} is a holomorphic line bundle over $\mathbb{CP}^m$.

Define L to be the dual of L^{-1}. Then L is a holomorphic line bundle over $\mathbb{CP}^m$, called the *hyperplane bundle*. So, L^k is a holomorphic line bundle over $\mathbb{CP}^m$ for each $k \in \mathbb{Z}$. It can be shown [132, p. 145] that every holomorphic line bundle over $\mathbb{CP}^m$ is isomorphic to L^k for some $k \in \mathbb{Z}$. There is an isomorphism $H^2(\mathbb{CP}^m, \mathbb{Z}) \cong \mathbb{Z}$, and making this identification we find that $c_1(L^k) = k \in \mathbb{Z}$. Thus $c_1 : \mathcal{P}_{\mathbb{CP}^m} \to H^2(\mathbb{CP}^m, \mathbb{Z}) \cong \mathbb{Z}$ is a group isomorphism. The canonical bundle $K_{\mathbb{CP}^m}$ of $\mathbb{CP}^m$ is isomorphic to L^{-m-1}.

The sheaf of holomorphic sections $\mathcal{O}_{L^k}$ of L^k over $\mathbb{CP}^m$ is written $\mathcal{O}(k)$. (Also, by an abuse of notation, $\mathcal{O}(k)$ often denotes the line bundle L^k.) The vector space of holomorphic sections of L^k is $H^0(\mathbb{CP}^m, \mathcal{O}(k))$, in the notation of sheaf cohomology. If $k < 0$ then $H^0(\mathbb{CP}^m, \mathcal{O}(k))$ is zero, and if $k \geqslant 0$ it is canonically identified with the set of homogeneous polynomials of degree k on $\mathbb{C}^{m+1}$, which is a vector space of dimension $\binom{m+k}{m}$.

Let M be a compact complex manifold, and L a holomorphic line bundle over M. Let V be the vector space $H^0(M, \mathcal{O}_L)$ of holomorphic sections of L over M. Then V is a finite-dimensional vector space over $\mathbb{C}$, of dimension $m + 1$, say. For each point $p \in M$, define a map $\phi_p : V \to L_p$ by $\phi_p(s) = s(p)$. Then ϕ_p is linear, so that $\phi_p \in V^* \otimes L_p$. Define p to be a *base point* of L if $\phi_p = 0$, and let $B \subset M$ be the set of base points of L. Then if $p \in M \setminus B$, then ϕ_p is nonzero in $V^* \otimes L_p$, and thus $[\phi_p] \in P(V^* \otimes L_p)$.

But L is a line bundle, so $L_p \cong \mathbb{C}$ as complex vector spaces. Therefore the projective spaces $P(V^*)$ and $P(V^* \otimes L_p)$ are naturally isomorphic, and we can regard $[\phi_p]$ as a point in $P(V^*)$. So, define a map $\iota_L : M \setminus B \to P(V^*)$ by $\iota_L(p) = [\phi_p]$. Now $P(V^*)$ is a complex projective space $\mathbb{CP}^m$, and thus a complex manifold, and B is closed in M, so that $M \setminus B$ is also a complex manifold. It is easy to show that $\iota_L : M \setminus B \to \mathbb{CP}^m$ is a holomorphic map of complex manifolds. When $L = K_M^r$, a power of the canonical bundle, the maps $\iota_{K_M^r}$ are called the *pluricanonical maps*, and are important in algebraic geometry.

A line bundle L over a compact complex manifold M is called *very ample* if L has no base points in M, and the map $\iota_L : M \to \mathbb{CP}^m$ is an *embedding* of M in $\mathbb{CP}^m$. Also, L is called *ample* if L^k is very ample for some $k > 0$. Thus, if L is very ample, then ι_L identifies M with a complex submanifold of $\mathbb{CP}^m$, its image $\iota_L(M)$. Now by Chow's Theorem [132, p. 167], every compact complex submanifold of $\mathbb{CP}^m$ is a nonsingular projective variety. Thus, if M is a compact complex manifold with an ample line bundle, then M is a projective variety.

The remarkable *Kodaira Embedding Theorem* [132, p. 181] gives a simple criterion for a holomorphic line bundle L over a compact complex manifold M to be ample. A line bundle L is called *positive* if its first Chern class $c_1(L)$ can be represented, as a de Rham cohomology class, by a closed $(1, 1)$-form α which is positive in the sense of §5.4. The Kodaira Embedding Theorem says that L is ample if and only if it is positive. Therefore, a compact complex manifold with a positive line bundle is a projective variety. Because of this, many problems on compact complex and Kähler manifolds become problems about projective varieties, and can be attacked algebraically.

5.10.1 Divisors

Now we shall explore the connections between line bundles and divisors, which are formal sums of hypersurfaces in complex manifolds.

Definition 5.10.2 Let M be a complex manifold. A closed subset $N \subset M$ is said to be a *hypersurface* in M if for each $p \in N$ there is an open neighbourhood U of p in M and a nonzero holomorphic function $f : U \rightarrow \mathbb{C}$, such that $N \cap U = \{u \in U : f(u) = 0\}$. A hypersurface $N \subset M$ is called *irreducible* if it is not the union of two hypersurfaces N_1, N_2 with $N_1, N_2 \neq N$.

In general, a hypersurface $N \subset M$ is a singular submanifold of M, of codimension one. Every hypersurface in M can be written uniquely as a union of irreducible hypersurfaces, and if M is compact then this union is finite. Suppose now that M is a complex manifold, L a holomorphic line bundle over M, and s a nonzero holomorphic section of L. Define $N \subset M$ to be the set $\{m \in M : s(m) = 0\}$. Then N is a hypersurface in M. Thus, there is a link between holomorphic line bundles, holomorphic sections, and hypersurfaces. To render this link more explicit, we make another definition.

Definition 5.10.3 Let M be a complex manifold. An irreducible hypersurface $N \subset M$ is called a *prime divisor* on M. A *divisor* D on M is a locally finite formal linear combination $D = \sum_i a_i N_i$, where $a_i \in \mathbb{Z}$, and each N_i is a prime divisor. Here 'locally finite' means that each compact subset of M meets only a finite number of the hypersurfaces N_i. The divisor D is called *effective* if $a_i \geqslant 0$ for all i.

Suppose as before that M is a complex manifold, L a holomorphic line bundle over M, and s a nonzero holomorphic section of L. Let N be the hypersurface $N = \{m \in M : s(m) = 0\}$. Then N may be written in a unique way as a locally finite union $N = \bigcup_i N_i$, where the N_i are prime divisors. For each i, there is a unique positive integer a_i, such that s *vanishes to order* a_i *along* N_i. Define $D = \sum_i a_i N_i$. Then D is an effective divisor. In this way, whenever we have a nonzero holomorphic section of a holomorphic line bundle over M, we construct an effective divisor on M.

This construction is reversible, in the following sense. Suppose that L_1, L_2 are holomorphic line bundles over M, and s_1, s_2 are nonzero holomorphic sections of L_1, L_2 respectively. Let D_1, D_2 be the effective divisors constructed from s_1, s_2. It can be shown that $D_1 = D_2$ if and only if there exists an isomorphism $\phi : L_1 \rightarrow L_2$ of holomorphic line bundles, such that $\phi(s_1) = s_2$, and the isomorphism ϕ is then unique. Moreover, if D is an effective divisor on M, then there exists a holomorphic line bundle L over M, and a nonzero holomorphic section s of L, that yields the divisor D.

Thus there is a 1-1 correspondence between effective divisors on M, and isomorphism classes of holomorphic line bundles equipped with nonzero holomorphic sections. In the same way, there is a 1-1 correspondence between divisors on M, and isomorphism classes of line bundles equipped with nonzero *meromorphic* sections. In this case, the divisor $\sum_i a_i N_i$ corresponds to a section s with a *zero* of order a_i along N_i if $a_i > 0$, and a *pole* of order $-a_i$ along N_i if $a_i < 0$.

Let M be a compact complex manifold of complex dimension m, and L a holomorphic line bundle over M, with a nonzero meromorphic section associated to a divisor D. Then $D = \sum_i a_i N_i$ defines a homology class $[D] = \sum_i a_i [N_i] \in H_{2m-2}(M, \mathbb{Z})$. Under the natural isomorphism $H^2(M, \mathbb{Z}) \cong H_{2m-2}(M, \mathbb{Z})$, this homology class is identified with $c_1(L)$, the first Chern class of L. This gives one way to understand $c_1(L)$.

Here is a result on the topology of a hypersurface N in M associated to a *positive* line bundle L.

Theorem 5.10.4. (Lefschetz Hyperplane Theorem) *Let M be a compact complex m-manifold, N a nonsingular hypersurface in M, and L the holomorphic line bundle over M associated to the divisor N. Suppose L is positive. Then*

(a) *the map $H^k(M, \mathbb{C}) \to H^k(N, \mathbb{C})$ induced by the inclusion $N \hookrightarrow M$ is an isomorphism for $0 \leqslant k \leqslant m - 2$ and injective for $k = m - 1$, and*

(b) *the map of homotopy groups $\pi_k(N) \to \pi_k(M)$ induced by the inclusion $N \hookrightarrow M$ is an isomorphism for $0 \leqslant k \leqslant m - 2$ and surjective for $k = m - 1$.*

The result also holds if M and N are orbifolds instead of manifolds, and N is a nonsingular hypersurface in the orbifold sense.

This is known as the *Lefschetz Hyperplane Theorem*, as we can take M to be a submanifold of $\mathbb{CP}^n$ and L the restriction to M of the line bundle $\mathcal{O}(1)$ over $\mathbb{CP}^n$, and then N is the intersection of M with a hyperplane H in $\mathbb{CP}^n$. Part (a) is proved in Griffiths and Harris [132, p. 156] for complex manifolds, and rather more general and complicated versions of (b) are proved by Goresky and MacPherson [129, p. 153] and Hamm [147], in which M, N can be singular complex varieties, not just orbifolds.

6

The Calabi Conjecture

Let (M, J) be a compact, complex manifold, and g a Kähler metric on M, with Ricci form ρ. From §5.6 we know that ρ is a closed (1,1)-form and $[\rho] = 2\pi c_1(M)$ in $H^2(M, \mathbb{R})$. It is natural to ask which closed (1,1)-forms can be the Ricci forms of a Kähler metric on M. The Calabi Conjecture [67,68] answers this question.

The Calabi Conjecture. *Let (M, J) be a compact, complex manifold, and g a Kähler metric on M, with Kähler form ω. Suppose that ρ' is a real, closed $(1,1)$-form on M with $[\rho'] = 2\pi c_1(M)$. Then there exists a unique Kähler metric g' on M with Kähler form ω', such that $[\omega'] = [\omega] \in H^2(M, \mathbb{R})$, and the Ricci form of g' is ρ'.*

The conjecture was posed by Calabi in 1954, who also showed that if g' exists it must be unique. It was eventually proved by Yau in 1976, [344,345]. Before this, Aubin [15] had made significant progress towards a proof. In this chapter we will give a proof of the Calabi Conjecture that broadly follows Yau's own proof, with some differences. My main references were Yau's paper [345], and the treatment given in Aubin's book [16, §7]. The proof is also explained, in French, by Bourguignon et al. [42].

In §6.1 the Calabi Conjecture is reformulated as a nonlinear, elliptic partial differential equation in a real function ϕ. Section 6.2 states four results, Theorems C1–C4, and then proves the Calabi Conjecture assuming these theorems. After some preparatory work in §6.3, Theorems C1–C4 are proved in §6.4–§6.7 respectively, and the proof of the Calabi Conjecture is complete. Finally, section 6.8 discusses some analytic issues from the proof.

The proof of the Calabi Conjecture is very important in the subject of Riemannian holonomy groups, for the following reason. Suppose M is a compact Kähler manifold with $c_1(M) = 0$. Then we may choose the 2-form ρ' in the Calabi Conjecture to be zero, and so the proof of the conjecture guarantees the existence of a Kähler metric g' on M with zero Ricci form. Thus, we construct families of *Ricci-flat Kähler metrics* on compact complex manifolds.

Now from §3.4.1, a generic Kähler metric g has $\mathrm{Hol}^0(g) = \mathrm{U}(m)$, but if g is Ricci-flat then $\mathrm{Hol}^0(g) \subseteq \mathrm{SU}(m)$. If g is irreducible, Berger's Theorem implies that either $\mathrm{Hol}^0(g) = \mathrm{SU}(m)$, or $m = 2k$ and $\mathrm{Hol}^0(g) = \mathrm{Sp}(k)$. Therefore, the Calabi Conjecture proof yields examples of compact Riemannian manifolds with holonomy $\mathrm{SU}(m)$ and $\mathrm{Sp}(k)$. These manifolds, called *Calabi–Yau manifolds* and *hyperkähler manifolds* respectively, will be the subjects of Chapters 7 and 10.

There are other applications of the proof of the Calabi Conjecture which we shall not discuss; for instance, it can be used to find Kähler metrics with positive or negative definite Ricci curvature on some compact complex manifolds M, and this has consequences for the fundamental group $\pi_1(M)$, and the group of biholomorphisms of M. There are also results on the existence of Kähler–Einstein metrics on complex manifolds that are closely related to the Calabi Conjecture proof. For more information on these topics, see Besse [30, §11].

6.1 Reformulating the Calabi Conjecture

We shall rewrite the Calabi Conjecture in terms of a partial differential equation. Let (M, J) be a compact, complex manifold, g a Kähler metric on M with Kähler form ω, and ρ the Ricci form of g. Let ρ' be a real, closed (1,1)-form on M with $[\rho'] = 2\pi c_1(M)$. To solve the Calabi Conjecture we must find a Kähler metric g', with Kähler form ω', such that $[\omega] = [\omega']$ and g' has Ricci form ρ'.

As $[\rho'] = 2\pi c_1(M) = [\rho]$ we have $[\rho' - \rho] = 0$ in $H^2(M, \mathbb{R})$, so by the proof of Lemma 5.5.1 there exists a smooth real function f on M, unique up to addition of a constant, such that $\rho' = \rho - \frac{1}{2}\mathrm{dd}^c f$. Define a smooth, positive function F on M by $(\omega')^m = F \cdot \omega^m$. Using eqns (5.6) and (5.7) of §5.6 we deduce that $\frac{1}{2}\mathrm{dd}^c(\log F) = \rho - \rho' = \frac{1}{2}\mathrm{dd}^c f$. Thus $\mathrm{dd}^c(f - \log F) = 0$, so that $f - \log F$ is constant on M.

Define $A > 0$ by $f - \log F = -\log A$. Then $F = A\mathrm{e}^f$, and g' must satisfy

$$(\omega')^m = A\mathrm{e}^f \omega^m. \tag{6.1}$$

As $[\omega'] = [\omega] \in H^2(M, \mathbb{R})$, and M is compact, we see that $\int_M (\omega')^m = \int_M \omega^m$. Substituting (6.1) in and applying (5.4), we deduce that

$$A \int_M \mathrm{e}^f \mathrm{d}V_g = \int_M \mathrm{d}V_g = \mathrm{vol}_g(M), \tag{6.2}$$

where $\mathrm{d}V_g$ is the volume form on M induced by g, and $\mathrm{vol}_g(M)$ the volume of M with this volume form. This determines the constant A.

Note that in this book, all manifolds are by definition assumed to be *connected*. If M were not connected then we would have to choose a different constant A for each connected component of M. We have shown that the Calabi Conjecture is equivalent to the following:

The Calabi Conjecture (second version). *Let (M, J) be a compact, complex manifold, and g a Kähler metric on M, with Kähler form ω. Let f be a smooth real function on M, and define $A > 0$ by $A \int_M \mathrm{e}^f \mathrm{d}V_g = \mathrm{vol}_g(M)$. Then there exists a unique Kähler metric g' on M with Kähler form ω', such that $[\omega'] = [\omega] \in H^2(M, \mathbb{R})$, and $(\omega')^m = A\mathrm{e}^f \omega^m$.*

Here is a way to understand this. The conjecture is about the existence of metrics with *prescribed volume forms*. Every volume form on M may be written as $F\mathrm{d}V_g$, for F a smooth real function. We impose two conditions on this volume form: firstly that it should be positive, so that $F > 0$, and secondly that it should have the same total volume as $\mathrm{d}V_g$, so that $\int_M F\mathrm{d}V_g = \int_M \mathrm{d}V_g$. Then the Calabi Conjecture says that there is a unique Kähler metric g' with the same Kähler class, such that $dV_{g'} = F dV_g$, that is, with the chosen volume form.

This is in fact a considerable simplification. The first statement of the conjecture prescribed the Ricci curvature of g', which depends on g' and its second derivatives, and was in effect m^2 real equations on g'. But this second statement depends only on g', not on its derivatives, and imposes only one real equation on g'.

Next, observe that as $[\omega'] = [\omega]$, by Lemma 5.5.1 there exists a smooth real function ϕ on M, unique up to addition of a constant, such that

$$\omega' = \omega + dd^c \phi. \tag{6.3}$$

Suppose also that ϕ satisfies the equation $\int_M \phi \, dV_g = 0$. This then specifies ϕ uniquely. So, we deduce that the Calabi Conjecture is equivalent to the following:

The Calabi Conjecture (third version). *Let (M, J) be a compact, complex manifold, and g a Kähler metric on M, with Kähler form ω. Let f be a smooth real function on M, and define $A > 0$ by $A \int_M e^f dV_g = \mathrm{vol}_g(M)$. Then there exists a unique smooth real function ϕ such that*

  (i) $\omega + dd^c \phi$ *is a positive $(1,1)$-form,*
 (ii) $\int_M \phi \, dV_g = 0$, *and*
(iii) $(\omega + dd^c \phi)^m = A e^f \omega^m$ *on M.*

Moreover, part (iii) is equivalent to the following:

(iii)′ *Choose holomorphic coordinates $z_1, \ldots, z_m$ on an open set U in M. Then $g_{\alpha\bar\beta}$ may be interpreted as an $m \times m$ Hermitian matrix indexed by $\alpha, \bar\beta = 1, 2, \ldots, m$ in U. The condition on ϕ is*

$$\det\left(g_{\alpha\bar\beta} + \frac{\partial^2 \phi}{\partial z_\alpha \partial \bar z_{\bar\beta}} \right) = A e^f \det\left(g_{\alpha\bar\beta} \right). \tag{6.4}$$

For part (iii)′, eqn (6.3) gives $g'_{\alpha\bar\beta} = g_{\alpha\bar\beta} + \partial_\alpha \bar\partial_{\bar\beta} \phi$, and the result follows from eqn (5.6) of §5.6. Equation (6.4) is a nonlinear, elliptic, second-order partial differential equation in ϕ, of a kind known as a *Monge–Ampère equation*. We have reduced the Calabi Conjecture to a problem in analysis, that of showing that a particular p.d.e. has a unique, smooth solution.

The difficulty of the Calabi Conjecture, and the reason it took twenty years to complete, is that nonlinear equations in general are difficult to solve, and the nonlinearities of (6.4) are of a particularly severe kind, as they are nonlinear in the derivatives of highest order.

In fact part (i) follows from part (iii), as the following lemma shows.

Lemma 6.1.1 *Let (M, J) be a compact, complex manifold, and g a Kähler metric on M, with Kähler form ω. Let $f \in C^0(M)$, and define A by $A \int_M e^f dV_g = \mathrm{vol}_g(M)$. Suppose $\phi \in C^2(M)$ satisfies the equation $(\omega + dd^c \phi)^m = A e^f \omega^m$ on M. Then $\omega + dd^c \phi$ is a positive $(1,1)$-form.*

Proof Choose holomorphic coordinates $z_1, \ldots, z_m$ on a connected open set U in M. Then in U, the new metric g' is $g'_{\alpha\bar\beta} = g_{\alpha\bar\beta} + \partial^2 \phi / \partial z_\alpha \partial \bar z_{\bar\beta}$. As usual, we may interpret

$g'_{\alpha\bar\beta}$ as an $m \times m$ Hermitian matrix indexed by $\alpha, \bar\beta = 1, 2, \ldots, m$ in U. A Hermitian matrix has real eigenvalues. From §5.4, $\omega + \mathrm{dd}^c\phi$ is a positive (1,1)-form if and only if g' is a Hermitian metric, that is, if and only if the eigenvalues of the matrix $g'_{\alpha\bar\beta}$ are all *positive*.

But from (6.4), $\det(g'_{\alpha\bar\beta}) > 0$ on U, so $g'_{\alpha\bar\beta}$ has no zero eigenvalues. Therefore by continuity of $g'_{\alpha\bar\beta}$, if the eigenvalues of $g'_{\alpha\bar\beta}$ are positive at some point $p \in U$ then they are positive everywhere in U. So by covering M with such open sets U and using the connectedness of M, we can show that if $\omega + \mathrm{dd}^c\phi$ is positive at some point $p \in M$, then it is positive on all of M.

Since M is compact and ϕ is continuous, ϕ has a minimum on M. Let $p \in M$ be a minimum point of ϕ, and U a coordinate patch containing p. It is easy to show that at p the matrix $\partial^2\phi/\partial z_\alpha\partial\bar z_{\bar\beta}$ has nonnegative eigenvalues, and so $g'_{\alpha\bar\beta}$ has positive eigenvalues at p. Thus $\omega + \mathrm{dd}^c\phi$ is positive at p, and everywhere on M.   □

6.2   Overview of the proof of the Calabi Conjecture

We begin by stating four results, Theorems C1–C4, which will be proved later in the chapter. These are the four main theorems which make up our proof of the Calabi Conjecture. After this, we will prove the Calabi Conjecture assuming Theorems C1–C4, and make some comments on the proof.

Theorem C1 *Let (M, J) be a compact, complex manifold, and g a Kähler metric on M, with Kähler form ω. Let $Q_1 \geqslant 0$. Then there exist $Q_2, Q_3, Q_4 \geqslant 0$ depending only on M, J, g and Q_1, such that the following holds.*
Suppose $f \in C^3(M)$, $\phi \in C^5(M)$ and $A > 0$ satisfy the equations

$$\|f\|_{C^3} \leqslant Q_1, \quad \int_M \phi \,\mathrm{d}V_g = 0, \quad \text{and} \quad (\omega + \mathrm{dd}^c\phi)^m = Ae^f\omega^m.$$

Then $\|\phi\|_{C^0} \leqslant Q_2$, $\|\mathrm{dd}^c\phi\|_{C^0} \leqslant Q_3$ and $\|\nabla\mathrm{dd}^c\phi\|_{C^0} \leqslant Q_4$.

Theorem C2 *Let (M, J) be a compact, complex manifold, and g a Kähler metric on M, with Kähler form ω. Let $Q_1, \ldots, Q_4 \geqslant 0$ and $\alpha \in (0, 1)$. Then there exists $Q_5 \geqslant 0$ depending only on $M, J, g, Q_1, \ldots, Q_4$ and α, such that the following holds.*
Suppose $f \in C^{3,\alpha}(M)$, $\phi \in C^5(M)$ and $A > 0$ satisfy $(\omega + \mathrm{dd}^c\phi)^m = Ae^f\omega^m$ and the inequalities

$$\|f\|_{C^{3,\alpha}} \leqslant Q_1, \quad \|\phi\|_{C^0} \leqslant Q_2, \quad \|\mathrm{dd}^c\phi\|_{C^0} \leqslant Q_3 \quad \text{and} \quad \|\nabla\mathrm{dd}^c\phi\|_{C^0} \leqslant Q_4.$$

Then $\phi \in C^{5,\alpha}(M)$ and $\|\phi\|_{C^{5,\alpha}} \leqslant Q_5$. Also, if $f \in C^{k,\alpha}(M)$ for $k \geqslant 3$ then $\phi \in C^{k+2,\alpha}(M)$, and if $f \in C^\infty(M)$ then $\phi \in C^\infty(M)$.

Theorem C3 *Let (M, J) be a compact complex manifold, and g a Kähler metric on M, with Kähler form ω. Fix $\alpha \in (0, 1)$, and suppose that $f' \in C^{3,\alpha}(M)$, $\phi' \in C^{5,\alpha}(M)$ and $A' > 0$ satisfy the equations*

$$\int_M \phi' \,\mathrm{d}V_g = 0 \quad \text{and} \quad (\omega + \mathrm{dd}^c\phi')^m = A'e^{f'}\omega^m.$$

Then whenever $f \in C^{3,\alpha}(M)$ and $\|f - f'\|_{C^{3,\alpha}}$ is sufficiently small, there exist $\phi \in C^{5,\alpha}(M)$ and $A > 0$ such that

$$\int_M \phi \, \mathrm{d}V_g = 0 \quad \text{and} \quad (\omega + \mathrm{dd}^c\phi)^m = A\mathrm{e}^f\omega^m.$$

Theorem C4 *Let (M, J) be a compact complex manifold, and g a Kähler metric on M, with Kähler form ω. Let $f \in C^1(M)$. Then there is at most one function $\phi \in C^3(M)$ such that $\int_M \phi \, \mathrm{d}V_g = 0$ and $(\omega + \mathrm{dd}^c\phi)^m = A\mathrm{e}^f\omega^m$ on M.*

Here are some remarks on these results. The positive constants A, A' above are determined entirely by f, f' using (6.2). Theorem C1 is due to Yau [345, §2]. Results of this type are called *a priori estimates*, because it tells us in advance (a priori) that any solution to a given equation must satisfy a certain bound. Finding such a priori estimates was the most difficult part of the Calabi Conjecture, and was Yau's biggest contribution to the proof.

As eqn (6.4) is a nonlinear, elliptic p.d.e., we can draw on the fruit of decades of hard work on the properties of solutions of elliptic equations. Theorem C2 uses results about the differentiability of solutions of elliptic equations, and Theorem C3 uses results on the existence of solutions of elliptic equations. Theorem C4 shows that if ϕ exists, then it is unique. It has an elementary proof found by Calabi [68, p. 86]. Broadly speaking, Theorems C1–C3 concern the *existence* of the function ϕ, Theorem C2 is about the *smoothness* of ϕ, and Theorem C4 is about the *uniqueness* of ϕ. Using Theorems C1–C4 we now prove the Calabi Conjecture.

6.2.1   The proof of the Calabi Conjecture

We start with a definition, the purpose of which will become clear soon.

Definition 6.2.1 Let (M, J) be a compact complex manifold, and g a Kähler metric on M, with Kähler form ω. Fix $\alpha \in (0, 1)$ and $f \in C^{3,\alpha}(M)$. Define S to be the set of all $t \in [0, 1]$ for which there exists $\phi \in C^{5,\alpha}(M)$ with $\int_M \phi \, \mathrm{d}V_g = 0$ and $A > 0$, such that $(\omega + \mathrm{dd}^c\phi)^m = A\mathrm{e}^{tf}\omega^m$ on M.

Now, using Theorems C1 and C2 we will show that this set S is *closed*, and using Theorem C3 we will show that S is *open*.

Theorem 6.2.2 *In Definition 6.2.1, the set S is a closed subset of $[0, 1]$.*

Proof It must be shown that S contains its limit points, and therefore is closed. Let $\{t_j\}_{j=0}^{\infty}$ be a sequence in S, which converges to some $t' \in [0, 1]$. We will prove that $t' \in S$. Since $t_j \in S$, by definition there exists $\phi_j \in C^{5,\alpha}(M)$ and $A_j > 0$ such that

$$\int_M \phi_j \, \mathrm{d}V_g = 0 \quad \text{and} \quad (\omega + \mathrm{dd}^c\phi_j)^m = A_j\mathrm{e}^{t_j f}\omega^m. \tag{6.5}$$

Define Q_1 by $Q_1 = \|f\|_{C^{3,\alpha}}$. Let Q_2, Q_3, Q_4 be the constants given by Theorem C1, which depend on Q_1, and Q_5 the constant given by Theorem C2, which depends on $Q_1, \ldots, Q_4$.

As $t_j \in [0, 1]$, $\|t_j f\|_{C^3} \leqslant Q_1$. So, applying Theorem C1 with ϕ_j in place of ϕ and $t_j f$ in place of f, we see that $\|\phi_j\|_{C^0} \leqslant Q_2$, $\|\mathrm{dd}^c\phi_j\|_{C^0} \leqslant Q_3$ and $\|\nabla\mathrm{dd}^c\phi_j\|_{C^0} \leqslant Q_4$

for all j. Thus, by Theorem C2, $\phi_j \in C^{5,\alpha}(M)$ and $\|\phi_j\|_{C^{5,\alpha}} \leqslant Q_5$ for all j. Now the Kondrakov Theorem, Theorem 1.2.3, says that the inclusion $C^{5,\alpha}(M) \to C^5(M)$ is *compact*, in the sense of Definition 1.2.2. It follows that as the sequence $\{\phi_j\}_{j=0}^\infty$ is bounded in $C^{5,\alpha}(M)$, it lies in a compact subset of $C^5(M)$. Therefore there exists a subsequence $\{\phi_{i_j}\}_{j=0}^\infty$ which converges in $C^5(M)$. Let $\phi' \in C^5(M)$ be the limit of this subsequence.

Define $A' > 0$ by $A' \int_M e^{t'f} dV_g = \text{vol}_g(M)$. Then $A_{i_j} \to A'$ as $j \to \infty$, because $t_{i_j} \to t'$ as $j \to \infty$. Since $\{\phi_{i_j}\}_{j=0}^\infty$ converges in C^2 we may take the limit in (6.5), giving

$$\int_M \phi' \, dV_g = 0 \quad \text{and} \quad (\omega + dd^c\phi')^m = A'e^{t'f}\omega^m. \tag{6.6}$$

Theorems C1 and C2 then show that $\phi' \in C^{5,\alpha}(M)$. Therefore $t' \in S$. So S contains its limit points, and is closed. $\qquad\square$

Theorem 6.2.3 *In Definition 6.2.1, the set S is an open subset of $[0,1]$.*

Proof Suppose $t' \in S$. Then by definition there exist $\phi' \in C^{5,\alpha}(M)$ with $\int_M \phi' \, dV_g = 0$ and $A' > 0$, such that $(\omega + dd^c\phi')^m = A'e^{t'f}\omega^m$ on M. Apply Theorem C3, with $t'f$ in place of f', and tf in place of f, for $t \in [0,1]$. The theorem shows that whenever $|t - t'| \cdot \|f\|_{C^{3,\alpha}}$ is sufficiently small, there exist $\phi \in C^{5,\alpha}(M)$ and $A > 0$ such that

$$\int_M \phi \, dV_g = 0 \quad \text{and} \quad (\omega + dd^c\phi)^m = Ae^{tf}\omega^m.$$

But then $t \in S$. Thus, if $t \in [0,1]$ is sufficiently close to t' then $t \in S$, and S contains an open neighbourhood in $[0,1]$ of each t' in S. So S is open. $\qquad\square$

Using Lemma 6.1.1 and Theorems 6.2.2 and 6.2.3 we shall prove an existence result for the function ϕ. Notice that parts (i)–(iii) come from the third version of the Calabi Conjecture.

Theorem 6.2.4 *Let (M, J) be a compact complex manifold, and g a Kähler metric on M with Kähler form ω. Choose $\alpha \in (0,1)$, and let $f \in C^{3,\alpha}(M)$. Then there exist $\phi \in C^{5,\alpha}(M)$ and $A > 0$ such that*

(i) $\omega + dd^c\phi$ *is a positive $(1,1)$-form,*

(ii) $\int_M \phi \, dV_g = 0$, *and*

(iii) $(\omega + dd^c\phi)^m = Ae^f\omega^m$ *on M.*

Proof Theorems 6.2.2 and 6.2.3 imply that S is an open and closed subset of $[0,1]$. Since $[0,1]$ is connected, either $S = \emptyset$ or $S = [0,1]$. But when $t = 0$, the function $\phi \equiv 0$ satisfies the conditions in Definition 6.2.1, so that $0 \in S$. Thus S cannot be empty, and $S = [0,1]$. It follows that $1 \in S$. So, setting $t = 1$, there exist $\phi \in C^{5,\alpha}(M)$ with $\int_M \phi \, dV_g = 0$ and $A > 0$ such that $(\omega + dd^c\phi)^m = Ae^f\omega^m$ on M. Therefore parts (ii) and (iii) of the theorem hold for ϕ. By Lemma 6.1.1, part (i) holds as well. This completes the proof. $\qquad\square$

Finally, using Theorem 6.2.4 and Theorems C2 and C4, we show:

Theorem 6.2.5 *The Calabi Conjecture is true.*

Proof   Suppose (M, J) is a compact, complex manifold, and g a Kähler metric on M, with Kähler form ω. Let $f \in C^\infty(M)$. Then Theorem 6.2.4 constructs $\phi \in C^{5,\alpha}(M)$ and $A > 0$ for which conditions (i)–(iii) of the third version of the Calabi Conjecture hold. Theorems C2 and C4 show that ϕ is smooth and unique. This proves the third version of the Calabi Conjecture.   □

6.2.2   The continuity method

The idea used in the proofs above is known as the *continuity method*, and it works like this. The goal is to prove that a particular nonlinear equation, in our case the equation

$$(\omega + \mathrm{dd}^c\phi)^m = Ae^f\omega^m,$$

has a solution ϕ. The first step is to think of a similar equation which we already know has a solution. In this case we choose the equation

$$(\omega + \mathrm{dd}^c\phi)^m = \omega^m,$$

which has the obvious solution $\phi = 0$.

The second step is to write down a 1-parameter family of equations depending continuously on $t \in [0, 1]$, such that when $t = 0$ the equation is the one we know has a solution, and when $t = 1$ the equation is the one which must be solved. In our case this family of equations is

$$(\omega + \mathrm{dd}^c\phi_t)^m = A_t e^{tf}\omega^m.$$

To complete the proof one must show that the set S of $t \in [0, 1]$ for which the corresponding equation has a solution ϕ_t, is both open and closed in $[0, 1]$. For then, as the equation is soluble when $t = 0$, it is also soluble when $t = 1$ by the argument in Theorem 6.2.4, which is what we want.

Here are two standard arguments that are used to show S is open and closed. To show S is open, suppose that $t' \in S$, so a solution $\phi_{t'}$ exists. Then, one tries to show that when t is close to t' in $[0, 1]$, there is a solution ϕ_t that is close to $\phi_{t'}$ (in some Banach space). To do this it is usually enough to consider the *linearization* of the equation about $\phi_{t'}$, which simplifies the problem.

To show S is closed, one shows that S contains its limit points. Suppose $\{t_j\}_{j=0}^\infty$ is a sequence in S that converges to t'. Then there is a corresponding sequence of solutions $\{\phi_{t_j}\}_{j=0}^\infty$. Now by establishing *a priori bounds* on all solutions ϕ_t in some Banach norm, it may be possible to show that they lie in some *compact* subset in a Banach space. If this is so, the sequence $\{\phi_{t_j}\}_{j=0}^\infty$ contains a convergent subsequence. One then shows that the limit of this subsequence is a solution $\phi_{t'}$ for $t = t'$, so $t' \in S$, and S is closed. This is the continuity method.

6.3   Calculations at a point

Let (M, J) be a compact, complex manifold of dimension m, and g a Kähler metric on M with Kähler form ω. Let $f \in C^0(M)$, $\phi \in C^2(M)$ and $A > 0$. Set $\omega' = \omega + \mathrm{dd}^c\phi$, and suppose $(\omega')^m = Ae^f\omega^m$ on M. Lemma 6.1.1 then shows that ω' is a real, positive $(1,1)$-form, which therefore determines a Kähler metric g'. Let p be a point in M. In this section we shall find expressions for ω and ω' at p, and derive several inequalities that will be useful later.

Lemma 6.3.1 *In the situation above there are holomorphic coordinates $z_1, \ldots, z_m$ on M near p, such that g, g', ω and ω' are given at p by*

$$g_p = 2|\mathrm{d}z_1|^2 + \cdots + 2|\mathrm{d}z_m|^2, \quad g'_p = 2a_1|\mathrm{d}z_1|^2 + \cdots + 2a_m|\mathrm{d}z_m|^2, \qquad (6.7)$$

$$\omega_p = i(\mathrm{d}z_1 \wedge \mathrm{d}\bar{z}_1 + \cdots + \mathrm{d}z_m \wedge \mathrm{d}\bar{z}_m),$$

$$\text{and} \quad \omega'_p = i(a_1\mathrm{d}z_1 \wedge \mathrm{d}\bar{z}_1 + \cdots + a_m\mathrm{d}z_m \wedge \mathrm{d}\bar{z}_m), \qquad (6.8)$$

where $a_1, \ldots, a_m$ are positive real numbers.

Proof Since $T_p^{(1,0)}M$ is isomorphic to $\mathbb{C}^m$ as a complex vector space, if we fix a basis for $T_p^{(1,0)}M$ over $\mathbb{C}$, then we may regard $(g_p)_{\alpha\bar{\beta}}$ and $(g'_p)_{\alpha\bar{\beta}}$ as invertible, Hermitian $m \times m$ complex matrices. By elementary linear algebra, using simultaneous diagonalization, one can choose a basis $(v_1, \ldots, v_m)$ for T_pM over $\mathbb{C}$ with respect to which

$$(g_p)_{\alpha\bar{\beta}} = \begin{cases} 1 & \text{if } \alpha = \bar{\beta}, \\ 0 & \text{if } \alpha \neq \bar{\beta}, \end{cases} \quad \text{and} \quad (g'_p)_{\alpha\bar{\beta}} = \begin{cases} a_j & \text{if } \alpha = \bar{\beta} = j, \\ 0 & \text{if } \alpha \neq \bar{\beta}, \end{cases} \qquad (6.9)$$

where $a_1, \ldots, a_m$ are real numbers, and $a_j > 0$ for $j = 1, \ldots, m$ as g' is a metric.

Clearly, it is possible to find holomorphic coordinates $z_1, \ldots, z_m$ on M near p, such that $v_j = \partial/\partial z_j$ at p. Equations (6.7) and (6.8) then follow immediately from (6.9) and the equation $\omega_{ac} = J_a^b g_{bc}$. $\qquad \square$

Next, we relate $a_1, \ldots, a_m$ to Ae^f and $\Delta\phi$. In order to be consistent with [345] and [16], we define the Laplacian Δ on a Kähler manifold by $\Delta\phi = -g^{\alpha\bar{\beta}}\partial_\alpha\partial_{\bar{\beta}}\phi$. Be warned that this is equal to half of the usual d-Laplacian on a Riemannian manifold.

Lemma 6.3.2 *In the situation of the previous lemma, we have*

$$\prod_{j=1}^m a_j = Ae^{f(p)}, \quad \frac{\partial^2\phi}{\partial z_j \partial \bar{z}_j}(p) = a_j - 1 \quad \text{and} \quad (\Delta\phi)(p) = m - \sum_{j=1}^m a_j. \qquad (6.10)$$

Proof From (6.8) we see that $\omega_p^m = i^m m! \, \mathrm{d}z_1 \wedge \mathrm{d}\bar{z}_1 \wedge \cdots \wedge \mathrm{d}z_m \wedge \mathrm{d}\bar{z}_m$ and $(\omega'_p)^m = \prod_{j=1}^m a_j \cdot i^m m! \, \mathrm{d}z_1 \wedge \mathrm{d}\bar{z}_1 \wedge \cdots \wedge \mathrm{d}z_m \wedge \mathrm{d}\bar{z}_m$. But $(\omega')^m = Ae^f \omega^m$, and therefore $\prod_{j=1}^m a_j = Ae^{f(p)}$, the first equation of (6.10). As $\omega' = \omega + \mathrm{dd}^c\phi$, we have

$$(g'_p)_{\alpha\bar{\beta}} = (g_p)_{\alpha\bar{\beta}} + \frac{\partial^2\phi}{\partial z_\alpha \partial \bar{z}_\beta}.$$

Putting $\alpha = \bar{\beta} = j$ and substituting (6.9) in gives the second equation of (6.10). Now $g^{\alpha\bar{\beta}}$ is the inverse of $g_{\alpha\bar{\beta}}$ as an $m \times m$ matrix. Hence by (6.9), $g_p^{\alpha\bar{\beta}}$ is 1 if $\alpha = \bar{\beta}$, and 0 otherwise. So, as $\Delta\phi = -g^{\alpha\bar{\beta}}\partial_\alpha\partial_{\bar{\beta}}\phi$, the third equation of (6.10) follows from the second. $\qquad \square$

Let $T = T^{a_1 \cdots a_k}_{c_1 \cdots c_l}$ be a tensor and g a Riemannian metric on M. Define $|T|_g$ by

$$|T|^2_g = T^{a_1 \cdots a_k}_{c_1 \cdots c_l} T^{b_1 \cdots b_k}_{d_1 \cdots d_l} g_{a_1 b_1} \cdots g_{a_k b_k} g^{c_1 d_1} \cdots g^{c_l d_l}. \qquad (6.11)$$

Using this notation, and the material of the previous two lemmas, the following result is very easy, so we omit the proof.

Lemma 6.3.3 *In the situation of Lemma 6.3.1, at p we have*

$$|\mathrm{dd}^c\phi|^2_g = 2\sum_{j=1}^m (a_j - 1)^2, \quad |g'_{ab}|^2_g = 2\sum_{j=1}^m a_j^2 \quad \text{and} \quad |g'^{ab}|^2_g = 2\sum_{j=1}^m a_j^{-2}.$$

Here g'^{ab} is the matrix inverse of g'_{ab} in coordinates. Now we shall prove some inequalities that will be useful in the next few sections.

Proposition 6.3.4 *Let (M, J) be a compact, complex manifold and g a Kähler metric on M, with Kähler form ω. Let $f \in C^0(M)$, $\phi \in C^2(M)$ and $A > 0$. Set $\omega' = \omega + \mathrm{dd}^c\phi$, suppose $(\omega')^m = A\mathrm{e}^f \omega^m$, and let g' be the metric with Kähler form ω'. Then*

$$\Delta\phi \leqslant m - mA^{1/m}\mathrm{e}^{f/m} < m, \qquad (6.12)$$

and there are constants c_1, c_2 and c_3 depending only on m and upper bounds for $\|f\|_{C^0}$ and $\|\Delta\phi\|_{C^0}$, such that

$$\|g'_{ab}\|_{C^0} \leqslant c_1, \quad \|g'^{ab}\|_{C^0} \leqslant c_2 \quad \text{and} \quad \|\mathrm{dd}^c\phi\|_{C^0} \leqslant c_3. \qquad (6.13)$$

Here all norms are with respect to the metric g.

Proof Inequality (6.12) follows immediately from the first and last equations of (6.10), and the fact that the geometric mean $(a_1 \cdots a_m)^{1/m}$ is less than or equal to the arithmetic mean $\frac{1}{m}(a_1 + \cdots + a_m)$. As $\Delta\phi(p) = m - \sum_{j=1}^m a_j$ by (6.10), and the a_j are positive, from Lemma 6.3.3 one can show that at p, $|\mathrm{dd}^c\phi|^2_g \leqslant 2m + 2(m - \Delta\phi)^2$, and $|g'_{ab}|^2_g \leqslant 2(m - \Delta\phi)^2$. As these hold for all $p \in M$, the first and last inequalities of (6.13) hold with constants c_1, c_3 depending only on m and $\|\Delta\phi\|_{C^0}$.

Now if $\log A > -\inf_M f$ then $A\mathrm{e}^f > 1$ on M, which contradicts the equation $\int_M A\mathrm{e}^f \mathrm{d}V_g = \int_M \mathrm{d}V_g$. Similarly $\log A < -\sup_M f$ leads to a contradiction. Hence, $-\sup_M f \leqslant \log A \leqslant -\inf_M f$, and $|\log A| \leqslant \|f\|_{C^0}$. It follows that

$$\mathrm{e}^{-2\|f\|_{C^0}} \leqslant A\mathrm{e}^f \leqslant \mathrm{e}^{2\|f\|_{C^0}} \qquad (6.14)$$

on M. Since $\prod_{j=1}^m a_j = A\mathrm{e}^{f(p)}$ by (6.10), we see that

$$a_j^{-1} = A^{-1}\mathrm{e}^{-f(p)} \prod_{1 \leqslant k \leqslant m, \, j \neq k} a_k.$$

Using (6.14) to estimate $A^{-1}\mathrm{e}^{-f(p)}$ and the inequality $a_k \leqslant m - \Delta\phi(p)$ derived from the third equation of (6.10), we find that $a_j^{-2} \leqslant \mathrm{e}^{4\|f\|_{C^0}} (m - \Delta\phi(p))^{2m-2}$. From this and Lemma 6.3.3 the second inequality of (6.13) follows for some c_2 depending on $m, \|f\|_{C^0}$ and $\|\Delta\phi\|_{C^0}$, and the proof is complete. □

This shows that an a priori bound for $\Delta\phi$ yields a priori bounds for g' and $\mathrm{dd}^c\phi$.

Lemma 6.3.5 *In the situation of Proposition 6.3.4, we have*

$$\mathrm{d}\phi \wedge \mathrm{d}^c\phi \wedge \omega^{m-1} = \tfrac{1}{m}\,|\nabla\phi|_g^2\,\omega^m \quad \text{and} \quad \mathrm{d}\phi \wedge \mathrm{d}^c\phi \wedge \omega^{m-j-1} \wedge (\omega')^j = F_j\,\omega^m$$

for $j = 1, 2, \ldots, m-1$, where F_j is a nonnegative real function on M.

Proof We have $\mathrm{d}\phi \wedge \mathrm{d}^c\phi \wedge \omega^{m-1} = \big(\mathrm{d}\phi \wedge \mathrm{d}^c\phi, *(\omega^{m-1})\big)\,\mathrm{d}V_g$ by properties of the Hodge star. But $*(\omega^{m-1}) = (m-1)!\,\omega$, and $\omega^m = m!\,\mathrm{d}V_g$ by (5.4). Therefore

$$\mathrm{d}\phi \wedge \mathrm{d}^c\phi \wedge \omega^{m-1} = \frac{(m-1)!}{m!}\big(\mathrm{d}\phi \wedge \mathrm{d}^c\phi, \omega\big) \cdot \omega^m. \tag{6.15}$$

Now $(\mathrm{d}^c\phi)_a = -J_a^b(\mathrm{d}\phi)_b$, so $(\mathrm{d}\phi \wedge \mathrm{d}^c\phi, \omega) = -(\mathrm{d}\phi)_a J_b^e(\mathrm{d}\phi)_e \omega_{cd} g^{ac} g^{bd} = |\nabla\phi|_g^2$, since $-J_b^e \omega_{cd} g^{ac} g^{bd} = g^{ae}$. Substituting this into (6.15) gives the first equation.

The reason $\mathrm{d}\phi \wedge \mathrm{d}^c\phi \wedge \omega^{m-1}$ is a nonnegative multiple of ω^m is that $*(\omega^{m-1})$ is a *positive* (1,1)-form, in the sense of §5.4. In the same way, if we can show $*\big(\omega^{m-j-1} \wedge (\omega')^j\big)$ is a positive (1,1)-form, it will follow that $\mathrm{d}\phi \wedge \mathrm{d}^c\phi \wedge \omega^{m-j-1} \wedge (\omega')^j$ is a nonnegative multiple of ω^m, which is what we have to prove. But using (6.8) one can readily show that $*\big(\omega^{m-j-1} \wedge (\omega')^j\big)$ is positive, and the proof is finished. $\square$

6.4   The proof of Theorem C1

We will now prove Theorem C1 of §6.2.

Theorem C1 *Let (M, J) be a compact, complex manifold, and g a Kähler metric on M, with Kähler form ω. Let $Q_1 \geqslant 0$. Then there exist $Q_2, Q_3, Q_4 \geqslant 0$ depending only on M, J, g and Q_1, such that the following holds.*

Suppose $f \in C^3(M)$, $\phi \in C^5(M)$ and $A > 0$ satisfy the equations

$$\|f\|_{C^3} \leqslant Q_1, \quad \int_M \phi\,\mathrm{d}V_g = 0, \quad \text{and} \quad (\omega + \mathrm{dd}^c\phi)^m = A\mathrm{e}^f \omega^m.$$

Then $\|\phi\|_{C^0} \leqslant Q_2$, $\|\mathrm{dd}^c\phi\|_{C^0} \leqslant Q_3$ and $\|\nabla\mathrm{dd}^c\phi\|_{C^0} \leqslant Q_4$.

The result was first proved by Yau [345, Prop. 2.1]. Parts of the proof (in particular, the a priori estimate of $\|\phi\|_{C^0}$ in §6.4.1) were simplified by Kazhdan, Bourguignon and Aubin, and the treatment we will give is based on that in Aubin's book [16, p. 260–267]. The way the proof works is to make a sequence of a priori estimates of ϕ in different Banach norms. Each estimate depends on the previous estimate, and each estimate is slightly stronger than the last.

First an estimate of $\|\phi\|_{L^2}$ is found. Then we show that an estimate of $\|\phi\|_{L^p}$ leads to an estimate of $\|\phi\|_{L^{p\epsilon}}$, where $\epsilon = \frac{m}{m-1}$, and so by an induction argument we estimate $\|\phi\|_{L^p}$ for all p. Taking the limit as $p \to \infty$ yields an estimate for $\|\phi\|_{C^0}$, proving the first part of the theorem. Using this estimate we are then able to bound $\|\Delta\phi\|_{C^0}$, and from this we bound $\|\mathrm{dd}^c\phi\|_{C^0}$. Finally we bound $\|\nabla\mathrm{dd}^c\phi\|_{C^0}$ and the theorem follows.

For the rest of this section, we study the following situation. Let M be a compact Kähler manifold, with complex structure J, Kähler metric g and Kähler form ω. Let

$f \in C^3(M)$ be a fixed function satisfying $\|f\|_{C^3} \leqslant Q_1$, where Q_1 is a given constant. For simplicity we shall assume, by adding a constant to f if necessary, that $\int_M \mathrm{e}^f \mathrm{d}V_g = \mathrm{vol}_g(M)$, so that the constant A is 1. (As A can be estimated in terms of $\|f\|_{C^0}$, this assumption does not matter.) Let $\phi \in C^5(M)$ satisfy the equations

$$\int_M \phi \, \mathrm{d}V_g = 0 \quad \text{and} \quad (\omega + \mathrm{d}\mathrm{d}^c \phi)^m = \mathrm{e}^f \omega^m,$$

define $\omega' = \omega + \mathrm{d}\mathrm{d}^c \phi$, and let g' be the associated Kähler metric.

6.4.1   Estimates of order zero

We begin with the following proposition.

Proposition 6.4.1 *Let $p > 1$ be a real number. Then in the situation above,*

$$\int_M \left| \nabla |\phi|^{p/2} \right|_g^2 \mathrm{d}V_g \leqslant \frac{mp^2}{4(p-1)} \int_M (1 - \mathrm{e}^f) \phi |\phi|^{p-2} \mathrm{d}V_g. \tag{6.16}$$

Proof As $(\omega')^m = \mathrm{e}^f \omega^m$ and $\omega - \omega' = -\mathrm{d}\mathrm{d}^c \phi$, we see that

$$(1 - \mathrm{e}^f)\omega^m = \omega^m - (\omega')^m = -\mathrm{d}\mathrm{d}^c \phi \wedge \left(\omega^{m-1} + \cdots + (\omega')^{m-1} \right). \tag{6.17}$$

Now M is compact, and so Stokes' Theorem shows that

$$\int_M \mathrm{d}\left[\phi |\phi|^{p-2} \mathrm{d}^c \phi \wedge \left(\omega^{m-1} + \omega^{m-2} \wedge \omega' + \cdots + (\omega')^{m-1} \right) \right] = 0.$$

Expanding this equation, remembering that $\mathrm{d}\omega = \mathrm{d}\omega' = 0$, and substituting in (6.17) and the equation $\mathrm{d}(\phi |\phi|^{p-2}) = (p-1)|\phi|^{p-2} \mathrm{d}\phi$, we find

$$\int_M \phi |\phi|^{p-2} (1 - \mathrm{e}^f)\omega^m = (p-1) \int_M |\phi|^{p-2} \mathrm{d}\phi \wedge \mathrm{d}^c \phi$$
$$\wedge \left(\omega^{m-1} + \omega^{m-2} \wedge \omega' + \cdots + (\omega')^{m-1} \right).$$

Now Lemma 6.3.5 gives expressions for $\mathrm{d}\phi \wedge \mathrm{d}^c \phi \wedge \omega^{m-1}$ and $\mathrm{d}\phi \wedge \mathrm{d}^c \phi \wedge \omega^{m-j-1} \wedge (\omega')^j$, and substituting these in shows that

$$\int_M (1 - \mathrm{e}^f)\phi |\phi|^{p-2} \omega^m = \frac{p-1}{m} \int_M |\phi|^{p-2} \left(|\nabla \phi|_g^2 + F_1 + \cdots + F_{m-1} \right)\omega^m,$$

where $F_1, \dots, F_{m-1}$ are nonnegative functions on M. Thus (5.4) yields

$$\int_M |\phi|^{p-2} \left(|\nabla \phi|_g^2 + F_1 + \cdots + F_{m-1} \right)\mathrm{d}V_g = \frac{m}{p-1} \int_M (1 - \mathrm{e}^f)\phi |\phi|^{p-2} \mathrm{d}V_g.$$

Combining this with the equation $\frac{1}{4}p^2 |\phi|^{p-2} |\nabla \phi|_g^2 = \left| \nabla |\phi|^{p/2} \right|_g^2$ and the inequality $0 \leqslant |\phi|^{p-2} F_j$ gives (6.16), as we have to show. $\square$

For the rest of this section, let $\epsilon = \frac{m}{m-1}$. Next we prove

Lemma 6.4.2 *There are constants C_1, C_2 depending on M, g, such that if $\psi \in L_1^2(M)$ then $\|\psi\|_{L^{2\epsilon}}^2 \leqslant C_1\big(\|\nabla\psi\|_{L^2}^2 + \|\psi\|_{L^2}^2\big)$, and if $\int_M \psi\,\mathrm{d}V_g = 0$ then $\|\psi\|_{L^2} \leqslant C_2\|\nabla\psi\|_{L^2}$.*

Proof By the Sobolev Embedding Theorem, Theorem 1.2.1, $L_1^2(M)$ is continuously embedded in $L^{2\epsilon}(M)$, and the first inequality of the lemma easily follows. Now consider the operator $\mathrm{d}^*\mathrm{d} : C^\infty(M) \to C^\infty(M)$. It is well-known that $\mathrm{Ker}(\mathrm{d}^*\mathrm{d})$ is the constant functions, and the nonzero eigenvalues of $\mathrm{d}^*\mathrm{d}$ are positive and form a discrete spectrum. Let $\lambda_1 > 0$ be the smallest positive eigenvalue of $\mathrm{d}^*\mathrm{d}$. If $\psi \in C^\infty(M)$ satisfies $\int_M \psi\,\mathrm{d}V_g = 0$, then ψ is L^2-orthogonal to $\mathrm{Ker}(\mathrm{d}^*\mathrm{d})$, so ψ is a sum of eigenvectors of $\mathrm{d}^*\mathrm{d}$ with eigenvalues not less than λ_1.

It follows that $\langle \psi, \mathrm{d}^*\mathrm{d}\psi \rangle \geqslant \lambda_1 \langle \psi, \psi \rangle$, so $\|\mathrm{d}\psi\|_{L^2}^2 \geqslant \lambda_1 \|\psi\|_{L^2}^2$ by integration by parts. As $C^\infty(M)$ is dense in $L_1^2(M)$, this inequality extends to $\psi \in L_1^2(M)$, by continuity. Thus if $\psi \in L_1^2(M)$ and $\int_M \psi\,\mathrm{d}V_g = 0$ then $\|\psi\|_{L^2} \leqslant \lambda_1^{-1/2}\|\nabla\psi\|_{L^2}$, giving the second inequality of the lemma with $C_2 = \lambda_1^{-1/2}$. $\qquad\square$

In the next two results we find a priori estimates of $\|\phi\|_{L^p}$ for $p \in [2, 2\epsilon]$ and $p \in [2, \infty)$ respectively.

Lemma 6.4.3 *There is a constant C_3 depending on M, g and Q_1 such that if $p \in [2, 2\epsilon]$ then $\|\phi\|_{L^p} \leqslant C_3$.*

Proof Putting $p = 2$ in Proposition 6.4.1 and using the inequality $|1 - \mathrm{e}^f| \leqslant \mathrm{e}^{Q_1}$, we see that $\|\nabla\phi\|_{L^2}^2 \leqslant m\,\mathrm{e}^{Q_1}\|\phi\|_{L^1}$. As $\int_M \phi\,\mathrm{d}V_g = 0$, Lemma 6.4.2 shows that $\|\phi\|_{L^2} \leqslant C_2\|\nabla\phi\|_{L^2}$, and $\|\phi\|_{L^1} \leqslant \mathrm{vol}_g(M)^{1/2}\|\phi\|_{L^2}$ by Hölder's inequality. Combining these three shows that

$$\|\nabla\phi\|_{L^2}^2 \leqslant m\,C_2\mathrm{e}^{Q_1}\mathrm{vol}_g(M)^{1/2} \cdot \|\nabla\phi\|_{L^2},$$

and so $\|\nabla\phi\|_{L^2} \leqslant c$, where $c = m\,C_2\mathrm{e}^{Q_1}\,\mathrm{vol}_g(M)^{1/2}$. Therefore

$$\|\phi\|_{L^2} \leqslant c\,C_2 \quad \text{and so} \quad \|\phi\|_{L^{2\epsilon}}^2 \leqslant C_1\big(c^2 + c^2C_2^2\big),$$

using Lemma 6.4.2. Define C_3 by $C_3 = \max\big(c\,C_2, c\,C_1^{1/2}(1+C_2^2)^{1/2}\big)$. Then $\|\phi\|_{L^2} \leqslant C_3$ and $\|\phi\|_{L^{2\epsilon}} \leqslant C_3$, so by Hölder's inequality if $p \in [2, 2\epsilon]$ then $\|\phi\|_{L^p} \leqslant C_3$. $\qquad\square$

Proposition 6.4.4 *There are constants Q_2, C_4 depending on M, g and Q_1 such that for each $p \geqslant 2$, we have $\|\phi\|_{L^p} \leqslant Q_2(C_4p)^{-m/p}$.*

Proof Define $C_4 = C_1\epsilon^{m-1}\big(m\,\mathrm{e}^{Q_1} + \frac{1}{2}\big) > 0$, and choose $Q_2 > 0$ such that

$$\begin{aligned} Q_2 &\geqslant C_3(C_4p)^{m/p} \quad \text{for } 2 \leqslant p \leqslant 2\epsilon, \text{ and} \\ Q_2 &\geqslant (C_4p)^{m/p} \quad\;\; \text{for } 2 \leqslant p < \infty. \end{aligned} \tag{6.18}$$

Such a constant must exist, as $\lim_{p\to\infty}(C_4p)^{m/p} = 1$.

We will prove the proposition by a form of induction on p. The first step is that if $p \in [2, 2\epsilon]$, then $\|\phi\|_{L^p} \leqslant C_3$ by Lemma 6.4.3, and $C_3 \leqslant Q_2(C_4p)^{-m/p}$ by the first inequality of (6.18), and so $\|\phi\|_{L^p} \leqslant Q_2(C_4p)^{-m/p}$ as we want. For the inductive step, suppose that $\|\phi\|_{L^p} \leqslant Q_2(C_4p)^{-m/p}$ holds for all p with $2 \leqslant p \leqslant k$, where $k \geqslant 2\epsilon$. We shall show that $\|\phi\|_{L^q} \leqslant Q_2(C_4 q)^{-m/q}$ holds for all q with $2 \leqslant q \leqslant \epsilon k$, and therefore by induction, the inequality holds for all $p \in [2, \infty)$.

Let $p \in [2, k]$. Then $p^2/4(p-1) \leqslant p$ as $p \geqslant 2$, and since $|1 - e^f| \leqslant e^{Q_1}$ we see from Proposition 6.4.1 that $\left\|\nabla|\phi|^{p/2}\right\|_{L^2}^2 \leqslant mp\, e^{Q_1} \|\phi\|_{L^{p-1}}^{p-1}$. Applying Lemma 6.4.2 with $\psi = |\phi|^{p/2}$ gives $\|\phi\|_{L^{\epsilon p}}^p \leqslant C_1\big(\big\|\nabla|\phi|^{p/2}\big\|_{L^2}^2 + \|\phi\|_{L^p}^p\big)$. Combining these two equations shows that

$$\|\phi\|_{L^{\epsilon p}}^p \leqslant mp\, C_1 e^{Q_1} \|\phi\|_{L^{p-1}}^{p-1} + C_1 \|\phi\|_{L^p}^p.$$

Let $q = \epsilon p$. As $p \in [2, k]$ we have $\|\phi\|_{L^p} \leqslant Q_2(C_4 p)^{-m/p}$, and by the second part of (6.18) we see that $Q_2(C_4 p)^{-m/p} \geqslant 1$, and thus

$$\|\phi\|_{L^q}^p \leqslant Q_2^p (C_4 p)^{-m}\big(mp\, C_1 e^{Q_1} + C_1\big),$$
$$\text{and also}\quad \big(Q_2(C_4 q)^{-m/q}\big)^p = Q_2^p (C_4 p \epsilon)^{1-m}.$$

However, as $p \geqslant 2$ the inequality $mp\, C_1 e^{Q_1} + C_1 \leqslant C_4 p \epsilon^{1-m}$ follows from the definition of C_4, so comparing the right hand sides of the above equations we see that $\|\phi\|_{L^q}^p \leqslant \big(Q_2(C_4 q)^{-m/q}\big)^p$, and therefore $\|\phi\|_{L^q} \leqslant Q_2(C_4 q)^{-m/q}$. This holds for all $q \in [2\epsilon, \epsilon k]$, and the inductive step is complete. $\square$

Now we can prove the first part of Theorem C1.

Corollary 6.4.5 *The function ϕ satisfies $\|\phi\|_{C^0} \leqslant Q_2$.*

Proof As ϕ is continuous on a compact manifold, $\|\phi\|_{C^0} = \lim_{p \to \infty} \|\phi\|_{L^p}$. But $\|\phi\|_{L^p} \leqslant Q_2(C_4 p)^{-m/p}$ by Proposition 6.4.4, and $\lim_{p \to \infty} (C_4 p)^{-m/p} = 1$, so the result follows. $\square$

6.4.2 Second-order estimates

Here is some notation that will be used for the next calculations. We have two Kähler metrics g and g' on M. Let ∇ be the Levi-Civita connection of g. If T is a tensor on M, we will write $\nabla_{a_1 \cdots a_k} T$ in place of $\nabla_{a_1} \cdots \nabla_{a_k} T$, the kth derivative of T using ∇. This notation will be used together with the notation for complex tensors introduced in §5.2. Let $R^a{}_{bcd}$ be the Riemann curvature of g. Also, for $\psi \in C^2(M)$, let $\Delta\psi$ be the Laplacian of ψ with respect to g, and let $\Delta'\psi$ be the Laplacian of ψ with respect to g'. Then in this notation,

$$\Delta\psi = -g^{\alpha\bar{\beta}} \nabla_{\alpha\bar{\beta}} \psi \quad \text{and} \quad \Delta'\psi = -g'^{\alpha\bar{\beta}} \nabla_{\alpha\bar{\beta}} \psi.$$

The second formula defines the Laplacian with respect to g' using the Levi-Civita connection of g. It is valid because this component of $\nabla^2 \psi$ is independent of the choice of Kähler metric.

Using this notation we shall prove:

Proposition 6.4.6 *In the situation above, we have*

$$\begin{aligned}
\Delta'(\Delta\phi) = {}&-\Delta f + g^{\alpha\bar{\lambda}} g'^{\mu\bar{\beta}} g'^{\gamma\bar{\nu}} \nabla_{\alpha\bar{\beta}\gamma}\phi\, \nabla_{\bar{\lambda}\mu\bar{\nu}}\phi \\
&+ g'^{\alpha\bar{\beta}} g^{\gamma\bar{\delta}} \big(R^{\bar{\epsilon}}{}_{\bar{\delta}\gamma\bar{\beta}} \nabla_{\alpha\bar{\epsilon}}\phi - R^{\bar{\epsilon}}{}_{\bar{\beta}\alpha\bar{\delta}} \nabla_{\gamma\bar{\epsilon}}\phi\big).
\end{aligned} \tag{6.19}$$

Proof Taking the log of eqn (6.4) and applying ∇ gives $\nabla_{\bar{\lambda}} f = g'^{\mu\bar{\nu}} \nabla_{\bar{\lambda}\mu\bar{\nu}}\phi$. There-fore, as $\Delta f = -g^{\alpha\bar{\lambda}}\nabla_{\alpha\bar{\lambda}} f$, we have

$$\Delta f = -g^{\alpha\bar{\lambda}}(\nabla_{\alpha} g'^{\mu\bar{\nu}})\nabla_{\bar{\lambda}\mu\bar{\nu}}\phi - g^{\alpha\bar{\lambda}}g'^{\mu\bar{\nu}}\nabla_{\alpha\bar{\lambda}\mu\bar{\nu}}\phi. \qquad (6.20)$$

But $g'^{\mu\bar{\beta}}g'_{\bar{\beta}\gamma} = \delta^{\mu}_{\gamma}$, and $\nabla_{\alpha}g'_{\bar{\beta}\gamma} = \nabla_{\alpha\bar{\beta}\gamma}\phi$ as $g'_{\bar{\beta}\gamma} = g_{\bar{\beta}\gamma} + \nabla_{\bar{\beta}\gamma}\phi$, so that

$$0 = \nabla_{\alpha}\delta^{\mu}_{\gamma} = g'_{\bar{\beta}\gamma}\nabla_{\alpha}g'^{\mu\bar{\beta}} + g'^{\mu\bar{\beta}}\nabla_{\alpha\bar{\beta}\gamma}\phi.$$

Contracting with $g'^{\gamma\bar{\nu}}$ shows that $\nabla_{\alpha}g'^{\mu\bar{\nu}} = -g'^{\mu\bar{\beta}}g'^{\gamma\bar{\nu}}\nabla_{\alpha\bar{\beta}\gamma}\phi$. Substituting this into (6.20) yields

$$\Delta f = g^{\alpha\bar{\lambda}}g'^{\mu\bar{\beta}}g'^{\gamma\bar{\nu}}\nabla_{\alpha\bar{\beta}\gamma}\phi\nabla_{\bar{\lambda}\mu\bar{\nu}}\phi - g^{\alpha\bar{\lambda}}g'^{\mu\bar{\nu}}\nabla_{\alpha\bar{\lambda}\mu\bar{\nu}}\phi,$$

and rearranging and changing some indices gives

$$g'^{\alpha\bar{\beta}}g^{\gamma\bar{\delta}}\nabla_{\gamma\bar{\delta}\alpha\bar{\beta}}\phi = -\Delta f + g^{\alpha\bar{\lambda}}g'^{\mu\bar{\beta}}g'^{\gamma\bar{\nu}}\nabla_{\alpha\bar{\beta}\gamma}\phi\nabla_{\bar{\lambda}\mu\bar{\nu}}\phi. \qquad (6.21)$$

However, it can be shown that

$$g'^{\alpha\bar{\beta}}g^{\gamma\bar{\delta}}\nabla_{\alpha\bar{\beta}\gamma\bar{\delta}}\phi - g'^{\alpha\bar{\beta}}g^{\gamma\bar{\delta}}\nabla_{\gamma\bar{\delta}\alpha\bar{\beta}}\phi = g'^{\alpha\bar{\beta}}g^{\gamma\bar{\delta}}\left(R^{\bar{\epsilon}}{}_{\bar{\delta}\gamma\bar{\beta}}\nabla_{\alpha\bar{\epsilon}}\phi - R^{\bar{\epsilon}}{}_{\bar{\beta}\alpha\bar{\delta}}\nabla_{\gamma\bar{\epsilon}}\phi\right),$$

and this combines with (6.21) and $\Delta'\Delta\phi = g'^{\alpha\bar{\beta}}g^{\gamma\bar{\delta}}\nabla_{\alpha\bar{\beta}\gamma\bar{\delta}}\phi$ to give (6.19), as we want. $\qquad\square$

The next result gives the second part of Theorem C1.

Proposition 6.4.7 *There are constants c_1, c_2 and Q_3 depending on M, J, g and Q_1, such that*

$$\|g'_{ab}\|_{C^0} \leqslant c_1, \quad \|g'^{ab}\|_{C^0} \leqslant c_2 \quad and \quad \|dd^c\phi\|_{C^0} \leqslant Q_3. \qquad (6.22)$$

Proof Define a function F on M by $F = \log(m - \Delta\phi) - \kappa\phi$, where κ is a constant to be chosen later. Note that $m - \Delta\phi > 0$ by Proposition 6.3.4, so F is well-defined. We shall find an expression for $\Delta'F$. It is easy to show that

$$\Delta'F = -(m - \Delta\phi)^{-1}\Delta'\Delta\phi + (m - \Delta\phi)^{-2}g'^{\alpha\bar{\lambda}}g^{\mu\bar{\beta}}g^{\gamma\bar{\nu}}\nabla_{\alpha\bar{\beta}\gamma}\phi\,\nabla_{\bar{\lambda}\mu\bar{\nu}}\phi - \kappa\Delta'\phi.$$

Now $\Delta'\phi = -g'^{\alpha\bar{\beta}}\nabla_{\alpha\bar{\beta}}\phi = g'^{\alpha\bar{\beta}}(g_{\alpha\bar{\beta}} - g'_{\alpha\bar{\beta}}) = g'^{\alpha\bar{\beta}}g_{\alpha\bar{\beta}} - m$, so (6.19) and the equation above give

$$\Delta'F = (m - \Delta\phi)^{-1}\Delta f + \kappa\left(m - g'^{\alpha\bar{\beta}}g_{\alpha\bar{\beta}}\right) - (m - \Delta\phi)^{-1}(G + H), \qquad (6.23)$$

where G and H are defined by

$$G = g^{\alpha\bar{\lambda}}g'^{\mu\bar{\beta}}g'^{\gamma\bar{\nu}}\nabla_{\alpha\bar{\beta}\gamma}\phi\,\nabla_{\bar{\lambda}\mu\bar{\nu}}\phi - (m - \Delta\phi)^{-1}g'^{\alpha\bar{\lambda}}g^{\mu\bar{\beta}}g^{\gamma\bar{\nu}}\nabla_{\alpha\bar{\beta}\gamma}\phi\,\nabla_{\bar{\lambda}\mu\bar{\nu}}\phi,$$
$$H = g'^{\alpha\bar{\beta}}g^{\gamma\bar{\delta}}\left(R^{\bar{\epsilon}}{}_{\bar{\delta}\gamma\bar{\beta}}\nabla_{\alpha\bar{\epsilon}}\phi - R^{\bar{\epsilon}}{}_{\bar{\beta}\alpha\bar{\delta}}\nabla_{\gamma\bar{\epsilon}}\phi\right).$$

Now expanding the inequality

$$g^{\alpha\bar{\lambda}}g'^{\mu\bar{\beta}}g'^{\gamma\bar{\nu}}\left[(m - \Delta\phi)\nabla_{\alpha\bar{\beta}\gamma}\phi - g'_{\alpha\bar{\beta}}\nabla_{\gamma}\Delta\phi\right]\cdot\left[(m - \Delta\phi)\nabla_{\bar{\lambda}\mu\bar{\nu}}\phi - g'_{\bar{\lambda}\mu}\nabla_{\bar{\nu}}\Delta\phi\right] \geqslant 0$$

and dividing by $(m - \Delta\phi)^2$ shows that $G \geqslant 0$. Also, using the inequalities $|\nabla_{\alpha\bar{\beta}}\phi|_g \leqslant (m - \Delta\phi)$ and $|g'^{\alpha\bar{\beta}}|_g \leqslant g'^{\alpha\bar{\beta}}g_{\alpha\bar{\beta}}$ one can see that there is a constant $C_5 \geqslant 0$ depending on m and $\|R\|_{C^0}$ such that $|H| \leqslant C_5(m - \Delta\phi)g'^{\alpha\bar{\beta}}g_{\alpha\bar{\beta}}$. Substituting these inequalities and $\Delta f \leqslant Q_1$ into (6.23) gives

$$\Delta' F \leqslant (m - \Delta\phi)^{-1}Q_1 + \kappa(m - g'^{\alpha\bar{\beta}}g_{\alpha\bar{\beta}}) + C_5 g'^{\alpha\bar{\beta}}g_{\alpha\bar{\beta}}. \tag{6.24}$$

At a point p where F is maximum, $\Delta' F \geqslant 0$, and so by (6.24) at p we get

$$(\kappa - C_5)g'^{\alpha\bar{\beta}}g_{\alpha\bar{\beta}} \leqslant m\kappa + Q_1(m - \Delta\phi)^{-1}. \tag{6.25}$$

Rearranging eqn (6.12) of Proposition 6.3.4 gives $m - \Delta\phi \geqslant m\,\mathrm{e}^{f/m}$, and as $\|f\|_{C^0} \leqslant Q_1$ this shows that $(m - \Delta\phi)^{-1} \leqslant \frac{1}{m}\mathrm{e}^{Q_1/m}$. Now choose $\kappa = C_5 + 1$. Then (6.25) implies that $g'^{\alpha\bar{\beta}}g_{\alpha\bar{\beta}} \leqslant C_6$ at p, where $C_6 = m\kappa + \frac{1}{m}Q_1\mathrm{e}^{Q_1/m}$.

Let us apply Lemma 6.3.1 to find expressions for g and g' at p in terms of positive constants $a_1, \ldots, a_m$. In this notation, using Lemma 6.3.2 we have

$$m - \Delta\phi = \sum_{j=1}^{m} a_j, \quad g'^{\alpha\bar{\beta}}g_{\alpha\bar{\beta}} = \sum_{j=1}^{m} a_j^{-1}, \quad \text{and} \quad \prod_{j=1}^{m} a_j = \mathrm{e}^{f(p)}.$$

It easily follows that at p we have

$$m - \Delta\phi \leqslant \mathrm{e}^{f(p)}\big(g'^{\alpha\bar{\beta}}g_{\alpha\bar{\beta}}\big)^{m-1} \leqslant \mathrm{e}^{Q_1}C_6^{m-1}.$$

Therefore at a maximum of F, we see that $F \leqslant Q_1 + (m-1)\log C_6 - \kappa \inf \phi$, and as $\|\phi\|_{C^0} \leqslant Q_2$ by Corollary 6.4.5 this shows that $F \leqslant Q_1 + (m-1)\log C_6 + \kappa Q_2$ everywhere on M. Substituting for F and exponentiating gives

$$0 < m - \Delta\phi \leqslant C_6^{m-1}\exp(Q_1 + \kappa Q_2 + \kappa\phi) \leqslant C_6^{m-1}\exp(Q_1 + 2\kappa Q_2)$$

on M. This gives an a priori estimate for $\|\Delta\phi\|_{C^0}$. But Proposition 6.3.4 gives estimates for $\|g'_{ab}\|_{C^0}, \|g'^{ab}\|_{C^0}$ and $\|\mathrm{dd}^c\phi\|_{C^0}$ in terms of upper bounds for $\|f\|_{C^0}$ and $\|\Delta\phi\|_{C^0}$, so (6.22) holds for some constants c_1, c_2 and Q_3. All the constants in this proof, including c_1, c_2 and Q_3, depend only on M, J, g and Q_1. $\qquad\square$

6.4.3   Third-order estimates

Define a function $S \geqslant 0$ on M by $4S^2 = |\nabla\mathrm{dd}^c\phi|^2_{g'}$, so that in index notation

$$S^2 = g'^{\alpha\bar{\lambda}}g'^{\mu\bar{\beta}}g'^{\gamma\bar{\nu}}\nabla_{\alpha\bar{\beta}\gamma}\phi\,\nabla_{\bar{\lambda}\mu\bar{\nu}}\phi.$$

Our goal is to find an a priori upper bound for S, and this will be done by finding a formula for $\Delta'(S^2)$ and then using an argument similar to that used in Proposition 6.4.7. First, here is some notation that will be used for the next result. Suppose A, B, C are tensors on M. Let us write $P^{a,b,c}(A, B, C)$ for any polynomial in the tensors A, B, C alone, that is homogeneous of degree a in A, degree b in B and degree c in C.

Using this notation, we shall give in the next proposition an expression for $\Delta'(S^2)$. The proof is a straightforward but rather long and tedious calculation, and we will omit it. Readers can find it in [345, App. A] and [15, p. 410–411].

Proposition 6.4.8 *In the notation above, we have:*

$$
\begin{aligned}
-\Delta'\big(S^2\big) = {} & \big|\nabla_{\bar{\alpha}\beta\bar{\gamma}\delta}\phi - g'^{\lambda\bar{\mu}}\nabla_{\bar{\alpha}\lambda\bar{\gamma}}\phi\nabla_{\beta\bar{\mu}\delta}\phi\big|^2_{g'} \\
& + \big|\nabla_{\alpha\beta\bar{\gamma}\delta}\phi - g'^{\lambda\bar{\mu}}\nabla_{\alpha\bar{\gamma}\lambda}\phi\nabla_{\beta\bar{\mu}\delta}\phi - g'^{\lambda\bar{\mu}}\nabla_{\alpha\bar{\mu}\delta}\phi\nabla_{\lambda\bar{\gamma}\beta}\phi\big|^2_{g'} \\
& + P^{4,2,1}\big(g'^{\alpha\bar{\beta}},\, \nabla_{\alpha\bar{\beta}\gamma}\phi,\, \nabla_{\alpha\bar{\beta}}f\big) + P^{4,2,1}\big(g'^{\alpha\bar{\beta}},\, \nabla_{\alpha\bar{\beta}\gamma}\phi,\, R^a{}_{bcd}\big) \\
& + P^{3,1,1}\big(g'^{\alpha\bar{\beta}},\, \nabla_{\alpha\bar{\beta}\gamma}\phi,\, \nabla_{\bar{\alpha}\beta\bar{\gamma}}f\big) + P^{3,1,1}\big(g'^{\alpha\bar{\beta}},\, \nabla_{\alpha\bar{\beta}\gamma}\phi,\, \nabla_e R^a{}_{bcd}\big).
\end{aligned}
\tag{6.26}
$$

In (6.26), we use the notation defined in (6.11) that for T a tensor on M, $|T|_{g'}$ is the modulus of T calculated with respect to the metric g'. Also, the four polynomials $P^{a,b,c}$ in (6.26) are not the same but different polynomials. Now there is a certain similarity between Proposition 6.4.8 and Proposition 6.4.6. Here is one way to see it.

In (6.19), the left hand side $\Delta'(\Delta\phi)$ involves $\nabla^4\phi$. However, the right hand side involves only $\nabla^2\phi$ and $\nabla^3\phi$, together with g', Δf and R. Moreover, the terms on the right hand side involving $\nabla^3\phi$ are nonnegative. In the same way, the left hand side of (6.26) involves $\nabla^5\phi$, but right hand side involves only $\nabla^3\phi$ and $\nabla^4\phi$, and the terms on the right hand side involving $\nabla^4\phi$ are nonnegative.

So, both (6.19) and (6.26) express a derivative of ϕ in terms of lower derivatives of ϕ, and g', f and R, and the highest derivative of ϕ on the right hand side contributes only nonnegative terms. Now Proposition 6.4.7 used (6.19) to find an a priori bound for $\|\Delta\phi\|_{C^0}$. Because of the similarities between (6.19) and (6.26), we will be able to use the same method to find an a priori bound for S, and hence for $\|\nabla\mathrm{dd}^c\phi\|_{C^0}$.

It follows quickly from (6.26) that

Corollary 6.4.9 *There is a constant C_7 depending on Q_1, c_1, c_2 and $\|R^a{}_{bcd}\|_{C^1}$ with*

$$
\Delta'\big(S^2\big) \leqslant C_7\big(S^2 + S\big).
$$

Here c_1, c_2 are the constants from Proposition 6.4.7. To prove the corollary, observe that the first two terms on the right hand side of (6.26) are nonnegative, and can be neglected. Of the four terms of type $P^{a,b,c}$, the first two are quadratic in $\nabla_{\alpha\bar{\beta}\gamma}\phi$ and must be estimated by a multiple of S^2, and the second two are linear in $\nabla_{\alpha\bar{\beta}\gamma}\phi$ and must be estimated by a multiple of S. As $\|f\|_{C^3} \leqslant Q_1$, the factors of $\nabla_{\alpha\bar{\beta}}f$ and $\nabla_{\bar{\alpha}\beta\bar{\gamma}}f$ are estimated using Q_1, and by using the estimates (6.22) of Proposition 6.4.7 for g'_{ab} and g'^{ab}, the corollary quickly follows.

The next result, together with Corollary 6.4.5 and Proposition 6.4.7, completes the proof of Theorem C1.

Proposition 6.4.10 $\|\nabla\mathrm{dd}^c\phi\|_{C^0} \leqslant Q_4$ *for Q_4 depending only on M, J, g and Q_1.*

Proof Using the formulae for g' in §6.3, it is easy to show that

$$
g^{\alpha\bar{\lambda}}g'^{\mu\bar{\beta}}g'^{\gamma\bar{\nu}}\nabla_{\alpha\bar{\beta}\gamma}\phi\nabla_{\bar{\lambda}\mu\bar{\nu}}\phi \geqslant c_2^{-1}S^2,
$$

where $c_2 > 0$ comes from Proposition 6.4.7. So Proposition 6.4.6 gives

$$\Delta'(\Delta\phi) \geqslant c_2^{-1} S^2 - C_8, \tag{6.27}$$

where C_8 depends on M, J, g and Q_1. Consider the function $S^2 - 2c_2C_7\Delta\phi$ on M. From Corollary 6.4.9 and (6.27) it follows that

$$\Delta'(S^2 - 2c_2C_7\Delta\phi) \leqslant C_7(S^2 + S) - 2c_2C_7(c_2^{-1}S^2 - C_8)$$
$$= -C_7(S - \tfrac{1}{2})^2 + 2c_2C_7C_8 + \tfrac{1}{4}C_7.$$

At a maximum p of $S^2 - 2c_2C_7\Delta\phi$, we have $\Delta'(S^2 - 2c_2C_7\Delta\phi) \geqslant 0$, and thus

$$(S - \tfrac{1}{2})^2 \leqslant 2c_2C_8 + \tfrac{1}{4}$$

at p. So, there is a constant $C_9 > 0$ depending on c_2, C_8 and the a priori estimate for $\|\Delta\phi\|_{C^0}$ found in Proposition 6.4.7, such that $S^2 - 2c_2C_7\Delta\phi \leqslant C_9$ at p. As p is a maximum, $S^2 - 2c_2C_7\Delta\phi \leqslant C_9$ holds on M. Using the estimate on $\|\Delta\phi\|_{C^0}$ again, we find an a priori estimate for $\|S\|_{C^0}$.

Now $2S = |\nabla \mathrm{dd}^c\phi|_{g'}$. It can be seen that $|\nabla \mathrm{dd}^c\phi|_g \leqslant c_1^{3/2}|\nabla \mathrm{dd}^c\phi|_{g'}$, where c_1 comes from Proposition 6.4.7. Thus, an a priori bound for $\|S\|_{C^0}$ gives one for $\|\nabla \mathrm{dd}^c\phi\|_{C^0}$, and there is Q_4 depending on M, J, g, Q_1 with $\|\nabla \mathrm{dd}^c\phi\|_{C^0} \leqslant Q_4$. $\square$

6.5   The proof of Theorem C2

Here is the proof of Theorem C2 of §6.2.

Theorem C2 *Let (M, J) be a compact, complex manifold, and g a Kähler metric on M, with Kähler form ω. Let $Q_1, \ldots, Q_4 \geqslant 0$ and $\alpha \in (0, 1)$. Then there exists $Q_5 \geqslant 0$ depending only on $M, J, g, Q_1, \ldots, Q_4$ and α, such that the following holds.*

Suppose $f \in C^{3,\alpha}(M)$, $\phi \in C^5(M)$ and $A > 0$ satisfy $(\omega + \mathrm{dd}^c\phi)^m = Ae^f\omega^m$ and the inequalities

$$\|f\|_{C^{3,\alpha}} \leqslant Q_1, \quad \|\phi\|_{C^0} \leqslant Q_2, \quad \|\mathrm{dd}^c\phi\|_{C^0} \leqslant Q_3 \quad \text{and} \quad \|\nabla \mathrm{dd}^c\phi\|_{C^0} \leqslant Q_4.$$

Then $\phi \in C^{5,\alpha}(M)$ and $\|\phi\|_{C^{5,\alpha}} \leqslant Q_5$. Also, if $f \in C^{k,\alpha}(M)$ for $k \geqslant 3$ then $\phi \in C^{k+2,\alpha}(M)$, and if $f \in C^\infty(M)$ then $\phi \in C^\infty(M)$.

We shall continue to use most of the notation of §6.4. In particular, the metrics g and g' and the Levi-Civita connection ∇ of g will be the same, repeated derivatives using ∇ will be written $\nabla_{a_1 \ldots a_k}$, and the operators $\Delta = g^{\alpha\bar\beta}\nabla_{\alpha\bar\beta}$ and $\Delta' = g'^{\alpha\bar\beta}\nabla_{\alpha\bar\beta}$ will be used. All norms in this section will be with respect to g.

Let us begin by stating three elliptic regularity results for Δ and Δ', which come from §1.4. Lemma 6.5.1 follows from (1.7) of Theorem 1.4.1. Lemmas 6.5.2 and 6.5.3 follow from estimates (1.10) and (1.11) of Theorem 1.4.3 respectively. To prove Lemmas 6.5.2 and 6.5.3, we cover M by a finite number of overlapping coordinate patches identified with the unit ball in $\mathbb{R}^{2m}$, using the argument described in §1.4.1.

Lemma 6.5.1 *Let $k \geqslant 0$ and $\alpha \in (0, 1)$. Then there exists a constant $E_{k,\alpha} > 0$ depending on k, α, M and g, such that if $\psi \in C^2(M)$, $\xi \in C^{k,\alpha}(M)$ and $\Delta\psi = \xi$, then $\psi \in C^{k+2,\alpha}(M)$ and $\|\psi\|_{C^{k+2,\alpha}} \leqslant E_{k,\alpha}(\|\xi\|_{C^{k,\alpha}} + \|\psi\|_{C^0})$.*

Lemma 6.5.2 *Let* $\alpha \in (0,1)$. *Then there exists a constant* $E'_\alpha > 0$ *depending on* α, M, g *and the norms* $\|g'_{ab}\|_{C^0}$ *and* $\|g'^{ab}\|_{C^{0,\alpha}}$, *such that if* $\psi \in C^2(M)$, $\xi \in C^0(M)$ *and* $\Delta'\psi = \xi$, *then* $\psi \in C^{1,\alpha}(M)$ *and* $\|\psi\|_{C^{1,\alpha}} \leqslant E'_\alpha \big(\|\xi\|_{C^0} + \|\psi\|_{C^0}\big)$.

Lemma 6.5.3 *Let* $k \geqslant 0$ *be an integer, and* $\alpha \in (0,1)$. *Then there exists a constant* $E'_{k,\alpha} > 0$ *depending on* k, α, M, g *and the norms* $\|g'_{ab}\|_{C^0}$ *and* $\|g'^{ab}\|_{C^{k,\alpha}}$, *such that if* $\psi \in C^2(M)$, $\xi \in C^{k,\alpha}(M)$ *and* $\Delta'\psi = \xi$, *then* $\psi \in C^{k+2,\alpha}(M)$ *and* $\|\psi\|_{C^{k+2,\alpha}} \leqslant E'_{k,\alpha}\big(\|\xi\|_{C^{k,\alpha}} + \|\psi\|_{C^0}\big)$.

The proof of Theorem C2 is based on eqn (6.19) of Proposition 6.4.6, which we reproduce here.

$$
\begin{aligned}
\Delta'(\Delta\phi) = {}&-\Delta f + g^{\alpha\bar{\lambda}}g'^{\mu\bar{\beta}}g'^{\gamma\bar{\nu}}\nabla_{\alpha\bar{\beta}\gamma}\phi\,\nabla_{\bar{\lambda}\mu\bar{\nu}}\phi \\
&+ g'^{\alpha\bar{\beta}}g^{\gamma\bar{\delta}}\big(R^{\bar{\epsilon}}_{\bar{\delta}\gamma\bar{\beta}}\nabla_{\alpha\bar{\epsilon}}\phi - R^{\bar{\epsilon}}_{\bar{\beta}\alpha\bar{\delta}}\nabla_{\gamma\bar{\epsilon}}\phi\big).
\end{aligned}
\tag{6.28}
$$

Applying the three lemmas to this equation, we prove the theorem by an inductive process known as 'bootstrapping'. First we find an a priori estimate for $\|\phi\|_{C^{3,\alpha}}$.

Proposition 6.5.4 *There is a constant* D_1 *depending on* $M, g, J, Q_1, \ldots, Q_4$ *and* α *such that* $\|\phi\|_{C^{3,\alpha}} \leqslant D_1$.

Proof   In this proof, all estimates and all constants depend only on $M, g, J, Q_1, \ldots, Q_4$ and α. By Proposition 6.4.7, $\|\mathrm{dd}^c\phi\|_{C^0} \leqslant Q_3$ implies $\|g'_{ab}\|_{C^0} \leqslant c_1$ and $\|g'^{ab}\|_{C^0} \leqslant c_2$. Also, $\nabla g'^{ab}$ can be expressed in terms of $\nabla \mathrm{dd}^c\phi$, g'^{ab} and J, so that the estimates for $\|g'^{ab}\|_{C^0}$ and $\|\nabla \mathrm{dd}^c\phi\|_{C^0}$ yield an estimate for $\|\nabla g'^{ab}\|_{C^0}$. Combining the estimates for $\|g'^{ab}\|_{C^0}$ and $\|\nabla g'^{ab}\|_{C^0}$, we can find an estimate for $\|g'^{ab}\|_{C^{0,\alpha}}$.

Thus there are a priori estimates for $\|g'_{ab}\|_{C^0}$ and $\|g'^{ab}\|_{C^{0,\alpha}}$. Therefore, Lemma 6.5.2 holds with a constant E'_α depending on $M, g, J, Q_1, \ldots, Q_4$ and α. Put $\psi = \Delta\phi$, and let ξ be the right hand side of (6.28), so that $\Delta\psi = \xi$. Now, combining a priori estimates of the C^0 norms of g'^{ab}, $\nabla_{\alpha\bar{\beta}}\phi$ and $\nabla_{\alpha\bar{\beta}\gamma}\phi$, the inequality $\|f\|_{C^2} \leqslant Q_1$ and a bound for $R^a{}_{bcd}$, we can find a constant D_2 such that $\|\xi\|_{C^0} \leqslant D_2$.

So, by Lemma 6.5.2, $\Delta\phi = \psi$ lies in $C^{1,\alpha}(M)$ and $\|\Delta\phi\|_{C^{1,\alpha}} \leqslant E'_\alpha(D_2 + Q_3)$, as $\|\Delta\phi\|_{C^0} \leqslant \|\mathrm{dd}^c\phi\|_{C^0} \leqslant Q_3$. Therefore by Lemma 6.5.1, $\phi \in C^{3,\alpha}(M)$ and

$$
\|\phi\|_{C^{3,\alpha}} \leqslant E_{1,\alpha}\big(\|\Delta\phi\|_{C^{1,\alpha}} + \|\phi\|_{C^0}\big) \leqslant E_{1,\alpha}\big(E'_\alpha(D_2 + Q_3) + Q_2\big).
$$

Thus, putting $D_1 = E_{1,\alpha}(E'_\alpha(D_2 + Q_3) + Q_2)$, the proposition is complete.   □

Theorem C2 will follow from the next proposition.

Proposition 6.5.5 *For each* $k \geqslant 3$, *if* $f \in C^{k,\alpha}(M)$ *then* $\phi \in C^{k+2,\alpha}(M)$, *and there exists an a priori bound for* $\|\phi\|_{C^{k+2,\alpha}}(M)$ *depending on* $M, g, J, Q_1, \ldots, Q_4, k, \alpha$, *and a bound for* $\|f\|_{C^{k,\alpha}}$.

Proof   The proof is by induction on k. The result is stated for $k \geqslant 3$ only, because Theorem C1 uses $\|f\|_{C^3}$ to bound $\|\nabla \mathrm{dd}^c\phi\|_{C^0}$. However, it is convenient to start the induction at $k = 2$. In this proof, we say a constant 'depends on the k-data' if it depends only on $M, g, J, Q_1, \ldots, Q_4, k, \alpha$ and bounds for $\|f\|_{C^{k,\alpha}}$ and $\|f\|_{C^3}$. Our inductive hypothesis is that $f \in C^{k,\alpha}(M)$ and $\phi \in C^{k+1,\alpha}(M)$, and that there is an a priori bound for $\|\phi\|_{C^{k+1,\alpha}}$ depending on the $(k-1)$-data. By Proposition 6.5.4, this holds for $k = 2$, and this is the first step in the induction.

Write $\psi = \Delta\phi$, and let ξ be the right hand side of (6.28), so that $\Delta'\psi = \xi$. Now the term $-\Delta f$ on the right hand side of (6.28) is bounded in $C^{k-2,\alpha}(M)$ by $\|f\|_{C^{k,\alpha}}$, and hence in terms of the k-data. It is easy to see that every other term on the right hand side of (6.28) can be bounded in $C^{k-2,\alpha}(M)$ in terms of M, g, J and bounds for $\|g'^{ab}\|_{C^0}$ and $\|\phi\|_{C^{k+1,\alpha}}$. By the inductive hypothesis, these are all bounded in terms of the $(k-1)$-data. Therefore, we can find a bound $F_{k,\alpha}$ depending on the k-data, such that $\|\xi\|_{C^{k-2,\alpha}} \leqslant F_{k,\alpha}$.

Using the inductive hypothesis again, we may bound $\|g'_{ab}\|_{C^0}$ and $\|g'^{ab}\|_{C^{k,\alpha}}$ in terms of the $(k-1)$-data. Therefore we may apply Lemma 6.5.3 to $\psi = \Delta\phi$, which shows that $\Delta\phi \in C^{k,\alpha}(M)$ and

$$\|\Delta\phi\|_{C^{k,\alpha}} \leqslant E'_{k-2,\alpha}\big(\|\xi\|_{C^{k-2,\alpha}} + \|\Delta\phi\|_{C^0}\big) \leqslant E'_{k-2,\alpha}\big(F_{k,\alpha} + Q_3\big),$$

since $\|\Delta\phi\|_{C^0} \leqslant Q_3$, where $E'_{k-2,\alpha}$ depends on the $(k-1)$-data. Thus by Lemma 6.5.1, $\phi \in C^{k+2,\alpha}(M)$ as we have to prove, and

$$\|\phi\|_{C^{k+2,\alpha}} \leqslant E_{k,\alpha}\big(\|\Delta\phi\|_{C^{k,\alpha}} + \|\phi\|_{C^0}\big) \leqslant E_{k,\alpha}\big(E'_{k-2,\alpha}(F_{k,\alpha} + Q_3) + Q_2\big),$$

since $\|\phi\|_{C^0} \leqslant Q_2$. This is an a priori bound for $\|\phi\|_{C^{k+2,\alpha}}$ depending only on the k-data. Therefore by induction, the result holds for all k. $\qquad\square$

Now, to prove Theorem C2, first put $k = 3$ in Proposition 6.5.5. This shows that $\phi \in C^{5,\alpha}(M)$ and gives an a priori bound for $\|\phi\|_{C^{5,\alpha}}$. Let Q_5 be this bound. Then $\|\phi\|_{C^{5,\alpha}} \leqslant Q_5$, and Q_5 depends only on $M, J, g, Q_1, \ldots, Q_4$ and α, since Q_1 is a bound for $\|f\|_{C^{3,\alpha}}$. Also, if $k \geqslant 3$ and $f \in C^{k,\alpha}(M)$, Proposition 6.5.5 shows that $\phi \in C^{k+2,\alpha}(M)$. Since this holds for all k, if $f \in C^\infty(M)$ then $\phi \in C^\infty(M)$, and the theorem is proved.

6.6   The proof of Theorem C3

Now we prove Theorem C3 of §6.2.

Theorem C3 *Let (M, J) be a compact complex manifold, and g a Kähler metric on M, with Kähler form ω. Fix $\alpha \in (0, 1)$, and suppose that $f' \in C^{3,\alpha}(M)$, $\phi' \in C^{5,\alpha}(M)$ and $A' > 0$ satisfy the equations*

$$\int_M \phi' \, dV_g = 0 \quad \text{and} \quad (\omega + dd^c\phi')^m = A' e^{f'} \omega^m. \tag{6.29}$$

Then whenever $f \in C^{3,\alpha}(M)$ and $\|f - f'\|_{C^{3,\alpha}}$ is sufficiently small, there exist $\phi \in C^{5,\alpha}(M)$ and $A > 0$ such that

$$\int_M \phi \, dV_g = 0 \quad \text{and} \quad (\omega + dd^c\phi)^m = A e^f \omega^m. \tag{6.30}$$

Proof Define X to be the vector subspace of $\phi \in C^{5,\alpha}(M)$ for which $\int_M \phi \, dV_g = 0$. Let U be the subset of $\phi \in X$ such that $\omega + dd^c \phi$ is a *positive* (1,1)-form on M. Then U is open in X. Let Y be the Banach space $C^{3,\alpha}(M)$. Suppose that $\phi \in U$ and $a \in \mathbb{R}$. Then $\omega + dd^c \phi$ is a positive (1,1)-form, so that $(\omega + dd^c \phi)^m$ is a positive multiple of ω^m at each point. Therefore there exists a unique real function f on M such that $(\omega + dd^c \phi)^m = e^{a+f} \omega^m$ on M, and as $\phi \in C^{5,\alpha}(M)$, it follows that $f \in C^{3,\alpha}(M)$.

Define a function $F : U \times \mathbb{R} \to Y$ by $F(\phi, a) = f$, where $(\omega + dd^c \phi)^m = e^{a+f} \omega^m$ on M. We have just shown that F is well-defined, and it is easy to see that F is a smooth map of Banach spaces. Now let f', ϕ' and A' be as in the theorem, and let $a' = \log A'$. By (6.29), ϕ' lies in U and $F(\phi', a') = f'$. We shall evaluate the first derivative $dF_{(\phi', a')}$ of F at (ϕ', a'). Calculation shows that

$$\big(\omega + dd^c(\phi' + \epsilon\psi)\big)^m = \exp\big(a' + \epsilon b + f_\epsilon\big)\, \omega^m,$$

where $f_\epsilon = f' - \epsilon b - \epsilon \Delta' \psi + O(\epsilon^2)$. Here g' is the $C^{3,\alpha}$ Kähler metric with Kähler form $\omega + dd^c \phi'$, and $\Delta' = -g'^{\alpha\bar\beta} \nabla_{\alpha\bar\beta}$ is the Laplacian with respect to g'. Therefore $F(\phi' + \epsilon\psi, a' + \epsilon b) = f_\epsilon$. From the first order term in ϵ in the expression for f_ϵ, we see that the first derivative of F is

$$dF_{(\phi', a')} : X \times \mathbb{R} \to Y, \quad \text{given by} \quad dF_{(\phi', a')}(\psi, b) = -b - \Delta' \psi. \tag{6.31}$$

Now the operator $\Delta' : C^{5,\alpha}(M) \to C^{3,\alpha}(M)$ is a linear elliptic operator of order 2 with $C^{3,\alpha}$ coefficients. As M is connected, the kernel $\operatorname{Ker} \Delta'$ is the set of constant functions on M. Calculated with respect to the metric g, the dual $(\Delta')^*$ of Δ' is given by $(\Delta')^* \psi = e^{f'} \Delta'(e^{-f'} \psi)$. Therefore $\operatorname{Ker}(\Delta')^*$ is the set of constant multiples of $e^{-f'}$. It follows that $\psi \perp \operatorname{Ker} \Delta'$ if and only if $\int_M \psi \, dV_g = 0$, and $\chi \perp \operatorname{Ker}(\Delta')^*$ if and only if $\langle \chi, e^{-f'} \rangle = 0$.

Let us apply Theorem 1.5.4 of §1.5 to Δ'. The theorem shows that if $\chi \in C^{3,\alpha}(M)$, then there exists $\psi \in C^{5,\alpha}(M)$ with $\Delta' \psi = f$ if and only if $\langle \chi, e^{-f'} \rangle = 0$. Also, if in addition we require that $\int_M \psi \, dV_g = 0$, then ψ is unique. Let $\chi \in Y = C^{3,\alpha}(M)$. Then there is a unique $b \in \mathbb{R}$ such that $\langle \chi + b, e^{-f'} \rangle = 0$. Hence, there exists a unique $\psi \in C^{5,\alpha}(M)$ with $\int_M \psi \, dV_g = 0$, such that $\Delta' \psi = -b - \chi$. But then $\psi \in X$, and (6.31) shows that $dF_{(\phi', a')}(\psi, b) = \chi$.

Thus, if $\chi \in Y$, there exist unique $\psi \in X$ and $b \in \mathbb{R}$, such that $dF_{(\phi', a')}(\psi, b) = \chi$. So $dF_{(\phi', a')} : X \times \mathbb{R} \to Y$ is an *invertible* linear map. It is continuous and has continuous inverse, and thus is an isomorphism of $X \times \mathbb{R}$ and Y as both vector spaces and topological spaces. Therefore, applying the Inverse Mapping Theorem for Banach spaces, Theorem 1.2.4, there is an open neighbourhood $U' \subset U \times \mathbb{R}$ of (ϕ', a') in $X \times \mathbb{R}$ and an open neighbourhood $V' \subset Y$ of f' in Y, such that $F : U' \to V'$ is a homeomorphism.

So, whenever $f \in C^{3,\alpha}(M) = Y$ and $\|f - f'\|_{C^{3,\alpha}}$ is sufficiently small, we have $f \subset V'$, and there is a unique $(\phi, a) \in U'$ with $F(\phi, a) = f$. Since $\phi \in X$, the first equation of (6.30) holds, and $\phi \in C^{5,\alpha}(M)$. Putting $A = e^a$ gives $A > 0$, and as $F(\phi, a) = f$, the second equation of (6.30) holds by definition of F. This completes the proof of Theorem C3. $\qquad\square$

6.7   The proof of Theorem C4

We now prove Theorem C4 of §6.2, which completes our proof of the Calabi Conjecture. Note that the proof is very similar to that of Proposition 6.4.1.

Theorem C4 *Let (M, J) be a compact complex manifold, and g a Kähler metric on M, with Kähler form ω. Let $f \in C^1(M)$. Then there is at most one function $\phi \in C^3(M)$ such that $\int_M \phi \, dV_g = 0$ and $(\omega + dd^c\phi)^m = Ae^f\omega^m$ on M.*

Proof Suppose $\phi_1, \phi_2 \in C^3(M)$ satisfy $\int_M \phi_j \, dV_g = 0$ and $(\omega + dd^c\phi_j)^m = Ae^f\omega^m$ for $j = 1, 2$. We shall show that $\phi_1 = \phi_2$, so that any solution ϕ of these two equations is unique in $C^3(M)$. Write $\omega_1 = \omega + dd^c\phi_1$ and $\omega_2 = \omega + dd^c\phi_2$. By Lemma 6.1.1, both ω_1 and ω_2 are positive (1,1)-forms. Let g_1 and g_2 be the C^1 Kähler metrics associated to ω_1 and ω_2.

As $\omega_1^m = Ae^f\omega^m = \omega_2^m$ and $\omega_1 - \omega_2 = dd^c(\phi_1 - \phi_2)$, we see that

$$0 = \omega_1^m - \omega_2^m = dd^c(\phi_1 - \phi_2) \wedge \left(\omega_1^{m-1} + \omega_1^{m-2} \wedge \omega_2 + \cdots + \omega_2^{m-1}\right).$$

Now M is compact, and so Stokes' Theorem gives

$$\int_M d\left[(\phi_1 - \phi_2)d^c(\phi_1 - \phi_2) \wedge \left(\omega_1^{m-1} + \cdots + \omega_2^{m-1}\right)\right] = 0.$$

Combining the previous two equations and using $d\omega_1 = d\omega_2 = 0$, we find

$$\int_M d(\phi_1 - \phi_2) \wedge d^c(\phi_1 - \phi_2) \wedge \left(\omega_1^{m-1} + \cdots + \omega_2^{m-1}\right) = 0. \qquad (6.32)$$

Following the proof of Lemma 6.3.5, we can prove that

$$d(\phi_1 - \phi_2) \wedge d^c(\phi_1 - \phi_2) \wedge \omega_1^{m-1} = \frac{1}{m}\left|d(\phi_1 - \phi_2)\right|_{g_1}^2 \omega_1^m$$

$$\text{and} \quad d(\phi_1 - \phi_2) \wedge d^c(\phi_1 - \phi_2) \wedge \omega_1^{m-1-j} \wedge \omega_2^j = F_j\omega_1^m,$$

where F_j is a nonnegative function on M. Substituting these into (6.32), and using the fact that $d(\phi_1 - \phi_2)$ is continuous, we deduce that $d(\phi_1 - \phi_2) = 0$. Since M is connected, $\phi_1 - \phi_2$ is constant. But $\int_M \phi_1 \, dV_g = 0 = \int_M \phi_2 \, dV_g$, so $\phi_1 - \phi_2 = 0$, and $\phi_1 = \phi_2$. This completes the proof. □

6.8   A discussion of the proof

One question that preoccupied the author during the writing of this chapter is: can the proof be simplified? In particular, is it really necessary to find a priori estimates for the first three derivatives of ϕ, and must one assume that ϕ is five times differentiable, or can one work with fewer derivatives? Here are the reasons for using this many derivatives of ϕ and f, and for choosing the Banach spaces such as $C^{5,\alpha}(M)$ that appear in the proof.

Clearly ϕ must be at least twice differentiable, or else the eqn (6.4) that ϕ must satisfy makes no sense. So one must find a priori estimates for at least the first two derivatives of ϕ. However, it is less obvious that one really needs the third-order estimates of §6.4.3. There are results that show that if ϕ is a C^2 solution of a smooth, nonlinear, elliptic, second-order differential equation, then ϕ is C^∞. For example, this follows from Morrey [267, Th. 6.8.1]. These suggest that one could make do with only the second-order estimates in §6.4.2.

However, a careful examination shows that Morrey's results, and others of the same kind, tend to rely not only on bounds on $\nabla^2 \phi$, but also on a *modulus of continuity* for $\nabla^2 \phi$. Every continuous function on a compact manifold has a modulus of continuity, which is a real function that bounds how quickly the function varies from point to point. To apply these smoothness results in our situation would require an a priori bound for the modulus of continuity of $dd^c \phi$. Finding one is probably no easier than making the third-order estimates of §6.4.3, and having made them we are then able to complete the proof using more elementary results on linear elliptic equations.

Now, about the spaces used in the proofs. In Theorem C1, we need $\phi \in C^5(M)$ so that $\Delta'(S^2)$ should exist in Proposition 6.4.8 of §6.4.3. Theorems C2 and C3 require $\phi \in C^{5,\alpha}(M)$ for several reasons: ϕ must be in $C^5(M)$ for Theorem C1, Hölder spaces have good elliptic regularity properties, and the embedding $C^{5,\alpha}(M) \hookrightarrow C^5(M)$ is compact, which is used in Theorem 6.2.2. In Theorem C4, we take $\phi \in C^3(M)$ because in applying Stokes' Theorem in §6.7 we suppose that $d\omega_1 = d\omega_2 = 0$, which involves the third derivatives of ϕ.

Here is another issue. In §6.4.2, the bound computed for $\|dd^c \phi\|_{C^0}$ depends on $\|f\|_{C^2}$, that is, on C^0 bounds for f and Δf. However, if $\phi \in C^2(M)$, it is only necessary that $f \in C^0(M)$, and experience with linear elliptic equations suggests that if f is bounded in $C^{0,\alpha}(M)$, then ϕ should be bounded in $C^{2,\alpha}(M)$. So, the proof seems to use an unnecessarily strong bound on f. Similarly, the third-order estimates for ϕ in §6.4.3 involve $\|f\|_{C^3}$, when one would expect to use only $\|f\|_{C^{1,\alpha}}$. I do not know whether it is possible to make the a priori estimates of ϕ without using these strong bounds on f, but I suspect the answer may be no.

7
Calabi–Yau manifolds

In this chapter we study compact Riemannian manifolds with holonomy $SU(m)$. If (X, g) is a compact Riemannian $2m$-manifold with $\text{Hol}(g) \subseteq SU(m)$ then g is Ricci-flat, and X admits a complex structure J with (X, J, g) Kähler and a holomorphic $(m, 0)$-form θ with $|\theta| \equiv 2^{m/2}$. The quadruple (X, J, g, θ) is then called a *Calabi–Yau manifold* or *Calabi–Yau m-fold*. They have this name because one can use Yau's solution of the Calabi Conjecture to show that suitable compact complex manifolds (X, J) can be extended to Calabi–Yau manifolds (X, J, g, θ).

The holonomy groups $Sp(k)$ and $SU(m)$ are closely related, since $Sp(k)$ is a subgroup of $SU(2k)$. Thus any metric g with $\text{Hol}(g) = Sp(k)$ is also Kähler and Ricci-flat. The inclusion $Sp(k) \subseteq SU(2k)$ is proper for $k > 1$, but $SU(2) = Sp(1)$. The holonomy groups $Sp(k)$ will be covered in Chapter 10. We postpone discussion of $SU(2) = Sp(1)$ until then, as it fits in better with the material there. This means that much of the focus of this chapter is on the holonomy group $SU(3)$.

If (X, J, g) is a compact Kähler manifold with holonomy $SU(m)$ for $m \geqslant 3$, then the complex manifold (X, J) is projective with $c_1(X) = 0$. Thus the complex manifolds underlying Calabi–Yau manifolds are algebraic objects, and it is natural to study them using *complex algebraic geometry*. Projective manifolds exist in huge numbers, and the condition $c_1(X) = 0$ is comparatively easy to test for. Many examples of Calabi–Yau manifolds can be constructed simply by considering some class of projective manifolds, and calculating which ones satisfy $c_1(X) = 0$.

Although much of the chapter is algebraic geometry, a rather technical subject, we will try to present it in an elementary way and keep the technicalities to a minimum. One thing we will not say much about is what the metrics on the Calabi–Yau manifolds are like: here we are mostly interested in the underlying complex manifolds. In fact little is known about general Calabi–Yau metrics.

The subject of Calabi–Yau manifolds is too big to be properly described in one chapter of a book. We aim to cover three main areas in detail: firstly, the basic differential geometry of Calabi–Yau manifolds; secondly, orbifolds and crepant resolutions; and thirdly, ways of constructing Calabi–Yau manifolds. There are other important areas, such as rational curves on Calabi–Yau 3-folds, that we shall not have space to discuss.

Section 7.1 introduces the holonomy groups $SU(m)$, and the differential geometry of compact Ricci-flat Kähler manifolds. Then §7.2–§7.5 discuss a special class of resolutions of singular complex manifolds, called *crepant resolutions*. Their importance to Calabi–Yau geometry is that if X is a singular Calabi–Yau manifold and $(\tilde{X}, \pi)$ a

Kähler crepant resolution of X, then $\tilde{X}$ is a nonsingular Calabi–Yau manifold. This plays a part in a number of constructions of Calabi–Yau manifolds. We focus in particular on crepant resolutions of *complex orbifolds*.

Section 7.6 describes constructions of Calabi–Yau manifolds, and Calabi–Yau 3-folds in particular, using algebraic geometry. For example, a nonsingular hypersurface of degree $m+1$ in $\mathbb{CP}^m$ is a Calabi–Yau manifold. By replacing $\mathbb{CP}^m$ by a weighted projective space, or more generally by a compact toric variety, one can construct many examples of Calabi–Yau manifolds. Finally, §7.7 explains the deformation theory of the complex manifolds underlying Calabi–Yau manifolds.

For further reading, two books primarily about Calabi–Yau manifolds are Gross, Huybrechts and Joyce [138], and Hübsch [167], which is now rather out of date. Some books on mirror symmetry, which also cover Calabi–Yau manifolds, are Cox and Katz [80] and Voisin [335] (both written from a mathematical point of view), and the monumental Hori et al. [166] (which mixes mathematics and string theory).

7.1 Ricci-flat Kähler manifolds and Calabi–Yau manifolds

Identify $\mathbb{R}^{2m}$ with $\mathbb{C}^m$ with complex coordinates $(z_1, \ldots, z_m)$, and define a metric g, a real 2-form ω and a complex m-form θ on $\mathbb{C}^m$ by

$$g = |\mathrm{d}z_1|^2 + \cdots + |\mathrm{d}z_m|^2, \quad \omega = \tfrac{i}{2}(\mathrm{d}z_1 \wedge \mathrm{d}\bar{z}_1 + \cdots + \mathrm{d}z_m \wedge \mathrm{d}\bar{z}_m),$$
$$\text{and} \quad \theta = \mathrm{d}z_1 \wedge \cdots \wedge \mathrm{d}z_m. \tag{7.1}$$

The subgroup of $\mathrm{GL}(2m, \mathbb{R})$ preserving g, ω and θ is $\mathrm{SU}(m)$. Therefore, by the results of §2.5.1, every Riemannian manifold (X, g) with holonomy $\mathrm{SU}(m)$ admits natural forms ω and θ, constant under the Levi-Civita connection, such that g, ω and θ can be written in the form (7.1) at each point of X.

There is a unique complex structure J on X such that $\omega_{ac} = J_a^b g_{bc}$. Then g is a Kähler metric on (X, J), with Kähler form ω, as in Chapter 5. Also, θ is a holomorphic $(m, 0)$-form with respect to J, so we call θ the *holomorphic volume form* on X. Thus, every Riemannian manifold with holonomy $\mathrm{SU}(m)$ is a Kähler manifold with a constant holomorphic volume form. Conversely, if (X, J, g) is Kähler and θ is a holomorphic volume form on X with $\nabla\theta = 0$, then $\mathrm{Hol}(g) \subseteq \mathrm{SU}(m)$.

Now $\Lambda^{m,0}$ is the *canonical bundle* K_X of X, a holomorphic line bundle, and a holomorphic volume form θ is a nonvanishing holomorphic section of K_X. Such a θ exists if and only if K_X is *trivial*, that is, if it is isomorphic to the trivial holomorphic line bundle $X \times \mathbb{C}$ over X. Thus, if (X, J, g) is Kähler and $\mathrm{Hol}(g) \subseteq \mathrm{SU}(m)$, then K_X is trivial.

Also, since the *first Chern class* $c_1(X)$ of X is a characteristic class of K_X, we see that $c_1(X) = 0$ in $H^2(X, \mathbb{Z})$. Hence, any complex manifold X admitting Kähler metrics with holonomy in $\mathrm{SU}(m)$ must have $c_1(X) = 0$. Another important property of metrics with holonomy $\mathrm{SU}(m)$ is that they are *Ricci-flat*.

Proposition 7.1.1 *Let (X, J, g) be a Kähler manifold. Then $\mathrm{Hol}^0(g) \subset \mathrm{SU}(m)$ if and only if g is Ricci-flat.*

Proof As g is Kähler, the Levi-Civita connection ∇ of g induces a connection ∇^K on $K_X = \Lambda^{m,0}X$, which has holonomy group $\mathrm{Hol}(\nabla^K) \subseteq \mathrm{U}(1)$. Now if $A \in \mathrm{U}(m)$

acts on $\mathbb{C}^m$, the induced map on $\Lambda^{m,0}\mathbb{C}^m$ is multiplication by $\det A$. Therefore, the relationship between $\mathrm{Hol}^0(\nabla) = \mathrm{Hol}^0(g)$ and $\mathrm{Hol}^0(\nabla^\kappa)$ is given by $\mathrm{Hol}^0(\nabla^\kappa) = \det(\mathrm{Hol}^0(\nabla))$, where $\det : U(m) \to U(1)$ is the determinant map. It follows that $\mathrm{Hol}^0(\nabla^\kappa) = \{1\}$ if and only if $\mathrm{Hol}^0(g) \subseteq SU(m)$.

But by the Frobenius Theorem, $\mathrm{Hol}^0(\nabla^\kappa) = \{1\}$ if and only if ∇^κ is *flat*, that is, if the curvature of ∇^κ is identically zero. Since the gauge group of ∇^κ is $U(1)$, the curvature of ∇^κ is just a closed 2-form. It can be shown using (5.7) that this 2-form is exactly the *Ricci form* ρ of g. Thus, we have shown that $\mathrm{Hol}^0(g) \subseteq SU(m)$ if and only if $\rho \equiv 0$, that is, if and only if g is Ricci-flat. $\qquad\square$

One important consequence of the proof of the Calabi Conjecture in Chapter 6, is the existence of families of Ricci-flat Kähler metrics on suitable compact complex manifolds. As we explained in the introduction to Chapter 6, if X is Kähler and $c_1(X) = 0$ in $H^2(X, \mathbb{R})$ then the Calabi Conjecture yields a Kähler metric g' on X with Ricci form $\rho' = 0$, so that g' is Ricci-flat. We state this in the following theorem.

Theorem 7.1.2 *Let (X, J) be a compact complex manifold admitting Kähler metrics, with $c_1(X) = 0$. Then there is a unique Ricci-flat Kähler metric in each Kähler class on X. The Ricci-flat Kähler metrics on X form a smooth family of dimension $h^{1,1}(X)$, isomorphic to the Kähler cone $\mathcal{K}_X$ of X.*

Now, a generic Kähler metric g on a complex m-manifold X has holonomy group $\mathrm{Hol}(g) = U(m)$. But by Proposition 7.1.1 $\mathrm{Hol}^0(g) \subseteq SU(m)$ if and only if g is Ricci-flat. Thus Theorem 7.1.2 constructs metrics with special holonomy on compact manifolds. Our next three results pin down their holonomy groups more precisely.

By Theorem 3.5.5 any compact Ricci-flat Riemannian manifold (X, g) admits a finite cover isometric to $(T^n \times N, g' \times g'')$, where (T^n, g') is a flat Riemannian torus, and (N, g'') a compact, simply-connected Ricci-flat Riemannian manifold. If g is also Kähler then so are (T^n, g') and (N, g''), so that $n = 2l$. But N is simply-connected, and g'' is complete as N is compact. Therefore Theorem 3.2.7 applies, and we may write (N, g'') as a Riemannian product $(X_1 \times \cdots \times X_k, g_1 \times \cdots \times g_k)$, where each (X_j, g_j) is irreducible. If (N, g'') is Kähler then the factors (X_j, g_j) are Kähler. This gives:

Proposition 7.1.3 *Each compact Ricci-flat Kähler manifold (X, J, g) admits a finite cover isomorphic to a product Kähler manifold*

$$(T^{2l} \times X_1 \times \cdots \times X_k, J_0 \times \cdots \times J_k, g_0 \times \cdots \times g_k),$$

where (T^{2l}, J_0, g_0) is a flat Kähler torus and (X_j, J_j, g_j) a compact, simply-connected, irreducible, Ricci-flat Kähler manifold for $j = 1, \ldots, k$.

Now let (X, J, g) be a compact, simply-connected, irreducible, Ricci-flat Kähler manifold. As g is Ricci-flat, but not flat, it must be nonsymmetric. So X is simply-connected and g is nonsymmetric and irreducible, and Theorem 3.4.1 shows that $\mathrm{Hol}(g)$ lies on Berger's list. Because g is also Kähler, we see that $\mathrm{Hol}(g)$ must be one of $U(m)$, $SU(m)$ or $Sp(m/2)$. But $\mathrm{Hol}^0(g) \subseteq SU(m)$ as g is Ricci-flat, which excludes $U(m)$, so $\mathrm{Hol}(g) = SU(m)$ or $Sp(m/2)$.

On the other hand, if (X, J, g) is compact and Kähler and $\mathrm{Hol}(g)$ is $SU(m)$ or $Sp(m/2)$ then g is Ricci-flat as $\mathrm{Hol}^0(g) \subseteq SU(m)$, and irreducible as $\mathrm{Hol}^0(g)$ does not split. By the previous proposition X has a simply-connected finite cover, and so $\pi_1(X)$ is finite. Thus we have proved:

Proposition 7.1.4 *Let (X, J, g) be a compact, simply-connected, irreducible, Ricci-flat Kähler manifold, of dimension m. Then either $m \geqslant 2$ and $\mathrm{Hol}(g) = \mathrm{SU}(m)$, or $m \geqslant 4$ is even and $\mathrm{Hol}(g) = \mathrm{Sp}(m/2)$. Conversely, if (X, J, g) is a compact Kähler manifold and $\mathrm{Hol}(g)$ is $\mathrm{SU}(m)$ or $\mathrm{Sp}(m/2)$, then g is Ricci-flat and irreducible and X has finite fundamental group.*

Next we show that any closed $(p, 0)$-form on a compact Ricci-flat manifold is constant. The proof involves the 'Bochner argument' used to prove Theorem 3.5.4.

Proposition 7.1.5 *Suppose (X, J, g) is a compact Ricci-flat Kähler manifold and ξ a smooth $(p, 0)$-form on X. Then $\nabla \xi = 0$ if and only if $\mathrm{d}\xi = 0$ if and only if $\bar{\partial}\xi = 0$, where ∇ is the Levi-Civita connection of g. Hence $H^{p,0}(X)$ is isomorphic to the vector space of constant $(p, 0)$-forms ξ on X.*

Proof For the first part, we shall show $\nabla\xi = 0$ implies $\mathrm{d}\xi = 0$ implies $\bar{\partial}\xi = 0$ implies $\nabla\xi = 0$. The first two implications are obvious. For the third, as $\bar{\partial}^*$ is zero on $(p, 0)$-forms, $\bar{\partial}\xi = 0$ implies $\Delta_{\bar{\partial}}\xi = 0$, so $(\mathrm{dd}^* + \mathrm{d}^*\mathrm{d})\xi = 0$ by (5.9). Taking the inner product with ξ and integrating by parts then shows that $\mathrm{d}\xi = \mathrm{d}^*\xi = 0$, as X is compact, and thus ξ lies in the space $\mathscr{H}^{p,0}$ of harmonic $(p, 0)$-forms defined in §5.7.1, which is isomorphic to $H^{p,0}(X)$.

Now the Weitzenbock formula for p-forms (3.13) shows that $(\mathrm{dd}^* + \mathrm{d}^*\mathrm{d})\xi = \nabla^*\nabla\xi - 2\tilde{R}(\xi)$, where $\tilde{R}$ is defined in (3.14) and depends on the Riemann curvature $R^a{}_{bcd}$ of g. Combining the symmetries (3.2) and (3.3) of $R^a{}_{bcd}$ with the decomposition (5.5) of the curvature tensor of a Kähler metric, we may show that $R^{\alpha}{}_{\beta\bar{\gamma}\delta} = R^{\alpha}{}_{\delta\bar{\gamma}\beta}$, in the notation of §5.2.1.

Since ξ is a $(p, 0)$-form, this implies that the first term on the right hand side of (3.14) is zero. However, the second term depends on the Ricci curvature, and g is Ricci-flat. Thus $\tilde{R}(\xi) = 0$, so that $(\mathrm{dd}^* + \mathrm{d}^*\mathrm{d})\xi = \nabla^*\nabla\xi$ by the Weitzenbock formula. So $\nabla^*\nabla\xi = 0$, and integrating by parts shows that $\nabla\xi = 0$, as X is compact. This proves that if $\bar{\partial}\xi = 0$ then $\nabla\xi = 0$. The last part is immediate. $\square$

Applying the proposition to $(m, 0)$-forms when $\dim X = m$, we get:

Corollary 7.1.6 *Suppose (X, J, g) is a compact Ricci-flat Kähler m-manifold. Then $\mathrm{Hol}(g) \subseteq \mathrm{SU}(m)$ if and only if the canonical bundle K_X of X is trivial.*

Now as we explained in §2.5.1, the tensors constant under a connection ∇ are entirely determined by the holonomy group $\mathrm{Hol}(\nabla)$. Combining this with Proposition 7.1.5, we see that if (X, J, g) is a compact Ricci-flat Kähler manifold, then $H^{p,0}(X)$ is determined by $\mathrm{Hol}(g)$. That is, $\mathrm{Hol}(g)$ is (up to conjugation) a subgroup of $\mathrm{U}(m)$, which acts naturally on $\Lambda^{p,0}\mathbb{C}^m$, and $H^{p,0}(X)$ is isomorphic to the subspace of $\Lambda^{p,0}\mathbb{C}^m$ invariant under $\mathrm{Hol}(g)$. In particular, the group $\mathrm{SU}(m)$ fixes $\Lambda^{0,0}\mathbb{C}^m$ and $\Lambda^{m,0}\mathbb{C}^m$, but fixes no nonzero elements of $\Lambda^{p,0}\mathbb{C}^m$ for $0 < p < m$. Hence if $\mathrm{Hol}(g) = \mathrm{SU}(m)$ then $H^{p,0}(X)$ is $\mathbb{C}$ if $p = 0, m$ and zero otherwise. We have proved:

Proposition 7.1.7 *Suppose (X, J, g) is a compact Kähler m-manifold with holonomy $\mathrm{SU}(m)$ and Hodge numbers $h^{p,q}$. Then $h^{0,0} = h^{m,0} = 1$ and $h^{p,0} = 0$ for $p \neq 0, m$.*

Compact Kähler manifolds with holonomy $\mathrm{SU}(m)$ for $m \geqslant 3$ are *algebraic*.

Theorem 7.1.8 *Let* (X, J, g) *be a compact Kähler m-manifold with* $\mathrm{Hol}(g) = \mathrm{SU}(m)$ *for* $m \geqslant 3$. *Then* (X, J) *is projective, that is, it is isomorphic to a complex submanifold of* $\mathbb{CP}^N$ *for some* $N > m$, *and is an algebraic variety.*

Proof From Proposition 7.1.7 we see that $h^{2,0}(X) = h^{0,2}(X) = 0$. Thus $H^{1,1}(X) = H^2(X, \mathbb{C})$, and $H^{1,1}(X) \cap H^2(X, \mathbb{R}) = H^2(X, \mathbb{R})$. Now the Kähler cone $\mathcal{K}_X$ of X is a nonempty open subset of $H^{1,1}(X) \cap H^2(X, \mathbb{R})$, so that $\mathcal{K}_X$ is open in $H^2(X, \mathbb{R})$. But $H^2(X, \mathbb{Q})$ is dense in $H^2(X, \mathbb{R})$, so $\mathcal{K}_X \cap H^2(X, \mathbb{Q})$ is nonempty. Choose α in $\mathcal{K}_X \cap H^2(X, \mathbb{Q})$. Let k be a positive integer such that $k\alpha \in H^2(X, \mathbb{Z})$. Then $k\alpha \in \mathcal{K}_X$, as $\mathcal{K}_X$ is a cone, so there exists a closed positive (1,1)-form β on X such that $[\beta] = k\alpha$.

Using the theory of holomorphic line bundles explained in §5.10, one can show that there exists a holomorphic line bundle L over X with $c_1(L) = k\alpha$. Since $c_1(L)$ is represented by a positive (1,1)-form β, the Kodaira Embedding Theorem [132, p. 181] shows that L is *ample*. That is, for large n the holomorphic sections of L^n embed X as a complex submanifold of $\mathbb{CP}^N$. But by Chow's Theorem [132, p. 167], any complex submanifold of $\mathbb{CP}^N$ is an algebraic variety. $\qquad\qquad\square$

Wilson [339, 340] studies the *Kähler cone* of compact Kähler manifolds with holonomy $\mathrm{SU}(3)$. We summarize his results.

Theorem 7.1.9 *Let* (X, J) *be a compact complex 3-manifold admitting Kähler metrics with holonomy* $\mathrm{SU}(3)$. *Define the cubic cone* $W = \{D \in H^2(X, \mathbb{R}) : D^3 = 0\}$. *Then the closure* $\bar{\mathcal{K}}_X$ *in* $H^2(X, \mathbb{R})$ *of the Kähler cone* $\mathcal{K}_X$ *of* X *is locally rational polyhedral away from* W. *Furthermore, the codimension one faces of* $\bar{\mathcal{K}}_X$ *correspond to primitive birational contractions* $\pi : X \to Y$ *of* X.

Wilson also shows that such contractions $\pi : X \to Y$ have a coarse classification into Types I, II and III, depending on which curves or surfaces in X are contracted. He uses his results to show that $\bar{\mathcal{K}}_X$ is invariant under small deformations of X unless X contains a quasi-ruled elliptic surface E.

We now come to define Calabi–Yau manifolds. Here are five *inequivalent* definitions the author found in the literature:

- a compact complex manifold (X, J) with $c_1(X) = 0$ admitting Kähler metrics,
- a projective manifold X with $c_1(X) = 0$,
- a compact Ricci-flat Kähler manifold,
- a compact Kähler manifold (X, J, g) with $\mathrm{Hol}(g) \subseteq \mathrm{SU}(m)$, and
- a compact Kähler manifold (X, J, g) with $\mathrm{Hol}(g) = \mathrm{SU}(m)$.

Manifolds such as Enriques surfaces or complex tori satisfy some of these definitions, but not others. We adopt the following sixth definition. It is unusual in regarding a choice of holomorphic volume form θ as part of the structure. The reason for this is that for the special Lagrangian geometry of Chapters 8 and 9 we will always need a choice of θ.

Definition 7.1.10 Let $m \geqslant 2$. A *Calabi–Yau m-fold* is a quadruple (X, J, g, θ) such that (X, J) is a compact m-dimensional complex manifold, g a Kähler metric on (X, J)

with Kähler form ω, and θ a nonvanishing holomorphic $(m, 0)$-form on X called the *holomorphic volume form*, which satisfies

$$\omega^m / m! = (-1)^{m(m-1)/2} (i/2)^m \theta \wedge \bar{\theta}. \tag{7.2}$$

The constant factor in (7.2) is chosen to make $\mathrm{Re}\,\theta$ a *calibration*.

Here is the point of this. If (X, J, g) is a Kähler manifold and θ a nonvanishing holomorphic volume form on X then K_X is trivial, as it has a nonvanishing section θ, so $c_1(X) = 0$. Thus the Ricci form ρ of g is exact, and using (5.6) and (5.7) we can show that ρ is given explicitly by $\rho = \mathrm{dd}^c \log \det |\theta|$, where $|\theta|$ is computed using g. If ω, θ satisfy (7.2) then an easy calculation gives $|\theta| \equiv 2^{m/2}$, so $\rho \equiv 0$, and g is Ricci-flat. Proposition 7.1.5 and Corollary 7.1.6 then show that $\nabla\theta = 0$ and $\mathrm{Hol}(g) \subseteq \mathrm{SU}(m)$.

If (X, J, g) is a Kähler manifold with $\mathrm{Hol}(g) \subseteq \mathrm{SU}(m)$, as some authors define Calabi–Yau manifolds, then K_X is trivial and the vector space of constant $(m, 0)$-forms is isomorphic to $\mathbb{C}$. So we can choose a constant $(m, 0)$-form θ with $|\theta| \equiv 2^{m/2}$, which then satisfies (7.2). It is unique up to change of phase, $\theta \mapsto e^{i\psi}\theta$.

The rest of the chapter will focus on Calabi–Yau m-folds (X, J, g, θ) with $\mathrm{Hol}(g) = \mathrm{SU}(m)$, for $m \geqslant 3$. Then (X, J) is projective by Theorem 7.1.8, with K_X trivial. Thus we can study the underlying complex manifold (X, J) and the holomorphic volume form θ using *complex algebraic geometry*. We can also use complex algebraic geometry to describe the *Kähler cone* $\mathcal{K}_X$ of X, as in Theorem 7.1.9. Then Theorem 7.1.2 shows that the family of Calabi–Yau metrics g on (X, J) is isomorphic to $\mathcal{K}_X$. So the chapter will be mostly algebraic geometry.

7.2 Crepant resolutions, small resolutions, and flops

Let X be a singular complex algebraic variety, and $(\tilde{X}, \pi)$ a resolution of X, as in §5.9.1. We call $\tilde{X}$ a *crepant resolution* of X if $\pi^*(K_X) \cong K_{\tilde{X}}$, where K_X and $K_{\tilde{X}}$ are the *canonical bundles* of X and $\tilde{X}$. However, this statement is not as simple as it seems. What do we mean by a holomorphic line bundle such as K_X over a singular variety? How do we define the pull-back $\pi^*(K_X)$, and is it really a holomorphic line bundle on $\tilde{X}$, or some more singular object?

There is an extensive and complicated theory of line bundles over singular varieties, involving Weil divisors, Cartier divisors, and so on. But as this is not an algebraic geometry text we will bypass most of these ideas to simplify things. We shall interpret holomorphic line bundles on singular varieties in terms of *invertible sheaves*.

Definition 7.2.1 Let X be a complex algebraic variety, and $\mathcal{O}$ the sheaf of regular functions on X. An *invertible sheaf* $\mathcal{F}$ on X is a sheaf of $\mathcal{O}$-modules that is locally isomorphic to $\mathcal{O}$.

If X is nonsingular, L is a holomorphic line bundle over X, and $\mathcal{O}(L)$ is the sheaf of holomorphic sections of L, then $\mathcal{O}(L)$ is an invertible sheaf. Conversely, each invertible sheaf over a nonsingular variety comes from a line bundle. Thus, invertible sheaves are a natural generalization to singular varieties of the idea of holomorphic line bundle.

A useful property of invertible sheaves is that if $\mathcal{F}$ is an invertible sheaf over Y and $\pi : X \to Y$ is a morphism, then by [149, §II.5] one can define the *inverse image sheaf*

$\pi^*(\mathscr{F})$, which is an invertible sheaf on X. Hence, if X is a singular variety with $\mathcal{O}(K_X)$ invertible, and $(\tilde{X}, \pi)$ is a resolution of X, then $\pi^*(K_X)$ is a well-defined holomorphic line bundle on $\tilde{X}$.

Definition 7.2.2 Let X be a complex algebraic variety, and suppose that the sheaf of regular sections $\mathcal{O}(K_X)$ of K_X is an invertible sheaf. Let $(\tilde{X}, \pi)$ be a *resolution* of X. Then the inverse image sheaf $\pi^*(\mathcal{O}(K_X))$ is an invertible sheaf on $\tilde{X}$, and so represents a holomorphic line bundle on $\tilde{X}$, written $\pi^*(K_X)$.

We say that a prime divisor E in $\tilde{X}$ is *exceptional* if $\pi(E)$ has codimension at least 2 in X. Let the exceptional divisors be $E_1, \ldots, E_n$. Then there exist unique integers $a_1, \ldots, a_n$ such that in divisors on $\tilde{X}$, we have

$$[K_{\tilde{X}}] = [\pi^*(K_X)] + \textstyle\sum_{i=1}^n a_i E_i.$$

We call the divisor $\sum_{i=1}^n a_i E_i$ on $\tilde{X}$ the *discrepancy* of $(\tilde{X}, \pi)$.

(i) If $a_i \geqslant 0$ for all i, we say X has *canonical singularities*.

(ii) if $a_i > 0$ for all i, we say X has *terminal singularities*.

(iii) If $a_i = 0$ for all i, we say $\pi : \tilde{X} \to X$ is a *minimal resolution* or *crepant resolution*, because it has zero discrepancy. Then $K_{\tilde{X}} \cong \pi^*(K_X)$.

Canonical and terminal singularities were defined by Reid [284, 285]. It can be shown that their definition is independent of the choice of resolution $\pi : \tilde{X} \to X$. They are of interest because they are the singularities which occur in the pluricanonical models of varieties of general type.

Next we consider another special class of resolutions, called *small resolutions*.

Definition 7.2.3 Let X be a singular algebraic variety, and let $(\tilde{X}, \pi)$ be a resolution of X. We call $\tilde{X}$ a *small resolution* if it has no exceptional divisors. Here is another way to say this. Define the *exceptional set* E of the resolution to be the set of points $e \in \tilde{X}$ such that $\dim[\pi^{-1}(\pi(e))] \geqslant 1$. Usually $E = \pi^{-1}(S)$, where S is the singular set of X. Then $\tilde{X}$ is a small resolution if E is of codimension at least 2 in $\tilde{X}$.

Clearly, any small resolution is a *crepant resolution*, as the condition in Definition 7.2.2 that $a_i = 0$ for all exceptional prime divisors E_i holds trivially because there are no E_i. For the same reason, any singular manifold X which admits a small resolution has *terminal singularities*. Conversely, if X has terminal singularities then any crepant resolution of X must be a small resolution.

Because blowing up always introduces new exceptional divisors, small resolutions cannot be constructed by the usual strategy of blowing up singular points. This makes small resolutions difficult to find, for general singularities. One common situation in which small resolutions arise is if X is a complex 3-fold with isolated singularities, and $\tilde{X}$ is a resolution of X in which each singular point is replaced by a rational curve $\mathbb{CP}^1$, or a finite union of rational curves. Such singularities are called *double points*, and are studied in [112]. Here is an example.

Example 7.2.4 Define a hypersurface X in $\mathbb{C}^4$ by

$$X = \{(z_1, \ldots, z_4) \in \mathbb{C}^4 : z_1 z_2 = z_3 z_4\}.$$

Then X has a single, isolated singular point at 0, called an *ordinary double point*, or *node*. Define $\tilde{X}_1 \subset \mathbb{C}^4 \times \mathbb{CP}^1$ by

$$\tilde{X}_1 = \left\{ \left((z_1, \ldots, z_4), [x_1, x_2] \right) \in \mathbb{C}^4 \times \mathbb{CP}^1 : z_1 x_2 = z_4 x_1, \quad z_3 x_2 = z_2 x_1 \right\},$$

and define $\pi_1 : \tilde{X}_1 \to \mathbb{C}^4$ by $\pi_1 : \left((z_1, \ldots, z_4), [x_1, x_2] \right) \mapsto (z_1, \ldots, z_4)$. Now since x_1, x_2 are not both zero, the equations $z_1 x_2 = z_4 x_1, z_3 x_2 = z_2 x_1$ together imply that $z_1 z_2 = z_3 z_4$. Therefore π_1 maps $\tilde{X}_1$ to X. It is easy to see that π_1 is surjective, that $\pi_1^{-1}(0) = \mathbb{CP}^1$, and that π_1 is injective except at $0 \in X$. So $\pi_1 : \tilde{X}_1 \to X$ is a *small resolution* of X.

Similarly, define $\tilde{X}_2 \subset \mathbb{C}^4 \times \mathbb{CP}^1$ by

$$\tilde{X}_2 = \left\{ \left((z_1, \ldots, z_4), [y_1, y_2] \right) \in \mathbb{C}^4 \times \mathbb{CP}^1 : z_1 y_2 = z_3 y_1, \quad z_4 y_2 = z_2 y_1 \right\},$$

and define $\pi_2 : \tilde{X}_2 \to X$ as above. Then $\pi_2 : \tilde{X}_2 \to X$ is also a small resolution of X. It can be shown (using toric geometry) that $\tilde{X}_1$ and $\tilde{X}_2$ are the only two crepant resolutions of X, and that they are topologically distinct. Thus, a singularity can admit more than one crepant resolution.

This example is the basis of an important construction in the algebraic geometry of 3-folds, called a *flop*. Suppose that Y is a 3-fold with a single node. Then we can resolve the singularity with a small resolution in two different ways as above, to get two different nonsingular 3-folds Y_1, Y_2, each containing a rational curve $\mathbb{CP}^1$. Conversely, if we have a nonsingular 3-fold Y_1 containing a suitable $\mathbb{CP}^1$, with normal bundle $\mathcal{O}(-1) \oplus \mathcal{O}(-1)$, we can contract the $\mathbb{CP}^1$ to a point to get a 3-fold Y with a single node, and then resolve it in the other way to get a nonsingular 3-fold Y_2 that is different from Y_1.

This process of passing from Y_1 to Y_2 is called a *flop*. Now $c_1(Y_1) = 0$ if and only if $c_1(Y_2) = 0$. Thus, if Y_1 is a Calabi–Yau 3-fold and Y_2 is a flop of Y_1, then Y_2 will also be a Calabi–Yau 3-fold, provided that it is Kähler (which is not always the case). So given one Calabi–Yau 3-fold, one can often construct many more by flopping.

7.3 Crepant resolutions of quotient singularities

Let G be a nontrivial finite subgroup of $\mathrm{GL}(m, \mathbb{C})$. Then G acts on $\mathbb{C}^m$, and the quotient $\mathbb{C}^m/G$ is a singular complex manifold called a *quotient singularity*. It can be made into an algebraic variety, using the algebra of G-invariant polynomials on $\mathbb{C}^m$. If $x \in \mathbb{C}^m$, then xG is a singular point of $\mathbb{C}^m/G$ if and only if the stabilizer subgroup $\mathrm{Stab}(x) = \{\gamma \in G : \gamma \cdot x = x\}$ of x is nontrivial. Thus 0 is always a singular point of $\mathbb{C}^m/G$. If G acts freely on $\mathbb{C}^m \setminus \{0\}$, then 0 is the unique singular point of $\mathbb{C}^m/G$.

For reasons to be explained in §7.5, we are interested in crepant resolutions of quotient singularities $\mathbb{C}^m/G$. Now $\gamma \in G$ acts on $\Lambda^{m,0}\mathbb{C}^m$ by multiplication by $\det \gamma$. Thus the canonical bundle of $\mathbb{C}^m/G$ is only well-defined at 0 if $\det \gamma = 1$ for all $\gamma \in G$, that is, if $G \subset \mathrm{SL}(m, \mathbb{C})$. It follows that $\mathbb{C}^m/G$ can have a crepant resolution only if $G \subset \mathrm{SL}(m, \mathbb{C})$.

Therefore we will restrict our attention to finite subgroups $G \subset \mathrm{SL}(m, \mathbb{C})$. Since any such G is conjugate to a subgroup of $\mathrm{SU}(m)$, we may take $G \subset \mathrm{SU}(m)$ if we wish. In §7.3.1 we discuss crepant resolutions of $\mathbb{C}^2/G$. Then §7.3.2 summarizes what is known about the existence of crepant resolutions of $\mathbb{C}^m/G$ for $m \geqslant 3$ and $G \subset \mathrm{SL}(m, \mathbb{C})$ a finite subgroup. Finally §7.3.3 explains the McKay correspondence, which relates the topology of a crepant resolution of $\mathbb{C}^m/G$ to the group theory of G.

7.3.1   The Kleinian singularities $\mathbb{C}^2/G$ and their resolutions

The quotient singularities $\mathbb{C}^2/G$, for G a finite subgroup of $\mathrm{SU}(2)$, were first classified by Klein in 1884 and are called *Kleinian singularities;* they are also called *Du Val surface singularities*, or *rational double points*. The theory of these singularities and their resolutions (see for instance Slodowy [315]) is very rich, and has many connections to other areas of mathematics. There is a 1-1 correspondence between nontrivial finite subgroups $G \subset \mathrm{SU}(2)$ and the *Dynkin diagrams* of type A_r $(r \geqslant 1)$, D_r $(r \geqslant 4)$, E_6, E_7 and E_8. Let Γ be the Dynkin diagram associated to G. These diagrams appear in the classification of Lie groups, and each one corresponds to a unique compact, simple Lie group; they are the set of such diagrams containing no double or triple edges.

Each singularity $\mathbb{C}^2/G$ admits a unique crepant resolution (X, π). The preimage $\pi^{-1}(0)$ of the singular point is a union of a finite number of *rational curves* in X. These curves correspond naturally to the vertices of Γ. They all have self-intersection -2, and two curves intersect transversely at one point if and only if the corresponding vertices are joined by an edge in the diagram; otherwise the curves do not intersect.

These curves give a basis for the homology group $H_2(X, \mathbb{Z})$, which may be identified with the *root lattice* of the diagram, and the intersection form with respect to this basis is the negative of the Cartan matrix of Γ. Define Δ to be $\{\delta \in H_2(X, \mathbb{Z}) : \delta \cdot \delta = -2\}$. Then Δ is the *set of roots* of the diagram. There are also 1-1 correspondences between the curves and the nonidentity conjugacy classes in G, and also the nontrivial representations of G; it makes sense to regard the nonidentity conjugacy classes as a basis for $H_2(X, \mathbb{Z})$, and the nontrivial representations as a basis for $H^2(X, \mathbb{Z})$.

This correspondence between the Kleinian singularities $\mathbb{C}^2/G$, Dynkin diagrams, and other areas of mathematics became known as the *McKay correspondence*, after John McKay, who pointed it out [258].

7.3.2   Crepant resolutions of $\mathbb{C}^m/G$ for $m \geqslant 3$

We start with a brief introduction to *toric geometry*. Let $\mathbb{C}^*$ be $\mathbb{C} \setminus \{0\}$, regarded as a complex Lie group, with multiplication as the group operation. A *toric variety* is a normal complex algebraic variety X of dimension m, equipped with a holomorphic action of $(\mathbb{C}^*)^m$, and with a dense open subset $T \subset X$ upon which $(\mathbb{C}^*)^m$ acts freely and transitively. For an introduction to toric varieties, see Fulton [119] or Oda [270]. Toric geometry is the geometry of toric varieties.

Each toric variety is the union of a finite number of orbits of $(\mathbb{C}^*)^m$. All of the information about these orbits, and the way they fit together, is represented in a finite collection of combinatorial data called a *fan* [119, §1.4], and the toric variety can be reconstructed from its fan. The importance of toric varieties is that they are very well understood, they are easy to work with and to compute invariants for, and they provide a large family of examples of varieties that can be used to test ideas on.

Now suppose $G \subset \mathrm{SL}(m, \mathbb{C})$ is a finite *abelian* group. Then we can choose a coordinate system $(z_1, \ldots, z_m)$ on $\mathbb{C}^m$ such that all elements of G are represented by *diagonal* matrices. The group of all invertible diagonal matrices is isomorphic to $(\mathbb{C}^*)^m$ and commutes with G, and so $(\mathbb{C}^*)^m$ acts on $\mathbb{C}^m/G$. This makes $\mathbb{C}^m/G$ into a toric variety, and its structure is described by a fan.

Any resolution of $\mathbb{C}^m/G$ that is also a toric variety is described by a *subdivision* of the fan of $\mathbb{C}^m/G$. There is a simple condition on this subdivision that determines whether or not the resolution is crepant. Moreover, any crepant resolution of $\mathbb{C}^m/G$ must be a toric variety. Thus, if $G \subset \mathrm{SL}(m, \mathbb{C})$ is a finite abelian group, then toric geometry gives a simple method for finding all the crepant resolutions of $\mathbb{C}^m/G$.

This method was described independently by Roan [288], and Markushevich [253, App.]. Both of them proved [288, p. 528], [253, p. 273] that if $m = 2$ or 3, then a toric crepant resolution of $\mathbb{C}^m/G$ always exists. For examples of such resolutions, see [253, p. 269–271]. When $m = 2$ the resolution is unique, but for $m \geqslant 3$ there can be finitely many different crepant resolutions of $\mathbb{C}^m/G$. If $G \subset \mathrm{SL}(m, \mathbb{C})$ is abelian and $m \geqslant 4$ then $\mathbb{C}^m/G$ may or may not admit a crepant resolution, depending on the fan of $\mathbb{C}^m/G$. For example, we will show below that $\mathbb{C}^4/\{\pm 1\}$ has no crepant resolution.

For more general subgroups of $\mathrm{SL}(3, \mathbb{C})$, Roan [289, Th. 1] proves

Theorem 7.3.1 *Let G be any finite subgroup of $\mathrm{SL}(3, \mathbb{C})$. Then the quotient singularity $\mathbb{C}^3/G$ admits a crepant resolution.*

Roan's proof is by explicit construction, using the classification of finite subgroups of $\mathrm{SU}(3)$, and it relies on previous work by Ito [174, 175] and Markushevich. In dimension four and above, singularities are less well understood. However, there are simple criteria to determine when a quotient singularity $\mathbb{C}^m/G$ is terminal, which we now give. To state them we first define the *age grading* on G, following Reid [286, §2].

Definition 7.3.2 Let $G \subset \mathrm{SL}(m, \mathbb{C})$ be a finite group. Then each $\gamma \in G$ has m eigenvalues $e^{2\pi i a_1}, \ldots, e^{2\pi i a_m}$, where $a_1, \ldots, a_m \in [0, 1)$ are uniquely defined up to order. Define the *age* of γ to be $\mathrm{age}(\gamma) = a_1 + \cdots + a_m$. Then $\mathrm{age}(\gamma)$ is well-defined with $0 \leqslant \mathrm{age}(\gamma) < m$. Since $\det(\gamma) = 1$, we see that $e^{2\pi i\, \mathrm{age}(\gamma)} = 1$, so $\mathrm{age}(\gamma)$ is an integer. Thus $\mathrm{age}(\gamma) \in \{0, 1, \ldots, m-1\}$, and we have defined a mapping $\mathrm{age} : G \to \{0, 1, \ldots, m-1\}$.

The next result (see [269, Th. 2.3]) is due to Reid.

Theorem 7.3.3 *Let $G \subset \mathrm{SL}(m, \mathbb{C})$ be a finite subgroup. Then $\mathbb{C}^m/G$ is a terminal singularity if and only if $\mathrm{age}(\gamma) \neq 1$ for all $\gamma \in G$.*

But terminal quotient singularities have no crepant resolutions:

Proposition 7.3.4 *Let G be a nontrivial subgroup of $\mathrm{SL}(m, \mathbb{C})$, and suppose $\mathbb{C}^m/G$ is a terminal singularity. Then $\mathbb{C}^m/G$ admits no crepant resolution.*

Proof From §7.2, any crepant resolution of a terminal singularity must be a *small resolution*. Thus it is enough to show that $\mathbb{C}^m/G$ admits no small resolutions. Suppose first that $\mathbb{C}^m/G$ has only an isolated singularity at 0, and that (X, π) is a small resolution of $\mathbb{C}^m/G$. Then $\pi^{-1}(0)$ is a finite union of compact algebraic varieties of dimension at least one. By constructing complex curves in $\pi^{-1}(0)$ one can show that $b_2(X) \geqslant 1$.

Regard X as a compact manifold with boundary $\mathcal{S}^{2m-1}/G$. Poincaré duality for manifolds with boundary gives $b_2(X) = b_{2m-2}(X)$, and so $b_{2m-2}(X) \geqslant 1$. But X contracts onto $\pi^{-1}(0)$, and $\dim \pi^{-1}(0) < m-1$ since X is a small resolution, and together these imply that $b_{2m-2}(X) = 0$, a contradiction. If the singularities of $\mathbb{C}^m/G$ are not isolated, then the generic singular point looks locally like $(\mathbb{C}^k/H) \times \mathbb{C}^{m-k}$, where $H \subset \mathrm{SL}(k, \mathbb{C})$ and $\mathbb{C}^k/H$ has an isolated singularity at 0. Using this we reduce to the previous case. $\square$

By combining Theorem 7.3.3 and Proposition 7.3.4 we can show some singularities $\mathbb{C}^m/G$ with $G \subset \mathrm{SL}(m, \mathbb{C})$ have no crepant resolutions, as in the next example.

Example 7.3.5 Let G be the group $\{\pm 1\} \subset \mathrm{SL}(4, \mathbb{C})$. Then $\mathrm{age}(1) = 0$ and $\mathrm{age}(-1) = 2$, so $\mathrm{age}(\gamma) \neq 1$ for all $\gamma \in G$. Thus $\mathbb{C}^4/\{\pm 1\}$ is a terminal singularity by Theorem 7.3.3, and does not admit any crepant resolution.

Here is a summary of the above. Let $G \subset \mathrm{SL}(m, \mathbb{C})$ be a nontrivial finite subgroup. If $m = 2$, there is a unique crepant resolution of $\mathbb{C}^2/G$. If $m = 3$ there is a crepant resolution of $\mathbb{C}^3/G$, but it may not be unique. If $m \geqslant 4$ then $\mathbb{C}^m/G$ may or may not have a crepant resolution, which then may or may not be unique. If G is abelian, one can calculate whether or not a crepant resolution exists using toric geometry. Also, Theorem 7.3.3 gives a criterion to determine whether $\mathbb{C}^m/G$ is terminal, and if it is then no crepant resolution exists.

7.3.3   The McKay correspondence

We now discuss some conjectures and results which aim to describe the topology and geometry of crepant resolutions (X, π) of $\mathbb{C}^m/G$ in terms of the group theory of G. The main idea is given in the following conjecture, from Ito and Reid [176, p. 1–2]. Note that in Definition 7.3.2, the age grading $\mathrm{age}(\gamma)$ is unchanged under conjugation, and is therefore an invariant of the conjugacy class of γ.

Conjecture 7.3.6 *Let G be a finite subgroup of $\mathrm{SL}(m, \mathbb{C})$, and (X, π) a crepant resolution of $\mathbb{C}^m/G$. Then there exists a basis of $H^*(X, \mathbb{Q})$ consisting of algebraic cycles in 1-1 correspondence with conjugacy classes of G, such that conjugacy classes with age k correspond to basis elements of $H^{2k}(X, \mathbb{Q})$.*

Reid calls this conjecture the *McKay correspondence*, because it generalizes the McKay correspondence for subgroups of $\mathrm{SL}(2, \mathbb{C})$ mentioned in §7.3.1 to higher dimensions. For a good survey on it, see Reid [286]. In the case $m = 2$ the conjecture is already known. A partial proof of Conjecture 7.3.6 for $m \geqslant 3$ is given by Ito and Reid [176, Cor. 1.5], in the following result.

Theorem 7.3.7 *Let G be a finite subgroup of $\mathrm{SL}(m, \mathbb{C})$, and (X, π) a crepant resolution of $\mathbb{C}^m/G$. Then there is a 1-1 correspondence between exceptional prime divisors in X, which form a basis for $H^2(X, \mathbb{Q})$, and elements of G with age 1.*

Ito and Reid then deduce that Conjecture 7.3.6 is true for $m = 3$, using Poincaré duality to relate $H^4(X, \mathbb{Q})$ and $H_c^2(X, \mathbb{Q})$. Batyrev and Dais [22, Th. 5.4] prove Conjecture 7.3.6 for arbitrary m when G is *abelian*, using toric geometry, and also give their own proof for the $m = 3$ case [22, Prop. 5.6]. Batyrev [21] and Denef and Loeser [94] prove the following corollary of Conjecture 7.3.6 using *motivic integration*:

Theorem 7.3.8 *Let G be a finite subgroup of* $\mathrm{SL}(m, \mathbb{C})$, *and* (X, π) *a crepant resolution of* $\mathbb{C}^m/G$. *Then the Betti number* $b^{2k}(X)$ *is the number of conjugacy classes of G with age k, and* $b^{2k+1}(X) = 0$, *so the Euler characteristic $\chi(X)$ is the number of conjugacy classes in G.*

This means we can work out the Betti numbers of a crepant resolution X of $\mathbb{C}^m/G$ without knowing anything about the resolution. Although singularities $\mathbb{C}^m/G$ for $m \geqslant 3$ often admit several different crepant resolutions, Theorem 7.3.8 shows that they must all have the same Betti numbers. In the case $m = 3$ this is true because all the different crepant resolutions of $\mathbb{C}^3/G$ are related by *flops*, as in §7.2, and a flop does not change the Betti numbers.

7.3.4 Deformations of $\mathbb{C}^m/G$

Section 5.9 defined two ways to desingularize a singular variety: *resolution* and *deformation*. Having discussed crepant resolutions of the quotient singularities $\mathbb{C}^m/G$, we now briefly consider their deformations. In particular we would like to understand the smoothings X_t of $\mathbb{C}^m/G$ with $c_1(X_t) = 0$ for $G \subset \mathrm{SL}(m, \mathbb{C})$. Such smoothings are analogues of crepant resolutions of $\mathbb{C}^m/G$.

The deformation theory of the Kleinian singularities $\mathbb{C}^2/G$ of §7.3.1 is very well understood (see for instance Slodowy [315]). In studying deformations of $\mathbb{C}^m/G$ for $m \geqslant 3$, it turns out that the *codimension* of the singularities of $\mathbb{C}^m/G$ is important. If $G \subset \mathrm{SL}(m, \mathbb{C})$ then $\mathbb{C}^m/G$ cannot have singularities of codimension one, since no nonidentity element of $\mathrm{SL}(m, \mathbb{C})$ can fix a subspace $\mathbb{C}^{m-1} \subset \mathbb{C}^m$. Thus the singularities of $\mathbb{C}^m/G$ are of codimension at least two. However, the *Schlessinger Rigidity Theorem* [299] shows that if the singularities are of codimension three or more, then $\mathbb{C}^m/G$ has no deformations.

Theorem 7.3.9. (Schlessinger) *Let G be a finite subgroup of* $\mathrm{GL}(m, \mathbb{C})$ *for some $m \geqslant 3$, that acts freely on* $\mathbb{C}^m \setminus \{0\}$. *Then the singularity $\mathbb{C}^m/G$ is rigid, that is, it admits no nontrivial deformations. More generally, if G is a finite subgroup of* $\mathrm{GL}(m, \mathbb{C})$ *and the singularities of $\mathbb{C}^m/G$ are of codimension at least three, then $\mathbb{C}^m/G$ is rigid.*

Thus, if $G \subset \mathrm{SL}(m, \mathbb{C})$ then $\mathbb{C}^m/G$ can have nontrivial deformations X_t only if the singularities of $\mathbb{C}^m/G$ are of codimension two. It turns out that $c_1(X_t) = 0$ holds automatically in this case.

7.4 Complex orbifolds

Orbifolds are a special class of singular manifolds.

Definition 7.4.1 An *orbifold* is a singular real manifold X of dimension n whose singularities are locally isomorphic to quotient singularities $\mathbb{R}^n/G$ for finite subgroups $G \subset \mathrm{GL}(n, \mathbb{R})$, such that if $1 \neq \gamma \in G$ then the subspace V_γ of $\mathbb{R}^n$ fixed by γ has dim $V_\gamma \leqslant n - 2$.

For each singular point $x \in X$ there is a finite subgroup $G_x \subset \mathrm{GL}(n, \mathbb{R})$, unique up to conjugation, such that open neighbourhoods of x in X and 0 in $\mathbb{R}^n / G_x$ are homeomorphic (and, in a suitable sense, diffeomorphic). We call x an *orbifold point* of X, and G_x the *orbifold group* or *isotropy group* of x.

Orbifolds are studied in detail by Satake [298], who calls them *V-manifolds*. The condition on γ means that the singularities of the orbifold have real codimension at least two. This makes orbifolds behave like manifolds in many respects. For instance, compact orbifolds satisfy Poincaré duality, but this fails if we allow singularities of codimension one.

Here is an easy method for constructing orbifolds. Let M be a manifold, and G a finite group acting smoothly on M, with nonidentity fixed point sets of codimension at least two. Then the quotient M/G is an orbifold. The following proposition describes the singular set of M/G. The proof is elementary, and we omit it.

Proposition 7.4.2 *Let M be an oriented manifold and G a finite group acting smoothly and faithfully on M preserving orientation. Then M/G is an orbifold. For each $x \in M$, define the stabilizer subgroup of x to be $\mathrm{Stab}(x) = \{g \in G : g \cdot x = x\}$. If $\mathrm{Stab}(x) = \{1\}$ then xG is a nonsingular point of M/G. If $\mathrm{Stab}(x) \neq \{1\}$ then xG is a singular point, with orbifold group $\mathrm{Stab}(x)$. Thus the singular set of M/G is*

$$S = \{xG \in M/G : x \in M \text{ and } g \cdot x = x \text{ for some } 1 \neq g \in G\}.$$

Here the condition that G preserves orientation eliminates the possibility that M/G could have singularities in codimension one, which is not allowed by Definition 7.4.1. In a similar way, we define complex orbifolds.

Definition 7.4.3 A *complex orbifold* is a singular complex manifold of dimension m whose singularities are all locally isomorphic to quotient singularities $\mathbb{C}^m / G$, for finite subgroups $G \subset \mathrm{GL}(m, \mathbb{C})$. *Orbifold points* x and *orbifold groups* $G_x \subset \mathrm{GL}(m, \mathbb{C})$ are defined as above.

Clearly, any complex orbifold is also a real orbifold. Notice that the singular points of orbifolds do not need to be isolated. For example, a complex orbifold of dimension m can have singularities locally modelled on $(\mathbb{C}^k / G) \times \mathbb{C}^{m-k}$, where G is a finite subgroup of $\mathrm{GL}(k, \mathbb{C})$. If G acts freely on $\mathbb{C}^k \setminus \{0\}$ then the singular set of $(\mathbb{C}^k / G) \times \mathbb{C}^{m-k}$ is a copy of $\mathbb{C}^{m-k}$.

The singular set of a complex orbifold is itself a locally finite union of complex orbifolds of lower dimension. If M is a complex manifold and G a finite group acting holomorphically on M then M/G is a complex orbifold, as in Proposition 7.4.2. The *weighted projective spaces* are a special class of complex orbifolds.

Definition 7.4.4 Let $m \geqslant 1$ be an integer, and let $a_0, a_1, \ldots, a_m$ be positive integers with highest common factor 1. Let $\mathbb{C}^{m+1}$ have complex coordinates $(z_0, \ldots, z_m)$, and define an action of the complex Lie group $\mathbb{C}^*$ on $\mathbb{C}^{m+1}$ by

$$(z_0, \ldots, z_m) \overset{u}{\longmapsto} (u^{a_0} z_0, \ldots, u^{a_m} z_m), \qquad \text{for } u \in \mathbb{C}^*. \tag{7.3}$$

Define the *weighted projective space* $\mathbb{CP}^m_{a_0,\dots,a_m}$ to be $(\mathbb{C}^{m+1} \setminus \{0\})/\mathbb{C}^*$, where $\mathbb{C}^*$ acts on $\mathbb{C}^{m+1} \setminus \{0\}$ with the action (7.3). Then $\mathbb{CP}^m_{a_0,\dots,a_m}$ is compact and Hausdorff, and has the structure of a *complex orbifold*.

Note that $\mathbb{CP}^m_{1,\dots,1}$ is the usual complex projective space $\mathbb{CP}^m$. Thus the weighted projective spaces $\mathbb{CP}^m_{a_0,\dots,a_m}$ are a large family of complex orbifolds that generalize the complex manifolds $\mathbb{CP}^m$. For each $(z_0,\dots,z_m) \in \mathbb{C}^{m+1} \setminus \{0\}$, define $[z_0,\dots,z_m] \in \mathbb{CP}^m_{a_0,\dots,a_m}$ to be the orbit of $(z_0,\dots,z_m)$ under $\mathbb{C}^*$.

Here is why $\mathbb{CP}^m_{a_0,\dots,a_m}$ is a complex orbifold. Consider the point $[1,0,\dots,0]$. Under the action of $u \in \mathbb{C}^*$ the point $(1,0,\dots,0)$ is taken to $(u^{a_0},0,\dots,0)$. Therefore the stabilizer of $(1,0,\dots,0)$ in $\mathbb{C}^*$ is $G = \{u \in \mathbb{C}^* : u^{a_0} = 1\}$, which is a finite group isomorphic to $\mathbb{Z}_{a_0}$. It can be shown that the open set $U_0 = \{[z_0,\dots,z_m] \in \mathbb{CP}^m_{a_0,\dots,a_m} : z_0 \neq 0\}$ is naturally isomorphic to $\mathbb{C}^m/G$, where $\mathbb{C}^m$ has complex coordinates $(z_1,\dots,z_m)$ and $u \in G$ acts on $\mathbb{C}^m$ by

$$(z_1,\dots,z_m) \overset{u}{\longmapsto} (u^{a_1}z_1,\dots,u^{a_m}z_m).$$

Thus, if $a_0 > 1$ then $[1,0,\dots,0]$ is an orbifold point of $\mathbb{CP}^m_{a_0,\dots,a_m}$ with orbifold group $\mathbb{Z}_{a_0}$. In the same way, one can prove the following. Let $[z_0,\dots,z_m]$ be a point in $\mathbb{CP}^m_{a_0,\dots,a_m}$, and let k be the highest common factor of the set of those a_j for which $z_j \neq 0$. If $k = 1$ then $[z_0,\dots,z_m]$ is nonsingular, and if $k > 1$ then $[z_0,\dots,z_m]$ is an orbifold point with orbifold group $\mathbb{Z}_k$.

Note that we cannot write $\mathbb{CP}^m_{a_0,\dots,a_m}$ as M/G for M a complex manifold and G a *finite* group, and so not all orbifolds are of the form M/G. The construction of weighted projective spaces is an example of a more general phenomenon. Suppose a complex Lie group K acts holomorphically on a complex manifold M, such that the stabilizers of points in M are always finite subgroups of K. Then the quotient M/K is a complex orbifold, provided it is Hausdorff. Because of this, orbifold singularities occur naturally in moduli spaces and other geometrical problems.

7.4.1 Kähler and Calabi–Yau orbifolds

There is a natural notion of Kähler metric on complex orbifolds.

Definition 7.4.5 We say that g is a *Kähler metric* on a complex orbifold X if g is Kähler in the usual sense on the nonsingular part of X, and wherever X is locally isomorphic to $\mathbb{C}^m/G$, we can identify g with the quotient of a G-invariant Kähler metric defined near 0 in $\mathbb{C}^m$. A *Kähler orbifold* (X,J,g) is a complex orbifold (X,J) equipped with a Kähler metric g.

Examples of Kähler orbifolds are easy to find. For instance, all the weighted projective spaces $\mathbb{CP}^m_{a_0,\dots,a_m}$ admit Kähler metrics, generalizing the Fubini–Study metric on $\mathbb{CP}^m$. Also, suppose (M,J,g) is a Kähler manifold, and G is a finite group acting holomorphically on M, not necessarily preserving the Kähler metric g. Then $g' = \frac{1}{|G|}\sum_{\alpha \in G} \alpha^*(g)$ is a G-invariant Kähler metric on M, and so $(M/G,J,g')$ is a Kähler orbifold. Thus, if M/G is a complex orbifold and M is Kähler, then M/G is Kähler too.

Because orbifold points are quite a mild form of singularity, orbifolds share many of the good properties of manifolds. Many definitions and results about manifolds can be very easily generalized to definitions and results about orbifolds, such as the definition of orbifold Kähler metrics above. In particular, the ideas of smooth k-forms and (p,q)-forms make sense on complex orbifolds. De Rham and Dolbeault cohomology are well-defined on orbifolds and have nearly all of their usual properties.

Another result of interest to us is that the *Calabi Conjecture* holds for compact Kähler orbifolds. To interpret this, we first need to know what we mean by the first Chern class $c_1(X)$ when X is a complex orbifold. Now $c_1(X)$ is a characteristic class of the canonical bundle K_X of X. If all the orbifold groups of singular points in X lie in $\mathrm{SL}(m,\mathbb{C})$, then K_X is a genuine line bundle over X, and $c_1(X)$ is a well-defined element of $H^2(X,\mathbb{Z})$ in the usual way.

However, if the orbifold groups do not lie in $\mathrm{SL}(m,\mathbb{C})$ then K_X is a singular bundle with fibre $\mathbb{C}$ over nonsingular points of X, but with fibres $\mathbb{C}/\mathbb{Z}_k$ over orbifold points of X, for $k \geqslant 1$. It can be shown that in this case $c_1(X)$ is still well-defined, but exists in $H^2(X,\mathbb{Q})$ rather than $H^2(X,\mathbb{Z})$. So in both cases $c_1(X) \in H^2(X,\mathbb{R})$, and if X is Kähler with Ricci form ρ then $[\rho] = 2\pi\, c_1(X)$ in $H^2(X,\mathbb{R})$.

Thus the statement of the Calabi Conjecture in the introduction to Chapter 6 does at least make sense in the category of complex orbifolds. Moreover, the proof of the conjecture also works for orbifolds, with only cosmetic changes. Since an orbifold is locally the quotient of a manifold by a finite group G, locally one can lift any problem on an orbifold up to a G-invariant problem on a manifold, and using this principle one can adapt many proofs in geometry and analysis to the orbifold case. So, by analogy with Theorem 7.1.2 we may prove:

Theorem 7.4.6 *Let X be a compact complex orbifold with $c_1(X) = 0$, admitting Kähler metrics. Then there is a unique Ricci-flat Kähler metric in every Kähler class on X.*

Let X be a real or complex orbifold with singular set S, and g a Riemannian or Kähler metric on X. We define the *holonomy group* $\mathrm{Hol}(g)$ to be the holonomy group (in the usual sense) of the restriction of g to $X \setminus S$. With this definition, holonomy groups on orbifolds have most of the good properties of the manifold case. However, $X \setminus S$ may not be simply-connected if X is simply-connected. Thus, if X is a simply-connected orbifold and g a metric on X then $\mathrm{Hol}(g)$ may not be connected, in contrast to the manifold case.

Let (X, J, g) be a Kähler manifold, and G a finite group of holomorphic isometries of X. Then $(X/G, J, g)$ is a Kähler orbifold. Write $g_X, g_{X/G}$ for the metrics g on X and X/G. Then it is easy to show that the restricted holonomy groups satisfy $\mathrm{Hol}^0(g_{X/G}) = \mathrm{Hol}^0(g_X)$. Furthermore, $\mathrm{Hol}(g_X)$ is the normal subgroup of $\mathrm{Hol}(g_{X/G})$ consisting of parallel transport around those based loops in X/G that lift to based loops in X. There is a natural, surjective group homomorphism $\rho : G \to \mathrm{Hol}(g_{X/G})/\mathrm{Hol}(g_X)$. So $\mathrm{Hol}(g_{X/G})$ is a finite extension of $\mathrm{Hol}(g_X)$ by a quotient group of G.

By analogy with Definition 7.1.10, we define a *Calabi–Yau orbifold* (X, J, g, θ) to be a compact Kähler orbifold (X, J, g) with a holomorphic volume form θ satisfying (7.2). Since from §7.1 a Kähler metric g is Ricci-flat if and only if $\mathrm{Hol}^0(g) \subseteq \mathrm{SU}(m)$, we can use Theorem 7.4.6 to construct metrics with holonomy $SU(m)$ on suitable compact complex orbifolds, making them into Calabi–Yau orbifolds. Also, if (X, J, g, θ) is a Calabi–Yau manifold, and G a finite group of holomorphic isometries of X preserving θ, then $(X/G, J, g, \theta)$ is a Calabi–Yau orbifold.

If (X, J, g) is a Kähler orbifold and an orbifold point x in X has orbifold group G, then there is a natural inclusion

$$G \subseteq \mathrm{Hol}(g) \subseteq \mathrm{U}(m) \subset \mathrm{GL}(m, \mathbb{C}).$$

In particular, G is a subgroup of $\mathrm{Hol}(g)$. So if (X, J, g, θ) is a Calabi–Yau orbifold then $G \subset \mathrm{SU}(m) \subset \mathrm{SL}(m, \mathbb{C})$. This shows that if (X, J, g, θ) is a Calabi–Yau orbifold then the orbifold groups of singular points of X all lie in $\mathrm{SL}(m, \mathbb{C})$.

Let G be a finite subgroup of $\mathrm{SL}(m, \mathbb{C})$. Choose $\alpha \in G$ with $\alpha \neq 1$, and let V_α be the vector subspace of $\mathbb{C}^m$ fixed by α. Then $\dim V_\alpha < m$ as $\alpha \neq 1$. If $\dim V_\alpha = m - 1$ then α has exactly one eigenvalue that is not 1, contradicting $\det(\alpha) = 1$. So $\dim V_\alpha \leqslant m - 2$. But the singular set of $\mathbb{C}^m/G$ is the image in $\mathbb{C}^m/G$ of the V_α for $\alpha \neq 1$. Thus the singular set of $\mathbb{C}^m/G$ has codimension at least two. Hence the singularities of a Calabi–Yau orbifold are of complex codimension at least two.

7.5 Crepant resolutions of orbifolds

Let X be a complex orbifold of dimension m. We shall consider *crepant resolutions* $(\tilde{X}, \pi)$ of X. Since the singularities of X are locally isomorphic to quotient singularities $\mathbb{C}^m/G$ for finite $G \subset \mathrm{GL}(m, \mathbb{C})$, any crepant resolution $(\tilde{X}, \pi)$ of X is locally isomorphic to crepant resolutions of $\mathbb{C}^m/G$. But from §7.3 we already understand the crepant resolutions of $\mathbb{C}^m/G$ quite well, especially when $m = 2$ or 3. Thus, we can use our knowledge of crepant resolutions of $\mathbb{C}^m/G$ to study crepant resolutions of general complex orbifolds.

From §7.3, a necessary condition for $\mathbb{C}^m/G$ to admit crepant resolutions is that $G \subset \mathrm{SL}(m, \mathbb{C})$. Thus, a complex orbifold X can admit a crepant resolution only if all its orbifold groups lie in $\mathrm{SL}(m, \mathbb{C})$. (From §7.4, this condition holds automatically for Calabi–Yau orbifolds.) So let X be a complex orbifold with all orbifold groups in $\mathrm{SL}(m, \mathbb{C})$. For each singular point there are a finite number of possible crepant resolutions (which may be zero if $m \geqslant 4$). If G is an orbifold group of X and $\mathbb{C}^m/G$ admits no crepant resolution, then clearly X has no crepant resolution either. So suppose that $\mathbb{C}^m/G$ has at least one crepant resolution for all orbifold groups G of X.

The obvious way to construct a crepant resolution of X is to choose a crepant resolution of $\mathbb{C}^m/G$ for each singular point of X with orbifold group G, and then try to patch these together to form a crepant resolution $(\tilde{X}, \pi)$ of X. If the singularities of X are *isolated* then we can independently choose any crepant resolution of each singular point, and fit them together in a unique way to get a crepant resolution of X.

For nonisolated singularities things are more complicated, as choosing a crepant resolution for one singular point x uniquely determines the choice of resolution of all other singular points in an open neighbourhood of x. The choice of resolution must vary continuously over the singular set, in an appropriate sense. However, in the case $m = 3$ this imposes no restrictions, as we show in the proof of the next result, due to Roan [289, p. 493].

Theorem 7.5.1 *Let X be a complex 3-orbifold with orbifold groups in* $\mathrm{SL}(3,\mathbb{C})$. *Then X admits a crepant resolution.*

Proof If X is a complex 3-orbifold with orbifold groups in $\mathrm{SL}(3,\mathbb{C})$ then the singular points of X divide into two types:

(a) singular points modelled on $(\mathbb{C}^2/H) \times \mathbb{C}$, for H a finite subgroup of $\mathrm{SU}(2)$, and

(b) singular points not of type (a).

The singular points of type (a) form a complex 1-manifold in X, but the singular points of type (b) are a *discrete* set of isolated points.

Now singular points of type (a) have a *unique* crepant resolution, by §7.3.1. So we only have more than one possible choice of resolution at singular points of type (b). But these are isolated from one another, and so there are no compatibility conditions between the choices. By Theorem 7.3.1 there is at least one possible crepant resolution for each singular point of type (b). Making an arbitrary choice in each case, we can patch the resolutions together in a unique way to get a crepant resolution of X. $\qquad\square$

It can happen that if X is a Kähler orbifold with nonisolated singularities, then some of the crepant resolutions $(\tilde{X}, \pi)$ of X are not Kähler. This is because, for reasons of global topology, it may not be possible to choose a class in $H^{1,1}(\tilde{X})$ which is simultaneously positive on homology classes of all the rational curves in $\tilde{X}$ introduced by the resolution, and this is a necessary condition for the existence of a Kähler class. However, extending the argument of Theorem 7.5.1 one can show that a Kähler 3-orbifold admitting crepant resolutions has at least one Kähler crepant resolution.

7.5.1 Crepant resolutions of quotients of complex tori

If X is a Calabi–Yau orbifold and $(\tilde{X}, \pi)$ is a Kähler crepant resolution of X, then $\tilde{X}$ has a family of Ricci-flat Kähler metrics which make it into a Calabi–Yau manifold. This gives a method of constructing Calabi–Yau manifolds. We start with a compact Kähler orbifold (X, J, g) with $\mathrm{Hol}(g) \subseteq \mathrm{SU}(m)$. If $(\tilde{X}, \pi)$ is any Kähler crepant resolution of X, then $\tilde{X}$ has a family of Ricci-flat Kähler metrics $\tilde{g}$. It can be shown that $\mathrm{Hol}(g) \subseteq \mathrm{Hol}(\tilde{g}) \subseteq \mathrm{SU}(m)$. Often we find that although $\mathrm{Hol}(g)$ may be a proper subgroup of $\mathrm{SU}(m)$, yet $\mathrm{Hol}(\tilde{g}) = \mathrm{SU}(m)$, and so $\tilde{X}$ is a Calabi–Yau manifold.

In particular we can take X to be T^{2m}/G, where T^{2m} is a flat Kähler torus and G a finite group. Let $\mathbb{C}^m$ have its standard complex structure J, Euclidean metric g, and holomorphic volume form θ, and let Λ be a *lattice* in $\mathbb{C}^m$. Then $\mathbb{C}^m/\Lambda$ is a compact torus T^{2m}, with a flat Kähler structure (J, g) and holomorphic volume form θ.

Let G be a finite group of automorphisms of T^{2m} preserving g, J and θ. Then $(T^{2m}/G, J, g)$ is a compact Kähler orbifold with orbifold groups in $\mathrm{SL}(m, \mathbb{C})$. Suppose $(\tilde{X}, \pi)$ is a crepant resolution of T^{2m}/G. Under good conditions, $\tilde{X}$ turns out to be a Calabi–Yau manifold.

The simplest case of this is the *Kummer construction*, in which $T^4/\{\pm 1\}$ is resolved to give a $K3$ surface. It will be described in Examples 10.3.2 and 10.3.14. The method was studied in higher dimensions by Roan [288]. Here is a useful result in the 3-dimensional case.

Theorem 7.5.2 *Let (T^6, J, g) be a flat Kähler torus with a holomorphic volume form θ, and suppose G is a finite group of automorphisms of T^6 preserving J, g and θ. Then $X = T^6/G$ is a compact complex orbifold with at least one Kähler crepant resolution $(\tilde{X}, \pi)$. There exist Ricci-flat Kähler metrics $\tilde{g}$ on $\tilde{X}$ with $\mathrm{Hol}(\tilde{g}) \subseteq \mathrm{SU}(3)$, and $\mathrm{Hol}(\tilde{g}) = \mathrm{SU}(3)$ if and only if $\pi_1(T^6/G)$ is finite.*

Proof As G preserves θ the orbifold groups of T^6/G all lie in $\mathrm{SL}(3, \mathbb{C})$, so by Theorem 7.5.1 and the discussion after it, T^6/G has at least one Kähler crepant resolution $(\tilde{X}, \pi)$, which has trivial canonical bundle. This $\tilde{X}$ admits Ricci-flat Kähler metrics $\tilde{g}$ with $\mathrm{Hol}(\tilde{g}) \subseteq \mathrm{SU}(3)$ by Theorem 7.1.2 and Corollary 7.1.6.

By the classification of holonomy groups, $\mathrm{Hol}^0(\tilde{g})$ must be $\{1\}$, $\mathrm{SU}(2)$ or $\mathrm{SU}(3)$. Propositions 7.1.3 and 7.1.4 then show that $\mathrm{Hol}(\tilde{g}) = \mathrm{SU}(3)$ if and only if $\pi_1(\tilde{X})$ is finite. But $\pi_1(\tilde{X}) \cong \pi_1(T^6/G)$, as crepant resolutions of orbifolds replace each singular point by a simply-connected set. Therefore $\mathrm{Hol}(\tilde{g}) = \mathrm{SU}(3)$ if and only if $\pi_1(T^6/G)$ is finite, as we have to prove. $\square$

Here is a simple example.

Example 7.5.3 Let $\zeta = -\frac{1}{2} + i\frac{\sqrt{3}}{2}$, so that $\zeta^3 = 1$, and define a lattice Λ in $\mathbb{C}^3$ by

$$\Lambda = \big\{ (a_1 + b_1\zeta, a_2 + b_2\zeta, a_3 + b_3\zeta) : a_j, b_j \in \mathbb{Z} \big\}.$$

Then $T^6 = \mathbb{C}^3/\Lambda$ is a complex torus, with a natural metric g and holomorphic volume form θ. Define a map $\alpha : T^6 \to T^6$ by

$$\alpha : (z_1, z_2, z_3) + \Lambda \longmapsto (\zeta z_1, \zeta z_2, \zeta z_3) + \Lambda.$$

Then α is well-defined, preserves g and θ, and α^3 is the identity. Hence $G = \{1, \alpha, \alpha^2\}$ is a group of automorphisms of T^6 isomorphic to $\mathbb{Z}_3$, and T^6/G is a Kähler orbifold.

The fixed points of α on T^6 are the 27 points

$$\big\{ (c_1, c_2, c_3) + \Lambda : c_1, c_2, c_3 \in \{0, \tfrac{i}{\sqrt{3}}, \tfrac{2i}{\sqrt{3}}\} \big\}.$$

Thus T^6/G has 27 isolated fixed points modelled on $\mathbb{C}^3/\mathbb{Z}_3$, where the action of $\mathbb{Z}_3$ on $\mathbb{C}^3$ is generated by $(z_1, z_2, z_3) \longmapsto (\zeta z_1, \zeta z_2, \zeta z_3)$. Now $\mathbb{C}^3/\mathbb{Z}_3$ has a unique crepant resolution, the blow-up of the singular point, in which the singular point is replaced by a copy of $\mathbb{CP}^2$. Therefore T^6/G has a unique crepant resolution Z, made by blowing up the 27 singular points.

Calculation shows that $\pi_1(T^6/G) = \{1\}$, so Z is a Calabi–Yau 3-fold by Theorem 7.5.2. Let us compute the Hodge numbers $h^{p,q}$ of Z. Since

$$h^{p,q} = h^{q,p} = h^{3-p,3-q} = h^{3-q,3-p} \quad \text{and} \quad h^{0,0} = h^{3,0} = 1, \ h^{1,0} = h^{2,0} = 0$$

by (5.10) and Proposition 7.1.7, it is enough to find $h^{1,1}$ and $h^{2,1}$. Now the forms $dz_j \wedge d\bar{z}_k$ for $j, k = 1, 2, 3$ are a natural basis for $H^{1,1}(T^6)$, so that $h^{1,1}(T^6) = 9$. The action of $\alpha \in G$ multiplies dz_j by ζ and $d\bar{z}_k$ by $\bar{\zeta}$, so that $dz_j \wedge d\bar{z}_k$ is multiplied by $\zeta\bar{\zeta} = 1$. Thus G acts trivially on $H^{1,1}(T^6)$, and so $H^{1,1}(T^6/G) \cong H^{1,1}(T^6)$, and

$h^{1,1}(T^6/G) = 9$. The resolution of each singular point adds 1 to $h^{1,1}$. Thus $h^{1,1}(Z) = 9 + 27 = 36$.

Similarly, a basis for $H^{2,1}(T^6)$ is $dz_j \wedge dz_k \wedge d\bar{z}_l$ for $j, k, l = 1, 2, 3$ and $j < k$. The action of $\alpha \in G$ multiplies $dz_j \wedge dz_k \wedge d\bar{z}_l$ by $\zeta^2 \bar{\zeta} = \zeta$. Thus the G-invariant part of $H^{2,1}(T^6)$ is $\{0\}$, and $h^{2,1}(T^6/G) = 0$. The resolution of the singular points does not change $h^{2,1}$, and so $h^{2,1}(Z) = 0$. Therefore the Hodge numbers of Z are $h^{1,1} = 36$ and $h^{2,1} = 0$. An interesting feature of this example is that $h^{2,1} = 0$. As we will see in §7.7, this implies that the complex 3-fold Z is *rigid*, and has no deformations.

Many other examples of Calabi–Yau manifolds can be constructed in this way using other finite groups G acting on T^{2m}, especially in the case $m = 3$. But one can show the number of distinct manifolds arising in this way in any one dimension is finite.

7.6 Complete intersections

Let $\mathbb{CP}^m$ have homogeneous coordinates $[z_0, \ldots, z_m]$, let $f(z_0, \ldots, z_m)$ be a nonzero homogeneous polynomial of degree d, and define X to be

$$\left\{ [z_0, \ldots, z_m] \in \mathbb{CP}^m : f(z_0, \ldots, z_m) = 0 \right\}.$$

Then we call X a *hypersurface of degree* d. Suppose X is nonsingular, which is true for generic f, so that X is a compact complex manifold of dimension $m - 1$. Now, under what conditions do we have $c_1(X) = 0$?

The *adjunction formula* for complex hypersurfaces [132, p. 147] gives that

$$K_X = (K_{\mathbb{CP}^m} \otimes L_X)|_X,$$

where L_X is the line bundle associated to the divisor $[X]$. In the notation of Example 5.10.1, any line bundle over $\mathbb{CP}^m$ is of the form $\mathcal{O}(k)$ for some integer k. It can be shown that $K_{\mathbb{CP}^m} = \mathcal{O}(-m-1)$ and $L_X = \mathcal{O}(d)$, and thus $K_X = \mathcal{O}(d-m-1)|_X$. Therefore K_X is trivial, so that $c_1(X) = 0$, if and only if $d = m+1$.

We have shown that any nonsingular hypersurface X in $\mathbb{CP}^m$ of degree $m + 1$ has $c_1(X) = 0$. But as X is also compact and Kähler, Theorem 7.1.2 shows X has a family of Ricci-flat Kähler metrics. In fact these metrics have holonomy $SU(m-1)$, and X is a *Calabi–Yau manifold* for $m \geqslant 3$. This is perhaps the simplest known method of finding Calabi–Yau manifolds. Now all nonsingular hypersurfaces in $\mathbb{CP}^m$ of degree $m + 1$ are diffeomorphic, and thus this method yields only one smooth manifold admitting Calabi–Yau structures in each dimension. We shall now describe three ways of generalizing this idea that give Calabi–Yau structures on many more manifolds.

7.6.1 Complete intersections in $\mathbb{CP}^m$

An algebraic variety X in $\mathbb{CP}^m$ is a *complete intersection* if $X = H_1 \cap \cdots \cap H_k$, where $H_1, \ldots, H_k$ are hypersurfaces in $\mathbb{CP}^m$ which intersect transversely along X, so that $\dim X = m - k$. Suppose that X is a complete intersection of hypersurfaces $H_1, \ldots, H_k$, and let d_j be the degree of H_j. Finding an expression for $c_1(X)$ as above, one easily proves that $c_1(X) = 0$ if and only if $d_1 + \cdots + d_k = m + 1$, and in this case X is a Calabi–Yau manifold.

Since complete intersections with fixed m and $d_1, \ldots, d_k$ are all equivalent under deformation, X depends as a smooth manifold only on m and $d_1, \ldots, d_k$. By [131, Th. 1], if hypersurfaces H_j in $\mathbb{CP}^m$ of degree d_j are chosen generically then $X = H_1 \cap \cdots \cap H_k$ is a nonsingular complete intersection. Now if $d_j = 1$ for any j then X may be regarded as the intersection of $k-1$ hypersurfaces in $\mathbb{CP}^{m-1}$ of degrees $d_1, \ldots, d_k$, omitting d_j.

Thus Calabi–Yau manifolds of dimension $m - k$ which are complete intersections are classified, as smooth manifolds, by integers m and $d_1, \ldots, d_k$, where $d_j \geqslant 2$ for $j = 1, \ldots, k$ and $d_1 + \cdots + d_k = m + 1$. There are only a finite number of possibilities for m and $d_1, \ldots, d_k$ in each dimension. For example, using the notation $(m \mid d_1, \ldots, d_k)$, the five complete intersections giving Calabi–Yau 3-folds are

$$(4 \mid 5), \quad (5 \mid 2, 4), \quad (5 \mid 3, 3), \quad (6 \mid 2, 2, 3) \quad \text{and} \quad (7 \mid 2, 2, 2, 2).$$

By applying the Lefschetz Hyperplane Theorem, Theorem 5.10.4, we see that if X is a complete intersection of dimension $m - k$ in $\mathbb{CP}^m$ then $H^j(X, \mathbb{C}) \cong H^j(\mathbb{CP}^m, \mathbb{C})$ for $0 \leqslant j < m - k$. This gives the Betti numbers $b^j(X)$ of X, except in the middle dimension $m - k$. To determine $b^{m-k}(X)$ we calculate the Euler characteristic of X, using a formula for the Chern classes of X.

7.6.2 Hypersurfaces in weighted projective spaces and toric varieties

The construction above may be generalized by replacing $\mathbb{CP}^m$ by a *weighted projective space* $\mathbb{CP}^m_{a_0, \ldots, a_m}$, as in Definition 7.4.4. This was studied in depth by Candelas, Lynker and Schimmrigk [71] in the case $m = 4$, and we now summarize their ideas. The weighted projective space $\mathbb{CP}^m_{a_0, \ldots, a_m}$ is the quotient of $\mathbb{C}^{m+1} \setminus \{0\}$ by the $\mathbb{C}^*$-action

$$(z_0, \ldots, z_m) \stackrel{u}{\longmapsto} (u^{a_0} z_0, \ldots, u^{a_m} z_m) \quad \text{for } u \in \mathbb{C}^*.$$

We call a nonzero polynomial $f(z_0, \ldots, z_m)$ *weighted homogeneous of degree d* if

$$f(u^{a_0} z_0, \ldots, u^{a_m} z_m) = u^d f(z_0, \ldots, z_m) \quad \text{for all } u, z_0, \ldots, z_m \in \mathbb{C}.$$

Let f be such a polynomial, and define a hypersurface X in $\mathbb{CP}^m_{a_0, \ldots, a_m}$ by

$$X = \big\{ [z_0, \ldots, z_m] \in \mathbb{CP}^m_{a_0, \ldots, a_m} : f(z_0, \ldots, z_m) = 0 \big\}.$$

Then we call X a *hypersurface of degree d* in $\mathbb{CP}^m_{a_0, \ldots, a_m}$.

Now $\mathbb{CP}^m_{a_0, \ldots, a_m}$ is an orbifold. Usually the hypersurface X intersects the singularities of $\mathbb{CP}^m_{a_0, \ldots, a_m}$, and at these points X itself is singular. We don't want to exclude all such X, and therefore we cannot restrict our attention to nonsingular X. Instead, we define the polynomial f to be *transverse* if $f(z_0, \ldots, z_m) = 0$ and $\mathrm{d}f(z_0, \ldots, z_m) = 0$ have no common solutions in $\mathbb{C}^{m+1} \setminus \{0\}$. If f is transverse then the only singular points of X are also singular points of $\mathbb{CP}^m_{a_0, \ldots, a_m}$, and in fact X is an *orbifold*, all of whose orbifold groups are cyclic.

So we restrict our attention to hypersurfaces X defined by transverse polynomials f. Note that for given weights $a_0, \ldots, a_m$ and degree d, there may not exist any transverse polynomials f. Any such f must be the sum of monomials $z_0^{b_0} \cdots z_m^{b_m}$, where

$b_0, \ldots, b_m$ are nonnegative integers with $a_0 b_0 + \cdots + a_m b_m = d$. If there are not enough suitable solutions $\{b_j\}$ to this equation, then there are no transverse f. For example, by [71, p. 389] a necessary (but not sufficient) criterion for there to exist a transverse polynomial f of degree d is that for each $i = 0, \ldots, m$ there exists a j such that a_i divides $d - a_j$.

Let X be a hypersurface in $\mathbb{CP}^m_{a_0, \ldots, a_m}$ of degree d, defined by a transverse polynomial f. Then since X is an orbifold, the first Chern class $c_1(X)$ is well-defined. It can be shown that $c_1(X) = 0$ if and only if $d = a_0 + \cdots + a_m$. Moreover, in this case the canonical bundle K_X of X is trivial, and this implies that the orbifold groups of X lie in $\mathrm{SL}(m-1, \mathbb{C})$. Therefore X is a Calabi–Yau orbifold, and if $(\tilde{X}, \pi)$ is a Kähler crepant resolution of X, then $\tilde{X}$ is a Calabi–Yau manifold.

In particular, when $m = 4$ the dimension of X is 3, and so by Theorem 7.5.1 and the discussion after it, X admits at least one Kähler crepant resolution $\tilde{X}$, which is then a Calabi–Yau 3-fold. This gives a method for constructing Calabi–Yau 3-folds:

- First choose a weighted projective space $\mathbb{CP}^4_{a_0, \ldots, a_4}$, where $a_0, \ldots, a_4$ are positive integers with highest common factor 1.
- If possible, find a hypersurface X in $\mathbb{CP}^4_{a_0, \ldots, a_4}$ defined by a transverse polynomial f of degree $a_0 + \cdots + a_4$.
- This X is a Calabi–Yau orbifold, whose orbifold groups are cyclic subgroups of $\mathrm{SL}(3, \mathbb{C})$. Let $(\tilde{X}, \pi)$ be a Kähler crepant resolution of X. There is at least one such resolution. Then $\tilde{X}$ is a Calabi–Yau 3-fold.

The two big advantages of this method are that, firstly, there are many possibilities for $a_0, \ldots, a_4$ and so the construction yields many Calabi–Yau 3-folds, and secondly, the calculations involved are sufficiently mechanical that the construction can be implemented on a computer.

In particular, once $a_0, \ldots, a_4$ are fixed, all hypersurfaces X defined by transverse f of degree $a_0 + \cdots + a_4$ are deformation equivalent. Thus the topology of the orbifold X depends only on $a_0, \ldots, a_4$. Although X may admit several different crepant resolutions $\tilde{X}_1, \ldots, \tilde{X}_k$, the Hodge numbers of $\tilde{X}_j$ depend only on the topology of X, and hence only on $a_0, \ldots, a_4$.

Candelas et al. [71, §3] explained how to calculate the Hodge numbers of $\tilde{X}$ from $a_0, \ldots, a_4$. They then used a computer program to search for quintuples $(a_0, \ldots, a_4)$ for which a suitable transverse polynomial f exists, and to calculate the Hodge numbers of the corresponding Calabi–Yau 3-folds. In this way they constructed some 6000 examples of Calabi–Yau 3-folds, which realized 2339 distinct pairs of Hodge numbers $(h^{1,1}, h^{2,1})$. This was many more examples than were known at the time.

When Candelas et al. plotted the Hodge numbers $(h^{1,1}, h^{2,1})$ of their examples on a graph [71, Fig. 1, p. 384], they found that their graph had an approximate, but very persuasive symmetry: for nearly every Calabi–Yau 3-fold with $h^{1,1} = x$ and $h^{2,1} = y$ in their examples, there was another Calabi–Yau 3-fold with $h^{1,1} = y$ and $h^{2,1} = x$. This was one of the first pieces of experimental evidence supporting the idea of *Mirror Symmetry* for Calabi–Yau 3-folds, which will be the subject of Chapter 9.

Batyrev [20] made significant progress in explaining this by studying Calabi–Yau manifolds which are hypersurfaces in *compact toric varieties*. As weighted projective spaces are toric this includes the examples of [71], and also some others. Batyrev showed that nonsingular toric 4-folds T containing Calabi–Yau 3-folds X as anticanonical divisors are classified by *reflexive polytopes* Δ in $\mathbb{R}^4$, with vertices in $\mathbb{Z}^4$. Each such Δ has a dual reflexive polytope $\hat{\Delta}$, and if $X, \hat{X}$ are Calabi–Yau 3-fold divisors in $T, \hat{T}$ then $h^{1,1}(X) = h^{2,1}(\hat{X})$ and $h^{2,1}(X) = h^{1,1}(\hat{X})$. Thus, this construction automatically yields Calabi–Yau 3-folds in mirror pairs.

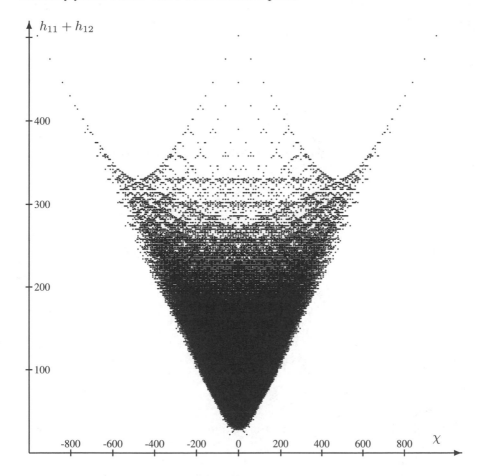

Fig. 7.1 $h_{11} + h_{12}$ versus Euler number $\chi = 2(h_{11} - h_{12})$ of Calabi–Yau 3-folds.

Reflexive polytopes Δ in $\mathbb{R}^4$ are specified by a finite amount of data, the set of vertices of Δ in $\mathbb{Z}^4$. Kreuzer and Skarke [225] used a computer to produce a complete list of such polytopes, consisting of 473,800,776 examples. They also calculated the Hodge numbers of the corresponding Calabi–Yau 3-fold hypersurfaces, which realize 30,108 distinct pairs of Hodge numbers $(h^{1,1}, h^{2,1})$. A graph of these Hodge numbers is given in Figure 7.1, reproduced from [225] by kind permission of Maximilian Kreuzer. Its symmetry under reflection in the vertical axis is a striking illustration of mirror symmetry. The patterns in the graph, and the nature of the curve bounding the inhabited region from above, have never been explained.

7.7 Deformations of Calabi–Yau manifolds

The Kodaira–Spencer–Kuranishi deformation theory for compact complex manifolds was explained in §5.9.2. It turns out that when (X, J, g, θ) is a Calabi–Yau manifold, the deformation theory of the underlying complex manifold (X, J) is particularly simple. Our next result is due independently to Tian [326, Th. 1] and Todorov [329, Th. 1].

Theorem 7.7.1 *Let X be a compact Kähler m-manifold with trivial canonical bundle. Then the local moduli space of deformations of the complex structure of X is a complex manifold of dimension $h^{m-1,1}(X)$. All the complex structures in this local moduli space are also Kähler with trivial canonical bundle.*

Here is why the moduli space has dimension $h^{m-1,1}(X)$. The deformation theory of a compact complex manifold X depends on the sheaf cohomology groups $H^*(X, \Theta_X)$, which are the cohomology of the complex

$$0 \to C^\infty(T^{1,0}X) \xrightarrow{\bar{\partial}} C^\infty(T^{1,0}X \otimes \Lambda^{0,1}X) \xrightarrow{\bar{\partial}} C^\infty(T^{1,0}X \otimes \Lambda^{0,2}X) \xrightarrow{\bar{\partial}} \cdots.$$

But since X has trivial canonical bundle there exists a holomorphic volume form θ on X, which is a section of $\Lambda^{m,0}$. Contraction with θ gives an isomorphism between $T^{1,0}X$ and $\Lambda^{m-1,0}X$. Thus

$$T^{1,0} \otimes \Lambda^{0,q}X \cong \Lambda^{m-1,0}X \otimes \Lambda^{0,q}X \cong \Lambda^{m-1,q}X.$$

Therefore the complex above is isomorphic to the complex

$$0 \to C^\infty(\Lambda^{m-1,0}X) \xrightarrow{\bar{\partial}} C^\infty(\Lambda^{m-1,1}X) \xrightarrow{\bar{\partial}} C^\infty(\Lambda^{m-1,2}X) \xrightarrow{\bar{\partial}} \cdots.$$

But this is part of the *Dolbeault complex* of X, as in §5.2.2 and §5.7.1. So $H^q(X, \Theta_X)$ is isomorphic to $H^{m-1,q}(X)$.

Thus in the Kodaira–Spencer–Kuranishi deformation theory of §5.9.2, the space of *infinitesimal deformations* of X is $H^{m-1,1}(X)$, and the space of *obstructions* to deforming X is $H^{m-1,2}(X)$. Now Tian and Todorov show that even though $H^{m-1,2}(X)$ may be nonzero, the obstructions are ineffective, and every infinitesimal deformation of X lifts to an actual deformation. Hence the base space of the *Kuranishi family* of deformations of X is an open set in $H^{m-1,1}(X)$. They also show that the Kuranishi family is *universal*. So the local moduli space of deformations of the complex structure of X is isomorphic to an open set in $H^{m-1,1}(X)$, and is a complex manifold with dimension $h^{m-1,1}(X)$.

Corollary 7.7.2 *Let (X, J, g, θ) be a Calabi–Yau m-fold of dimension m. Then the local moduli space of deformations of the Calabi–Yau structure (J, g, θ) of X is a smooth real manifold of dimension $h^{1,1}(X) + 2h^{m-1,1}(X) + 1$.*

Proof From Theorem 7.7.1, the local deformations of J form a manifold of complex dimension $h^{m-1,1}(X)$, and thus of real dimension $2h^{m-1,1}(X)$. But from §7.1, the family of Calabi–Yau metrics g on the fixed complex manifold (X, J) is a real manifold of dimension $h^{1,1}(X)$ isomorphic to the Kähler cone $\mathcal{K}_X$ of (X, J). This also holds on (X, J_t) for any small deformation J_t of J. Having chosen J, g, the holomorphic volume form θ is unique up to phase $\theta \mapsto e^{i\psi}\theta$, which contributes one more real parameter. So the local moduli space of Calabi–Yau structures on X is a real manifold of dimension $h^{1,1}(X) + 2h^{m-1,1}(X) + 1$. $\qquad\square$

Now let X be a compact real manifold of dimension $2m$, let $\mathscr{A}_X$ be the set of integrable complex structures J on X extending to Calabi–Yau m-folds (X, J, g, θ), and let $\mathscr{D}_X$ be the group of diffeomorphisms of X isotopic to the identity map. Then $\mathscr{M}_X = \mathscr{A}_X/\mathscr{D}_X$ is the moduli space of *Calabi–Yau complex structures* on X.

Note that $\mathscr{A}_X$ and $\mathscr{D}_X$ are both infinite-dimensional. However, by Theorem 7.7.1 we see that $\mathscr{M}_X$ is a nonsingular complex manifold of dimension $h^{m-1,1}(X)$, where the Hodge number is a locally constant function of complex structure in $\mathscr{M}_X$. We would like to study the *global* geometry of $\mathscr{M}_X$, and hence to get some information about the entire collection of Calabi–Yau structures on a fixed real manifold. Here is a tool that helps us to do this.

For each $J \in \mathscr{A}_X$, write X_J for the complex manifold (X, J). Then $H^m(X, \mathbb{C})$ splits into a direct sum of Dolbeault groups $H^{p,q}(X_J)$ for $p + q = m$. In particular, $H^{m,0}(X_J)$ is a 1-dimensional subspace of $H^m(X, \mathbb{C})$, and so it defines a point in the complex projective space $P(H^m(X, \mathbb{C}))$. Clearly, this point depends only on the equivalence class of J in $\mathscr{M}_X$. Define a map $\Phi : \mathscr{M}_X \to P(H^m(X, \mathbb{C}))$ by $\Phi(J \cdot \mathscr{D}_X) = H^{m,0}(X_J)$. We call Φ the *period map* of X.

We showed above how to use a holomorphic volume form θ on X_J to define an isomorphism between $H^{m-1,1}(X_J)$ and the tangent space to $\mathscr{M}_X$ at the equivalence class of J. This tangent space is isomorphic to $H^{m,0}(X_J)^* \otimes H^{m-1,1}(X_J)$, without making any choice of θ. However, $H^{m,0}(X_J)^* \otimes H^{m-1,1}(X_J)$ is also isomorphic to a subspace of the tangent space to $P(H^m(X, \mathbb{C}))$ at the point $H^{m,0}(X_J)$.

Thus there is a natural isomorphism between the tangent space to $\mathscr{M}_X$ at $J \cdot \mathscr{D}_X$, and a subspace of the tangent space to $P(H^m(X, \mathbb{C}))$ at $\Phi(J \cdot \mathscr{D}_X)$, and $d\Phi$ induces exactly this isomorphism. This has two important consequences: firstly, Φ is a holomorphic map of complex manifolds, and secondly, Φ is an immersion. Hence we have proved:

Theorem 7.7.3 *Let X be a compact real $2m$-manifold, $\mathscr{M}_X$ be the moduli space of Calabi–Yau complex structures on X, and $\Phi : \mathscr{M}_X \to P(H^m(X, \mathbb{C}))$ the map defined above. Then $\mathscr{M}_X$ has the structure of a complex manifold of dimension $h^{m-1,1}(X)$, and Φ is a holomorphic immersion of complex manifolds.*

Now in good cases, we can hope that Φ is an *embedding* rather than just an immersion, and its image $\text{Im}\,\Phi$ can be explicitly identified in $P(H^m(X, \mathbb{C}))$. We can then give a very precise description of the moduli space of Calabi–Yau complex structures $\mathscr{M}_X$ on X. This happens for $K3$ surfaces, which are Calabi–Yau manifolds of dimension 2, as we will explain in §10.3. One can also describe the moduli space $\mathscr{M}_X$ when X is a Calabi–Yau m-fold for $m \geqslant 3$ defined as a complete intersection in some $\mathbb{CP}^n$ using this method, as all small deformations of X are also complete intersections in $\mathbb{CP}^n$.

8

Special Lagrangian geometry

Special Lagrangian submanifolds (or SL m-folds for short) in $\mathbb{C}^m$, or in a Calabi–Yau m-fold or almost Calabi–Yau m-fold (X, J, g, θ), are real m-dimensional submanifolds in $\mathbb{C}^m$ or X calibrated by the real part $\operatorname{Re}\theta$ of the holomorphic volume form θ. They were invented by Harvey and Lawson [151, §III], who concentrated on SL m-folds in $\mathbb{C}^m$. For a long time, essentially the only nontrivial result on SL m-folds in Calabi–Yau m-folds was McLean's beautiful theorem on the deformation theory of compact SL m-folds in §8.4.1 below.

However, in 1996 Strominger, Yau and Zaslow [317] put forward the SYZ Conjecture explaining mirror symmetry of Calabi–Yau 3-folds $X, \hat{X}$ in terms of fibrations of $X, \hat{X}$ by SL 3-folds, including singular fibres, as we shall describe in Chapter 9. This generated considerable interest in special Lagrangian geometry amongst mathematicians and physicists, and there are now many more examples known, and a fairly well-developed theory of singularities of SL m-folds.

Sections 8.1–8.3 concern SL m-folds in $\mathbb{C}^m$, covering the basic theory, constructions of examples, and more detail on SL cones and Asymptotically Conical SL m-folds. Then §8.4 discusses SL m-folds in (almost) Calabi–Yau m-folds, focussing on compact, nonsingular SL m-folds, and §8.5 surveys what is known about singular SL m-folds in (almost) Calabi–Yau m-folds.

8.1 Special Lagrangian submanifolds in $\mathbb{C}^m$

Here is the basic definition.

Definition 8.1.1 Let $\mathbb{C}^m \cong \mathbb{R}^{2m}$ have complex coordinates $(z_1, \ldots, z_m)$, and as in (7.1) define a metric g, Kähler form ω and complex volume form θ on $\mathbb{C}^m$ by

$$g = |\mathrm{d}z_1|^2 + \cdots + |\mathrm{d}z_m|^2, \quad \omega = \tfrac{i}{2}(\mathrm{d}z_1 \wedge \mathrm{d}\bar{z}_1 + \cdots + \mathrm{d}z_m \wedge \mathrm{d}\bar{z}_m),$$
$$\text{and} \quad \theta = \mathrm{d}z_1 \wedge \cdots \wedge \mathrm{d}z_m. \tag{8.1}$$

Then $\operatorname{Re}\theta$ and $\operatorname{Im}\theta$ are real m-forms on $\mathbb{C}^m$, both calibrations. Let L be an oriented real m-submanifold of $\mathbb{C}^m$. We call L a *special Lagrangian submanifold* of $\mathbb{C}^m$, or *SL m-fold* for short, if L is calibrated with respect to $\operatorname{Re}\theta$, in the sense of §4.1.

More generally, if $\psi \in \mathbb{R}$ we call L *special Lagrangian with phase* $e^{i\psi}$ if it is calibrated w.r.t. $\cos\psi \operatorname{Re}\theta + \sin\psi \operatorname{Im}\theta$. When we refer to SL m-folds L without specifying a phase we mean phase 1, that is, calibrated w.r.t. $\operatorname{Re}\theta$ as above. Since L is an

SL m-fold with phase $e^{i\psi}$ in $\mathbb{C}^m$ if and only if $e^{-i\psi/m}L$ is an SL m-fold with phase 1, studying SL m-folds with phase 1 tells us about SL m-folds with arbitrary phase.

Here [151, Cor. III.1.11] is an alternative characterization of SL m-folds.

Proposition 8.1.2 *Let L be a real m-dimensional submanifold of $\mathbb{C}^m$. Then L admits an orientation making it into an SL m-fold if and only if $\omega|_L \equiv 0$ and $\operatorname{Im}\theta|_L \equiv 0$. More generally, it admits an orientation making it into an SL m-fold with phase $e^{i\psi}$ if and only if $\omega|_L \equiv 0$ and $(\cos\psi\,\operatorname{Im}\theta - \sin\psi\,\operatorname{Re}\theta)|_L \equiv 0$.*

In practice, the condition $\omega|_L \equiv \operatorname{Im}\theta|_L \equiv 0$ is a more useful definition of SL m-folds than being calibrated with respect to $\operatorname{Re}\theta$. Also, an m-submanifold L in $\mathbb{C}^m$ is called *Lagrangian* if $\omega|_L \equiv 0$. (This is a term from symplectic geometry, and ω is a symplectic structure.) Thus special Lagrangian submanifolds are Lagrangian submanifolds satisfying the extra condition $\operatorname{Im}\theta|_L \equiv 0$, which is how they get their name.

Using Proposition 8.1.2 it is easy to determine the family $\mathcal{F}$ of SL m-planes in $\mathbb{C}^m$:

Proposition 8.1.3 *The family $\mathcal{F}$ of oriented real m-planes V in $\mathbb{C}^m$ with $\operatorname{Re}\theta|_V = \operatorname{vol}_V$ is isomorphic to $\mathrm{SU}(m)/\mathrm{SO}(m)$, with dimension $\frac{1}{2}(m^2 + m - 2)$.*

As special Lagrangian submanifolds in $\mathbb{C}^m$ are calibrated, they are minimal. Harvey and Lawson [151, Th. III.2.7] use this to show that they are real analytic:

Theorem 8.1.4 *Let L be an SL m-fold in $\mathbb{C}^m$. Then L is real analytic wherever it is nonsingular.*

Examples are known [151, p. 97] of special Lagrangian singularities which are not real analytic. Harvey and Lawson also prove [151, Th. III.5.5]:

Theorem 8.1.5 *Let P be a real analytic $(m-1)$-submanifold in $\mathbb{C}^m$ with $\omega|_P \equiv 0$. Then there exists a locally unique SL m-fold L in $\mathbb{C}^m$ containing P.*

They assume P is real analytic because their proof uses the *Cartan–Kähler Theorem*, from the subject of exterior differential systems, and this only works in the real analytic category. The submanifold L is defined by a kind of Taylor series, which converges in a small neighbourhood of P. One can use the same methods to show that SL m-folds in $\mathbb{C}^m$ 'depend on 2 functions of $m-1$ variables', in the sense of exterior differential systems. Thus, there are very many special Lagrangian submanifolds in $\mathbb{C}^m$.

8.1.1   Special Lagrangian 2-folds in $\mathbb{C}^2$ and the quaternions

The smallest interesting dimension, $m = 2$, is a special case. Let $\mathbb{C}^2$ have complex coordinates (z_1, z_2), complex structure I, and metric g, Kähler form ω and holomorphic 2-form θ as in (8.1). Define real coordinates (x_0, x_1, x_2, x_3) on $\mathbb{C}^2 \cong \mathbb{R}^4$ by $z_0 = x_0 + ix_1$, $z_1 = x_2 + ix_3$. Then

$$g = dx_0^2 + \cdots + dx_3^2, \qquad\qquad \omega = dx_0 \wedge dx_1 + dx_2 \wedge dx_3,$$
$$\operatorname{Re}\theta = dx_0 \wedge dx_2 - dx_1 \wedge dx_3 \qquad \text{and} \quad \operatorname{Im}\theta = dx_0 \wedge dx_3 + dx_1 \wedge dx_2.$$

Now define a *different* set of complex coordinates (w_1, w_2) on $\mathbb{C}^2 = \mathbb{R}^4$ by $w_1 = x_0 + ix_2$ and $w_2 = x_1 - ix_3$. Then $\omega - i\operatorname{Im}\theta = dw_1 \wedge dw_2$.

But by Proposition 8.1.2, a real 2-submanifold $L \subset \mathbb{R}^4$ is special Lagrangian if and only if $\omega|_L \equiv \operatorname{Im} \theta|_L \equiv 0$. Thus, L is special Lagrangian if and only if $(\mathrm{d}w_1 \wedge \mathrm{d}w_2)|_L \equiv 0$. But this holds if and only if L is a *holomorphic curve* with respect to the complex coordinates (w_1, w_2).

Here is another way to say this. There are *two different* complex structures I and J involved in this problem, associated to the two different complex coordinate systems (z_1, z_2) and (w_1, w_2) on $\mathbb{R}^4$. In the coordinates $(x_0, \ldots, x_3)$, I and J are given by

$$I\left(\tfrac{\partial}{\partial x_0}\right) = \tfrac{\partial}{\partial x_1}, \quad I\left(\tfrac{\partial}{\partial x_1}\right) = -\tfrac{\partial}{\partial x_0}, \quad I\left(\tfrac{\partial}{\partial x_2}\right) = \tfrac{\partial}{\partial x_3}, \quad I\left(\tfrac{\partial}{\partial x_3}\right) = -\tfrac{\partial}{\partial x_2},$$

$$J\left(\tfrac{\partial}{\partial x_0}\right) = \tfrac{\partial}{\partial x_2}, \quad J\left(\tfrac{\partial}{\partial x_1}\right) = -\tfrac{\partial}{\partial x_3}, \quad J\left(\tfrac{\partial}{\partial x_2}\right) = -\tfrac{\partial}{\partial x_0}, \quad J\left(\tfrac{\partial}{\partial x_3}\right) = \tfrac{\partial}{\partial x_1}.$$

The usual complex structure on $\mathbb{C}^2$ is I, but a 2-fold L in $\mathbb{C}^2$ is special Lagrangian if and only if it is holomorphic w.r.t. the alternative complex structure J. This means that special Lagrangian 2-folds are already very well understood, so we generally focus our attention on dimensions $m \geqslant 3$.

We can express all this in terms of the *quaternions* $\mathbb{H}$. The complex structures I, J anticommute, so that $IJ = -JI$, and $K = IJ$ is also a complex structure on $\mathbb{R}^4$, and $\langle 1, I, J, K \rangle$ is an algebra of automorphisms of $\mathbb{R}^4$ isomorphic to $\mathbb{H}$.

8.1.2   Special Lagrangian submanifolds in $\mathbb{C}^m$ as graphs

In symplectic geometry, there is a well-known way of manufacturing *Lagrangian* submanifolds of $\mathbb{R}^{2m} \cong \mathbb{C}^m$, which works as follows. Let $f : \mathbb{R}^m \to \mathbb{R}$ be a smooth function, and define

$$\Gamma_f = \left\{ \left(x_1 + i\tfrac{\partial f}{\partial x_1}(x_1, \ldots, x_m), \ldots, x_m + i\tfrac{\partial f}{\partial x_m}(x_1, \ldots, x_m) \right) : x_1, \ldots, x_m \in \mathbb{R} \right\}.$$

Then Γ_f is a smooth real m-dimensional submanifold of $\mathbb{C}^m$, with $\omega|_{\Gamma_f} \equiv 0$. Identifying $\mathbb{C}^m \cong \mathbb{R}^{2m} \cong \mathbb{R}^m \times (\mathbb{R}^m)^*$, we may regard Γ_f as the graph of the 1-form $\mathrm{d}f$ on $\mathbb{R}^m$, so that Γ_f is the *graph of a closed 1-form*. Locally, but not globally, every Lagrangian submanifold arises from this construction.

Now by Proposition 8.1.2, a special Lagrangian m-fold in $\mathbb{C}^m$ is a Lagrangian m-fold L satisfying the additional condition that $\operatorname{Im} \theta|_L \equiv 0$. We shall find the condition for Γ_f to be a special Lagrangian m-fold. Define the *Hessian* $\operatorname{Hess} f$ of f to be the $m \times m$ matrix $\left(\tfrac{\partial^2 f}{\partial x_i \partial x_j} \right)_{i,j=1}^m$ of real functions on $\mathbb{R}^m$. Then it is easy to show that $\operatorname{Im} \theta|_{\Gamma_f} \equiv 0$ if and only if

$$\operatorname{Im} \det{}_{\mathbb{C}} \left(I_m + i \operatorname{Hess} f \right) \equiv 0 \quad \text{on } \mathbb{R}^m, \tag{8.2}$$

where I_m is the $m \times m$ identity matrix. This is a *nonlinear second-order elliptic partial differential equation* upon the function $f : \mathbb{R}^m \to \mathbb{R}$.

It is known that if $f : \mathbb{R}^m \to \mathbb{R}$ is a global solution of (8.2) satisfying one of several extra conditions, to do with convexity or order of growth, then f must be a quadratic polynomial, so that Γ_f is a real affine m-plane in $\mathbb{C}^m$. For more details see Yuan [346].

8.1.3   Local discussion of special Lagrangian deformations

Suppose L_0 is a special Lagrangian submanifold in $\mathbb{C}^m$ (or, more generally, in some Calabi–Yau m-fold). What can we say about the family of *special Lagrangian deformations* of L_0, that is, the set of special Lagrangian m-folds L that are 'close to L_0'

in a suitable sense? Essentially, deformation theory is one way of thinking about the question 'how many special Lagrangian submanifolds are there in $\mathbb{C}^m$?'.

Locally (that is, in small enough open sets), every special Lagrangian m-fold looks quite like $\mathbb{R}^m$ in $\mathbb{C}^m$. Therefore deformations of special Lagrangian m-folds should look like special Lagrangian deformations of $\mathbb{R}^m$ in $\mathbb{C}^m$. So, we would like to know what special Lagrangian m-folds L in $\mathbb{C}^m$ close to $\mathbb{R}^m$ look like.

As $\mathbb{R}^m$ is the graph Γ_f associated to the function $f \equiv 0$, a graph Γ_f will be close to $\mathbb{R}^m$ if the function f and its derivatives are small. But then Hess f is small, so we can approximate eqn (8.2) by its *linearization*. Now

$$\operatorname{Im} \det_{\mathbb{c}}\left(I_m + i \operatorname{Hess} f\right) = \operatorname{Tr} \operatorname{Hess} f + \text{higher order terms}.$$

Thus, when the second derivatives of f are small, eqn (8.2) reduces approximately to $\operatorname{Tr} \operatorname{Hess} f \equiv 0$. But $\operatorname{Tr} \operatorname{Hess} f = \frac{\partial^2 f}{(\partial x_1)^2} + \cdots + \frac{\partial^2 f}{(\partial x_m)^2} = -\Delta f$, where Δ is the *Laplacian* on $\mathbb{R}^m$.

Hence, the small special Lagrangian deformations of $\mathbb{R}^m$ in $\mathbb{C}^m$ are approximately parametrized by small *harmonic functions* on $\mathbb{R}^m$. Actually, because adding a constant to f has no effect on Γ_f, this parametrization is degenerate. We can get round this by parametrizing instead by df, which is a closed and coclosed 1-form. This justifies the following:

Principle. *Small special Lagrangian deformations of a special Lagrangian m-fold L are approximately parametrized by closed and coclosed 1-forms α on L.*

This is the idea behind McLean's Theorem, Theorem 8.4.5 below.

We have seen using (8.2) that the deformation problem for special Lagrangian m-folds can be written as an *elliptic equation*. In particular, there are the same number of equations as functions, so the problem is neither overdetermined nor underdetermined. Therefore we do not expect special Lagrangian m-folds to be very few and very rigid (as would be the case if (8.2) were overdetermined), nor to be very abundant and very flabby (as would be the case if (8.2) were underdetermined).

If we think about Proposition 8.1.3 for a while, this may seem surprising. For the set $\mathcal{F}$ of special Lagrangian m-planes in $\mathbb{C}^m$ has dimension $\frac{1}{2}(m^2 + m - 2)$, but the set of all real m-planes in $\mathbb{C}^m$ has dimension m^2. So the special Lagrangian m-planes have codimension $\frac{1}{2}(m^2 - m + 2)$ in the set of all m-planes.

This means that the condition for a real m-submanifold L in $\mathbb{C}^m$ to be special Lagrangian is $\frac{1}{2}(m^2 - m + 2)$ real equations on each tangent space of L. However, the freedom to vary L is the sections of its normal bundle in $\mathbb{C}^m$, which is m real functions. When $m \geqslant 3$, there are more equations than functions, so we would expect the deformation problem to be *overdetermined*.

The explanation is that because ω is a *closed* 2-form, submanifolds L with $\omega|_L \equiv 0$ are much more abundant than would otherwise be the case. So the closure of ω is a kind of integrability condition necessary for the existence of many special Lagrangian submanifolds, just as the integrability of an almost complex structure is a necessary condition for the existence of many complex submanifolds of dimension greater than 1 in a complex manifold.

8.2   Constructing examples of SL m-folds in $\mathbb{C}^m$

We now survey constructions of SL m-folds in $\mathbb{C}^m$. This has been a very active area in recent years, and we give many references. The same examples have often been discovered independently by different people, and described in different terminology. For instance, special Lagrangian cones in $\mathbb{C}^m$, and special Legendrian or minimal Legendrian $(m-1)$-folds in $\mathcal{S}^{2m-1}$, and minimal Lagrangian $(m-1)$-folds in $\mathbb{CP}^{m-1}$, are all essentially the same thing, but appear in different branches of the literature.

Our principal interest in these constructions is that they yield many examples of *singular* SL m-folds in $\mathbb{C}^m$, and so hopefully will help in understanding what general singularities of SL m-folds in Calabi–Yau m-folds are like. In particular, later we shall be interested in examples of SL cones and asymptotically conical SL m-folds in $\mathbb{C}^m$.

8.2.1   Special Lagrangian m-folds with large symmetry groups

Here is a method used by Harvey and Lawson [151, §III.3], Haskins [155], Goldstein [127, 128] and the author [196] to construct examples of SL m-folds in $\mathbb{C}^m$. The group $\mathrm{SU}(m) \ltimes \mathbb{C}^m$ acts on $\mathbb{C}^m$ preserving all the structure g, ω, θ, so that it takes SL m-folds to SL m-folds in $\mathbb{C}^m$. Let G be a Lie subgroup of $\mathrm{SU}(m) \ltimes \mathbb{C}^m$ with Lie algebra $\mathfrak{g}$, and N a connected G-invariant SL m-fold in $\mathbb{C}^m$.

Since G preserves the symplectic form ω on $\mathbb{C}^m$, one can show that it has a *moment map* $\mu : \mathbb{C}^m \to \mathfrak{g}^*$. As N is Lagrangian, one can show that μ is constant on N, that is, $\mu \equiv c$ on N for some $c \in Z(\mathfrak{g}^*)$, the *centre* of $\mathfrak{g}^*$.

If the orbits of G in N are of codimension 1 (that is, dimension $m - 1$), then N is a 1-parameter family of G-orbits $\mathcal{O}_t$ for $t \in \mathbb{R}$. After reparametrizing the variable t, it can be shown that the special Lagrangian condition is equivalent to an o.d.e. in t upon the orbits $\mathcal{O}_t$.

Thus, we can construct examples of cohomogeneity one SL m-folds in $\mathbb{C}^m$ by solving an o.d.e. in the family of $(m - 1)$-dimensional G-orbits $\mathcal{O}$ in $\mathbb{C}^m$ with $\mu|_{\mathcal{O}} \equiv c$, for fixed $c \in Z(\mathfrak{g}^*)$. This o.d.e. usually turns out to be *integrable*.

Now suppose N is a *special Lagrangian cone* in $\mathbb{C}^m$, invariant under a subgroup $G \subset \mathrm{SU}(m)$ which has orbits of dimension $m - 2$ in N. In effect the symmetry group of N is $G \times \mathbb{R}_+$, where $\mathbb{R}_+$ acts by *dilations*, as N is a cone. Thus, in this situation too the symmetry group of N acts with cohomogeneity one, and we again expect the problem to reduce to an o.d.e.

One can show that $N \cap \mathcal{S}^{2m-1}$ is a 1-parameter family of G-orbits $\mathcal{O}_t$ in $\mathcal{S}^{2m-1} \cap \mu^{-1}(0)$ satisfying an o.d.e. By solving this o.d.e. we construct SL cones in $\mathbb{C}^m$. When $G = \mathrm{U}(1)^{m-2}$, the o.d.e. has many *periodic solutions* yielding large families of distinct SL cones on T^{m-1}. In particular, this gives many examples of SL T^2-cones in $\mathbb{C}^3$, studied by Haskins [155], the author [196], and in different language by Castro and Urbano [73].

Bryant [62] proves a different kind of result, on SL m-folds in $\mathbb{C}^m$ invariant under $\mathrm{SO}(m-1)$ acting on $\mathbb{C} \times \mathbb{C}^{m-1}$ trivially on $\mathbb{C}$ and in the usual way on $\mathbb{C}^{m-1}$. He shows that for each real analytic curve C in $\mathbb{C} \times \{0\}$, there are locally exactly $m - 1$ distinct $\mathrm{SO}(m-1)$-invariant SL m-folds N in $\mathbb{C}^m$ whose fixed locus under $\mathrm{SO}(m-1)$ is C.

8.2.2   Evolution equations for special Lagrangian m-folds

The following method was used by the author in [190, 191] to construct examples of
SL m-folds in $\mathbb{C}^m$. A related but less general method was used by Lawlor [231], and
completed by Harvey [150, p. 139–143]. Let P be a real analytic $(m-1)$-dimensional
manifold, and χ a nonvanishing real analytic section of $\Lambda^{m-1}TP$. Let $\{\phi_t : t \in \mathbb{R}\}$ be
a 1-parameter family of real analytic maps $\phi_t : P \to \mathbb{C}^m$. Consider the o.d.e.

$$\left(\frac{d\phi_t}{dt}\right)^b = (\phi_t)_*(\chi)^{a_1 \cdots a_{m-1}}(\operatorname{Re}\theta)_{a_1 \cdots a_{m-1} a_m} g^{a_m b}, \qquad (8.3)$$

using the index notation for (real) tensors on $\mathbb{C}^m$, where g^{ab} is the inverse of the Euclidean metric g_{ab} on $\mathbb{C}^m$.

It is shown in [190, §3] that if the ϕ_t satisfy (8.3) and $\phi_0^*(\omega) \equiv 0$, then $\phi_t^*(\omega) \equiv 0$
for all t, and $N = \{\phi_t(p) : p \in P, t \in \mathbb{R}\}$ is an SL m-fold in $\mathbb{C}^m$ wherever it is
nonsingular. We think of (8.3) as an *evolution equation*, and N as the result of evolving
a 1-parameter family of $(m-1)$-submanifolds $\phi_t(P)$ in $\mathbb{C}^m$.

Here is one way to understand this result. Suppose we are given $\phi_t : P \to \mathbb{C}^m$ for
some t, and we want to find an SL m-fold N in $\mathbb{C}^m$ containing the $(m-1)$-submanifold
$\phi_t(P)$. As N is Lagrangian, a necessary condition for this is that $\omega|_{\phi_t(P)} \equiv 0$, and
hence $\phi_t^*(\omega) \equiv 0$ on P.

The effect of eqn (8.3) is to flow $\phi_t(P)$ in the direction in which $\operatorname{Re}\theta$ is 'largest'.
The result is that $\operatorname{Re}\theta$ is 'maximized' on N, given the initial conditions. But $\operatorname{Re}\theta$ is
maximal on N exactly when N is calibrated w.r.t. $\operatorname{Re}\theta$, that is, when N is special Lagrangian. The same technique also works for other calibrations, such as the associative
and coassociative calibrations on $\mathbb{R}^7$, and the Cayley calibration on $\mathbb{R}^8$.

Now (8.3) evolves amongst the infinite-dimensional family of real analytic maps
$\phi : P \to \mathbb{C}^m$ with $\phi^*(\omega) \equiv 0$, so it is an *infinite-dimensional* problem, and thus
difficult to solve explicitly. However, there are *finite-dimensional* families $\mathcal{C}$ of maps
$\phi : P \to \mathbb{C}^m$ such that evolution stays in $\mathcal{C}$. This gives a *finite-dimensional* o.d.e.,
which can hopefully be solved fairly explicitly. For example, if we take G to be a Lie
subgroup of $\mathrm{SU}(m) \ltimes \mathbb{C}^m$, P to be an $(m-1)$-dimensional homogeneous space G/H,
and $\phi : P \to \mathbb{C}^m$ to be G-equivariant, we recover the construction of §8.2.1.

But there are also other possibilities for $\mathcal{C}$ which do not involve a symmetry assumption. Suppose P is a submanifold of $\mathbb{R}^n$, and χ the restriction to P of a linear or affine
map $\mathbb{R}^n \to \Lambda^{m-1}\mathbb{R}^n$. (This is a strong condition on P and χ.) Then we can take $\mathcal{C}$ to
be the set of restrictions to P of linear or affine maps $\mathbb{R}^n \to \mathbb{C}^m$.

For instance, set $m = n$ and let P be a quadric in $\mathbb{R}^m$. Then one can construct SL
m-folds in $\mathbb{C}^m$ with few symmetries by evolving quadrics in Lagrangian planes $\mathbb{R}^m$
in $\mathbb{C}^m$. When P is a quadric cone in $\mathbb{R}^m$ this gives many SL cones on products of
spheres $\mathcal{S}^a \times \mathcal{S}^b \times \mathcal{S}^1$.

8.2.3   Ruled special Lagrangian 3-folds

A 3-submanifold N in $\mathbb{C}^3$ is called *ruled* if it is fibred by a 2-dimensional family $\mathcal{F}$
of real lines in $\mathbb{C}^3$. A *cone* N_0 in $\mathbb{C}^3$ is called *two-sided* if $N_0 = -N_0$. Two-sided
cones are automatically ruled. If N is a ruled 3-fold in $\mathbb{C}^3$, the *asymptotic cone* N_0 of

N is the two-sided cone fibred by the lines passing through 0 and parallel to those in $\mathcal{F}$. Nonsingular ruled SL 3-folds N are automatically *Asymptotically Conical* in the sense of §8.3.3, with cone N_0.

Ruled SL 3-folds are studied by Harvey and Lawson [151, §III.3.C, §III.4.B], Bryant [60, §3] and the author [195]. Each (oriented) real line in $\mathbb{C}^3$ is determined by its *direction* in $\mathcal{S}^5$ together with an orthogonal *translation* from the origin. Thus a ruled 3-fold N is determined by a 2-dimensional family of directions and translations.

The condition that N be special Lagrangian turns out [195, §5] to reduce to two equations, the first involving only the direction components, and the second *linear* in the translation components. Thus, if a ruled 3-fold N in $\mathbb{C}^3$ is special Lagrangian, so is its asymptotic cone N_0. And the ruled SL 3-folds N asymptotic to a given two-sided SL cone N_0 come from solutions of a linear equation, and so form a *vector space*.

Let N_0 be a two-sided SL cone, and let $\Sigma = N_0 \cap \mathcal{S}^5$. Then Σ is a *Riemann surface*. Holomorphic vector fields on Σ give solutions to the linear equation (though not all solutions) [195, §6], and so yield new ruled SL 3-folds. In particular, each SL T^2-cone gives a 2-dimensional family of ruled SL 3-folds, which are generically diffeomorphic to $T^2 \times \mathbb{R}$ as immersed 3-submanifolds.

8.2.4   Integrable systems

Let N_0 be a special Lagrangian cone in $\mathbb{C}^3$, and set $\Sigma = N_0 \cap \mathcal{S}^5$. As N_0 is calibrated, it is minimal in $\mathbb{C}^3$, and so Σ is minimal in $\mathcal{S}^5$. That is, Σ is a *minimal Legendrian surface* in $\mathcal{S}^5$. Let $\pi : \mathcal{S}^5 \to \mathbb{CP}^2$ be the Hopf projection. One can also show that $\pi(\Sigma)$ is a *minimal Lagrangian surface* in $\mathbb{CP}^2$.

Regard Σ as a *Riemann surface*. Then the inclusions $\iota : \Sigma \to \mathcal{S}^5$ and $\pi \circ \iota : \Sigma \to \mathbb{CP}^2$ are *conformal harmonic maps*. Now harmonic maps from Riemann surfaces into $\mathcal{S}^n$ and $\mathbb{CP}^m$ are an *integrable system*. There is a complicated theory for classifying them in terms of algebro-geometric 'spectral data', and finding 'explicit' solutions. In principle, this gives all harmonic maps from T^2 into $\mathcal{S}^n$ and $\mathbb{CP}^m$. So, the field of integrable systems offers the hope of a *classification* of all SL T^2-cones in $\mathbb{C}^3$.

For a good general introduction to this field, see Fordy and Wood [109]. Sharipov [309] and Ma and Ma [252] apply this integrable systems machinery to describe minimal Legendrian tori in $\mathcal{S}^5$, and minimal Lagrangian tori in $\mathbb{CP}^2$, respectively, giving explicit formulae in terms of Prym theta functions. McIntosh [257] provides a more recent, readable, and complete discussion of special Lagrangian cones in $\mathbb{C}^3$ from the integrable systems perspective.

Carberry and McIntosh [72] use integrable systems to prove that for every $n \geqslant 1$ there exists a smooth n-dimensional family of SL T^2-cones in $\mathbb{C}^3$, with spectral curves of genus $2n + 4$, which are pairwise non-congruent under the SU(3) action. This is surprising, as on naïve geometric grounds one would expect special Lagrangian cones with isolated singularities to have no nontrivial deformations, and it shows that any classification of SL singularities will involve continuous as well as discrete parameters.

Haskins [154] estimates geometric quantities associated to an SL T^2-cone N_0, such as the area of Σ, in terms of the spectral genus of Σ. The families of SL T^2-cones constructed by U(1)-invariance in §8.2.1, and by evolving quadrics in §8.2.2, are both part of a more general, explicit, 'integrable systems' family of conformal harmonic maps $\mathbb{R}^2 \to \mathcal{S}^5$ with Legendrian image, involving two commuting, integrable o.d.e.s., described in [193].

8.2.5   Exterior differential systems

The theory of *exterior differential systems*, or *Cartan–Kähler theory*, can be used to study SL m-folds in $\mathbb{C}^m$ satisfying various extra conditions. The standard text on exterior differential systems is Bryant et al. [63], and a gentler introduction is Ivey and Landsberg [177]. Here are some papers using these techniques. Bryant [60] and Ionel [172] study and largely classify special Lagrangian submanifolds in $\mathbb{C}^3$ and $\mathbb{C}^4$ respectively, whose second fundamental form at a generic point has nontrivial stabilizer group in SO(3) or SO(4). Also Bryant [62] studies SO$(m-1)$-invariant SL m-folds in $\mathbb{C}^m$ with fixed loci, and [59] shows that any compact, oriented, real analytic Riemannian 3-manifold can be embedded as an SL 3-fold in a noncompact Calabi–Yau 3-fold.

8.2.6   Analytic construction of U(1)-invariant SL 3-folds in $\mathbb{C}^3$

Next we summarize the author's three papers [205–207], which study SL 3-folds N in $\mathbb{C}^3$ invariant under the U(1)-action

$$
e^{i\psi} : (z_1, z_2, z_3) \mapsto (e^{i\psi} z_1, e^{-i\psi} z_2, z_3) \quad \text{for } e^{i\psi} \in U(1). \tag{8.4}
$$

These three papers are briefly surveyed in [198]. Locally we can write N in the form

$$
N = \big\{ (z_1, z_2, z_3) \in \mathbb{C}^3 : z_1 z_2 = v(x, y) + iy, \quad z_3 = x + iu(x, y), \\
|z_1|^2 - |z_2|^2 = 2a, \quad (x, y) \in S \big\}, \tag{8.5}
$$

where S is a domain in $\mathbb{R}^2$, $a \in \mathbb{R}$ and $u, v : S \to \mathbb{R}$ are continuous.

Here we may take $|z_1|^2 - |z_2|^2 = 2a$ to be one of the equations defining N as $|z_1|^2 - |z_2|^2$ is the *moment map* of the U(1)-action (8.4), and so $|z_1|^2 - |z_2|^2$ is constant on any U(1)-invariant Lagrangian 3-fold in $\mathbb{C}^3$. Effectively (8.5) just means that we are choosing $x = \mathrm{Re}\, z_3$ and $y = \mathrm{Im}(z_1 z_2)$ as local coordinates on the 2-manifold $N/U(1)$. Then we find [205, Prop. 4.1]:

Proposition 8.2.1 *Let* S, a, u, v *and* N *be as above. Then*

(a) *If* $a = 0$, *then* N *is a (possibly singular) SL 3-fold in* $\mathbb{C}^3$ *if* u, v *are differentiable and satisfy*

$$
\frac{\partial u}{\partial x} = \frac{\partial v}{\partial y} \quad \text{and} \quad \frac{\partial v}{\partial x} = -2(v^2 + y^2)^{1/2} \frac{\partial u}{\partial y}, \tag{8.6}
$$

except at points $(x, 0)$ *in* S *with* $v(x, 0) = 0$, *where* u, v *need not be differentiable. The singular points of* N *are those of the form* $(0, 0, z_3)$, *where* $z_3 = x + iu(x, 0)$ *for* $(x, 0) \in S$ *with* $v(x, 0) = 0$.

(b) *If* $a \neq 0$, *then* N *is a nonsingular SL 3-fold in* $\mathbb{C}^3$ *if and only if* u, v *are differentiable in* S *and satisfy*

$$
\frac{\partial u}{\partial x} = \frac{\partial v}{\partial y} \quad \text{and} \quad \frac{\partial v}{\partial x} = -2(v^2 + y^2 + a^2)^{1/2} \frac{\partial u}{\partial y}. \tag{8.7}
$$

Now (8.6) and (8.7) are *nonlinear Cauchy–Riemann equations*. Thus, we may treat $u + iv$ as like a holomorphic function of $x + iy$. Many of the results in [205–207] are analogues of well-known results in elementary complex analysis.

In [205, Prop. 7.1] we show that solutions $u, v \in C^1(S)$ of (8.7) come from a potential $f \in C^2(S)$ satisfying a second-order quasilinear elliptic equation.

Proposition 8.2.2 *Let S be a domain in $\mathbb{R}^2$ and $u, v \in C^1(S)$ satisfy (8.7) for $a \neq 0$. Then there exists $f \in C^2(S)$ with $\frac{\partial f}{\partial y} = u$, $\frac{\partial f}{\partial x} = v$ and*

$$P(f) = \left(\left(\frac{\partial f}{\partial x} \right)^2 + y^2 + a^2 \right)^{-1/2} \frac{\partial^2 f}{\partial x^2} + 2 \frac{\partial^2 f}{\partial y^2} = 0. \tag{8.8}$$

This f is unique up to addition of a constant, $f \mapsto f + c$. Conversely, all solutions of (8.8) yield solutions of (8.7).

In the following result, a condensation of [205, Th. 7.6] and [206, Th.s 9.20 & 9.21], we prove existence and uniqueness for the *Dirichlet problem* for (8.8).

Theorem 8.2.3 *Suppose S is a strictly convex domain in $\mathbb{R}^2$ invariant under $(x, y) \mapsto (x, -y)$, and $\alpha \in (0, 1)$. Let $a \in \mathbb{R}$ and $\phi \in C^{3,\alpha}(\partial S)$. Then if $a \neq 0$ there exists a unique solution f of (8.8) in $C^{3,\alpha}(S)$ with $f|_{\partial S} = \phi$. If $a = 0$ there exists a unique $f \in C^1(S)$ with $f|_{\partial S} = \phi$, which is twice weakly differentiable and satisfies (8.8) with weak derivatives. Furthermore, the map $C^{3,\alpha}(\partial S) \times \mathbb{R} \to C^1(S)$ taking $(\phi, a) \mapsto f$ is continuous.*

Here a domain S in $\mathbb{R}^2$ is *strictly convex* if it is convex and the curvature of ∂S is nonzero at each point. Also domains are by definition compact, with smooth boundary, and $C^{3,\alpha}(\partial S)$ and $C^{3,\alpha}(S)$ are *Hölder spaces* of functions on ∂S and S, as in §1.2.2. For more details see [205, 206].

Combining Propositions 8.2.1 and 8.2.2 and Theorem 8.2.3 gives existence and uniqueness for a large class of $U(1)$-invariant SL 3-folds in $\mathbb{C}^3$, with boundary conditions, and including *singular* SL 3-folds. It is interesting that this existence and uniqueness is *entirely unaffected* by singularities appearing in S°.

Here are some other areas covered in [205–207]. Examples of solutions u, v of (8.6) and (8.7) are given in [205, §5]. In [206] we give more precise statements on the regularity of singular solutions of (8.6) and (8.8). In [205, §6] and [207, §7] we consider the zeroes of $(u_1, v_1) - (u_2, v_2)$, where (u_j, v_j) are (possibly singular) solutions of (8.6) and (8.7).

We show that if $(u_1, v_1) \not\equiv (u_2, v_2)$ then the zeroes of $(u_1, v_1) - (u_2, v_2)$ in S° are *isolated*, with a positive integer *multiplicity*, and that the zeroes of $(u_1, v_1) - (u_2, v_2)$ in S° can be counted with multiplicity in terms of boundary data on ∂S. In particular, under some boundary conditions we can show $(u_1, v_1) - (u_2, v_2)$ has no zeroes in S°, so that the corresponding SL 3-folds do not intersect. This will be important in constructing $U(1)$-invariant SL fibrations in §9.4.5.

In [207, §9–§10] we study singularities of solutions u, v of (8.6). We show that either $u(x, -y) \equiv u(x, y)$ and $v(x, -y) \equiv -v(x, y)$, so that u, v are singular all along the x-axis, or else the singular points of u, v in S° are all *isolated*, with a positive integer *multiplicity*, and one of two *types*. We also show that singularities exist with every multiplicity and type, and multiplicity n singularities occur in codimension n in the family of all $U(1)$-invariant SL 3-folds.

A partial extension of these results to higher dimensions is given by Castro and Urbano [74], who consider SL m-folds in $\mathbb{C}^m$ invariant under $SO(a) \times SO(b)$ for $a, b \geq 1$ with $a + b = m$, and show that such SL m-folds can be written locally in terms of solutions u, v of a nonlinear Cauchy–Riemann equation generalizing (8.6). When $a = 2, b = 1$ and $m = 3$ their result reduces to Proposition 8.2.1(a). However, Theorem 8.2.3 and other analytic results of [205–207] have not yet been generalized to this context.

Section 9.4.5 will apply these results to *special Lagrangian fibrations* and the *SYZ Conjecture*.

8.2.7 SL cones on higher genus surfaces in $\mathbb{C}^3$

In an elegant but difficult paper, Haskins and Kapouleas [156] construct examples of SL cones C in $\mathbb{C}^3$ whose intersection with the unit sphere $\Sigma = C \cap \mathcal{S}^5$ is a compact Riemann surface of genus g, for all odd $g \geq 1$. Their method is analytic, and involves gluing together pieces of several SL cones in $\mathbb{C}^3$ invariant under different $U(1)$ subgroups of $SU(3)$ to get an approximately special Lagrangian cone, and then deforming to an exactly special Lagrangian one. The construction involves a large integer parameter, and Σ has large area, so the 'stability index' s-ind(C) of §8.3.1 below is also large.

See also Wang [337] for another analytic construction of SL cones C on higher genus surfaces Σ in $\mathbb{C}^3$, which works by minimizing the area of Legendrian surfaces Σ in $\mathcal{S}^5$ invariant under a finite group $\Gamma \subset SU(3)$, so Σ has small area. However, the author is unsure whether the proof is yet complete. Note too that the integrable systems ideas of §8.2.4 presently give no information on SL cones on surfaces Σ of genus $g > 1$.

8.2.8 Examples of singular special Lagrangian 3-folds in $\mathbb{C}^3$

We finish by describing four families of SL 3-folds in $\mathbb{C}^3$, as examples of the material of §8.2.1–§8.2.4. They have been chosen to illustrate different kinds of singular behaviour of SL 3-folds, and also to show how nonsingular SL 3-folds can converge to a singular SL 3-fold, to serve as a preparation for our discussion of singularities of SL m-folds in §8.5. Our first example derives from Harvey and Lawson [151, §III.3.A], and is discussed in detail in [194, §3] and [197, §4].

Example 8.2.4 Define a subset L_0 in $\mathbb{C}^3$ by

$$L_0 = \{ (re^{i\psi_1}, re^{i\psi_2}, re^{i\psi_3}) : r \geq 0, \ \psi_1, \psi_2, \psi_3 \in \mathbb{R}, \ \psi_1 + \psi_2 + \psi_3 = 0 \}. \quad (8.9)$$

Then L_0 is a *special Lagrangian cone* on T^2. An alternative definition is

$$L_0 = \{ (z_1, z_2, z_3) \in \mathbb{C}^3 : |z_1| = |z_2| = |z_3|, \ \mathrm{Im}(z_1 z_2 z_3) = 0, \ \mathrm{Re}(z_1 z_2 z_3) \geq 0 \}.$$

Let $a > 0$, write $\mathcal{S}^1 = \{ e^{i\psi} : \psi \in \mathbb{R} \}$, and define $\phi_a : \mathcal{S}^1 \times \mathbb{C} \to \mathbb{C}^3$ by

$$\phi_a : (e^{i\psi}, z) \mapsto \left((|z|^2 + 2a)^{1/2} e^{i\psi}, z, e^{-i\psi} \bar{z} \right).$$

Then ϕ_a is an *embedding*. Define $L_a = \text{Image}\,\phi_a$. Then L_a is a nonsingular special Lagrangian 3-fold in $\mathbb{C}^3$ diffeomorphic to $\mathcal{S}^1 \times \mathbb{R}^2$. An equivalent definition is

$$L_a = \big\{ (z_1, z_2, z_3) \in \mathbb{C}^3 : |z_1|^2 - 2a = |z_2|^2 = |z_3|^2,$$
$$\text{Im}(z_1 z_2 z_3) = 0, \quad \text{Re}(z_1 z_2 z_3) \geqslant 0 \big\}.$$

As $a \to 0_+$, the nonsingular SL 3-fold L_a converges to the singular SL cone L_0. Note that L_a is asymptotic to L_0 at infinity, so it is *asymptotically conical*, and that $L_a = a^{1/2} L_1$ for $a > 0$, so that the L_a for $a > 0$ are all homothetic to each other. Also, each L_a for $a \geqslant 0$ is invariant under the T^2 subgroup of SU(3) acting by

$$(z_1, z_2, z_3) \longmapsto (e^{i\psi_1} z_1, e^{i\psi_2} z_2, e^{i\psi_3} z_3) \text{ for } \psi_1, \psi_2, \psi_3 \in \mathbb{R} \text{ with } \psi_1 + \psi_2 + \psi_3 = 0,$$

and so fits into the framework of §8.2.1. By [205, Th. 5.1] the L_a may also be written in the form (8.5) for continuous $u, v : \mathbb{R}^2 \to \mathbb{R}$, as in §8.2.6.

Our second example is taken from [196, Ex. 9.4 & Ex. 9.5].

Example 8.2.5 Let a_1, a_2 be positive, coprime integers, and set $a_3 = -a_1 - a_2$. Let $c \in \mathbb{R}$, and define

$$L_c^{a_1, a_2} = \big\{ (e^{ia_1\psi} x_1, e^{ia_2\psi} x_2, ie^{ia_3\psi} x_3) : \psi \in \mathbb{R}, \ x_j \in \mathbb{R}, \ a_1 x_1^2 + a_2 x_2^2 + a_3 x_3^2 = c \big\}.$$

Then $L_c^{a_1, a_2}$ is a special Lagrangian 3-fold, which comes from the 'evolving quadrics' construction of §8.2.2. It is also symmetric under the U(1)-action

$$(z_1, z_2, z_3) \longmapsto (e^{ia_1\psi} z_1, e^{ia_2\psi} z_2, ie^{ia_3\psi} z_3) \quad \text{for } \psi \in \mathbb{R},$$

but this is not a necessary feature of the construction; these are just the easiest examples to write down.

When $c = 0$ and a_3 is odd, $L_0^{a_1, a_2}$ is an embedded special Lagrangian cone on T^2, with one singular point at 0. When $c = 0$ and a_3 is even, $L_0^{a_1, a_2}$ is two opposite embedded SL T^2-cones with one singular point at 0.

When $c > 0$ and a_3 is odd, $L_c^{a_1, a_2}$ is an embedded 3-fold diffeomorphic to a nontrivial real line bundle over the Klein bottle. When $c > 0$ and a_3 is even, $L_c^{a_1, a_2}$ is an embedded 3-fold diffeomorphic to $T^2 \times \mathbb{R}$. In both cases, $L_c^{a_1, a_2}$ is a *ruled* SL 3-fold, as in §8.2.3, since it is fibred by hyperboloids of one sheet in $\mathbb{R}^3$, which are ruled in two different ways.

When $c < 0$ and a_3 is odd, $L_c^{a_1, a_2}$ is an immersed copy of $\mathcal{S}^1 \times \mathbb{R}^2$. When $c < 0$ and a_3 is even, $L_c^{a_1, a_2}$ is two immersed copies of $\mathcal{S}^1 \times \mathbb{R}^2$. In every case with $c \neq 0$, $L_c^{a_1, a_2}$ is asymptotically conical as in §8.3.3, with cone $L_0^{a_1, a_2}$.

Our third example is adapted from Harvey and Lawson [151, §III.3.B].

Example 8.2.6 For each $t > 0$, define

$$L_t = \big\{ (e^{i\psi} x_1, e^{i\psi} x_2, e^{i\psi} x_3) : x_j \in \mathbb{R}, \quad \psi \in (0, \pi/3),$$
$$x_1^2 + x_2^2 + x_3^2 = t^2 (\sin 3\psi)^{-2/3} \big\}.$$

Then L_t is a nonsingular embedded SL 3-fold in $\mathbb{C}^3$ diffeomorphic to $\mathcal{S}^2 \times \mathbb{R}$. As $t \to 0_+$ it converges to the singular union L_0 of the two SL 3-planes

$$\Pi_1 = \big\{(x_1, x_2, x_3) : x_j \in \mathbb{R}\big\} \text{ and } \Pi_2 = \big\{(e^{i\pi/3}x_1, e^{i\pi/3}x_2, e^{i\pi/3}x_3) : x_j \in \mathbb{R}\big\},$$

which intersect at 0. Note that L_t is invariant under the action of the Lie subgroup $SO(3)$ of $SU(3)$, acting on $\mathbb{C}^3$ in the obvious way, so again this comes from the method of §8.2.1. Also L_t is asymptotically conical, with asymptotic cone L_0.

All the singular SL 3-folds we have seen so far have been *cones* in $\mathbb{C}^3$. Our final example, taken from [191], has more complicated singularities which are not cones. They are difficult to describe in a simple way, so we will not say much about them. For more details, see [191].

Example 8.2.7 In [191, §5] the author constructed a family of smooth maps $\Phi : \mathbb{R}^3 \to \mathbb{C}^3$ with special Lagrangian image $N = \text{Image}\,\Phi$. It is shown in [191, §6] that generic Φ in this family are immersions, so that N is nonsingular as an immersed SL 3-fold, but in codimension 1 in the family they develop isolated singularities.

Here is a rough description of these singularities, taken from [191, §6]. Taking the singular point to be at $\Phi(0, 0, 0) = 0$, one can write Φ as

$$\begin{aligned}
\Phi(x, y, t) = &\big(x + \tfrac{1}{4}g(\mathbf{u}, \mathbf{v})t^2\big)\,\mathbf{u} + \big(y^2 - \tfrac{1}{4}|\mathbf{u}|^2 t^2\big)\,\mathbf{v} \\
&+ 2yt\,\mathbf{u} \times \mathbf{v} + O\big(x^2 + |xy| + |xt| + |y|^3 + |t|^3\big),
\end{aligned} \tag{8.10}$$

where $\mathbf{u}, \mathbf{v}$ are linearly independent vectors in $\mathbb{C}^3$ with $\omega(\mathbf{u}, \mathbf{v}) = 0$, and $\times : \mathbb{C}^3 \times \mathbb{C}^3 \to \mathbb{C}^3$ is defined by

$$(r_1, r_2, r_3) \times (s_1, s_2, s_3) = \tfrac{1}{2}(\bar{r}_2 \bar{s}_3 - \bar{r}_3 \bar{s}_2, \bar{r}_3 \bar{s}_1 - \bar{r}_1 \bar{s}_3, \bar{r}_1 \bar{s}_2 - \bar{r}_2 \bar{s}_1).$$

The next few terms in the expansion (8.10) can also be given very explicitly, but we will not write them down as they are rather complex, and involve further choices of vectors $\mathbf{w}, \mathbf{x}, \ldots$.

What is going on here is that the lowest order terms in Φ are a *double cover* of the special Lagrangian plane $\langle \mathbf{u}, \mathbf{v}, \mathbf{u} \times \mathbf{v} \rangle_{\mathbb{R}}$ in $\mathbb{C}^3$, *branched* along the real line $\langle \mathbf{u} \rangle_{\mathbb{R}}$. The branching occurs when $y = t = 0$. Higher order terms deviate from the 3-plane $\langle \mathbf{u}, \mathbf{v}, \mathbf{u} \times \mathbf{v} \rangle_{\mathbb{R}}$, and make the singularity isolated.

8.3 SL cones and Asymptotically Conical SL m-folds

We now discuss *special Lagrangian cones* in $\mathbb{C}^m$, SL m-folds C in $\mathbb{C}^m$ invariant under dilations $C \mapsto tC$ for $t > 0$ which are generally singular at their vertex 0, and *Asymptotically Conical special Lagrangian m-folds (AC SL m-folds)* L in $\mathbb{C}^m$, nonsingular SL m-folds which are asymptotic to a special Lagrangian cone C at infinity in $\mathbb{C}^m$. There are two main reasons we are interested in these.

Firstly, it is natural to ask if there are any attractive distinguished classes of SL m-folds in $\mathbb{C}^m$. There are no compact SL m-folds L without boundary in $\mathbb{C}^m$, since then $\text{vol}(L) = \int_L \text{Re}\,\Omega = 0$ as $\text{Re}\,\Omega$ is exact on $\mathbb{C}^m$, a contradiction. Thus we must impose some kind of boundary condition on L, either in the interior or at infinity in $\mathbb{C}^m$. To be asymptotically conical is a simple boundary condition at infinity in $\mathbb{C}^m$, and we will see that AC SL m-folds behave like compact SL m-folds in Calabi–Yau m-folds.

Secondly, SL cones and AC SL m-folds are important in the theory of *singular* SL m-folds N in (almost) Calabi–Yau m-folds M, which will be discussed in §8.5. As in §4.4, geometric measure theory defines a large class of singular SL m-folds called special Lagrangian *integral currents* N in M, and each singular point x of N has a special Lagrangian *tangent cone* C in T_xM, which describes N near x to leading order. Thus, SL cones are the local models for general singularities of SL m-folds. Also, as in §8.5.3, AC SL m-folds L with cone C provide local models for how to desingularize SL m-folds in (almost) Calabi–Yau m-folds with singularities modelled on C.

Some topics included here really relate to later sections. In particular, stable and rigid SL cones C and the stability index s-ind(C) in §8.3.1 are important in the theory of SL m-folds with conical singularities in §8.5, and the deformation theory in Theorem 8.3.10 is modelled on McLean's Theorem for compact SL m-folds, Theorem 8.4.5 below, and makes more sense if read in parallel with §8.4.1. But it seemed better to collect material on SL m-folds in $\mathbb{C}^m$ together in §8.1–§8.3, as we have done.

8.3.1 Preliminaries on special Lagrangian cones

We define *special Lagrangian cones*, and some notation.

Definition 8.3.1 A (singular) SL m-fold C in $\mathbb{C}^m$ is called a *cone* if $C = tC$ for all $t > 0$, where $tC = \{t\mathbf{x} : \mathbf{x} \in C\}$. Let C be a closed SL cone in $\mathbb{C}^m$ with an isolated singularity at 0. Then $\Sigma = C \cap \mathcal{S}^{2m-1}$ is a compact, nonsingular $(m-1)$-submanifold of $\mathcal{S}^{2m-1}$. Let g_Σ be the restriction of g to Σ, where g is as in (8.1).

Set $C' = C \setminus \{0\}$. Define $\iota : \Sigma \times (0, \infty) \to \mathbb{C}^m$ by $\iota(\sigma, r) = r\sigma$. Then ι has image C'. By an abuse of notation, *identify* C' with $\Sigma \times (0, \infty)$ using ι. The *cone metric* on $C' \cong \Sigma \times (0, \infty)$ is $g = \iota^*(g) = \mathrm{d}r^2 + r^2 g_\Sigma$.

For $\alpha \in \mathbb{R}$, we say that a function $u : C' \to \mathbb{R}$ is *homogeneous of order* α if $u \circ t \equiv t^\alpha u$ for all $t > 0$. Equivalently, u is homogeneous of order α if $u(\sigma, r) \equiv r^\alpha v(\sigma)$ for some function $v : \Sigma \to \mathbb{R}$.

In [200, Lem. 2.3] we study *homogeneous harmonic functions* on C'.

Lemma 8.3.2 *In the situation of Definition 8.3.1, let $u(\sigma, r) \equiv r^\alpha v(\sigma)$ be a homogeneous function of order α on $C' = \Sigma \times (0, \infty)$, for $v \in C^2(\Sigma)$. Then*

$$\Delta u(\sigma, r) = r^{\alpha-2}\big(\Delta_\Sigma v - \alpha(\alpha + m - 2)v\big)(\sigma),$$

where Δ, Δ_Σ are the Laplacians on (C', g) and (Σ, g_Σ). Hence, u is harmonic on C' if and only if v is an eigenfunction of Δ_Σ with eigenvalue $\alpha(\alpha + m - 2)$.

Following [200, Def. 2.5], we define:

Definition 8.3.3 In Definition 8.3.1, suppose $m > 2$ and define

$$\mathcal{D}_\Sigma = \big\{\alpha \in \mathbb{R} : \alpha(\alpha + m - 2) \text{ is an eigenvalue of } \Delta_\Sigma\big\}. \tag{8.11}$$

Then $\mathcal{D}_\Sigma$ is a countable, discrete subset of $\mathbb{R}$. By Lemma 8.3.2, an equivalent definition is that $\mathcal{D}_\Sigma$ is the set of $\alpha \in \mathbb{R}$ for which there exists a nonzero homogeneous harmonic function u of order α on C'.

Define $m_\Sigma : \mathcal{D}_\Sigma \to \mathbb{N}$ by taking $m_\Sigma(\alpha)$ to be the multiplicity of the eigenvalue $\alpha(\alpha + m - 2)$ of Δ_Σ, or equivalently the dimension of the vector space of homogeneous harmonic functions u of order α on C'. Define $N_\Sigma : \mathbb{R} \to \mathbb{Z}$ by

$$N_\Sigma(\delta) = - \sum_{\alpha \in \mathcal{D}_\Sigma \cap (\delta, 0)} m_\Sigma(\alpha) \text{ if } \delta < 0, \text{ and } N_\Sigma(\delta) = \sum_{\alpha \in \mathcal{D}_\Sigma \cap [0, \delta]} m_\Sigma(\alpha) \text{ if } \delta \geqslant 0. \quad (8.12)$$

Then N_Σ is monotone increasing and upper semicontinuous, and is discontinuous exactly on $\mathcal{D}_\Sigma$, increasing by $m_\Sigma(\alpha)$ at each $\alpha \in \mathcal{D}_\Sigma$. As the eigenvalues of Δ_Σ are nonnegative, we see that $\mathcal{D}_\Sigma \cap (2 - m, 0) = \emptyset$ and $N_\Sigma \equiv 0$ on $(2 - m, 0)$.

We define the *stability index* of C, and *stable* and *rigid* cones [201, Def. 3.6]. These definitions will be important in the conical singularities material of §8.5.

Definition 8.3.4 Let C be an SL cone in $\mathbb{C}^m$ for $m > 2$ with an isolated singularity at 0, let G be the Lie subgroup of $\mathrm{SU}(m)$ preserving C, and use the notation of Definitions 8.3.1 and 8.3.3. Then [201, eqn (8)] shows that

$$m_\Sigma(0) = b^0(\Sigma), \quad m_\Sigma(1) \geqslant 2m \quad \text{and} \quad m_\Sigma(2) \geqslant m^2 - 1 - \dim G. \quad (8.13)$$

Define the *stability index* s-ind(C) to be

$$\text{s-ind}(C) = N_\Sigma(2) - b^0(\Sigma) - m^2 - 2m + 1 + \dim G.$$

Then s-ind$(C) \geqslant 0$ by (8.13), as $N_\Sigma(2) \geqslant m_\Sigma(0) + m_\Sigma(1) + m_\Sigma(2)$ by (8.12). We call C *stable* if s-ind$(C) = 0$.

Following [200, Def. 6.7], we call C *rigid* if $m_\Sigma(2) = m^2 - 1 - \dim G$. As

$$\text{s-ind}(C) \geqslant m_\Sigma(2) - (m^2 - 1 - \dim G) \geqslant 0,$$

we see that *if C is stable, then C is rigid.*

We shall see in §8.5.2 that s-ind(C) is the dimension of an obstruction space to deforming an SL m-fold N with a conical singularity with cone C, and that if C is *stable* then the deformation theory of N simplifies. An SL cone C is *rigid* if all infinitesimal deformations of C as an SL cone come from $\mathrm{SU}(m)$ rotations of C.

Haskins [154] uses integrable systems techniques as in §8.2.4 to study the stability index s-ind(C) of special Lagrangian T^2-cones C in $\mathbb{C}^3$. Amongst other results, he derives a lower bound for s-ind(C) in terms of the spectral genus of $\Sigma = C \cap \mathcal{S}^5$, and proves that the Clifford torus cone L_0 of (8.9) is up to $\mathrm{SU}(3)$ isomorphism the unique stable SL T^2-cone C in $\mathbb{C}^3$.

8.3.2 Examples of special Lagrangian cones

In our first example, generalizing Example 8.2.4, we can compute the data of §8.3.1 very explicitly.

Example 8.3.5 Here is a family of special Lagrangian cones constructed by Harvey and Lawson [151, §III.3.A]. For $m \geqslant 3$, define

$$ C_{\mathrm{HL}}^m = \{(z_1, \ldots, z_m) \in \mathbb{C}^m : i^{m+1} z_1 \cdots z_m \in [0, \infty), \quad |z_1| = \cdots = |z_m|\}. \quad (8.14) $$

Then C_{HL}^m is a *special Lagrangian cone* in $\mathbb{C}^m$ with an isolated singularity at 0, and $\Sigma_{\mathrm{HL}}^m = C_{\mathrm{HL}}^m \cap \mathcal{S}^{2m-1}$ is an $(m-1)$-torus T^{m-1}. Both C_{HL}^m and Σ_{HL}^m are invariant under the $\mathrm{U}(1)^{m-1}$ subgroup of $\mathrm{SU}(m)$ acting by

$$ (z_1, \ldots, z_m) \mapsto (e^{i\psi_1} z_1, \ldots, e^{i\psi_m} z_m) \text{ for } \psi_j \in \mathbb{R} \text{ with } \psi_1 + \cdots + \psi_m = 0. \quad (8.15) $$

In fact $\pm C_{\mathrm{HL}}^m$ for m odd, and $C_{\mathrm{HL}}^m, iC_{\mathrm{HL}}^m$ for m even, are the unique SL cones in $\mathbb{C}^m$ invariant under (8.15), which is how Harvey and Lawson constructed them.

The metric on $\Sigma_{\mathrm{HL}}^m \cong T^{m-1}$ is flat, so it is easy to compute the eigenvalues of $\Delta_{\Sigma_{\mathrm{HL}}^m}$. This was done by Marshall [254, §6.3.4]. There is a 1-1 correspondence between $(n_1, \ldots, n_{m-1}) \in \mathbb{Z}^{m-1}$ and eigenvectors of $\Delta_{\Sigma_{\mathrm{HL}}^m}$ with eigenvalue

$$ m \sum_{i=1}^{m-1} n_i^2 - \sum_{i,j=1}^{m-1} n_i n_j. \quad (8.16) $$

Using (8.16) and a computer we can find the eigenvalues of $\Delta_{\Sigma_{\mathrm{HL}}^m}$ and their multiplicities. The Lie subgroup G_{HL}^m of $\mathrm{SU}(m)$ preserving C_{HL}^m has identity component the $\mathrm{U}(1)^{m-1}$ of (8.15), so that $\dim G_{\mathrm{HL}}^m = m - 1$. Thus we can calculate $\mathcal{D}_{\Sigma_{\mathrm{HL}}^m}$, $m_{\Sigma_{\mathrm{HL}}^m}$, $N_{\Sigma_{\mathrm{HL}}^m}$, and the stability index s-ind(C_{HL}^m). This was done by Marshall [254, Table 6.1] and the author [201, §3.2]. Table 8.1 gives the data m, $N_{\Sigma_{\mathrm{HL}}^m}(2)$, $m_{\Sigma_{\mathrm{HL}}^m}(2)$ and s-ind(C_{HL}^m) for $3 \leqslant m \leqslant 12$.

Table 8.1 Data for $\mathrm{U}(1)^{m-1}$-invariant SL cones C_{HL}^m in $\mathbb{C}^m$

m	3	4	5	6	7	8	9	10	11	12
$N_{\Sigma_{\mathrm{HL}}^m}(2)$	13	27	51	93	169	311	331	201	243	289
$m_{\Sigma_{\mathrm{HL}}^m}(2)$	6	12	20	30	42	126	240	90	110	132
s-ind(C_{HL}^m)	0	6	20	50	112	238	240	90	110	132

One can also prove that

$$ N_{\Sigma_{\mathrm{HL}}^m}(2) = 2m^2 + 1 \text{ and } m_{\Sigma_{\mathrm{HL}}^m}(2) = \text{s-ind}(C_{\mathrm{HL}}^m) = m^2 - m \text{ for } m \geqslant 10. \quad (8.17) $$

As C_{HL}^m is *stable* when s-ind$(C_{\mathrm{HL}}^m) = 0$ we see from Table 8.1 and (8.17) that C_{HL}^3 is a *stable* cone in $\mathbb{C}^3$, but C_{HL}^m is *unstable* for $m \geqslant 4$. Also C_{HL}^m is *rigid* when $m_{\Sigma_{\mathrm{HL}}^m}(2) = m^2 - m$, as $\dim G_{\mathrm{HL}}^m = m - 1$. Thus C_{HL}^m is *rigid* if and only if $m \neq 8, 9$, by Table 8.1 and (8.17).

Here is an example from [196, Ex. 9.4], generalizing Example 8.2.5.

Example 8.3.6 Let $a_1, \ldots, a_m \in \mathbb{Z}$ with $a_1 + \cdots + a_m = 0$ and highest common factor 1, such that $a_1, \ldots, a_k > 0$ and $a_{k+1}, \ldots, a_m < 0$ for $0 < k < m$. Define

$$L_0^{a_1, \ldots, a_m} = \{ \left(i e^{i a_1 \psi} x_1, e^{i a_2 \psi} x_2, \ldots, e^{i a_m \psi} x_m \right) : \psi \in [0, 2\pi), $$
$$x_1, \ldots, x_m \in \mathbb{R}, \qquad a_1 x_1^2 + \cdots + a_m x_m^2 = 0 \}.$$

Then $L_0^{a_1, \ldots, a_m}$ is an *immersed SL cone* in $\mathbb{C}^m$, with an isolated singularity at 0.

Define $C^{a_1, \ldots, a_m} = \{ (x_1, \ldots, x_m) \in \mathbb{R}^m : a_1 x_1^2 + \cdots + a_m x_m^2 = 0 \}$. Then $C^{a_1, \ldots, a_m}$ is a quadric cone on $\mathcal{S}^{k-1} \times \mathcal{S}^{m-k-1}$ in $\mathbb{R}^m$, and $L_0^{a_1, \ldots, a_m}$ is the image of an immersion $\Phi : C^{a_1, \ldots, a_m} \times \mathcal{S}^1 \to \mathbb{C}^m$, which is generically 2:1. Therefore $L_0^{a_1, \ldots, a_m}$ is an immersed SL cone on $(\mathcal{S}^{k-1} \times \mathcal{S}^{m-k-1} \times \mathcal{S}^1)/\mathbb{Z}_2$.

Ohnita [273] studies the symmetric spaces $\mathrm{SU}(p), \mathrm{SU}(p)/\mathrm{SO}(p), \mathrm{SU}(2p)/\mathrm{Sp}(p)$ for $p \geqslant 3$ and E_6/F_4, which may be embedded as homogeneous minimal Legendrian $(m-1)$-folds Σ in $\mathcal{S}^{2m-1}$, so that the cone C_Σ on Σ in $\mathbb{C}^m$ is special Lagrangian. Ohnita uses the representation theory of Lie groups to compute the spectrum of the Laplacian on Σ, and so determine the stability index and rigidity of the cone C_Σ. He shows that all these cones C_Σ are *rigid*, and that C_Σ is *stable* if and only if $\Sigma = \mathrm{SU}(3), \mathrm{SU}(3)/\mathrm{SO}(3), \mathrm{SU}(6)/\mathrm{Sp}(3)$ or E_6/F_4. This gives new examples of stable SL cones in $\mathbb{C}^m$ for $m = 6, 9, 15$ and 27. Many other examples of special Lagrangian cones have been constructed by the methods described in §8.2.

8.3.3 Asymptotically conical special Lagrangian submanifolds in $\mathbb{C}^m$

We now discuss *Asymptotically Conical* SL m-folds L in $\mathbb{C}^m$, [200, Def. 7.1].

Definition 8.3.7 Let C be a closed SL cone in $\mathbb{C}^m$ with isolated singularity at 0 for $m > 2$, and let $\Sigma = C \cap \mathcal{S}^{2m-1}$, so that Σ is a compact, nonsingular $(m-1)$-manifold, not necessarily connected. Let g_Σ be the metric on Σ induced by the metric g on $\mathbb{C}^m$ in (8.1), and r the radius function on $\mathbb{C}^m$. Define $\iota : \Sigma \times (0, \infty) \to \mathbb{C}^m$ by $\iota(\sigma, r) = r\sigma$. Then the image of ι is $C \setminus \{0\}$, and $\iota^*(g) = r^2 g_\Sigma + \mathrm{d}r^2$ is the cone metric on $C \setminus \{0\}$.

Let L be a closed, nonsingular SL m-fold in $\mathbb{C}^m$. We call L *asymptotically conical* (AC) with *rate* $\lambda < 2$ and *cone* C if there exists a compact subset $K \subset L$ and a diffeomorphism $\varphi : \Sigma \times (T, \infty) \to L \setminus K$ for $T > 0$, such that

$$\left| \nabla^k (\varphi - \iota) \right| = O(r^{\lambda - 1 - k}) \quad \text{as } r \to \infty \text{ for } k = 0, 1, \text{ and} \tag{8.18}$$
$$(\phi - \iota)(\sigma, r) \perp T_{\iota(\sigma, r)} C \quad \text{in } \mathbb{C}^m \text{ for all } (\sigma, r) \in \Sigma \times (T, \infty). \tag{8.19}$$

Here $\nabla, |.|$ are computed using the cone metric $\iota^*(g)$.

Actually (8.19) is not part of the original definition, but we show in [200, Th. 7.4] that if L satisfies all of Definition 8.3.7 except (8.19), and $T > 0$ is large enough, we can choose ϕ uniquely so that (8.19) holds. In [200, Ths 7.7 and 7.11] we study the asymptotic behaviour of L at infinity, showing that we can extend (8.18) to all $k \geqslant 0$, and also vary λ in $\mathbb{R} \setminus \mathcal{D}_\Sigma$.

Theorem 8.3.8 *Suppose L is an AC SL m-fold in $\mathbb{C}^m$ with cone C and rate λ, and use the notation of Definitions 8.3.1, 8.3.3 and 8.3.7. If either $\lambda = \lambda'$, or λ, λ' lie in the* *same connected component of* $\mathbb{R} \setminus \mathcal{D}_\Sigma$, *then* L *is an AC SL* m-*fold with rate* λ' *and* $\left|\nabla^k(\varphi - \iota)\right| = O(r^{\lambda' - 1 - k})$ *for all* $k \geqslant 0$. *Here* $\nabla, |.|$ *are computed using the cone metric* $\iota^*(g)$ *on* $\Sigma \times (T, \infty)$.

This is proved by showing that $\phi - \iota$ in Definition 8.3.7 satisfies a nonlinear elliptic equation, and then using elliptic regularity results to deduce asymptotic bounds for $\phi - \iota$ and all its derivatives, using the theories of analysis on manifolds with cylindrical ends developed by Lockhart and McOwen [245] or Melrose. The deformation theory of asymptotically conical SL m-folds in $\mathbb{C}^m$ has been studied independently by Pacini [274] and Marshall [254]. Pacini's results are earlier, but Marshall's are more complete.

Definition 8.3.9 Suppose L is an asymptotically conical SL m-fold in $\mathbb{C}^m$ with cone C and rate $\lambda < 2$, as in Definition 8.3.7. Define the *moduli space* $\mathcal{M}_L^\lambda$ *of deformations of* L *with rate* λ to be the set of AC SL m-folds $\hat{L}$ in $\mathbb{C}^m$ with cone C and rate λ, such that $\hat{L}$ is diffeomorphic to L and isotopic to L as an asymptotically conical submanifold of $\mathbb{C}^m$. One can define a natural *topology* on $\mathcal{M}_L^\lambda$.

Note that if L is an AC SL m-fold with rate λ, then it is also an AC SL m-fold with rate λ' for any $\lambda' \in [\lambda, 2)$. Thus we have defined a 1-parameter family of moduli spaces $\mathcal{M}_L^{\lambda'}$ for L, and not just one. The following result can be deduced from Marshall [254, Th. 6.2.15] and [254, Table 5.1]. (See also Pacini [274, Th.s 2 & 3].) It implies conjectures by the author in [194, Conj. 2.12] and [138, §10.2].

Theorem 8.3.10 *Let* L *be an asymptotically conical SL* m-*fold in* $\mathbb{C}^m$ *with cone* C *and rate* $\lambda < 2$, *and let* $\mathcal{M}_L^\lambda$ *be as in Definition 8.3.9. Set* $\Sigma = C \cap \mathcal{S}^{2m-1}$, *and let* $\mathcal{D}_\Sigma, N_\Sigma$ *be as in §8.3.1 and* $b^k(L), b^k_{\mathrm{cs}}(L)$ *be the Betti numbers in ordinary and compactly-supported cohomology. Then*

(a) *If* $\lambda \in (0, 2) \setminus \mathcal{D}_\Sigma$ *then* $\mathcal{M}_L^\lambda$ *is smooth with* $\dim \mathcal{M}_L^\lambda = b^1(L) - b^0(L) + N_\Sigma(\lambda)$.
   Note that if $0 < \lambda < \min(\mathcal{D}_\Sigma \cap (0, \infty))$ *then* $N_\Sigma(\lambda) = b^0(\Sigma)$.
(b) *If* $\lambda \in (2 - m, 0)$ *then* $\mathcal{M}_L^\lambda$ *is smooth with* $\dim \mathcal{M}_L^\lambda = b^1_{\mathrm{cs}}(L) = b^{m-1}(L)$.

To get a feel for this theorem, readers are advised to first read §8.4.1 on the deformation theory of compact SL m-folds in Calabi–Yau m-folds. The theorem says that if $\lambda \in (2 - m, 2) \setminus \mathcal{D}_\Sigma$ then the deformation theory for L with rate λ is *unobstructed* and $\mathcal{M}_L^\lambda$ is a *smooth manifold* with a given dimension, which is the sum of a topological contribution depending on L (roughly $b^1(L)$, as in Theorem 8.4.5), and an analytic contribution depending on the cone C and rate λ, involving an eigenvalue count over Σ.

For rates $\lambda < 2 - m$ the deformation theory will in general be obstructed, and similar to the conical singularities case of §8.5.1. Theorem 8.3.10 is proved in a similar way to Theorem 8.4.5, but using the theory of analysis on manifolds with cylindrical ends developed by Lockhart and McOwen [245].

Following [200, Def. 7.2] we define *cohomological invariants* $Y(L), Z(L)$ of L.

Definition 8.3.11 Let L be an AC SL m-fold in $\mathbb{C}^m$ with cone C, and let $\Sigma = C \cap \mathcal{S}^{2m-1}$. As ω, $\mathrm{Im}\,\theta$ in (8.1) are closed forms with $\omega|_L \equiv \mathrm{Im}\,\theta|_L \equiv 0$, they define classes in the relative de Rham cohomology groups $H^k(\mathbb{C}^m; L, \mathbb{R})$ for $k = 2, m$. Since Σ is in effect the boundary of L there is a long exact sequence

$$\cdots \to H_{\mathrm{cs}}^k(L, \mathbb{R}) \to H^k(L, \mathbb{R}) \to H^k(\Sigma, \mathbb{R}) \to H_{\mathrm{cs}}^{k+1}(L, \mathbb{R}) \to \cdots . \qquad (8.20)$$

But for $k > 1$ we have the exact sequence

$$0 = H^{k-1}(\mathbb{C}^m, \mathbb{R}) \to H^{k-1}(L, \mathbb{R}) \xrightarrow{\cong} H^k(\mathbb{C}^m; L, \mathbb{R}) \to H^k(\mathbb{C}^m, \mathbb{R}) = 0.$$

Let $Y(L) \in H^1(\Sigma, \mathbb{R})$ be the image of $[\omega]$ in $H^2(\mathbb{C}^m; L, \mathbb{R}) \cong H^1(L, \mathbb{R})$ under $H^1(L, \mathbb{R}) \to H^1(\Sigma, \mathbb{R})$ in (8.20), and $Z(L) \in H^{m-1}(\Sigma, \mathbb{R})$ be the image of $[\mathrm{Im}\,\theta]$ in $H^m(\mathbb{C}^m; L, \mathbb{R}) \cong H^{m-1}(L, \mathbb{R})$ under $H^{m-1}(L, \mathbb{R}) \to H^{m-1}(\Sigma, \mathbb{R})$ in (8.20).

Here are some conditions for $Y(L)$ or $Z(L)$ to be zero, [200, Prop. 7.3].

Proposition 8.3.12 *Let L be an AC SL m-fold in $\mathbb{C}^m$ with cone C and rate λ, and let $\Sigma = C \cap \mathcal{S}^{2m-1}$. If $\lambda < 0$ or $b^1(L) = 0$ then $Y(L) = 0$. If $\lambda < 2 - m$ or $b^0(\Sigma) = 1$ then $Z(L) = 0$.*

8.3.4   Examples of asymptotically conical SL m-folds

Many of the examples of SL m-folds in $\mathbb{C}^m$ constructed by methods described in §8.2 turn out to be asymptotically conical. Nearly all the known examples (up to translations) have minimum rate λ either 0 or $2 - m$, which are topologically significant values by Proposition 8.3.12. The only explicit, nontrivial examples known to the author with $\lambda \neq 0, 2 - m$ are in [191, Th. 11.6], and have $\lambda = \frac{3}{2}$.

We shall give three families of examples of AC SL m-folds L in $\mathbb{C}^m$ explicitly. The first is adapted from Harvey and Lawson [151, §III.3.A], and generalizes Example 8.2.4.

Example 8.3.13 Let C_{HL}^m be the SL cone in $\mathbb{C}^m$ of Example 8.3.5. We shall define a family of AC SL m-folds in $\mathbb{C}^m$ with cone C_{HL}^m. Let $a_1, \ldots, a_m \geqslant 0$ with exactly two of the a_j zero and the rest positive. Write $\mathbf{a} = (a_1, \ldots, a_m)$, and define

$$L_{\mathrm{HL}}^{\mathbf{a}} = \left\{ (z_1, \ldots, z_m) \in \mathbb{C}^m : i^{m+1} z_1 \cdots z_m \in [0, \infty), \right. \\ \left. |z_1|^2 - a_1 = \cdots = |z_m|^2 - a_m \right\}. \qquad (8.21)$$

Then $L_{\mathrm{HL}}^{\mathbf{a}}$ is an AC SL m-fold in $\mathbb{C}^m$ diffeomorphic to $T^{m-2} \times \mathbb{R}^2$, with cone C_{HL}^m and rate 0. It is invariant under the $\mathrm{U}(1)^{m-1}$ group (8.15). It is surprising that equations of the form (8.21) should define a nonsingular submanifold of $\mathbb{C}^m$ *without boundary*, but in fact they do.

Now suppose for simplicity that $a_1, \ldots, a_{m-2} > 0$ and $a_{m-1} = a_m = 0$. As $\Sigma_{\mathrm{HL}}^m \cong T^{m-1}$ we have $H^1(\Sigma_{\mathrm{HL}}^m, \mathbb{R}) \cong \mathbb{R}^{m-1}$, and calculation shows that $Y(L_{\mathrm{HL}}^{\mathbf{a}}) = (\pi a_1, \ldots, \pi a_{m-2}, 0) \in \mathbb{R}^{m-1}$ in the natural coordinates. Since $L_{\mathrm{HL}}^{\mathbf{a}} \cong T^{m-2} \times \mathbb{R}^2$ we have $H^1(L_{\mathrm{HL}}^{\mathbf{a}}, \mathbb{R}) = \mathbb{R}^{m-2}$, and $Y(L_{\mathrm{HL}}^{\mathbf{a}})$ lies in the image $\mathbb{R}^{m-2} \subset \mathbb{R}^{m-1}$ of $H^1(L_{\mathrm{HL}}^{\mathbf{a}}, \mathbb{R})$ in $H^1(\Sigma_{\mathrm{HL}}^m, \mathbb{R})$, as in Definition 8.3.11. As $b^0(\Sigma_{\mathrm{HL}}^m) = 1$, Proposition 8.3.12 shows that $Z(L_{\mathrm{HL}}^{\mathbf{a}}) = 0$.

Take $C = C_{\mathrm{HL}}^m$, $\Sigma = \Sigma_{\mathrm{HL}}^m$ and $L = L_{\mathrm{HL}}^{\mathbf{a}}$ in Theorem 8.3.10, and let $0 < \lambda < \min(\mathcal{D}_\Sigma \cap (0, \infty))$. Then $b^1(L) = m - 2$, $b^0(L) = 1$ and $N_\Sigma(\lambda) = b^0(\Sigma) = 1$, so Theorem 8.3.10(a) shows that $\dim \mathscr{M}_L^\lambda = m - 2$. This is consistent with the fact that L depends on $m - 2$ real parameters $a_1, \ldots, a_{m-2} > 0$.

The family of all $L_{\mathrm{HL}}^{\mathbf{a}}$ has $\frac{1}{2}m(m-1)$ connected components, indexed by which two of $a_1, \ldots, a_m$ are zero. Using the theory of §8.5.3, these can give many *topologically distinct* ways to desingularize SL m-folds with conical singularities with these cones.

Our second family, from [196, Ex. 9.4], generalizes Example 8.2.5.

Example 8.3.14 Let $m, a_1, \ldots, a_m, k$ and $L_0^{a_1, \ldots, a_m}$ be as in Example 8.3.6. For $0 \neq c \in \mathbb{R}$ define

$$L_c^{a_1, \ldots, a_m} = \big\{ \big(i e^{i a_1 \psi} x_1, e^{i a_2 \psi} x_2, \ldots, e^{i a_m \psi} x_m \big) : \psi \in [0, 2\pi),$$

$$x_1, \ldots, x_m \in \mathbb{R}, \qquad a_1 x_1^2 + \cdots + a_m x_m^2 = c \big\}.$$

Then $L_c^{a_1, \ldots, a_m}$ is an AC SL m-fold in $\mathbb{C}^m$ with rate 0 and cone $L_0^{a_1, \ldots, a_m}$. It is diffeomorphic as an immersed SL m-fold to $(\mathcal{S}^{k-1} \times \mathbb{R}^{m-k} \times \mathcal{S}^1)/\mathbb{Z}_2$ if $c > 0$, and to $(\mathbb{R}^k \times \mathcal{S}^{m-k-1} \times \mathcal{S}^1)/\mathbb{Z}_2$ if $c < 0$.

Our third family generalizes Example 8.2.6. It was first found by Lawlor [231], made more explicit by Harvey [150, p. 139–140], and discussed from a different point of view by the author in [190, §5.4(b)]. Our treatment is based on that of Harvey.

Example 8.3.15 Let $m > 2$ and $a_1, \ldots, a_m > 0$, and define polynomials p, P by

$$p(x) = (1 + a_1 x^2) \cdots (1 + a_m x^2) - 1 \quad \text{and} \quad P(x) = \frac{p(x)}{x^2}.$$

Define real numbers $\psi_1, \ldots, \psi_m$ and A by

$$\psi_k = a_k \int_{-\infty}^{\infty} \frac{\mathrm{d}x}{(1 + a_k x^2)\sqrt{P(x)}} \quad \text{and} \quad A = \omega_m (a_1 \cdots a_m)^{-1/2}, \qquad (8.22)$$

where ω_m is the volume of the unit sphere in $\mathbb{R}^m$. Clearly $\psi_k, A > 0$. But writing $\psi_1 + \cdots + \psi_m$ as one integral gives

$$\psi_1 + \cdots + \psi_m = \int_0^{\infty} \frac{p'(x)\mathrm{d}x}{(p(x)+1)\sqrt{p(x)}} = 2 \int_0^{\infty} \frac{\mathrm{d}w}{w^2 + 1} = \pi,$$

making the substitution $w = \sqrt{p(x)}$. So $\psi_k \in (0, \pi)$ and $\psi_1 + \cdots + \psi_m = \pi$. This yields a 1-1 correspondence between m-tuples $(a_1, \ldots, a_m)$ with $a_k > 0$, and $(m+1)$-tuples $(\psi_1, \ldots, \psi_m, A)$ with $\psi_k \in (0, \pi)$, $\psi_1 + \cdots + \psi_m = \pi$ and $A > 0$.

For $k = 1, \ldots, m$ and $y \in \mathbb{R}$, define a function $z_k : \mathbb{R} \to \mathbb{C}$ by

$$z_k(y) = e^{i\phi_k(y)} \sqrt{a_k^{-1} + y^2}, \quad \text{where} \quad \phi_k(y) = a_k \int_{-\infty}^{y} \frac{\mathrm{d}x}{(1 + a_k x^2)\sqrt{P(x)}}.$$

Now write $\psi = (\psi_1, \ldots, \psi_m)$, and define a submanifold $L^{\psi, A}$ in $\mathbb{C}^m$ by

$$L^{\psi, A} = \big\{ (z_1(y)x_1, \ldots, z_m(y)x_m) : y \in \mathbb{R}, \ x_k \in \mathbb{R}, \ x_1^2 + \cdots + x_m^2 = 1 \big\}.$$

Then $L^{\psi, A}$ is closed, embedded, and diffeomorphic to $\mathcal{S}^{m-1} \times \mathbb{R}$, and Harvey [150, Th. 7.78] shows that $L^{\psi, A}$ is *special Lagrangian*. One can also show that $L^{\psi, A}$ is *asymptotically conical*, with rate $2 - m$ and cone the union $\Pi^0 \cup \Pi^\psi$ of two special Lagrangian m-planes Π^0, Π^ψ in $\mathbb{C}^m$ given by

$$\Pi^0 = \big\{ (x_1, \dots, x_m) : x_j \in \mathbb{R} \big\} \quad \text{and} \quad \Pi^\psi = \big\{ (\mathrm{e}^{i\psi_1} x_1, \dots, \mathrm{e}^{i\psi_m} x_m) : x_j \in \mathbb{R} \big\}. \quad (8.23)$$

As $\lambda = 2 - m < 0$ we have $Y(L^{\psi,A}) = 0$ by Proposition 8.3.12. Now $L^{\psi,A} \cong \mathcal{S}^{m-1} \times \mathbb{R}$ so $H^{m-1}(L^{\psi,A}, \mathbb{R}) \cong \mathbb{R}$, and $\Sigma = (\Pi^0 \cup \Pi^\psi) \cap \mathcal{S}^{2m-1}$ is the disjoint union of two unit $(m-1)$-spheres $\mathcal{S}^{m-1}$, so $H^{m-1}(\Sigma, \mathbb{R}) \cong \mathbb{R}^2$. The image of $H^{m-1}(L^{\psi,A}, \mathbb{R})$ in $H^{m-1}(\Sigma, \mathbb{R})$ is $\big\{ (x, -x) : x \in \mathbb{R} \big\}$ in the natural coordinates. Calculation shows that $Z(L^{\psi,A}) = (A, -A) \in H^{m-1}(\Sigma, \mathbb{R})$, which is why we defined A this way in (8.22).

Apply Theorem 8.3.10 with $L = L^{\psi,A}$ and $\lambda \in (2 - m, 0)$. As $L \cong \mathcal{S}^{m-1} \times \mathbb{R}$ we have $b^1_{\mathrm{cs}}(L) = 1$, so part (b) of Theorem 8.3.10 shows that $\dim \mathcal{M}^\lambda_L = 1$. This is consistent with the fact that when ψ is fixed, $L^{\psi,A}$ depends on one real parameter $A > 0$. Here ψ is fixed in $\mathcal{M}^\lambda_L$ as the cone $C = \Pi^0 \cup \Pi^\psi$ of L depends on ψ, and all $\hat{L} \in \mathcal{M}^\lambda_L$ have the same cone C, by definition.

8.4 SL m-folds in (almost) Calabi–Yau m-folds

Here is how we define SL m-folds in Calabi–Yau m-folds.

Definition 8.4.1 Let (X, J, g, θ) be a Calabi–Yau m-fold. Then $\mathrm{Re}\,\theta$ is a *calibration* on (X, g). An oriented real m-submanifold N in X is called a *special Lagrangian submanifold (SL m-fold)* if it is calibrated with respect to $\mathrm{Re}\,\theta$.

More generally, if $\psi \in \mathbb{R}$ we call N *special Lagrangian with phase* $\mathrm{e}^{i\psi}$ if it is calibrated with respect to $\cos\psi\,\mathrm{Re}\,\theta + \sin\psi\,\mathrm{Im}\,\theta$. When we refer to SL m-folds L without specifying a phase we mean phase 1, that is, calibrated w.r.t. $\mathrm{Re}\,\theta$ as above.

From Proposition 8.1.2 we deduce an *alternative definition* of SL m-folds. It is often more useful than Definition 8.4.1.

Proposition 8.4.2 *Let* (X, J, g, θ) *be a Calabi–Yau m-fold, with Kähler form* ω, *and* N *a real m-dimensional submanifold in* X. *Then* N *admits an orientation making it into an SL m-fold in* X *if and only if* $\omega|_N \equiv 0$ *and* $\mathrm{Im}\,\theta|_N \equiv 0$. *More generally, it admits an orientation making it into an SL m-fold with phase* $\mathrm{e}^{i\psi}$ *if and only if* $\omega|_N \equiv 0$ *and* $(\cos\psi\,\mathrm{Im}\,\theta - \sin\psi\,\mathrm{Re}\,\theta)|_N \equiv 0$.

We also define SL m-folds in the much larger class of *almost Calabi–Yau m-folds*. The idea of extending special Lagrangian geometry to almost Calabi–Yau manifolds appears in the work of Goldstein [128, §3.1], Bryant [60, §1], who uses the term 'special Kähler' instead of 'almost Calabi–Yau', and the author [197].

Definition 8.4.3 Let $m \geqslant 2$. An *almost Calabi–Yau m-fold*, or *ACY m-fold* for short, is a quadruple (X, J, g, θ) such that (X, J, g) is a compact m-dimensional Kähler manifold, and θ is a non-vanishing holomorphic $(m, 0)$-form on X.

The difference between this and Definition 7.1.10 is that we do not require θ and the Kähler form ω of g to satisfy eqn (7.2), and hence g need not be Ricci-flat, nor have holonomy in $\mathrm{SU}(m)$.

Definition 8.4.4 Let (X, J, g, θ) be an ACY m-fold with Kähler form ω, and N a real m-dimensional submanifold of X. We call N a *special Lagrangian submanifold (SL m-fold)* if $\omega|_N \equiv \operatorname{Im}\theta|_N \equiv 0$. Then $\operatorname{Re}\theta|_N$ is a nonvanishing m-form on N, so N is orientable, with a unique orientation in which $\operatorname{Re}\theta|_N$ is positive.

More generally, we call N *special Lagrangian with phase* $e^{i\psi}$ if $\omega|_N \equiv 0$ and $(\cos\psi \operatorname{Im}\theta - \sin\psi \operatorname{Re}\theta)|_N \equiv 0$, and then N has a unique orientation in which $(\cos\psi \operatorname{Re}\theta + \sin\psi \operatorname{Im}\theta)|_N$ is positive.

Definitions 8.4.1 and 8.4.4 are equivalent when (X, J, g, θ) is a Calabi–Yau m-fold, by Proposition 8.4.2. We can also express SL m-folds in ACY m-folds as calibrated submanifolds, as follows. Suppose (X, J, g, θ) is an almost Calabi–Yau m-fold, with Kähler form ω. Let $f : X \to (0, \infty)$ be the unique smooth function such that

$$f^{2m}\omega^m/m! = (-1)^{m(m-1)/2}(i/2)^m \theta \wedge \bar{\theta}, \tag{8.24}$$

and define $\tilde{g}$ to be the conformally equivalent metric $f^2 g$ on X. Then $\operatorname{Re}\theta$ is a *calibration* on the Riemannian manifold $(X, \tilde{g})$, and SL m-folds N in (X, J, g, θ) are calibrated with respect to it, so that they are minimal with respect to $\tilde{g}$. If (X, J, g, θ) is Calabi–Yau then $f \equiv 1$ by (7.2), so $\tilde{g} = g$, and N is special Lagrangian if and only if it is calibrated with respect to $\operatorname{Re}\theta$ on (X, g), as in Definition 8.4.1.

Here are two reasons why working with almost Calabi–Yau m-folds is a good idea. Firstly, many examples of almost Calabi–Yau m-folds can be written down very explicitly—for instance, a quintic in $\mathbb{CP}^4$ with the Fubini–Study metric is an almost Calabi–Yau 3-fold. However, the *only* compact Calabi–Yau m-folds for which the metric is known explicitly are finite quotients of flat tori T^{2m}. For other examples we know only the existence of the metric, using Theorem 7.1.2. Problems such as constructing examples of special Lagrangian fibrations will probably be more feasible upon an explicit almost Calabi–Yau m-fold than on an inexplicit Calabi–Yau m-fold.

Secondly, SL m-folds should have much stronger *genericness properties* in almost Calabi–Yau m-folds than in Calabi–Yau m-folds. There are many situations in geometry in which one uses a genericity assumption to control singular behaviour. For instance, pseudo-holomorphic curves in an arbitrary almost complex manifold may have bad singularities, but the possible singularities in a generic almost complex manifold are much simpler. In the same way, one might hope that in a generic Calabi–Yau m-fold, compact SL m-folds may be better behaved than in an arbitrary Calabi–Yau m-fold.

But because Calabi–Yau manifolds come in only finite-dimensional families by Corollary 7.7.2, choosing a generic Calabi–Yau structure is a fairly weak assumption, and probably will not help very much. However, almost Calabi–Yau manifolds come in *infinite-dimensional* families, because of the freedom to change the metric using a Kähler potential as in §5.5, so choosing a generic almost Calabi–Yau structure is a much more powerful thing to do, and will probably simplify the singular behaviour of compact SL m-folds considerably. We will return to this idea in §8.5.2 and §8.5.5.

8.4.1 Deformations of compact special Lagrangian m-folds

The *deformation theory* of compact SL m-folds was studied by McLean [259, §3], who proved the following result in the Calabi–Yau case. The extension to the almost Calabi–Yau case is described in [138, §9.5].

Theorem 8.4.5 *Let N be a compact special Lagrangian m-fold in an almost Calabi–Yau m-fold (X, J, g, θ). Then the moduli space $\mathcal{M}_N$ of special Lagrangian deformations of N is a smooth manifold of dimension $b^1(N)$.*

Sketch proof Suppose for simplicity that N is an embedded submanifold. There is a natural orthogonal decomposition $TX|_N = TN \oplus \nu$, where $\nu \to N$ is the normal bundle of N in X. As N is Lagrangian, the complex structure $J : TX \to TX$ gives an isomorphism $J : \nu \to TN$. But the metric g gives an isomorphism $TN \cong T^*N$. Composing these two gives an isomorphism $\nu \cong T^*N$.

Let T be a small *tubular neighbourhood* of N in X. Then we can identify T with a neighbourhood of the zero section in ν. Using the isomorphism $\nu \cong T^*N$, we have an identification between T and a neighbourhood of the zero section in T^*N. This can be chosen to identify the Kähler form ω on T with the natural symplectic structure on T^*N. Let $\pi : T \to N$ be the obvious projection.

Under this identification, submanifolds N' in $T \subset X$ which are C^1 close to N are identified with the graphs of small smooth sections α of T^*N. That is, submanifolds N' of X close to N are identified with 1-forms α on N. We need to know: which 1-forms α are identified with special Lagrangian submanifolds N'?

Well, N' is special Lagrangian if $\omega|_{N'} \equiv \operatorname{Im} \theta|_{N'} \equiv 0$. Now $\pi|_{N'} : N' \to N$ is a diffeomorphism, so we can push $\omega|_{N'}$ and $\operatorname{Im} \theta|_{N'}$ down to N, and regard them as functions of α. Calculation shows that

$$\pi_*\big(\omega|_{N'}\big) = \mathrm{d}\alpha \quad \text{and} \quad \pi_*\big(\operatorname{Im}\theta|_{N'}\big) = F(\alpha, \nabla\alpha),$$

where F is a nonlinear function of its arguments. Thus, the moduli space $\mathcal{M}_N$ is locally isomorphic to the set of small 1-forms α on N such that $\mathrm{d}\alpha \equiv 0$ and $F(\alpha, \nabla\alpha) \equiv 0$.

Now it turns out that F satisfies $F(\alpha, \nabla\alpha) \approx \mathrm{d}(*\alpha)$ when α is small. Therefore $\mathcal{M}_N$ is locally approximately isomorphic to the vector space of 1-forms α with $\mathrm{d}\alpha = \mathrm{d}(*\alpha) = 0$. But by Hodge theory, this is isomorphic to the de Rham cohomology group $H^1(N, \mathbb{R})$, and is a manifold with dimension $b^1(N)$.

To carry out this last step rigorously requires some technical machinery: one must work with certain *Banach spaces* of sections of T^*N, $\Lambda^2 T^*N$ and $\Lambda^m T^*N$, use *elliptic regularity results* to prove that $\alpha \mapsto \big(\mathrm{d}\alpha, F(\alpha, \nabla\alpha)\big)$ has *closed image* in these Banach spaces, and then use the *Implicit Mapping Theorem for Banach spaces*, Theorem 1.2.5, to show that the kernel of the map is what we expect. $\qquad\square$

8.4.2 Obstructions to the existence of compact SL m-folds

Let (X, J, g, θ) be an almost Calabi–Yau m-fold, with Kähler form ω, and N an immersed real m-submanifold in X, with immersion $\iota : N \to X$. If N is special Lagrangian then $\omega|_N \equiv \operatorname{Im}\theta|_N \equiv 0$, so $[\omega|_N] = [\operatorname{Im}\theta|_N] = 0$ in de Rham cohomology $H^*(N, \mathbb{R})$. Now $[\omega|_N]$ and $[\operatorname{Im}\theta|_N]$ are unchanged under continuous variations of ι. Thus, $[\omega|_N] = [\operatorname{Im}\theta|_N] = 0$ is necessary not just for N to be special Lagrangian, but also for any isotopic submanifold N' in X to be special Lagrangian. That is, $[\omega|_N]$ and $[\operatorname{Im}\theta|_N]$ are *obstructions* to the existence of isotopic SL m-folds N'. This proves:

Proposition 8.4.6 *Let* (X, J, g, θ) *be an almost Calabi–Yau* m-*fold, and* N *a compact real* m-*submanifold in* X. *Then a necessary condition for* N *to be isotopic to an SL* m-*fold* N' *in* X *is that* $[\omega|_N] = 0$ *in* $H^2(N, \mathbb{R})$ *and* $[\operatorname{Im} \theta|_N] = 0$ *in* $H^m(N, \mathbb{R})$.

Now let $\big\{(X, J_t, g_t, \theta_t) : t \in (-\epsilon, \epsilon)\big\}$ be a smooth 1-parameter family of almost Calabi–Yau m-folds with Kähler forms ω_t. Suppose N_0 is a special Lagrangian m-fold in (X, J_0, g_0, θ_0). When can we extend N_0 to a smooth family of special Lagrangian m-folds N_t in (X, J_t, g_t, θ_t) for $t \in (-\epsilon, \epsilon)$? From above, a necessary condition is that $[\omega_t|_{N_0}] = [\operatorname{Im} \theta_t|_{N_0}] = 0$ for all t. Locally, this is also a *sufficient* condition.

Theorem 8.4.7 *Let* $\big\{(X, J_t, g_t, \theta_t) : t \in (-\epsilon, \epsilon)\big\}$ *be a smooth 1-parameter family of almost Calabi–Yau* m-*folds, with Kähler forms* ω_t. *Let* N_0 *be a compact SL* m-*fold in* (X, J_0, g_0, θ_0), *and suppose that* $[\omega_t|_{N_0}] = 0$ *in* $H^2(N_0, \mathbb{R})$ *and* $[\operatorname{Im} \theta_t|_{N_0}] = 0$ *in* $H^m(N_0, \mathbb{R})$ *for all* $t \in (-\epsilon, \epsilon)$. *Then* N_0 *extends to a smooth 1-parameter family* $\big\{N_t : t \in (-\delta, \delta)\big\}$, *where* $0 < \delta \leqslant \epsilon$ *and* N_t *is a compact SL* m-*fold in* (X, J_t, g_t, θ_t).

This can be proved using similar techniques to Theorem 8.4.5, though McLean did not prove it. Note that the condition $[\operatorname{Im} \theta_t|_{N_0}] = 0$ for all t can be satisfied by choosing the phases of the θ_t appropriately, and if the image of $H_2(N, \mathbb{R})$ in $H_2(X, \mathbb{R})$ is zero, then the condition $[\omega|_N] = 0$ holds automatically. Thus, the obstructions $[\omega_t|_{N_0}] = [\operatorname{Im} \theta_t|_{N_0}] = 0$ in Theorem 8.4.7 are actually fairly mild restrictions, and special Lagrangian m-folds should be thought of as pretty stable under small deformations of the almost Calabi–Yau structure.

8.4.3 Natural coordinates on the moduli space $\mathcal{M}_N$

Let N be a compact SL m-fold in an almost Calabi–Yau m-fold (X, J, g, θ). Theorem 8.4.5 shows that the moduli space $\mathcal{M}_N$ has dimension $b^1(N)$. By Poincaré duality $b^1(N) = b^{m-1}(N)$. Thus $\mathcal{M}_N$ has the same dimension as the de Rham cohomology groups $H^1(N, \mathbb{R})$ and $H^{m-1}(N, \mathbb{R})$.

We shall construct natural local diffeomorphisms $\Phi : \mathcal{M}_N \to H^1(N, \mathbb{R})$ and $\Psi : \mathcal{M}_N \to H^{m-1}(N, \mathbb{R})$. These induce two natural *affine structures* on $\mathcal{M}_N$, and can be thought of as two *natural coordinate systems* on $\mathcal{M}_N$. The material of this section can be found in Hitchin [162, §4].

Here is how to define Φ, Ψ. Let U be a connected and simply-connected open neighbourhood of N in $\mathcal{M}_N$. We will construct smooth maps $\Phi : U \to H^1(N, \mathbb{R})$ and $\Psi : U \to H^{m-1}(N, \mathbb{R})$ with $\Phi(N) = \Psi(N) = 0$, which are local diffeomorphisms.

Let $N' \in U$. Then as U is connected, there exists a smooth path $\gamma : [0, 1] \to U$ with $\gamma(0) = N$ and $\gamma(1) = N'$, and as U is simply-connected, γ is unique up to isotopy. Now γ parametrizes a family of submanifolds of X diffeomorphic to N, which we can lift to a smooth map $\Gamma : N \times [0, 1] \to X$ with $\Gamma(N \times \{t\}) = \gamma(t)$.

Consider the 2-form $\Gamma^*(\omega)$ on $N \times [0, 1]$. As each fibre $\gamma(t)$ is Lagrangian, we have $\Gamma^*(\omega)|_{N \times \{t\}} \equiv 0$ for each $t \in [0, 1]$. Therefore we may write $\Gamma^*(\omega) = \alpha_t \wedge \mathrm{d}t$, where α_t is a closed 1-form on N for $t \in [0, 1]$. Define $\Phi(N') = \big[\int_0^1 \alpha_t \, \mathrm{d}t\big] \in H^1(N, \mathbb{R})$. That is, we integrate the 1-forms α_t with respect to t to get a closed 1-form $\int_0^1 \alpha_t \, \mathrm{d}t$ on N, and then take its cohomology class.

Similarly, write $\Gamma^*(\operatorname{Im}\theta) = \beta_t \wedge \mathrm{d}t$, where β_t is a closed $(m-1)$-form on N for $t \in [0,1]$, and define $\Psi(N') = \left[\int_0^1 \beta_t \, \mathrm{d}t\right] \in H^{m-1}(N,\mathbb{R})$. Then Φ and Ψ are independent of choices made in the construction. We need to restrict to a simply-connected subset U of $\mathcal{M}_N$ so that γ is unique up to isotopy. Alternatively, one can define Φ and Ψ on the universal cover $\widetilde{\mathcal{M}}_N$ of $\mathcal{M}_N$.

8.4.4   Examples of SL *m*-folds in (almost) Calabi–Yau *m*-folds

Here is a method for finding examples of compact SL m-folds in compact Calabi–Yau m-folds. It is used by Bryant [58], for example, to produce examples of special Lagrangian 3-tori in Calabi–Yau 3-folds. It relies on the following proposition, which is well-known and easy to prove.

Proposition 8.4.8 *Let (X, J, g, θ) be an (almost) Calabi–Yau m-fold, and $\sigma : X \to X$ be an involution with $\sigma^*(J) = -J, \sigma^*(g) = g$ and $\sigma^*(\theta) = \bar{\theta}$. Then the fixed point set N of σ is a compact, nonsingular SL m-fold in X.*

Calabi–Yau m-folds (X, J, g, θ) with such involutions σ are easy to construct. One uses algebraic geometry to find a complex manifold (X, J) with K_X trivial and an antiholomorphic involution $\sigma : X \to X$. For instance, (X, J) could be a quintic in $\mathbb{CP}^4$ defined by a polynomial with real coefficients, and σ the restriction to X of complex conjugation $[z_0, \ldots, z_4] \mapsto [\bar{z}_0, \ldots, \bar{z}_4]$ on $\mathbb{CP}^4$. Theorem 7.1.2 implies that any Kähler class invariant under σ contains a unique Ricci-flat Kähler metric g with $\sigma^*(g) = g$. Then there exists a holomorphic volume θ, unique up to sign, such that (X, J, g, θ) is a Calabi–Yau m-fold and $\sigma^*(\theta) = \bar{\theta}$. Theorem 8.4.5 then shows that N in Proposition 8.4.8 lies in a smooth family of compact SL m-folds in X with dimension $b^1(N)$.

This is at present almost the only known successful method for producing compact SL m-folds in compact Calabi–Yau m-folds with holonomy $\mathrm{SU}(m)$ for $m \geqslant 3$. However, there are very good conjectural reasons for believing that compact SL m-folds in (almost) Calabi–Yau m-folds are actually very abundant. Schoen and Wolfson [300, 301] propose that given a Calabi–Yau m-fold (X, J, g, θ) and a suitable class of compact Lagrangian submanifolds in X, then by minimizing volume in this class one should be able to construct a minimal Lagrangian integral current, which is a finite sum of special Lagrangian integral currents with different phases $e^{i\psi}$.

Thomas and Yau [324, 325] express a similar idea in terms of Lagrangian mean curvature flow, and regard SL m-folds as representing (semi)stable objects w.r.t. a stability condition induced by θ on the *derived Fukaya category* $D^b(F(X))$ of (X, ω), which will be discussed in §9.3. In the author's opinion, Thomas and Yau's notion of stability should be replaced by Bridgeland's concept of stability condition on a triangulated category [51]. Overall, the rough expectation is that every graded Lagrangian in X with unobstructed Floer homology should be Hamiltonian equivalent to a connected sum of finitely many (possibly singular) SL m-folds in X with different phases, in a unique way. So X should have enough SL m-folds to generate the whole Fukaya category.

Many examples are known of SL m-folds in noncompact Calabi–Yau m-folds. Bryant [59] proves that any compact, oriented, real analytic Riemannian 3-manifold can be embedded as an SL 3-fold in a noncompact Calabi–Yau 3-fold, as the fixed points of an involution σ as above. Explicit examples of SL m-folds in explicit noncompact

Calabi–Yau m-folds such as $T^*\mathcal{S}^m$ and $\mathcal{O}(-1) \oplus \mathcal{O}(-1) \to \mathbb{CP}^1$ have been found by several authors using the ideas of §8.2, in particular the symmetries method of §8.2.1.

8.5 SL m-folds with isolated conical singularities

Finally we shall discuss *singular* SL m-folds in (almost) Calabi–Yau m-folds. As in §4.4, geometric measure theory gives one approach to this. However, GMT is too general for many purposes: it studies all kinds of singularities at once, including the most horrible. As a result almost nothing is known about the singularities of a general special Lagrangian integral current, except that they have Hausdorff codimension two.

An alternative approach is to define some class of fairly simple singularities of SL m-folds, and study such singular SL m-folds in depth. The author did this in a series of papers [200–204] on SL m-folds with *isolated conical singularities* in almost Calabi–Yau m-folds, whose singularities are locally modelled on SL cones with isolated singular points. We now describe the main results. For further discussion of special Lagrangian singularities, see [199]. Here [200, Def. 3.6 and Th. 4.4] is the basic definition.

Definition 8.5.1 Let (X, J, g, θ) be an almost Calabi–Yau m-fold for $m > 2$ with Kähler form ω, and define $f : X \to (0, \infty)$ as in (8.24). Suppose N is a compact singular SL m-fold in X with singularities at distinct points $x_1, \ldots, x_n \in N$, and no other singularities.

Fix isomorphisms $v_i : \mathbb{C}^m \to T_{x_i}X$ for $i = 1, \ldots, n$ such that $v_i^*(\omega) = \omega'$ and $v_i^*(\theta) = f(x_i)^m \theta'$, where ω', θ' are the forms on $\mathbb{C}^m$ from (8.1). Let $C_1, \ldots, C_n$ be SL cones in $\mathbb{C}^m$ with isolated singularities at 0. For $i = 1, \ldots, n$ let $\Sigma_i = C_i \cap \mathcal{S}^{2m-1}$, and let $\mu_i \in (2, 3)$ with $(2, \mu_i] \cap \mathcal{D}_{\Sigma_i} = \emptyset$, where $\mathcal{D}_{\Sigma_i}$ is defined in (8.11). Then we say that N has a *conical singularity* or *conical singular point* at x_i, with *rate* μ_i and *cone* C_i for $i = 1, \ldots, n$, if the following holds.

By *Darboux's Theorem* [256, Th. 3.15] there exist embeddings $\Upsilon_i : B_R \to X$ for $i = 1, \ldots, n$ satisfying $\Upsilon_i(0) = x_i$, $d\Upsilon_i|_0 = v_i$ and $\Upsilon_i^*(\omega) = \omega$, where B_R is the open ball of radius R about 0 in $\mathbb{C}^m$ for some small $R > 0$. Define $\iota_i : \Sigma_i \times (0, R) \to B_R$ by $\iota_i(\sigma, r) = r\sigma$ for $i = 1, \ldots, n$.

Define $N' = N \setminus \{x_1, \ldots, x_n\}$. Then there should exist a compact subset $K \subset N'$ such that $N' \setminus K$ is a union of open sets $S_1, \ldots, S_n$ with $S_i \subset \Upsilon_i(B_R)$, whose closures $\bar{S}_1, \ldots, \bar{S}_n$ are disjoint in N. For $i = 1, \ldots, n$ and some $R' \in (0, R]$ there should exist a smooth $\phi_i : \Sigma_i \times (0, R') \to B_R$ such that $\Upsilon_i \circ \phi_i : \Sigma_i \times (0, R') \to X$ is a diffeomorphism $\Sigma_i \times (0, R') \to S_i$, and

$$\left|\nabla^k(\phi_i - \iota_i)\right| = O(r^{\mu_i - 1 - k}) \quad \text{as } r \to 0 \text{ for } k = 0, 1, \text{ and} \tag{8.25}$$

$$(\phi_i - \iota_i)(\sigma, r) \perp T_{\iota_i(\sigma, r)}C_i \quad \text{in } \mathbb{C}^m \text{ for all } (\sigma, r) \in \Sigma_i \times (0, R'). \tag{8.26}$$

Here ∇ is the Levi-Civita connection of the cone metric $\iota_i^*(g)$ on $\Sigma_i \times (0, R')$, $|\,.\,|$ is computed using $\iota_i^*(g)$. If the cones $C_1, \ldots, C_n$ are *stable* in the sense of Definition 8.3.4, then we say that N has *stable conical singularities*.

This is similar to Definition 8.3.7, and in fact there are strong parallels between the theories of SL m-folds with conical singularities and of asymptotically conical SL m-folds. Actually (8.26) is not part of the original definition, but we show in [200, Th. 4.4] that if N satisfies all of Definition 8.5.1 except (8.26), we can choose the ϕ_i essentially uniquely so that (8.26) holds.

Not all singularities of SL m-folds satisfy Definition 8.5.1. For example, if L is an SL a-fold in $\mathbb{C}^a$ for $a > 0$ and N is an SL b-fold in $\mathbb{C}^b$ with a singularity at $x \in \mathbb{C}^b$, then $L \times N$ is an SL $(a + b)$-fold in $\mathbb{C}^{a+b}$ with nonisolated singularities along $L \times \{x\}$. There are also isolated SL singularities which are not conical. For example, using Theorem 8.2.3 as in §8.2.6 one can construct U(1)-invariant SL 3-folds in $\mathbb{C}^3$ with isolated singularities of multiplicity k for all $k \geqslant 1$. When $k \geqslant 2$ these are not conical singularities, as their tangent cone C is two copies of $\mathbb{R}^3$ intersecting in $\mathbb{R}$, which does not have an isolated singularity at 0. Also, Example 8.2.7 describes SL 3-folds with isolated singularities that are not conical, whose tangent cone is an $\mathbb{R}^3$ with multiplicity 2.

8.5.1 The asymptotic behaviour of N near x_i

Let X be an almost Calabi–Yau m-fold and N an SL m-fold in X with conical singularities at $x_1, \ldots, x_n$, with identifications υ_i and cones C_i. In [200] we study how quickly N converges to the cone $\upsilon(C_i)$ in $T_{x_i}X$ near x_i, and [200, Th. 5.5] gives:

Theorem 8.5.2 *Let* (X, J, g, θ) *be an almost Calabi–Yau m-fold and N a compact SL m-fold in X with conical singularities, and use the notation of Definition 8.5.1. Suppose $\mu_i' \in (2, 3)$ with $(2, \mu_i'] \cap \mathcal{D}_{\Sigma_i} = \emptyset$ for $i = 1, \ldots, n$. Then*

$$\left| \nabla^k (\phi_i - \iota_i) \right| = O(r^{\mu_i' - 1 - k}) \qquad \text{for all } k \geqslant 0 \text{ and } i = 1, \ldots, n. \qquad (8.27)$$

Hence N has conical singularities at x_i with cone C_i and rate μ_i', for all possible rates μ_i' allowed by Definition 8.5.1. Therefore, the definition of conical singularities is essentially independent of the choice of rate μ_i.

Theorem 8.5.2 in effect *strengthens* the definition of SL m-folds with conical singularities, Definition 8.5.1, as it shows that (8.25) actually implies the much stronger condition (8.27) on all derivatives. It is proved by showing that $\phi_i - \iota_i$ in Definition 8.5.1 satisfies a *nonlinear elliptic equation*, and then using *elliptic regularity results* to deduce asymptotic bounds for $\phi_i - \iota_i$ and all its derivatives.

Our next result [200, Th. 6.8] uses *geometric measure theory*, as in §4.4.

Theorem 8.5.3 *Let* (X, J, g, θ) *be an almost Calabi–Yau m-fold and $x \in X$, and fix an isomorphism $\upsilon : \mathbb{C}^m \to T_x X$ as in Definition 8.5.1. Using the ideas of §4.4, suppose T is a special Lagrangian integral current in X with $x \in T^\circ$, and $\upsilon_*(C)$ is a multiplicity 1 tangent cone to T at x, where C is a rigid special Lagrangian cone in $\mathbb{C}^m$, in the sense of Definition 8.3.4. Then T has a conical singularity at x, in the sense of Definition 8.5.1.*

This is a *weakening* of Definition 8.5.1 for *rigid* cones C. It is proved by applying Theorem 4.4.5, and using rigidity of C to improve the weak asymptotic estimate (4.1) to the stronger estimate (8.25). Theorem 8.5.3 also holds for the larger class of *Jacobi integrable* SL cones C, defined in [200, Def. 6.7]. Basically, it shows that if a singular SL m-fold T in X is locally modelled on a rigid SL cone C in only a very weak sense, then it necessarily satisfies Definition 8.5.1.

Theorems 8.5.2 and 8.5.3 justify the apparent arbitrariness of Definition 8.5.1, as they show that almost any other sensible definition of SL m-folds with singularities modelled on (rigid) SL cones C is equivalent to Definition 8.5.1.

8.5.2   Moduli spaces of SL m-folds with conical singularities

In [201] we study the *deformation theory* of compact SL m-folds with conical singularities, generalizing the nonsingular case of §8.4.1. Here is [201, Def. 5.4].

Definition 8.5.4   Let (X, J, g, θ) be an almost Calabi–Yau m-fold and N a compact SL m-fold in X with conical singularities at $x_1, \ldots, x_n$ with identifications $\upsilon_i : \mathbb{C}^m \to T_{x_i}X$ and cones $C_1, \ldots, C_n$. Define the *moduli space* $\mathcal{M}_N$ of deformations of N to be the set of $\hat{N}$ such that

(i) $\hat{N}$ is a compact SL m-fold in X with conical singularities at $\hat{x}_1, \ldots, \hat{x}_n$ with cones $C_1, \ldots, C_n$, for some $\hat{x}_i$ and identifications $\hat{\upsilon}_i : \mathbb{C}^m \to T_{\hat{x}_i}X$.

(ii) There exists a homeomorphism $\hat{\iota} : N \to \hat{N}$ with $\hat{\iota}(x_i) = \hat{x}_i$ for $i = 1, \ldots, n$ such that $\hat{\iota}|_{N'} : N' \to \hat{N}'$ is a diffeomorphism and $\hat{\iota}$ and ι are isotopic as continuous maps $N \to X$, where $\iota : N \to X$ is the inclusion.

In [201, Def. 5.6] we define a *topology* on $\mathcal{M}_N$, and explain why it is the natural one.

In [201, Th. 6.10] we describe $\mathcal{M}_N$ near N, in terms of a smooth map Φ between the *infinitesimal deformation space* $\mathcal{I}_{N'}$ and the *obstruction space* $\mathcal{O}_{N'}$.

Theorem 8.5.5   *Suppose (X, J, g, θ) is an almost Calabi–Yau m-fold and N a compact SL m-fold in X with conical singularities at $x_1, \ldots, x_n$ and cones $C_1, \ldots, C_n$. Let $\mathcal{M}_N$ be the moduli space of deformations of N as an SL m-fold with conical singularities in X, as in Definition 8.5.4. Set $N' = N \setminus \{x_1, \ldots, x_n\}$.*

Then there exist natural finite-dimensional vector spaces $\mathcal{I}_{N'}, \mathcal{O}_{N'}$ such that $\mathcal{I}_{N'}$ is isomorphic to the image of $H^1_{cs}(N', \mathbb{R})$ in $H^1(N', \mathbb{R})$ and $\dim \mathcal{O}_{N'} = \sum_{i=1}^n$ s-ind(C_i), where s-ind(C_i) is the stability index of Definition 8.3.4. There exists an open neighbourhood U of 0 in $\mathcal{I}_{N'}$, a smooth map $\Phi : U \to \mathcal{O}_{N'}$ with $\Phi(0) = 0$, and a map $\Xi : \{u \in U : \Phi(u) = 0\} \to \mathcal{M}_N$ with $\Xi(0) = N$ which is a homeomorphism with an open neighbourhood of N in $\mathcal{M}_N$.

The rather complicated proof generalizes that of Theorem 8.4.5, using the theory of analysis on manifolds with cylindrical ends developed by Lockhart and McOwen [245]. Roughly speaking, the spaces $\mathcal{I}_{N'}, \mathcal{O}_{N'}$ appear as corrected versions of the kernel and cokernel of a Laplacian acting between weighted Sobolev spaces of functions on N'. If the C_i are *stable* then $\mathcal{O}_{N'} = \{0\}$ and we deduce [201, Cor. 6.11]:

Corollary 8.5.6   *Suppose (X, J, g, θ) is an almost Calabi–Yau m-fold and N a compact SL m-fold in X with stable conical singularities, and let $\mathcal{M}_N$ and $\mathcal{I}_{N'}$ be as in Theorem 8.5.5. Then $\mathcal{M}_N$ is a smooth manifold of dimension $\dim \mathcal{I}_{N'}$.*

This has clear similarities with Theorem 8.4.5. Here is another simple condition for $\mathcal{M}_N$ to be a manifold near N, [201, Def. 6.12].

Definition 8.5.7   Let (X, J, g, θ) be an almost Calabi–Yau m-fold and N a compact SL m-fold in X with conical singularities, and let $\mathcal{I}_{N'}, \mathcal{O}_{N'}, U$ and Φ be as in Theorem 8.5.5. We call N *transverse* if the linear map $d\Phi|_0 : \mathcal{I}_{N'} \to \mathcal{O}_{N'}$ is surjective.

If N is transverse then $\{u \in U : \Phi(u) = 0\}$ is a manifold near 0, so Theorem 8.5.5 yields [201, Cor. 6.13]:

Corollary 8.5.8 *Suppose* (X, J, g, θ) *is an almost Calabi–Yau* m-*fold and* N *a transverse compact SL* m-*fold in* X *with conical singularities, and let* $\mathcal{M}_N$, $\mathcal{I}_{N'}$ *and* $\mathcal{O}_{N'}$ *be as in Theorem 8.5.5. Then* $\mathcal{M}_N$ *is smooth of dimension* $\dim \mathcal{I}_{N'} - \dim \mathcal{O}_{N'}$ *near* N.

Now there are a number of well-known moduli space problems in geometry where in general moduli spaces are obstructed and singular, but after a generic perturbation they become smooth manifolds. For instance, moduli spaces of instantons on 4-manifolds can be made smooth by choosing a generic metric, and similar things hold for Seiberg–Witten equations, and pseudo-holomorphic curves in symplectic manifolds.

In [201, §9] we try (but do not quite succeed) to replicate this for moduli spaces of SL m-folds with conical singularities, by choosing a *generic Kähler metric* in a fixed Kähler class. This is the idea behind [201, Conj. 9.5], partially proved in [201, §9]:

Conjecture 8.5.9 *Let* (X, J, g, θ) *be an almost Calabi–Yau* m-*fold,* N *a compact SL* m-*fold in* X *with conical singularities, and* $\mathcal{I}_{N'}$, $\mathcal{O}_{N'}$ *be as in Theorem 8.5.5. Then for a second category subset of Kähler forms* $\breve{\omega}$ *in the Kähler class of* g, *the moduli space* $\breve{\mathcal{M}}_N$ *of compact SL* m-*folds* $\hat{N}$ *with conical singularities in* $(X, J, \breve{\omega}, \theta)$ *isotopic to* N *consists solely of transverse* $\hat{N}$, *and so is smooth of dimension* $\dim \mathcal{I}_{N'} - \dim \mathcal{O}_{N'}$.

Notice that Conjecture 8.5.9 constrains the topology and cones of SL m-folds N with conical singularities that can occur in a generic almost Calabi–Yau m-fold, as we must have $\dim \mathcal{I}_{N'} \geqslant \dim \mathcal{O}_{N'}$.

8.5.3 Desingularizing compact SL m-folds with conical singularities

We now discuss the work of [202, 203] on *desingularizing* compact SL m-folds with conical singularities. Here is the basic idea. Let (X, J, g, θ) be an almost Calabi–Yau m-fold, and N a compact SL m-fold in X with conical singularities $x_1, \ldots, x_n$ and cones $C_1, \ldots, C_n$. Suppose $L_1, \ldots, L_n$ are AC SL m-folds in $\mathbb{C}^m$ with the same cones $C_1, \ldots, C_n$ as N.

If $t > 0$ then $tL_i = \{t\mathbf{x} : \mathbf{x} \in L_i\}$ is also an AC SL m-fold with cone C_i. We construct a 1-parameter family of compact, nonsingular *Lagrangian* m-folds N^t in (X, ω) for $t \in (0, \delta)$ by gluing tL_i into N at x_i, using a partition of unity.

When t is small, N^t is *close to special Lagrangian* (its phase is nearly constant), but also *close to singular* (it has large curvature and small injectivity radius). We prove using analysis that for small $t \in (0, \delta)$ we can deform N^t to a *special Lagrangian* m-fold $\tilde{N}^t$ in X, using a small Hamiltonian deformation.

The proof involves a delicate balancing act, showing that the advantage of being close to special Lagrangian outweighs the disadvantage of being nearly singular. Doing this in full generality is rather complex. Here is our simplest desingularization result, [202, Th. 6.13].

Theorem 8.5.10 *Suppose* (X, J, g, θ) *is an almost Calabi–Yau* m *fold and* N *a compact SL* m-*fold in* X *with conical singularities at* $x_1, \ldots, x_n$ *and cones* $C_1, \ldots, C_n$. *Let* $L_1, \ldots, L_n$ *be AC SL* m-*folds in* $\mathbb{C}^m$ *with cones* $C_1, \ldots, C_n$ *and rates* $\lambda_1, \ldots, \lambda_n$. *Suppose* $\lambda_i < 0$ *for* $i = 1, \ldots, n$, *and* $N' = N \setminus \{x_1, \ldots, x_n\}$ *is connected.*

Then there exists $\epsilon > 0$ and a smooth family $\{\tilde{N}^t : t \in (0, \epsilon]\}$ of compact, non-singular SL m-folds in X, such that $\tilde{N}^t$ is constructed by gluing tL_i into N at x_i for $i = 1, \ldots, n$. In the sense of currents in geometric measure theory, $\tilde{N}^t \rightarrow N$ as $t \rightarrow 0$.

The theorem contains two *simplifying assumptions*: (a) that N' is connected, and (b) that $\lambda_i < 0$ for all i. These avoid two kinds of *obstructions* to desingularizing N using the L_i. In [202, Th. 7.10] we remove assumption (a), allowing N' not connected.

Theorem 8.5.11 *Suppose (X, J, g, θ) is an almost Calabi–Yau m-fold and N a compact SL m-fold in X with conical singularities at $x_1, \ldots, x_n$ and cones $C_1, \ldots, C_n$. Define $f : X \rightarrow (0, \infty)$ as in (8.24). Let $L_1, \ldots, L_n$ be AC SL m-folds in $\mathbb{C}^m$ with cones $C_1, \ldots, C_n$ and rates $\lambda_1, \ldots, \lambda_n$. Suppose $\lambda_i < 0$ for $i = 1, \ldots, n$. Write $N' = N \setminus \{x_1, \ldots, x_n\}$ and $\Sigma_i = C_i \cap \mathcal{S}^{2m-1}$.*

Set $q = b^0(N')$, and let $N'_1, \ldots, N'_q$ be the connected components of N'. For $i = 1, \ldots, n$ let $l_i = b^0(\Sigma_i)$, and let $\Sigma_i^1, \ldots, \Sigma_i^{l_i}$ be the connected components of Σ_i. Define $k(i, j) = 1, \ldots, q$ by $\Upsilon_i \circ \varphi_i \big(\Sigma_i^j \times (0, R') \big) \subset N'_{k(i,j)}$ for $i = 1, \ldots, n$ and $j = 1, \ldots, l_i$. Suppose that

$$\sum_{\substack{1 \leqslant i \leqslant n, \, 1 \leqslant j \leqslant l_i: \\ k(i,j)=k}} f(x_i)^m Z(L_i) \cdot [\Sigma_i^j] = 0 \quad \text{for all } k = 1, \ldots, q. \qquad (8.28)$$

Suppose also that the compact m-manifold N obtained by gluing L_i into N' at x_i for $i = 1, \ldots, n$ is connected. A sufficient condition for this to hold is that N and L_i for $i = 1, \ldots, n$ are connected.

Then there exists $\epsilon > 0$ and a smooth family $\{\tilde{N}^t : t \in (0, \epsilon]\}$ of compact, nonsingular SL m-folds in (X, J, g, θ) diffeomorphic to N, such that $\tilde{N}^t$ is constructed by gluing tL_i into N' at x_i for $i = 1, \ldots, n$. In the sense of currents, $\tilde{N}^t \rightarrow N$ as $t \rightarrow 0$.

The new issue here is that if N' is not connected then there is an *analytic obstruction* to deforming N^t to $\tilde{N}^t$, because the Laplacian Δ^t on functions on N^t has *small eigenvalues* of size $O(t^{m-2})$. As in §8.3.3 the L_i have *cohomological invariants* $Z(L_i)$ in $H^{m-1}(\Sigma_i, \mathbb{R})$ derived from the relative cohomology class of Im θ. It turns out that we can only deform N^t to $\tilde{N}^t$ if the $Z(L_i)$ satisfy (8.28). This arises by requiring the projection of an error term to the eigenspaces of Δ^t with small eigenvalues to be zero.

In [203, Th. 6.13] we remove assumption (b), extending Theorem 8.5.10 to the case $\lambda_i \leqslant 0$, and allowing $Y(L_i) \neq 0$.

Theorem 8.5.12 *Let (X, J, g, θ) be an almost Calabi–Yau m-fold for $2 < m < 6$, and N a compact SL m-fold in X with conical singularities at $x_1, \ldots, x_n$ and cones $C_1, \ldots, C_n$. Let $L_1, \ldots, L_n$ be AC SL m-folds in $\mathbb{C}^m$ with cones $C_1, \ldots, C_n$ and rates $\lambda_1, \ldots, \lambda_n$. Suppose that $\lambda_i \leqslant 0$ for $i = 1, \ldots, n$, that $N' = N \setminus \{x_1, \ldots, x_n\}$ is connected, and that there exists $\varrho \in H^1(N', \mathbb{R})$ such that $\big(Y(L_1), \ldots, Y(L_n) \big)$ is the image of ϱ under the natural map $H^1(N', \mathbb{R}) \rightarrow \bigoplus_{i=1}^n H^1(\Sigma_i, \mathbb{R})$, where $\Sigma_i = C_i \cap \mathcal{S}^{2m-1}$. Then there exists $\epsilon > 0$ and a smooth family $\{\tilde{N}^t : t \in (0, \epsilon]\}$ of compact, nonsingular SL m-folds in (X, J, g, θ), such that $\tilde{N}^t$ is constructed by gluing tL_i into N' at x_i for $i = 1, \ldots, n$. In the sense of currents, $\tilde{N}^t \rightarrow N$ as $t \rightarrow 0$.*

From §8.3.3, the L_i have *cohomological invariants* $Y(L_i)$ in $H^1(\Sigma_i, \mathbb{R})$ derived from the relative cohomology class of ω. The new issue in Theorem 8.5.12 is that if $Y(L_i) \neq 0$ then there are obstructions to the existence of N^t as a *Lagrangian m-fold*. That is, we can only define N^t if the $Y(L_i)$ satisfy an equation. This did not appear in Theorem 8.5.10, as $\lambda_i < 0$ implies that $Y(L_i) = 0$ by Proposition 8.3.12.

To define the N^t when $Y(L_i) \neq 0$ we must also use a more complicated construction. This introduces new errors. To overcome these errors when we deform N^t to $\tilde{N}^t$ we must assume that $m < 6$. There are also [203, Th. 6.12] a result combining the modifications of Theorems 8.5.11 and 8.5.12, and [203, §7–§8] extensions of all these results to *families* of almost Calabi–Yau m-folds $(X, J^s, \omega^s, \theta^s)$ for $s \in \mathcal{F}$, but for brevity we will not give them.

8.5.4 Connected sums of compact SL m-folds

The results of §8.5.3 were applied in [204, §9] to desingularize self-intersection points of *immersed* SL m-folds. Similar theorems were proved by also Adrian Butscher [66], Dan Lee [241] and Yng-Ing Lee [242]. Here is some notation for self-intersection points of immersed SL m-folds.

Definition 8.5.13 Let (X, J, g, θ) be an almost Calabi–Yau m-fold for $m > 2$. Suppose N is a compact, nonsingular, immersed SL m-fold in X. That is, N is a compact m-manifold (not necessarily connected) and $\iota : N \to X$ an immersion, with special Lagrangian image. Call $x \in X$ a *self-intersection point* of N if $\iota^*(x)$ is at least two points in N. Call such an x *transverse* if $\iota^*(x)$ is exactly two points x^+, x^- in N, and $\iota_*(T_{x^+}N) \cap \iota_*(T_{x^-}N) = \{0\}$ in $T_x X$.

Let x be a transverse self-intersection point of N, and $x^\pm$ as above. Then by a kind of simultaneous diagonalization, it can be shown that we can choose an isomorphism $\upsilon : \mathbb{C}^m \to T_x X$ satisfying the conditions $\upsilon^*(\omega) = \omega'$ and $\upsilon^*(\theta) = f(x)^m \theta'$ of Definition 8.5.1, such that $\upsilon^*\big(\iota_*(T_{x^+}N)\big) = \Pi^0$ and $\upsilon^*\big(\iota_*(T_{x^-}N)\big) = \Pi^\psi$, for some unique Π^0, Π^ψ defined in (8.23) using $\psi = (\psi_1, \ldots, \psi_m)$ with $0 < \psi_1 \leqslant \cdots \leqslant \psi_m < \pi$.

Since Π^ψ is special Lagrangian with some orientation it follows that $\psi_1 + \cdots + \psi_m = k\pi$ for some integer k, with $0 < k < m$ as $0 < \psi_i < \pi$. Define the *type* of x to be k. This depends on the order of x^+, x^-, and exchanging $x^\pm$ replaces k by $m - k$.

If N is an immersed SL m-fold in X and x is a transverse self-intersection point of *type one* then N has a conical singularity at x with cone $\Pi^0 \cup \Pi^\psi$, where $\psi_i \in (0, \pi)$ with $\psi_1 + \cdots + \psi_m = \pi$. In Example 8.3.15, for each $A > 0$ we constructed an AC SL m-fold $L^{\psi,A}$ in $\mathbb{C}^m$ with the same cone $\Pi^0 \cup \Pi^\psi$. So we can use the results of §8.5.3 to glue $L^{\psi,A}$ into N at x. Applying Theorem 8.5.11 as in [204, Th. 9.7] yields:

Theorem 8.5.14 Let (X, J, g, θ) be an almost Calabi–Yau m-fold for $m > 2$, and N a compact, immersed SL m-fold in X, not necessarily connected. Define $f : X \to (0, \infty)$ as in (8.24). Suppose $x_1, \ldots, x_n \in X$ are transverse self-intersection points of N with type 1, and let $x_i^\pm \in N$ be as in Definition 8.5.13. Set $q = b^0(N)$, and let $N_1, \ldots, N_q$ be the connected components of N. Suppose $A_1, \ldots, A_n > 0$ satisfy

$$\sum_{i=1,\ldots,n: \, x_i^+ \in N_k} f(x_i)^m A_i = \sum_{i=1,\ldots,n: \, x_i^- \in N_k} f(x_i)^m A_i \quad \text{for all } k = 1, \ldots, q. \quad (8.29)$$

Let $\tilde{N}$ be the oriented multiple connected sum of N with itself at the pairs of points x_i^+, x_i^- for $i = 1, \ldots, n$. Suppose $\tilde{N}$ is connected.

Then there exists $\epsilon > 0$ and a smooth family $\{\tilde{N}^t : t \in (0, \epsilon]\}$ of compact, immersed SL m-folds in (X, J, g, θ) diffeomorphic to $\tilde{N}$, such that $\tilde{N}^t$ is constructed by gluing an AC SL m-fold $L^{\psi_i, t^m A_i}$ from Example 8.3.15 into N at x_i for $i = 1, \ldots, n$. In the sense of currents in geometric measure theory, $\tilde{N}^t \to N$ as $t \to 0$. If $x_1, \ldots, x_n$ are the only self-intersection points of N then $\tilde{N}^t$ is embedded.

When $m = 3$ the only possible types of a transverse self-intersection point x of N are 1 and $2 = m - 1$, so swapping x_+, x_- if necessary we can take all self-intersection points of N to have type 1. If also N is connected, so that $q = 1$, then (8.29) holds automatically for all choices of $A_1, \ldots, A_n > 0$. This gives [204, Cor. 9.6]:

Corollary 8.5.15 *Let (X, J, g, θ) be an almost Calabi–Yau 3-fold and N a compact, immersed SL 3-fold with transverse self-intersection points in X, which is connected as an abstract 3-manifold. Then N is a limit of embedded SL 3-folds in X.*

These results also extend to connected sums of SL m-folds in *families* of almost Calabi–Yau m-folds (X, J^s, g^s, θ^s) for $s \in \mathcal{F}$. For details see [204, §9.3].

8.5.5   The index of singularities of SL m-folds

We now consider the *boundary* $\partial \mathcal{M}_{\tilde{N}}$ of a moduli space $\mathcal{M}_{\tilde{N}}$ of SL m-folds.

Definition 8.5.16 Let (X, J, g, θ) be an almost Calabi–Yau m-fold, $\tilde{N}$ a compact, nonsingular SL m-fold in X, and $\mathcal{M}_{\tilde{N}}$ the moduli space of deformations of $\tilde{N}$ in X. Then $\mathcal{M}_{\tilde{N}}$ is a smooth manifold of dimension $b^1(\tilde{N})$, in general noncompact. We can construct a natural *compactification* $\overline{\mathcal{M}}_{\tilde{N}}$ as follows.

Regard $\mathcal{M}_{\tilde{N}}$ as a moduli space of special Lagrangian *integral currents* in the sense of Geometric Measure Theory, as in §4.4. Let $\overline{\mathcal{M}}_{\tilde{N}}$ be the closure of $\mathcal{M}_{\tilde{N}}$ in the space of integral currents. As elements of $\mathcal{M}_{\tilde{N}}$ have uniformly bounded volume, $\overline{\mathcal{M}}_{\tilde{N}}$ is *compact*. Define the *boundary* $\partial \mathcal{M}_{\tilde{N}}$ to be $\overline{\mathcal{M}}_{\tilde{N}} \setminus \mathcal{M}_{\tilde{N}}$. Then elements of $\partial \mathcal{M}_{\tilde{N}}$ are *singular SL integral currents*.

In good cases, say if (X, J, g, θ) is suitably generic, it seems reasonable that $\partial \mathcal{M}_{\tilde{N}}$ should be divided into a number of *strata*, each of which is a moduli space of singular SL m-folds with singularities of a particular type, and is itself a manifold with singularities. In particular, some or all of these strata could be moduli spaces $\mathcal{M}_N$ of SL m-folds N with isolated conical singularities, as in §8.5.2.

Let $\mathcal{M}_{\tilde{N}}$ be a moduli space of compact, nonsingular SL m-folds $\tilde{N}$ in (X, J, g, θ), and $\mathcal{M}_N$ a moduli space of singular SL m-folds N in $\partial \mathcal{M}_{\tilde{N}}$ with singularities of a particular type, and $N \in \mathcal{M}_N$. Following [204, §8.3], we (loosely) define the *index* of the singularities of N to be $\mathrm{ind}(N) = \dim \mathcal{M}_{\tilde{N}} - \dim \mathcal{M}_N$, provided $\mathcal{M}_N$ is smooth near N. Note that $\mathrm{ind}(N)$ depends on the choice of desingularization type $\tilde{N}$ of N.

In [204, Th. 8.10] we use the results of [201–203] to compute $\mathrm{ind}(N)$ when N is *transverse* with conical singularities, in the sense of Definition 8.5.7. Here is a simplified version of the result, where we assume that $H^1_{\mathrm{cs}}(L_i, \mathbb{R}) \to H^1(L_i, \mathbb{R})$ is surjective to avoid a complicated correction term to $\mathrm{ind}(N)$ related to the obstructions to defining $\tilde{N}$ as a Lagrangian m-fold.

Theorem 8.5.17 *Let N be a compact, transverse SL m-fold in (X, J, g, θ) with conical singularities at $x_1, \ldots, x_n$ and rigid SL cones $C_1, \ldots, C_n$. Let $L_1, \ldots, L_n$ be AC SL m-folds in $\mathbb{C}^m$ with cones $C_1, \ldots, C_n$, such that the natural projection $H^1_{\mathrm{cs}}(L_i, \mathbb{R}) \to H^1(L_i, \mathbb{R})$ is surjective. Construct desingularizations $\tilde{N}$ of N by gluing AC SL m-folds $L_1, \ldots, L_n$ in at $x_1, \ldots, x_n$, as in §8.5.3. Then*

$$\mathrm{ind}(N) = 1 - b^0(N') + \textstyle\sum_{i=1}^n b^1_{\mathrm{cs}}(L_i) + \sum_{i=1}^n \text{s-ind}(C_i). \qquad (8.30)$$

If the cones C_i are not *rigid*, for instance if $C_i \setminus \{0\}$ is not connected, then (8.30) should be corrected, as in [204, §8.3]. If Conjecture 8.5.9 is true then for a *generic* Kähler form ω, *all* compact SL m-folds N with conical singularities are transverse, and so Theorem 8.5.17 and [204, Th. 8.10] allow us to calculate $\mathrm{ind}(N)$.

Now singularities with *small index* are the most commonly occurring, and so arguably the most interesting kinds of singularity. Also, as $\mathrm{ind}(N) \leqslant \dim \mathscr{M}_{\tilde{N}}$, for various problems, such as the SYZ Conjecture in Chapter 9, it will only be necessary to know about singularities with index up to a certain value. This motivates the following:

Problem 8.5.18 Classify types of singularities of SL 3-folds with *small index* in suitably generic almost Calabi–Yau 3-folds, say with index 1, 2 or 3.

Here we restrict to $m = 3$ to make the problem more feasible, though still difficult. Note, however, that we do *not* restrict to conical singularities, so a complete, rigorous answer will require a theory of more general kinds of singularities of SL 3-folds.

One can make some progress on this problem simply by studying the many known examples of singular SL 3-folds, calculating or guessing the index of each, and ruling out other kinds of singularity by plausible-sounding arguments. Using these techniques the author has a conjectural classification of index 1 singularities of SL 3-folds, which involves the SL T^2-cone L_0 of (8.9), and several different kinds of singularity whose tangent cone is two copies of $\mathbb{R}^3$, intersecting in 0, $\mathbb{R}$ or $\mathbb{R}^3$.

Coming from another direction, *integrable systems* techniques may yield rigorous classification results for SL T^2-cones by index. Haskins [154, Th. A] has used them to prove that the SL T^2-cone L_0 in $\mathbb{C}^3$ of (8.9) is up to SU(3) equivalence the *unique* SL T^2-cone C with s-ind$(C) = 0$. Now the index of a singularity modelled on C is at least s-ind$(C) + 1$, so this implies that L_0 is the unique SL T^2-cone with index 1 in Problem 8.5.18.

9

Mirror symmetry and the SYZ Conjecture

Mirror symmetry is a mysterious, non-classical relationship between pairs of Calabi–Yau 3-folds $X, \hat{X}$. It was discovered by physicists working in *string theory*, a branch of theoretical physics that aims to quantize gravity by modelling particles not as points but as 1-dimensional 'loops of string'. Special Lagrangian geometry is important in mirror symmetry, because of the *SYZ Conjecture*, which explains mirror symmetry in terms of dual fibrations $f : X \to B$, $\hat{f} : \hat{X} \to B$ with special Lagrangian fibres.

Readers are warned that this is the most flawed and unsatisfactory chapter in the book, due partly to the incompetence of the author, and partly to the nature of the subject. String theory is a huge endeavour, and difficult for mathematicians to penetrate. Mirror symmetry is not a clean mathematical statement, but a swamp of complicated, continuously evolving conjectures, which are slowly being turned into theorems. Each field would take a book far longer than this to explain properly, such as Hori et al. [166] on mirror symmetry.

We begin in §9.1 with a discussion of string theory and mirror symmetry aimed at mathematicians new to the subject. Section 9.2 describes the first forms of mirror symmetry to be discovered, namely the occurrence of Calabi–Yau 3-folds in pairs $X, \hat{X}$ with Hodge numbers satisfying $h^{p,q}(X) = h^{q,p}(\hat{X})$, and the prediction of numbers of rational curves on $\hat{X}$ in terms of the complex structure moduli space of its mirror X. Sections 9.3 and 9.4 discuss two rather different 'explanations' of mirror symmetry, the homological mirror symmetry conjecture, and the SYZ Conjecture.

9.1 String theory and mirror symmetry for dummies

I am now going to take the risk of offering a personal perspective on string theory, which is that of a puzzled mathematician on the outside looking in. I should stress that the dummy in question is me: although I have been talking to string theorists for most of my career, and have wanted to be allowed to play string theory with the big boys and girls for most of that time, I have never come close to understanding it, nor even to being able to follow an average string theory paper. Two good sources for mathematicians to learn some basic ideas in string theory are Douglas [99] and Cox and Katz [80, §1].

The greatest problem in modern physics is that our two most successful physical theories, quantum theory—describing the very small, atoms, particles, and so on—and general relativity—describing the very large, stars and galaxies—are incompatible. Ever since Einstein, the dream has been to find a theory of quantum gravity reconciling the two, perhaps giving quantum theory and general relativity as different limits of one more complicated theory; which might even be a theory of everything, exactly describing the whole universe. This has proved extraordinarily difficult.

String theory has for decades been a sexy area of theoretical physics, because its practitioners claim that it can quantize gravity. As far as I can tell, this claim rests on the fact that string theory has quantum theory and general relativity in it, and no-one has yet managed to persuade everyone else that string theory cannot possibly work. This is stronger evidence than it sounds, as most schemes for quantizing gravity cannot possibly work for reasons that become apparent very quickly. I am agnostic on whether this claim is true, or on whether string theory describes the universe in which we actually live. But I am sure that underlying string theory is some very major area of not yet understood mathematics, which one could call *quantum geometry*, that string theorists grasp at a heuristic, intuitive level.

A source of culture shock for mathematicians is that string theorists seem quite content to spend their lives working on things with as yet no mathematical definition: they often really cannot tell you what the things they are talking about actually are. Also, the parts of string theory we mathematicians might be able to understand if we try really hard, such as mirror symmetry, are actually only a tiny corner of the whole theory, obtained by taking several limits, setting various bells and whistles in the theory to zero, and throwing away almost all of the data.

In string theory, particles are modelled not as points but as 1-dimensional objects—'strings'—propagating in some background space-time M. String theorists aim to construct a *quantum theory* of the string's motion. The process of quantization is extremely complicated, and fraught with mathematical difficulties that are as yet still poorly understood. The basic recipe involves a *Feynman path integral* over all 'paths' of the system, that is, over all possible motions of the string over time in M. This is a huge infinite-dimensional space with no known good measure, so the integral makes no mathematical sense. But by treating it as if it did, and using techniques such as perturbative expansions of the integral, string theorists can do interesting physics.

One curious feature that emerges from the process of quantization is that the dimension of the background space time M must take a particular value, which is not 4. The most popular theoretical framework is *supersymmetric string theory*, which requires $\dim M = 10$. (Other flavours are *M-theory*, with $\dim M = 11$, *F-theory*, with $\dim M = 12$, and *non-supersymmetric string theory*, with $\dim M = 26$.) To account for the disparity between the 10 dimensions the theory requires, and the 4 space-time dimensions we observe, it is supposed that the space we live in looks locally like $M = \mathbb{R}^4 \times X$, where $\mathbb{R}^4$ is Minkowski space, and X is a compact Riemannian 6-manifold with radius of order 10^{-33}cm, the Planck length. Since the Planck length is so small, space then appears to macroscopic observers to be 4-dimensional.

It turns out that because of supersymmetry, which is to do with constant spinors, X has to be a *Calabi–Yau 3-fold*. Moreover, the geometry and topology of X then determines the laws of the low-energy physics we observe in our 4-dimensional world. Because of this, string theorists were interested in understanding and constructing Calabi–Yau 3-folds, in the hope of finding examples with the right properties to describe our own physics. However, the enormous multitude of examples now known has become an embarrassment, as they lead to so many different versions of 4-dimensional physics that they drain string theory of any predictive power.

One can consider either *closed strings*—strings without ends, circles S^1—or *open strings*—strings allowed to have ends, circles S^1 and intervals $[0, 1]$. Closed strings are simpler, and were studied first. Generalizing to open strings leads to the idea of *branes*. From one point of view, a brane is a boundary condition for an open string. When an open string moves in a background space-time M, its end points should be constrained to lie in a submanifold N. This submanifold is thought of as a p-dimensional membrane, or 'p-brane', a joke which a few pea-brained string theorists still find funny.

For the boundary condition to preserve some supersymmetry, N should be *calibrated* by one of the natural calibrations. So the classical limit of a brane in a Calabi–Yau 3-fold might be an SL 3-fold, or a holomorphic curve or surface. This is important for us, as it explains string theorists' interest in calibrated geometry. Actually, things are more complicated than this. At the least, N must be augmented with some extra data—something like a flat U(1)-connection for SL 3-folds, or a holomorphic vector bundle for complex submanifolds. But this is still a very partial description of branes.

String theorists expect that each Calabi–Yau 3-fold X should have a quantization, which is a *super-conformal field theory* (*SCFT*). I am not sure whether there is yet a rigorous mathematical definition of SCFT (though conformal field theories are well-defined), but they are certainly very complicated mathematical objects, including things like Hilbert spaces with large families of multilinear operations satisfying relations, and (for open strings) probably one or more triangulated categories whose objects are topological branes. Invariants of X such as the Dolbeault groups $H^{p,q}(X)$ and numbers of holomorphic curves in X translate to properties of the SCFT.

This leads to the idea of *mirror symmetry*. If X is a Calabi–Yau 3-fold, $H^{1,1}(X)$ and $H^{2,1}(X)$ can be recovered from the associated SCFT as eigenspaces of a certain operator. Furthermore, the only difference between the SCFT representations of $H^{1,1}(X)$ and $H^{2,1}(X)$ is the sign of their eigenvalue under a particular U(1)-action, and the choice of sign is only a matter of convention. This led several physicists to conjecture that there should exist a second Calabi–Yau 3-fold $\hat{X}$ which should have the same SCFT but with the opposite sign for the U(1)-action, so that $H^{1,1}(X) \cong H^{2,1}(\hat{X})$ and $H^{2,1}(X) \cong H^{1,1}(\hat{X})$. We call X and $\hat{X}$ a *mirror pair*. All this is very surprising from the mathematical point of view, but there is by now a great deal of evidence in favour of mirror symmetry and the existence of mirror pairs.

Although we shall concentrate on mirror symmetry for Calabi–Yau 3-folds, it is also studied in other contexts. One of these is mirror symmetry between Calabi–Yau m-folds for other $m \geqslant 1$. Another is mirror symmetry between Fano varieties and Landau–Ginzburg models. Here a *Fano variety* X is a nonsingular projective complex variety X with ample anticanonical bundle K_X^{-1}. A *Landau–Ginzburg model* $(\hat{X}, \hat{f})$ is roughly speaking a noncompact Kähler manifold $\hat{X}$ with a holomorphic function $\hat{f} : \hat{X} \rightarrow \mathbb{C}$, and one is primarily interested in the stationary points of $\hat{f}$.

9.2 Early mathematical formulations of mirror symmetry

We now explain two aspects of mirror symmetry of Calabi–Yau 3-folds $X, \hat{X}$, developed by Candelas and others in 1990–1, which were the first really persuasive evidence for mathematicians that mirror symmetry is true. These are mirror symmetry at the level of Hodge numbers, and closed string mirror symmetry, which is a relationship between the variation of Hodge structure of X and numbers of rational curves on $\hat{X}$.

9.2.1 Mirror symmetry at the level of Hodge numbers

Mirror symmetry suggests that for each Calabi–Yau 3-fold X there should be another Calabi–Yau 3-fold $\hat{X}$ whose Hodge numbers satisfy $h^{p,q}(X) = h^{q,p}(\hat{X})$, so that $h^{1,1}(X) = h^{2,1}(\hat{X})$ and $h^{2,1}(X) = h^{1,1}(\hat{X})$. This cannot be exactly true, since as in Example 7.5.3 there exist Calabi–Yau 3-folds X with $h^{2,1}(X) = 0$, but there cannot exist any mirror Calabi–Yau 3-fold $\hat{X}$ with $h^{1,1}(\hat{X}) = 0$, since $h^{1,1} \geqslant 1$ for any compact Kähler manifold. However, it could still be approximately true.

As in §7.6.2, one of the earliest pieces of mathematical evidence for mirror symmetry was found by Candelas, Lynker and Schimmrigk [71], who computed the Hodge numbers $(h^{1,1}, h^{2,1})$ of Calabi–Yau 3-folds obtained as crepant resolutions of hypersurfaces in weighted projective spaces $\mathbb{CP}^4_{a_0,\dots,a_4}$. When plotted on a graph these displayed an approximate, but very persuasive, symmetry under exchanging $h^{1,1}$ and $h^{2,1}$.

Later, Batyrev [20] showed that Calabi–Yau 3-folds that are hypersurfaces in compact toric 4-folds T naturally come in mirror pairs at the Hodge number level, since each allowed toric 4-fold T has a dual toric 4-fold $\hat{T}$, and if $X, \hat{X}$ are Calabi–Yau hypersurfaces in $T, \hat{T}$ then $h^{1,1}(X) = h^{2,1}(\hat{X})$ and $h^{2,1}(X) = h^{1,1}(\hat{X})$. The graph of the corresponding Hodge numbers, Figure 7.1, above, has an exact reflection symmetry when $h^{1,1}$ and $h^{2,1}$ are exchanged.

9.2.2 Closed string mirror symmetry and counting curves

Suppose $X, \hat{X}$ are a mirror pair of Calabi–Yau 3-folds, in the string theory sense. What does this mean mathematically, and how can we relate the geometry of X to the geometry of $\hat{X}$? The first answer to this was given by Candelas, de la Ossa, Green, and Parkes [70] in 1991, interpreted for mathematicians by Morrison [268]. We shall call this area *closed string mirror symmetry*, as it uses only closed rather than open string ideas. It is now a mature subject. Some good mathematical books taking this approach are Voisin [335], Gross [138, Part II], and the very thorough Cox and Katz [80].

Here is a very brief sketch. In §9.1 we said that a Calabi–Yau 3-fold (X, J, g, θ) determines a super-conformal field theory (SCFT), whatever that is. This is an oversimplification: we also need some extra data on X to determine the SCFT. The most important is a *B-field*, a closed 2-form B on X. This appears in the physics in expressions of the form $\exp\left(\int_\Sigma 2\pi i (B + i\omega)\right)$, for Σ a closed oriented 2-submanifold in X.

Clearly, this depends only on the cohomology class $[B] \in H^2(X, \mathbb{R})$, and is also unchanged by adding elements of $H^2(X, \mathbb{Z})$ to $[B]$, so it depends only on the projection $[B] \in H^2(X, \mathbb{R})/H^2(X, \mathbb{Z})$. Also, it depends holomorphically on the *complexified Kähler form* $\omega_c = B + i\omega$, and on the projection of its cohomology class $[\omega_c] \in H^2(X, \mathbb{C})/H^2(X, \mathbb{Z})$. Thus, the SCFT depends, in some sense holomorphically, on two complex cohomology classes, $[\omega_c] \in H^2(X, \mathbb{C})/H^2(X, \mathbb{Z})$ and $[\theta] \in H^3(X, \mathbb{C})$.

Now consider the *moduli space* $\mathcal{M}^X_{\text{SCFT}}$ of SCFT's coming from X in this way. It turns out that $\mathcal{M}^X_{\text{SCFT}}$ is a complex manifold locally parametrized by independent choices of $[\omega_{\text{C}}], [\theta]$. That is, $\mathcal{M}^X_{\text{SCFT}}$ is locally a product $\mathcal{M}^X_{\text{Käh}} \times \mathcal{M}^X_{\text{cx}}$ of the *complexified Kähler moduli space* $\mathcal{M}^X_{\text{Käh}}$ of X, parametrized by $[\omega_{\text{C}}]$, and the *complex structure moduli space* $\mathcal{M}^X_{\text{cx}}$ of X, parametrized by $[\theta]$. Here $[\omega_{\text{C}}]$ locally varies freely in $H^2(X, \mathbb{C})/H^2(X, \mathbb{Z})$, but $[\theta]$ is confined to a complex Lagrangian cone in $H^3(X, \mathbb{C})$, so that $\dim_{\mathbb{C}} \mathcal{M}^X_{\text{cx}} = h^{3,0}(X) + h^{2,1}(X) = \frac{1}{2} \dim_{\mathbb{C}} H^3(X, \mathbb{C})$.

Suppose $X, \hat{X}$ are a mirror pair. Then by definition their SCFT's are isomorphic under a sign change, so their SCFT moduli spaces must be isomorphic, that is, $\mathcal{M}^X_{\text{SCFT}} \cong \mathcal{M}^{\hat{X}}_{\text{SCFT}}$. This isomorphism turns out to preserve the local product structures but exchanges complex and Kähler factors. Thus we expect local isomorphisms

$$\mathcal{M}^X_{\text{cx}} \cong \mathcal{M}^{\hat{X}}_{\text{Käh}} \quad \text{and} \quad \mathcal{M}^X_{\text{Käh}} \cong \mathcal{M}^{\hat{X}}_{\text{cx}}. \tag{9.1}$$

These isomorphisms do not simply identify $[\theta]$ with $[\hat{\omega}_{\text{C}}]$ and $[\omega_{\text{C}}]$ with $[\hat{\theta}]$, as $[\theta]$ lies in a complex Lagrangian cone in $H^3(\hat{X}, \mathbb{C})$ and $[\hat{\omega}_{\text{C}}]$ in $H^2(\hat{X}, \mathbb{C})/H^2(\hat{X}, \mathbb{Z})$, which are very different spaces. Instead, they are given by *mirror maps* $m : \mathcal{M}^X_{\text{cx}} \to \mathcal{M}^{\hat{X}}_{\text{Käh}}$ and $\hat{m} : \mathcal{M}^{\hat{X}}_{\text{cx}} \to \mathcal{M}^X_{\text{Käh}}$, which amount to writing $[\hat{\omega}_{\text{C}}]$ and $[\omega_{\text{C}}]$ as some complicated holomorphic functions of $[\theta]$ and $[\hat{\theta}]$ respectively.

One of the important problems in the theory is to determine these mirror maps. It is done as follows. Roughly speaking, the moduli spaces $\mathcal{M}^X_{\text{cx}}, \mathcal{M}^{\hat{X}}_{\text{Käh}}$ admit partial compactifications $\overline{\mathcal{M}}^X_{\text{cx}}, \overline{\mathcal{M}}^{\hat{X}}_{\text{Käh}}$ including singular limit points. One such singular limit in $\overline{\mathcal{M}}^{\hat{X}}_{\text{Käh}}$ is the *large radius limit point* obtained by letting $\hat{\omega}$ go to infinity in the Kähler cone of $\hat{X}$, that is, by taking the limit of $[\hat{B} + it\hat{\omega}]$ as $t \to \infty$. The corresponding kind of singular limit in $\overline{\mathcal{M}}^X_{\text{cx}}$ is called the *large complex structure limit*, or a *maximally unipotent boundary point*, a limiting family of complex structures whose monodromy is maximally unipotent.

It turns out that near a large complex structure limit point in $\overline{\mathcal{M}}^X_{\text{cx}}$ and a large radius limit point in $\overline{\mathcal{M}}^{\hat{X}}_{\text{Käh}}$, one can define systems of *canonical coordinates* on $\mathcal{M}^X_{\text{cx}}$ and $\mathcal{M}^{\hat{X}}_{\text{Käh}}$, and argue that the mirror map $m : \mathcal{M}^X_{\text{cx}} \to \mathcal{M}^{\hat{X}}_{\text{Käh}}$ must identify these coordinates. So, after a lot of work, one can actually find the mirror map m *explicitly* near these limits.

Once we know the mirror map, we can use it to compare aspects of the geometry of X and $\hat{X}$ that are reflected in their common SCFT. For example, something that is easy to compute for X might translate to something difficult to compute for $\hat{X}$, so we could conjecture nontrivial new information about $\hat{X}$. This applies for the *Yukawa couplings*, which are holomorphic sections of the vector bundles $S^3 T^* \mathcal{M}^X_{\text{cx}}, S^3 T^* \mathcal{M}^{\hat{X}}_{\text{Käh}}$ over $\mathcal{M}^X_{\text{cx}}, \mathcal{M}^{\hat{X}}_{\text{Käh}}$ identified by the mirror map m.

On $\mathcal{M}^X_{\text{cx}}$, the Yukawa coupling is given exactly by a simple formula involving the third derivatives of the inclusion $\mathcal{M}^X_{\text{cx}} \hookrightarrow H^3(X, \mathbb{C})$ taking $[(X, J, \theta)] \mapsto [\theta]$, and is easy to compute if we know $\mathcal{M}^X_{\text{cx}}$. On $\mathcal{M}^{\hat{X}}_{\text{Käh}}$, the Yukawa coupling is shown by Feynman path integral techniques to be the sum over $\beta \in H_2(\hat{X}, \mathbb{Z})$ of a term involving a number n_β of rational curves in $\hat{X}$ with homology class β; in fact n_β is a *Gromov–Witten invariant*. Candelas et al. [70] used these ideas to predict values for the Gromov–Witten invariants of the quintic Calabi–Yau 3-fold. At the time this was remarkable, as noone knew how to compute these invariants, but the conjectures have since been proved by Givental and Lian, Liu and Yau.

The string theorists Bershadsky, Cecotti, Ooguri and Vafa [29] showed how to use similar ideas to compute Gromov–Witten invariants of $\hat{X}$ for curves of every genus $g \geqslant 0$, rather than just rational curves of genus 0. The story is more complicated, and involves nonholomorphic potential functions over $\mathscr{M}^X_{\mathrm{cx}}, \mathscr{M}^{\hat{X}}_{\mathrm{Käh}}$ satisfying a p.d.e.

We can now refine our notion of *mirror pair*. Rather than thinking of this as just a pair of Calabi–Yau 3-folds, it is better to think of mirror symmetry as identifying a large complex structure limit point of the complex moduli space $\mathscr{M}^X_{\mathrm{cx}}$ of one Calabi–Yau 3-fold X with the large radius limit of the Kähler moduli space $\mathscr{M}^{\hat{X}}_{\mathrm{Käh}}$ of another Calabi–Yau 3-fold $\hat{X}$. Better still, we can say that mirror symmetry identifies a large radius limit times a large complex structure limit in $\mathscr{M}^X_{\mathrm{SCFT}} \approx \mathscr{M}^X_{\mathrm{Käh}} \times \mathscr{M}^X_{\mathrm{cx}}$ with a large complex structure limit times a large radius limit in $\mathscr{M}^{\hat{X}}_{\mathrm{SCFT}} \approx \mathscr{M}^{\hat{X}}_{\mathrm{cx}} \times \mathscr{M}^{\hat{X}}_{\mathrm{Käh}}$.

We draw two conclusions from this. Firstly, $\mathscr{M}^X_{\mathrm{cx}}$ could have several large complex structure limit points, so that X (or rather, the family of Calabi–Yau 3-folds deformation equivalent to X) could have several mirrors $\hat{X}_1, \ldots, \hat{X}_k$. Or $\overline{\mathscr{M}}^X_{\mathrm{cx}}$ could have no large complex structure limit points, and no mirrors. This happens if $h^{2,1}(X) = 0$, agreeing with the comment in §9.2.1 that no mirror $\hat{X}$ can exist in this case.

Secondly, mirror symmetry is really about *limiting families* of Calabi–Yau 3-folds undergoing some kind of collapse, rather than just about pairs of Calabi–Yau 3-folds. When string theorists make mathematical conjectures, these limits are often suppressed for simplicity. But sometimes it is necessary to put the limits back in again to make the conjectures true.

9.3   Kontsevich's homological mirror symmetry proposal

Another major advance in the mirror symmetry story was due to Kontsevich [219] in 1994, and is known as *homological mirror symmetry* (*HMS*). At the time it was a visionary proposal, as this was before string theorists had really got started on branes. Kapustin and Orlov [211] give a survey on homological mirror symmetry, explaining the physics to mathematicians. Equation (9.1) illustrates the:

Principle 9.3.1 *If X and $\hat{X}$ are a mirror pair of Calabi–Yau 3-folds, then mirror symmetry relates the complex geometry of X to the symplectic geometry of $\hat{X}$, and the symplectic geometry of X to the complex geometry of $\hat{X}$.*

That is, mirror symmetry relates things on X depending on J, θ but independent of g, ω, B to things on $\hat{X}$ depending on $\hat{\omega}, \hat{B}$ but independent of $\hat{g}, \hat{J}, \hat{\theta}$, and vice versa. The Yukawa couplings discussed in §9.2.2 are an example of this. On $\mathscr{M}^X_{\mathrm{cx}}$ the Yukawa coupling clearly depends only on J, θ. On $\mathscr{M}^{\hat{X}}_{\mathrm{Käh}}$, the Yukawa coupling depends on numbers n_β of rational curves in $\hat{X}$, which are $\hat{J}$-holomorphic and so appear to depend on $\hat{J}$. But Gromov–Witten invariants are independent of the choice of almost complex structure $\hat{J}'$ compatible with $\hat{\omega}$, so the n_β are independent of $\hat{J}$ and depend only on $(\hat{X}, \omega)$.

A string theory explanation for Principle 9.3.1 is that the SCFT associated to a Calabi–Yau 3-fold X admits two *topological twistings*, which are *topological quantum field theories*, called the *A-model* and the *B-model*. These are simpler theories than the original SCFT, containing much less information. The A-model depends only on X, ω, B, and so only on the symplectic geometry of X, and the B-model depends only on X, J, θ, and so only on the complex geometry of X. Mirror symmetry identifies the A-model for X with the B-model for $\hat{X}$, and vice versa.

Kontsevich's *homological mirror symmetry conjecture* (*HMS Conjecture*) [219] says that if X and $\hat{X}$ are mirror Calabi–Yau 3-folds then there should be an *equivalence of triangulated categories* between the *(bounded) derived category* $D^b(\mathrm{coh}(X))$ *of the abelian category* $\mathrm{coh}(X)$ *of coherent sheaves on* X and the *(bounded) derived category* $D^b(F(\hat{X}))$ *of the Fukaya* A_∞*-category* $F(\hat{X})$ *of* $\hat{X}$. Here $\mathrm{coh}(X)$ and $D^b(\mathrm{coh}(X))$ capture the complex algebraic geometry of X, and depend only on the complex manifold (X, J). Also $F(\hat{X})$ and $D^b(F(\hat{X}))$ capture the symplectic geometry of $\hat{X}$, and depend (up to an appropriate notion of equivalence) only on the symplectic manifold $(\hat{X}, \hat{\omega})$ and the B-field $\hat{B}$. So the conjecture is another example of Principle 9.3.1.

A string theory explanation for the HMS Conjecture is that the data of an open string topological quantum field theory should include a triangulated category of branes (boundary conditions). For the B-model on X, this category of topological B-branes should be equivalent to $D^b(\mathrm{coh}(X))$. For the A-model on $\hat{X}$, this category of topological A-branes should be equivalent to $D^b(F(\hat{X}))$. Equating the open string A-model on X with the open string B-model on $\hat{X}$ thus implies the HMS Conjecture.

There is a lot of complicated mathematics going on here. The rest of the section gives brief descriptions of some of the ideas involved, with references.

9.3.1 Categories, and equivalence of categories

A *category* $\mathcal{C}$ consists of a set (or class) $\mathrm{Ob}\,\mathcal{C}$ of *objects*, and for all $X, Y \in \mathrm{Ob}\,\mathcal{C}$ a set $\mathrm{Hom}(X, Y)$ or $\mathrm{Hom}_{\mathcal{C}}(X, Y)$ of *morphisms* $f : X \to Y$, and for all $X, Y, Z \in \mathrm{Ob}\,\mathcal{C}$ a *composition map* $\circ : \mathrm{Hom}(Y, Z) \times \mathrm{Hom}(X, Y) \to \mathrm{Hom}(X, Z)$. Composition of morphisms is *associative*, that is, $(h \circ g) \circ f = h \circ (g \circ f)$. Each $X \in \mathrm{Ob}\,\mathcal{C}$ has an *identity morphism* $\mathrm{id}_X \in \mathrm{Hom}(X, X)$, with $f \circ \mathrm{id}_X = f = \mathrm{id}_Y \circ f$ for all $f : X \to Y$.

A *functor* $F : \mathcal{C} \to \mathcal{D}$ between categories $\mathcal{C}, \mathcal{D}$ consists of maps $F : \mathrm{Ob}\,\mathcal{C} \to \mathrm{Ob}\,\mathcal{D}$ and $F_{XY} : \mathrm{Hom}_{\mathcal{C}}(X, Y) \to \mathrm{Hom}_{\mathcal{D}}(F(X), F(Y))$ for all $X, Y \in \mathrm{Ob}\,\mathcal{C}$, which preserve identities and compositions, that is, $F_{XX}(\mathrm{id}_X) = \mathrm{id}_{F(X)}$ and $F_{XZ}(g \circ f) = F_{YZ}(g) \circ F_{XY}(f)$ for all $X, Y, Z \in \mathrm{Ob}\,\mathcal{C}$ and $f : X \to Y$, $g : Y \to Z$. This is the natural notion of map (morphism) between categories.

Categories are a universal structure occurring almost everywhere in mathematics. A good introduction is Gelfand and Manin [123, §II]. Whenever we have some class of mathematical structures (for example, topological spaces, or smooth manifolds), and some class of maps between them (for example, continuous maps, or smooth maps), these generally form the objects and morphisms of a category. But we can also use as morphisms things which are nothing like maps between the objects of the category, as we shall see in §9.3.7 for the Fukaya category. The abstract categorical point of view, in which one forgets about the nature of the objects and morphisms and regards them simply as points and arrows, can be very powerful.

A question that arises often is: when are two categories $\mathcal{C}, \mathcal{D}$ *the same*? The obvious notion of strict isomorphism, that there should exist a functor $F : \mathcal{C} \to \mathcal{D}$ with $F : \mathrm{Ob}\,\mathcal{C} \to \mathrm{Ob}\,\mathcal{D}$ and all maps F_{XY} bijections, turns out to be too strong, and more-or-less useless in applications. Instead, a functor $F : \mathcal{C} \to \mathcal{D}$ is called an *equivalence* if $F_{XY} : \mathrm{Hom}_{\mathcal{C}}(X, Y) \to \mathrm{Hom}_{\mathcal{D}}(F(X), F(Y))$ is a bijection for all $X, Y \in \mathrm{Ob}\,\mathcal{C}$, and in addition $F : \mathrm{Ob}\,\mathcal{C} \to \mathrm{Ob}\,\mathcal{D}$ induces a bijection between *isomorphism classes of objects* in $\mathrm{Ob}\,\mathcal{C}$ and $\mathrm{Ob}\,\mathcal{D}$. This allows $\mathrm{Ob}\,\mathcal{C}, \mathrm{Ob}\,\mathcal{D}$ to be very different sets.

9.3.2 Additive and linear categories

A category $\mathcal{C}$ is *additive* if for all $X, Y, Z \in \mathrm{Ob}\,\mathcal{C}$ the set of morphisms $\mathrm{Hom}(X, Y)$ has the structure of an abelian group, and composition $\circ : \mathrm{Hom}(Y, Z) \times \mathrm{Hom}(X, Y) \to \mathrm{Hom}(X, Z)$ is biadditive. It is $\mathbb{K}$-*linear* if the $\mathrm{Hom}(X, Y)$ are vector spaces over some field $\mathbb{K}$, and composition is bilinear. Linear categories are additive. All the categories we are interested in are $\mathbb{C}$-linear.

The identity in the abelian group $\mathrm{Hom}(X, Y)$ is the *zero morphism* $0 : X \to Y$. A *complex* in an additive category is a finite or infinite sequence of objects and morphisms $\cdots \to X_k \xrightarrow{d_k} X_{k+1} \xrightarrow{d_{k+1}} X_{k+2} \to \cdots$ with $d_{k+1} \circ d_k = 0$ for all k.

9.3.3 The exact category $\mathrm{Vect}(X)$ of holomorphic vector bundles on X

Before discussing coherent sheaves, we explain a simpler idea. Let (X, J) be a complex manifold. Then we can define *holomorphic vector bundles* $E \to X$, with fibre $\mathbb{C}^k$. A morphism $f : E \to F$ of vector bundles $E, F \to X$ is a holomorphic bundle map $f : E \to F$ over X which is linear on each fibre. Equivalently, it is a holomorphic section of the vector bundle $E^* \otimes F \to X$.

The category $\mathrm{Vect}(X)$ of holomorphic vector bundles is a $\mathbb{C}$-linear category, since $\mathrm{Hom}(E, F) = H^0(E^* \otimes F)$ is a complex vector space. A complex $\cdots \to E \to F \to G \to \cdots$ in $\mathrm{Vect}(X)$ is called *exact* if for all $x \in X$, the restriction to the fibres $\cdots \to E|_x \to F|_x \to G|_x \to \cdots$ is an exact sequence of complex vector spaces. With this notion of exactness, $\mathrm{Vect}(X)$ is an *exact category*, as in [123, p. 275]. The *zero object* $0 \in \mathrm{Vect}(X)$ is the vector bundle with fibre $\{0\}$ at each point.

For many purposes in homological algebra it would really useful if every morphism $f : E \to F$ in $\mathrm{Vect}(X)$ lay in an exact sequence $0 \to K \xrightarrow{i} E \xrightarrow{f} F \xrightarrow{\pi} C \to 0$. Then $i : K \to E$ is called the *kernel* and $\pi : F \to C$ the *cokernel* of f. But this is not true. The dimensions of the kernel and cokernel of the fibre maps $f|_x : E|_x \to F|_x$ need not be locally constant in x, and if they jump then f has no kernel or cokernel as a holomorphic vector bundle. To get round this we enlarge $\mathrm{Vect}(X)$ to $\mathrm{coh}(X)$.

9.3.4 The abelian category $\mathrm{coh}(X)$ of coherent sheaves on X

An *abelian category* is basically an exact category (an additive category with a good notion of exact sequences) in which every morphism $f : E \to F$ has a kernel and cokernel. They are the setting for much of homological algebra. An exact category $\mathcal{E}$ can be enlarged to an abelian category in an essentially unique way up to equivalence, by adding extra objects to be the kernels and cokernels of morphisms in $\mathcal{E}$.

The category $\mathrm{coh}(X)$ of *coherent sheaves* on X is the result of enlarging $\mathrm{Vect}(X)$ to an abelian category in this way. But coherent sheaves are sensible geometric objects on X, not just categorical abstractions. Sheaves are defined in Definition 5.8.8, and a coherent sheaf on a $\mathbb{C}$-scheme $(X, \mathcal{O})$ is a sheaf of $\mathcal{O}$-modules $\mathcal{F}$ such that X can be covered by open affine subsets U of the form $\mathrm{Spec}\,A$ for a $\mathbb{C}$-algebra A, with $\mathcal{F}|_U$

isomorphic to the sheaf of $\mathcal{O}$-modules on U induced by a finitely-generated A-module.

Roughly speaking, a coherent sheaf E on X is a generalization of a holomorphic vector bundle, that has a fibre $E|_x$ for each $x \in X$ which is a finite-dimensional complex vector space, but $\dim E|_x$ need not be constant in x. We actually define E in terms of the $\mathbb{C}$-vector space of holomorphic sections of E over each open set $U \subseteq X$. A typical example is a vector bundle $E \to W$ over a complex subvariety W in X. For more information see Hartshorne [149, §II.5].

9.3.5 Triangulated categories and derived categories

A *triangulated category* $\mathcal{T}$ is an additive category equipped with a *shift functor* $[+1]$: $\mathcal{T} \to \mathcal{T}$ and a class of *distinguished triangles* $X \xrightarrow{u} Y \xrightarrow{v} Z \xrightarrow{w} X[+1]$ in $\mathcal{T}$, satisfying a list of axioms given in Gelfand and Manin [123, §IV].

If $\mathcal{A}$ is an abelian category, the *(bounded) derived category* $D^b(\mathcal{A})$ is a triangulated category constructed from $\mathcal{A}$ by a complicated procedure. The objects of $D^b(\mathcal{A})$ are *bounded complexes* in $\mathcal{A}$, that is, complexes $\cdots \to X_k \xrightarrow{d_k} X_{k+1} \xrightarrow{d_{k+1}} X_{k+2} \to \cdots$ for $k \in \mathbb{Z}$ with $X_k = 0$ for $|k| \gg 0$. The morphisms in $D^b(\mathcal{A})$ are obtained by starting with chain maps on complexes and then inverting *quasi-isomorphisms*, which are chain maps on complexes that induce isomorphisms on the cohomology of the complexes.

Thomas [323] gives a helpful short introduction to derived categories. An excellent textbook on derived and triangulated categories is Gelfand and Manin [123, §III]. Thomas expresses the main idea of derived categories as *working with complexes rather than their (co)homology*. We shall explain the point of derived categories slightly differently, as a way of remedying problems caused by *failures of exactness*.

A functor $F : \mathcal{A} \to \mathcal{B}$ between abelian categories is called *exact* if it takes exact sequences to exact sequences. There are many natural functors between abelian categories which are *not* exact, and this causes problems. For example, let A, B, C, D be vector bundles or coherent sheaves over a $\mathbb{C}$-variety X, and $0 \to B \to C \to D \to 0$ an exact sequence in $\mathrm{coh}(X)$. Then we have a long exact sequence of $\mathbb{C}$-vector spaces

$$0 \to \mathrm{Hom}(A, B) \to \mathrm{Hom}(A, C) \to \mathrm{Hom}(A, D) \to \mathrm{Ext}^1(A, B) \to \cdots .$$

In particular, $0 \to \mathrm{Hom}(A, B) \to \mathrm{Hom}(A, C) \to \mathrm{Hom}(A, D) \to 0$ need *not* be exact at $\mathrm{Hom}(A, D)$. So the functor $\mathrm{Hom}(A, -)$ from $\mathrm{coh}(X)$ to $\mathbb{C}$-vector spaces is not exact. A functor $F : \mathcal{A} \to \mathcal{B}$ is *left exact* if $0 \to B \to C \to D \to 0$ exact in $\mathcal{A}$ implies $0 \to F(B) \to F(C) \to F(D)$ exact in $\mathcal{B}$, and *right exact* if it implies $F(B) \to F(C) \to F(D) \to 0$ exact in $\mathcal{B}$. Thus $\mathrm{Hom}(A, -)$ is left exact, but not right exact.

An advantage of working with derived categories rather than abelian categories is that we can usually fix these problems with exactness. If $F : \mathcal{A} \to \mathcal{B}$ is a left exact functor of abelian categories then we can define the *right derived functor* $\mathbf{R}F : D^b(\mathcal{A}) \to D^b(\mathcal{B})$ which is an exact functor of triangulated categories, in the sense that it takes distinguished triangles (which are the triangulated analogue of short exact sequences) to distinguished triangles. Similarly, if F is right exact we define the *left derived functor* $\mathbf{L}F : D^b(\mathcal{A}) \to D^b(\mathcal{B})$, which is exact. A consequence of this is that functors between derived categories are often much better behaved than their abelian counterparts.

There is a problem in the definition of triangulated categories, the 'nonfunctoriality of the cone' described by Gelfand and Manin [123, p. 245], which means that some constructions can be done up to isomorphism, but not up to canonical isomorphism. To fix this requires some modification to the definition of triangulated categories, and it seems likely that this modification should also be included in the HMS Conjecture. One possibility is the *enhanced triangulated categories* of Bondal and Kapranov [36], involving *dg-categories*. Diaconescu [95] and Lazaroiu [234] have used string theory to argue that topological D-branes should form an enhanced triangulated category. Perhaps *triangulated A_∞-categories* might be another suitable context.

9.3.6 The derived category of coherent sheaves $D^b(\mathrm{coh}(X))$

In recent years, algebraic geometers have begun to study the algebraic varieties X via their derived categories of coherent sheaves $D^b(\mathrm{coh}(X))$. An interesting survey on this is Bridgeland [52]. Bondal and Orlov [37] prove that if X is a smooth algebraic variety whose canonical bundle K_X or its inverse K_X^{-1} is ample, then X can be reconstructed from the triangulated category $D^b(\mathrm{coh}(X))$ uniquely up to isomorphism. Furthermore, any autoequivalence Φ of $D^b(\mathrm{coh}(X))$ is induced by an automorphism $\phi : X \to X$.

For Calabi–Yau m-folds X, with K_X trivial, things are very different. If X, Y are Calabi–Yau m-folds, there can exist equivalences of triangulated categories $\Phi :$ $D^b(\mathrm{coh}(X)) \to D^b(\mathrm{coh}(Y))$ which do not come from isomorphisms $\phi : X \to Y$, and X, Y need not be isomorphic. It is conjectured that if X, Y are birational Calabi–Yau m-folds then there is an equivalence $\Phi : D^b(\mathrm{coh}(X)) \to D^b(\mathrm{coh}(Y))$, and this is known for $m = 3$. Also, if X is a Calabi–Yau m-fold then there may exist many autoequivalences Φ of $D^b(\mathrm{coh}(X))$ not induced by any automorphism $\phi : X \to X$, and the autoequivalences form a large, interesting group $\mathrm{Aut}(D^b(\mathrm{coh}(X)))$.

Thus derived categories $D^b(\mathrm{coh}(X))$ are particularly interesting for Calabi–Yau m-folds X, and they can have many hidden, non-classical symmetries. This is parallel to the idea of *dualities* in string theory, which are basically isomorphisms between SCFT's which do not come from isomorphisms of the classical geometries used to construct the SCFT's, such as mirror symmetry. Since triangulated categories of topological branes are a small part of the data making up an SCFT, isomorphisms between SCFT's naturally imply equivalences between triangulated categories.

9.3.7 The Fukaya category $F(\hat{X})$

Fukaya categories were introduced informally by Fukaya [116]. Fukaya, Oh, Ohta and Ono [118] have undertaken the mammoth task of laying the foundations of the theory and making it rigorous. Seidel's book [306] is a good place to learn the ideas; Seidel avoids many of the technical problems by only considering noncompact, exact symplectic manifolds. We first explain them in a rather oversimplified way, to give the basic ideas, and then explain modifications to and difficulties with this scheme.

The Fukaya category is an A_∞-*category* $\mathcal{A}$, which has objects $\mathrm{Ob}\,\mathcal{A}$ and morphisms $\mathrm{Hom}(A, B)$ for $A, B \in \mathcal{A}$ as usual. However, we do not have associative composition or morphisms $\circ$, as in a usual category. Instead, we have families of multilinear composition maps $m_k : \mathrm{Hom}(A_1, A_2) \times \mathrm{Hom}(A_2, A_3) \times \cdots \times \mathrm{Hom}(A_k, A_{k+1}) \to$ $\mathrm{Hom}(A_1, A_{k+1})$ for all $k \geq 1$ and $A_1, \ldots, A_{k+1} \in \mathcal{A}$, satisfying many complicated identities. Roughly speaking, m_2 is composition of morphisms, but it is only associative up to a homotopy given by m_3.

The Fukaya category $F(\hat{X})$ of a symplectic manifold $(\hat{X}, \hat{\omega})$ is an A_∞-category defined as follows. The objects of $F(\hat{X})$ are compact Lagrangian submanifolds of $\hat{X}$. Let L_1, L_2 be Lagrangians intersecting transversely. Then we define $\mathrm{Hom}(L_1, L_2) = \mathbb{C}^{L_1 \cap L_2}$, that is, $\mathrm{Hom}(L_1, L_2)$ is the $\mathbb{C}$-vector space with basis $L_1 \cap L_2$, a finite set. The maps $m_k : \mathrm{Hom}(L_1, L_2) \times \cdots \times \mathrm{Hom}(L_k, L_{k+1}) \to \mathrm{Hom}(L_1, L_{k+1})$, for transversely intersecting Lagrangians $L_1, \ldots, L_{k+1}$, are defined by 'counting' pseudoholomorphic discs D with boundary in $(\hat{X}, \hat{\omega})$, with respect to some almost complex structure $\hat{J}'$ compatible with $\hat{\omega}$, such that the boundary ∂D lies in $L_1 \cup \cdots \cup L_{k+1}$, and includes prescribed points in $L_i \cap L_{i+1}$ for $i = 1, \ldots, k$ and $L_1 \cap L_{k+1}$. The identities on m_k come from gluing results for pseudoholomorphic discs.

Here are the ways in which this must be modified for use in the HMS Conjecture.

- Rather than Lagrangians L in the Calabi–Yau m-fold $\hat{X}$ we must consider *graded Lagrangians* $\tilde{L} = (L, \phi)$ as in Seidel [302], that is, Lagrangians L equipped with a *phase function* $\phi : L \to \mathbb{R}$ such that $\hat{\theta}|_L \equiv \mathrm{e}^{i\phi} \mathrm{vol}_L$.

  If $\tilde{L}_1, \tilde{L}_2$ are two graded Lagrangians intersecting transversely, to each $x \in L_1 \cap L_2$ we assign a *Maslov index* $I(\tilde{L}_1, \tilde{L}_2, x) \in \mathbb{Z}$, the sum of $\frac{1}{\pi}(\phi_1(x) - \phi_2(x))$ and a correction term depending on the Lagrangian subspaces $T_x L_1, T_x L_2$ in $T_x \hat{X}$. We use these to make $\mathrm{Hom}(\tilde{L}_1, \tilde{L}_2) = \mathbb{C}^{L_1 \cap L_2}$ into a *graded vector space*, where each basis vector $x \in L_1 \cap L_2$ has grading $I(\tilde{L}_1, \tilde{L}_2, x)$.

- Actually, the objects in $F(\hat{X})$ should be triples $(\tilde{L}, E, A)$, where $\tilde{L} = (L, \phi)$ is a graded Lagrangian in $\hat{X}$, E a complex line bundle over L, and A a U(1)-connection on E. In the absence of B-fields A should be a flat connection, but with B-fields we require $F_A \equiv -2\pi i \hat{B}|_L$, where F_A is the curvature of A, a closed 2-form. We set $\mathrm{Hom}\big((\tilde{L}_1, E_1, A_1), (\tilde{L}_2, E_2, A_2)\big) = \bigoplus_{x \in L_1 \cap L_2} \mathrm{Hom}(E_1|_x, E_2|_x)$ when L_1, L_2 intersect transversely. The contribution of a pseudoholomorphic disc D to m_k involves parallel translation along intervals $L_i \cap \partial D$ using A_i.

Here is an (incomplete) list of difficulties in the theory. It illustrates the fact the Fukaya categories, while a really clever idea, are also technically a complete nightmare.

- We define morphisms $\mathrm{Hom}(L_1, L_2)$ only when L_1, L_2 intersect transversely. In particular, we never define identity morphisms.

- The m_k are defined by 'counting' moduli spaces $\mathscr{M}$ of pseudoholomorphic discs D. We count only those moduli spaces of virtual dimension 0. The identities on the m_k are proved by considering moduli spaces $\mathscr{M}'$ of pseudoholomorphic discs D' with virtual dimension 1. These $\mathscr{M}'$ are basically collections of oriented intervals, whose boundaries are unions of pieces of the form $\mathscr{M}_1 \times \mathscr{M}_2$, where $\mathscr{M}_1, \mathscr{M}_2$ are virtual dimension 0 moduli spaces contributing to m_k, m_l. The identities follow from the fact that the number of boundary points of an oriented interval $[0, 1]$, counted with signs, is zero.

  However, there is a problem. If there exist holomorphic discs $\tilde{D}$ with boundary in some L_i, these can also contribute to the boundary of $\mathscr{M}'$ in an unwanted way, and this invalidates the proof of the identities on the m_k. If a Lagrangian L (or rather, a triple $(\tilde{L}, E, A)$) has such bad discs which do not cancel when counted with signs, it has *obstructed Lagrangian Floer homology*. A lot of effort in [118] goes into understanding these obstructions. Lagrangians with obstructed Floer homology must be excluded from the Fukaya category.

- The definition of m_k involves a sum over pseudoholomorphic discs D weighted by a factor $\exp\left(\int_D 2\pi i(\hat{B}+i\hat{\omega})\right)$. There are likely to be infinitely many such discs D, whose areas diverge to infinity. It is not known whether this infinite sum converges. To get round this it is usual to work not over $\mathbb{C}$ but over a ring Λ of formal power series known as the *Novikov ring*. Roughly speaking this is equivalent to working not with a single symplectic manifold $(\hat{X},\hat{\omega})$ but with the 1-parameter family of symplectic manifolds $(\hat{X},t\hat{\omega})$ for $t \gg 0$ in $\mathbb{R}$. Effectively this is taking a *large radius limit* as $t \to \infty$, as in §9.2.2.

  This is a problem, because if $D^b(F(\hat{X}))$ is defined over a Novikov ring Λ it cannot be equivalent to $D^b(\mathrm{coh}(X))$ defined over $\mathbb{C}$. To get round this, perhaps one should start with a Calabi–Yau 3-fold X defined over Λ rather than $\mathbb{C}$, so that $D^b(\mathrm{coh}(X))$ is defined over Λ. Effectively this means working with a 1-parameter family of Calabi–Yau 3-folds X_t for $t \gg 0$ in $\mathbb{R}$. Since $(\hat{X},t\hat{\omega})$ goes to the large radius limit as $t \to \infty$, the mirror X_t must go to the *large complex structure limit* as $t \to \infty$. Thus we recover aspects of the picture of §9.2.2.

- Kapustin and Orlov [210] use physics to argue that for the HMS Conjecture to be true on Calabi–Yau 3-folds with nonzero first Betti number, the Fukaya category must be enlarged to include extra objects they call *coisotropic A-branes*, which are not Lagrangian submanifolds. It is not yet clear how to form these into a category.

9.3.8 The derived Fukaya category $D^b(F(\hat{X}))$

As for abelian categories, given an A_∞-category $\mathcal{A}$ one can construct a triangulated category $D^b(\mathcal{A})$ called the *(bounded) derived category* of $\mathcal{A}$, which was introduced by Kontsevich [219, p. 133-4] and is described at greater length by Seidel [305, §2], [306, §I.3]. One first constructs a triangulated A_∞-category $\mathrm{Tw}(\mathcal{A})$ of *twisted complexes* in $\mathcal{A}$. The morphism groups $\mathrm{Hom}_{\mathrm{Tw}(\mathcal{A})}(X,Y)$ in $\mathrm{Tw}(\mathcal{A})$ have the structure of complexes of vector spaces, and we define $D^b(\mathcal{A})$ to be $H^0\big(\mathrm{Tw}(\mathcal{A})\big)$, that is, the triangulated category with the same objects as $\mathrm{Tw}(\mathcal{A})$, but with morphisms $\mathrm{Hom}_{D^b(\mathcal{A})}(X,Y) = H^0\big(\mathrm{Hom}_{\mathrm{Tw}(\mathcal{A})}(X,Y)\big)$, taking cohomology of the complex.

When this is applied to the Fukaya category $F(\hat{X})$ it gives a triangulated category $D^b(F(\hat{X}))$. For simplicity consider $F(\hat{X})$ to have objects Lagrangians L in $\hat{X}$, rather than triples $(\tilde{L},E,A)$. Then the Lagrangians L in $F(\hat{X})$ occur as objects in $D^b(F(\hat{X}))$, though $D^b(F(\hat{X}))$ also contains many more general objects which are twisted complexes rather than single Lagrangians. For Lagrangians L_1, L_2 in $D^b(F(\hat{X}))$, the morphism group $\mathrm{Hom}_{D^b(F(\hat{X}))}(L_1, L_2)$ is the *Floer homology group* $HF^0(L_1, L_2)$. Lagrangian Floer homology is studied by Fukaya et al. [118], and is the foundation of Fukaya categories. If L_1, L_2 are *Hamiltonian equivalent* Lagrangians in $\hat{X}$ then they are isomorphic objects in $D^b(F(\hat{X}))$.

9.3.9 Extensions and further developments of homological mirror symmetry

Polishchuk [280] strengthens the homological mirror symmetry proposal by giving $D^b(\mathrm{coh}(X))$ the structure of a *triangulated A_∞-category*, which is stronger than that of a triangulated category, and then conjecturing that $D^b(\mathrm{coh}(X))$ and $D^b(F(\hat{X}))$ should be equivalent as triangulated A_∞-categories, not just as triangulated categories. This implies that families of higher products on morphisms in $D^b(\mathrm{coh}(X))$ and $D^b(F(\hat{X}))$ have to agree. In a similar vein, Diaconescu [95] and Lazaroiu [234] have used String Theory to argue that topological D-branes should form an *enhanced triangulated category* in the sense of Bondal and Kapranov [36].

As in §9.1, one can study mirror symmetry between Calabi–Yau m-folds for other $m \geqslant 1$, and between Fano varieties and Landau–Ginzburg models, as well as for Calabi–Yau 3-folds. There are also HMS Conjectures in these contexts. The case of Calabi–Yau 1-folds, that is, elliptic curves T^2, was studied by Kontsevich [219], Polishchuk and Zaslow [281] and Polishchuk [280], and is well understood.

For Calabi–Yau 2-folds, that is, $K3$ surfaces, an important paper of Seidel [305] proves a form of homological mirror symmetry for quartic $K3$ surfaces in $\mathbb{CP}^3$. On the complex side, consider the 1-parameter family of quartic surfaces X_q in $\mathbb{CP}^3$ given by $z_0 z_1 z_2 z_3 + q(z_0^4 + z_1^4 + z_2^4 + z_3^4) = 0$, regarded as a single surface over a Novikov ring Λ_q of formal power series in q for small q. On the symplectic side, consider the 1-parameter family of symplectic manifolds $(\hat{X}, t\hat{\omega})$ for $t \gg 0$, regarded as a single symplectic manifold over a Novikov ring $\hat{\Lambda}_t$ of series in e^{-ct}. Then Seidel shows that for some unknown isomorphism $\psi : \Lambda_q \to \hat{\Lambda}_t$ there is an equivalence between the triangulated category $D^b(\mathrm{coh}(X_q))$ over Λ_q and the triangulated category $D^b(F(\hat{X}))$ over $\hat{\Lambda}_t$. The equivalence is found 'by hand', by comparing finite systems of generators for each side.

Homological mirror symmetry between *Fano varieties* and *Landau–Ginzburg models* is discussed in physical terms by Hori, Iqbal and Vafa [165]. On the complex side for the Fano variety X, the appropriate triangulated category is just the derived category of coherent sheaves $D^b(\mathrm{coh}(X))$. On the symplectic side for the Landau–Ginzburg model $(\hat{X}, \hat{f})$, the correct analogue of the derived Fukaya category is the *derived category of Lagrangian vanishing cycles*, which is rigorously defined by Seidel [303] when the stationary points of $\hat{f}$ are isolated and nondegenerate. Examples of homological mirror symmetry equivalences between categories of this kind of are proved by Seidel [304] when X is $\mathbb{CP}^2$, and by Auroux, Katzarkov and Orlov [17, 18] when X is a weighted projective line $\mathbb{CP}^1_{a_0,a_1}$ or plane $\mathbb{CP}^2_{a_0,a_1,a_2}$, or a Fano surface.

9.3.10   Relations between closed string and homological mirror symmetry

It is an interesting question how the closed string mirror symmetry of §9.2.2, and the homological mirror symmetry of §9.3, are related. An answer is provided by Costello [78, 79], using elegant but formidably difficult constructions in category theory and homological algebra. The starting point is that the triangulated categories $D^b(\mathrm{coh}(X))$, $D^b(F(\hat{X}))$ of the HMS Conjecture, enhanced to triangulated A_∞-categories as in Polishchuk [280], are examples of what Costello calls *Calabi–Yau A_∞-categories*.

Costello shows Calabi–Yau A_∞-categories are equivalent to what he calls *open topological conformal field theories (TCFTs)*, which are functors from a complicated category constructed using singular chains on moduli spaces of open Riemann surfaces, to a category of chain complexes. He proves that each open TCFT can be enhanced to an *open-closed TCFT*, and then restricted to a *closed TCFT*. But closed TCFTs exactly encode the structures of closed string mirror symmetry. Applied to the Fukaya category $D^b(F(\hat{X}))$, the closed TCFT encodes the Gromov–Witten invariants of $\hat{X}$, for curves of every genus $g \geqslant 0$. Thus, closed string mirror symmetry is a consequence of homological mirror symmetry, but in a rather hidden form.

9.4 The SYZ Conjecture

The most geometric formulation of mirror symmetry was proposed by Strominger, Yau and Zaslow [317] in 1996, and is known as the *SYZ Conjecture*. Strominger et al. expressed their ideas in physics terms, and did not make a mathematically precise conjecture. Sections 9.4.1 and 9.4.2 discuss ways to make it precise, and §9.4.3–§9.4.5 summarize some mathematical progress in this area.

9.4.1 The basic idea

Here is a first attempt to state the SYZ Conjecture, which we admit from the outset is inadequate. The maps $f, \hat{f}$ below are called *special Lagrangian fibrations*, and the set of singular fibres Δ is called the *discriminant*.

Conjecture 9.4.1. (SYZ Conjecture, first approximation) *Let X and $\hat{X}$ be mirror Calabi–Yau 3-folds. Then (under some additional conditions) there should exist a compact 3-manifold B and surjective, continuous maps $f : X \to B$ and $\hat{f} : \hat{X} \to B$, with fibres $T^b = f^{-1}(b)$ and $\hat{T}^b = \hat{f}^{-1}(b)$ for $b \in B$, and a closed set Δ in B with $B \setminus \Delta$ dense, such that:*

(i) *For each $b \in B \setminus \Delta$, the fibres T^b and $\hat{T}^b$ are nonsingular special Lagrangian 3-tori T^3 in X and $\hat{X}$, which are in some sense dual to one another.*

(ii) *For each $b \in \Delta$, the fibres T^b and $\hat{T}^b$ are singular special Lagrangian 3-folds in X and $\hat{X}$.*

There are two main problems with Conjecture 9.4.1. The first concerns the *singular fibres* of $f, \hat{f}$. As Strominger, Yau and Zaslow explain, some of their string theory arguments break down near singular fibres, as if $b \in B$ is close to Δ then there may exist holomorphic discs D in X with boundary ∂D in $f^{-1}(b)$ and with area(D) arbitrarily small, and these 'instantons' will correct the moduli space geometry in a poorly understood way. So the SYZ prediction becomes unreliable near the singular fibres.

As a mathematical confirmation of this, using local models for special Lagrangian fibrations the author [197] argued that the statement above cannot be true in general. The idea is that there are types of singularities expected to occur in mirror special Lagrangian fibrations $f, \hat{f}$ (if these exist at all) such that the discriminants $\Delta, \hat{\Delta}$ of $f, \hat{f}$ are not locally homeomorphic. So part (ii) above, saying that $f, \hat{f}$ should have the same discriminant in B, cannot hold. This does not mean that the SYZ Conjecture is false, only that we have not yet found the right statement.

The second problem is that we have not said what we mean by *dual tori* in part (i). On the topological level, we can define duality between two tori $T, \hat{T}$ to be a choice of isomorphism $H^1(T, \mathbb{Z}) \cong H_1(\hat{T}, \mathbb{Z})$. But Strominger et al. have in mind something stronger than this. There is a natural notion of duality between tori equipped with flat Riemannian metrics. Such a torus may be written $T = V/\Lambda$, where V is a Euclidean vector space and Λ a lattice in V. Then the dual torus $\hat{T}$ is V^*/Λ^*, where V^* is the dual Euclidean vector space and Λ^* the dual lattice. In the situation considered by Strominger et al., the metrics on nonsingular fibres $T^b, \hat{T}^b$ for b not close to Δ are expected to be approximately flat, and approximately dual in this sense.

We briefly explain the physical justification for the SYZ Conjecture, ignoring issues to do with singularities. Strominger et al. begin by assuming *quantum mirror symmetry*, that is, that the full quantum string theories of $X, \hat{X}$ are equivalent, not just their SCFT's. Now each point x in X is a 0-brane in X, so that X can be reconstructed from its quantum string theory as a moduli space $\mathcal{M}_X$ of 0-branes in X. Since the quantum string theories are equivalent, $\mathcal{M}_X$ is also a moduli space of 3-branes in $\hat{X}$. The classical limit of a 3-brane is an SL 3-fold $\hat{T}$ in $\hat{X}$ with a flat U(1) connection $\hat{A}$. To get a moduli space of real dimension 6 we need $b^1(\hat{T}) = 3$, by Theorem 8.4.5.

Thus we expect X to be diffeomorphic to a moduli space $\mathcal{M}_X$ of pairs $(\hat{T}, \hat{A})$, where $\hat{T}$ is an SL 3-fold in $\hat{X}$ with $b^1(\hat{T}) = 3$, and $\hat{A}$ a flat U(1) connection on $\hat{T}$. Each family $T \subset \mathcal{M}_X$ of pairs $(\hat{T}, \hat{A})$ with $\hat{T}$ fixed is naturally isomorphic to T^3, and corresponds to a 3-submanifold of X, which Strominger et al. argue is special Lagrangian. Let B be the set of SL 3-folds $\hat{T}$ occurring in $\mathcal{M}_X$, which has dimension 3 by Theorem 8.4.5. Then the projection $f : X \cong \mathcal{M}_X \to B$ taking $(\hat{T}, \hat{A}) \mapsto \hat{T}$ should be a *special Lagrangian fibration*, with fibres T diffeomorphic to T^3. Note that B corresponds to families of both SL 3-folds T in X, and SL 3-folds $\hat{T}$ in $\hat{X}$.

We can now exchange X and $\hat{X}$, and realize $\hat{X}$ as a moduli space $\mathcal{M}_{\hat{X}}$ of 0-branes in $\hat{X}$, and so of 3-branes in X, which in the classical limit are pairs (T, A) of an SL 3-fold T in X with $b^1(T) = 3$ and a flat U(1)-connection A on T, and the projection $\hat{f} : \hat{X} \cong \mathcal{M}_{\hat{X}} \to \hat{B}$ taking $(T, A) \mapsto T$ is a special Lagrangian fibration with fibres $\hat{T}$ diffeomorphic to T^3. Strominger et al. claim that B and $\hat{B}$ should be identified, since both correspond to families of both SL 3-folds $\hat{T}$ in $\hat{X}$, and T in X. Hence the SL 3-folds $\hat{T}, T$ with $b^1(\hat{T}) = b^1(T) = 3$ above must be 3-tori. This yields the SYZ Conjecture.

9.4.2   Rewriting the SYZ Conjecture as a limiting statement

Many mathematical claims made by string theorists implicitly involve taking various limits, and to make mathematically plausible statements it is often necessary to put these limits in explicitly. To improve Conjecture 9.4.1, we must pay attention to the notion of 'classical limit' of 3-branes in §9.4.1. The claim that the classical limit of a 3-brane in $\hat{X}$ is an SL 3-fold $\hat{T}$ in $\hat{X}$ with a flat U(1) connection $\hat{A}$ really means that $\hat{X}$ should approach the *large radius limit*, as in §9.2.2, and so its mirror X should approach the *large complex structure limit*. The corresponding classical limit in X requires X to approach the large radius limit, and $\hat{X}$ the large complex structure limit.

Thus, we should rewrite the SYZ Conjecture in terms of 1-parameter families of mirror Calabi–Yau 3-folds $X_t, \hat{X}_t$ for $t \gg 0$, which both approach both large radius and large complex structure limits as $t \to \infty$. Here is an attempt to do this.

Conjecture 9.4.2. (SYZ Conjecture, second approximation, optimistic) *Let $R > 0$ and $X_t, \hat{X}_t$ for $t \in (R, \infty)$ be smooth 1-parameter families of mirror Calabi–Yau 3-folds, which both approach both large radius and large complex structure limits as $t \to \infty$. Then (under some additional conditions) there should exist $S \geqslant R$, a compact 3-manifold B and surjective, continuous maps $f_t : X_t \to B$ and $\hat{f}_t : \hat{X}_t \to B$ for $t \in (S, \infty)$, with fibres $T_t^b = f_t^{-1}(b)$ and $\hat{T}_t^b = \hat{f}_t^{-1}(b)$ for $b \in B$, such that:*

(a) For $t \in (S, \infty)$ there are closed sets $\Delta_t, \hat{\Delta}_t$ in B with $B \setminus \Delta_t, B \setminus \hat{\Delta}_t$ dense, such that T_t^b is a nonsingular SL 3-torus in X_t if $b \in B \setminus \Delta_t$ and a singular SL 3-fold in X_t if $b \in \Delta_t$, and $\hat{T}_t^b$ is a nonsingular SL 3-torus in $\hat{X}_t$ if $b \in B \setminus \hat{\Delta}_t$ and a singular SL 3-fold in $\hat{X}_t$ if $b \in \hat{\Delta}_t$.

(b) There is a closed set Δ_∞ in B with $B \setminus \Delta_\infty$ dense, such that Δ_t and $\hat{\Delta}_t$ converge to Δ_∞ as $t \to \infty$, in an appropriate sense.

(c) If $b \in B \setminus \Delta_\infty$ then $T_t^b, \hat{T}_t^b$ are nonsingular SL 3-tori in $X_t, \hat{X}_t$ for sufficiently large t. There exist flat Riemannian 3-tori $T_\infty^b, \hat{T}_\infty^b$ which are dual in the sense above, such that $T_t^b \to T_\infty^b$ and $\hat{T}_t^b \to \hat{T}_\infty^b$ as $t \to \infty$, as Riemannian 3-manifolds.

Conjecture 9.4.2 is *optimistic* as it makes strong claims on the existence of SL fibrations of Calabi–Yau 3-folds $X_t, \hat{X}_t$ sufficiently close to the large radius and complex structure limits. These claims would only really be plausible if the singularities of SL 3-folds are very well behaved. There is some evidence for this from special Lagrangian geometry, and the author expects something similar to Conjecture 9.4.2 to be true.

In Conjecture 9.4.2 we have retained the idea from Conjecture 9.4.1 that actual Calabi–Yau 3-folds $X_t, \hat{X}_t$ should have special Lagrangian fibrations, without taking any limit. But the mirror relationship between $X_t, \hat{X}_t$ is described in (b),(c) only in the limit as $t \to \infty$. This may be too strong, and it is conceivable that Calabi–Yau 3-folds with holonomy $\mathrm{SU}(3)$ never admit special Lagrangian fibrations at all. So we give an alternative, weaker statement:

Conjecture 9.4.3. (SYZ Conjecture, second approximation, pessimistic) *Let $R > 0$ and $X_t, \hat{X}_t$ for $t \in (R, \infty)$ be smooth 1-parameter families of mirror Calabi–Yau 3-folds, which both approach both large radius and large complex structure limits as $t \to \infty$. Then (under some additional conditions) there should exist $S \geqslant R$, a compact 3-manifold B and surjective, continuous maps $f_t : X_t \to B$ and $\hat{f}_t : \hat{X}_t \to B$ for $t \in (S, \infty)$, with fibres $T_t^b = f_t^{-1}(b)$ and $\hat{T}_t^b = \hat{f}_t^{-1}(b)$ for $b \in B$, such that:*

(a) *For $t \in (S, \infty)$ there are closed sets $\Delta_t, \hat{\Delta}_t$ in B with $B \setminus \Delta_t, B \setminus \hat{\Delta}_t$ dense, such that T_t^b is a nonsingular 3-torus in X_t if $b \in B \setminus \Delta_t$ and a singular 3-fold in X_t if $b \in \Delta_t$, and $\hat{T}_t^b$ is a nonsingular 3-torus in $\hat{X}_t$ if $b \in B \setminus \hat{\Delta}_t$ and a singular 3-fold in $\hat{X}_t$ if $b \in \hat{\Delta}_t$. Furthermore, the fibres $T_t^b, \hat{T}_t^b$ are approximately special Lagrangian in $X_t, \hat{X}_t$, and converge to exactly special Lagrangian as $t \to \infty$.*

(b) *There is a closed set Δ_∞ in B with $B \setminus \Delta_\infty$ dense, such that Δ_t and $\hat{\Delta}_t$ converge to Δ_∞ as $t \to \infty$, in an appropriate sense.*

(c) *If $b \in B \setminus \Delta_\infty$ then $T_t^b, \hat{T}_t^b$ are nonsingular 3-tori in $X_t, \hat{X}_t$ for sufficiently large t. There exist flat Riemannian 3-tori $T_\infty^b, \hat{T}_\infty^b$ which are dual in the sense above, such that $T_t^b \to T_\infty^b$ and $\hat{T}_t^b \to \hat{T}_\infty^b$ as $t \to \infty$, as Riemannian 3-manifolds.*

Here we have not said what we mean by $T_t^b, \hat{T}_t^b$ being 'approximately special Lagrangian' in (a), or how they converge to exactly special Lagrangian. Since T_t^b is SL in X_t if $\omega_t|_{T_t^b} \equiv \mathrm{Im}\,\theta_t|_{T_t^b} \equiv 0$, one way to interpret this might be to require Banach norms of $\omega_t|_{T_t^b}$ and $\mathrm{Im}\,\theta_t|_{T_t^b}$ to converge to zero as $t \to \infty$, and similarly for $\hat{T}_t^b$.

Conjecture 9.4.3 is only a limiting statement: it says essentially nothing about the special Lagrangian geometry of any Calabi–Yau 3-fold, it just predicts how families of Calabi–Yau 3-folds can collapse to a simpler structure in a singular limit. As such, it is rather disappointing from the point of view of the special Lagrangian geometer. One could also formulate statements intermediate between Conjectures 9.4.2 and 9.4.3 which do make nontrivial claims on the special Lagrangian geometry of X_t, $\hat{X}_t$, without asserting the existence of special Lagrangian fibrations.

For example, an SL fibration f of a Calabi–Yau m-fold X is basically a moduli space of *disjoint* SL m-folds, with one passing through each point $x \in X$. Now two SL 2-folds in homology classes with zero intersection number must be disjoint, as from §8.1.1 SL 2-folds are holomorphic curves in disguise, and always intersect with positive intersection number. But for SL m-folds with $m \geqslant 3$ this does not apply, so there is no particular reason for a pair of SL m-folds to be disjoint. Thus, it might be a good idea to replace the idea of SL fibration of X by a moduli space $\mathcal{M}$ of SL m-folds in X, such that $\mathcal{M}$ is a compact manifold without boundary, and for generic $x \in X$ there is exactly one SL m-fold in $\mathcal{M}$ passing through x *when counted with signs*.

See §9.4.4 for a discussion of related SYZ-type conjectures on the limit $t \to \infty$.

9.4.3 The symplectic topological approach to SYZ of Gross and Ruan

Early mathematical work on the SYZ Conjecture generally took an approach we describe as *symplectic topological*. In this approach, we mostly forget about complex structures, and treat $X, \hat{X}$ just as *symplectic manifolds*. We mostly forget about the 'special' condition, and treat $f, \hat{f}$ just as *Lagrangian fibrations*. We also impose the condition that B is a *smooth* 3-manifold and $f : X \to B$ and $\hat{f} : \hat{X} \to B$ are *smooth maps*. (It is not clear that $f, \hat{f}$ can in fact be smooth at every point, though).

Under these simplifying assumptions, Gross [133–136], Ruan [290–293], and others have built up a beautiful, detailed picture of how dual SYZ fibrations work at the global topological level, in particular for examples such as the quintic and its mirror, and for Calabi–Yau 3-folds constructed as hypersurfaces in toric 4-folds, using combinatorial data. A good introduction to this can be found in Gross [138, §19].

Here are some of the basic ideas. Let $f : X \to B$ be a Lagrangian fibration of a Calabi–Yau 3-fold X, with discriminant Δ. Define $B_0 = B \setminus \Delta$ and $X_0 = f^{-1}(B_0)$. Regard X as a compact symplectic manifold (X, ω), and its dense open set X_0 as a noncompact symplectic manifold (X_0, ω). Then $f : X_0 \to B_0$ is a smooth fibration of (X_0, ω), all of whose fibres are Lagrangian T^3's. The aim is:

(a) to describe (X_0, ω) in terms of essentially combinatorial data on B_0, and

(b) to reconstruct the compactification (X, ω) from (X_0, ω) in some natural way.

For (a), the most basic topological invariant of the fibration $f : X_0 \to B_0$ is its *monodromy* μ. The fibres T^b of f are all diffeomorphic to T^3, so $H_1(T^b, \mathbb{Z}) \cong \mathbb{Z}^3$. Thus $b \mapsto H_1(T^b, \mathbb{Z})$ is a local system with fibre $\mathbb{Z}^3$. Fix some base-point $b_0 \in B_0$ and an identification $H_1(T^{b_0}, \mathbb{Z}) \cong \mathbb{Z}^3$. Then parallel transport around loops γ based at b_0 induces a group homomorphism $\mu : \pi_1(B_0, b_0) \to \mathrm{Aut}(H_1(T^{b_0}, \mathbb{Z})) = \mathrm{GL}(3, \mathbb{Z})$ called the monodromy. If f is an SL fibration its fibres are oriented, which implies that μ actually maps $\pi_1(B_0, b_0) \to \mathrm{SL}(3, \mathbb{Z})$. If the fibration $f : X_0 \to B_0$ admits a section $s : B_0 \to X_0$ then B_0, b_0 and μ determine $f : X_0 \to B_0$ up to isomorphism as a smooth T^3-bundle, though they do not determine (X_0, ω) as a symplectic manifold.

To reconstruct (X_0, ω) we need some extra data on B_0, which comes from the material on *affine structures* on moduli spaces of SL m-folds in §8.4.3. Given a base point $b_0 \in B_0$ and an identification $H_1(T^{b_0}, \mathbb{Z}) \cong \mathbb{Z}^3$ as above, if U is small open ball about b_0 in B_0 then as in §8.4.3 we can define local coordinates $\Phi : U \to H^1(T^{b_0}, \mathbb{Z}) = \mathbb{R}^3$ by integrating the symplectic form ω on X_0 along paths in U. The transition maps between two such Φ are locally affine transformations $\psi : \mathbb{R}^3 \to \mathbb{R}^3$ whose linear parts lie in $\mathrm{GL}(3, \mathbb{Z})$. Thus these Φ give B_0 the structure of an *integral affine 3-manifold*. We write this structure as $\mathcal{I}$, and an integral affine manifold as $(B_0, \mathcal{I})$.

This can also be expressed in the language of Chapter 2. An *affine n-manifold* is equivalent to an n-manifold M with a torsion-free flat connection ∇, and it is *integral* if the holonomy $\mathrm{Hol}(\nabla)$ lies in $\mathrm{GL}(n, \mathbb{Z}) \subset \mathrm{GL}(n, \mathbb{R})$. Alternatively, an integral affine n-manifold $(B_0, \mathcal{I})$ is an n-manifold with a torsion-free $\mathrm{GL}(n, \mathbb{Z})$-structure. The monodromy $\mu : \pi_1(B_0, b_0) \to \mathrm{GL}(3, \mathbb{Z})$ above comes from the holonomy of ∇. For Proposition 2.2.6 gives a surjective homomorphism $\phi : \pi_1(B_0, b_0) \to \mathrm{Hol}_{b_0}(\nabla)/\mathrm{Hol}^0_{b_0}(\nabla)$. But $\mathrm{Hol}^0_{b_0}(\nabla) = \{1\}$ as ∇ is flat, and $\mathrm{Hol}_{b_0}(\nabla) \subseteq \mathrm{GL}(3, \mathbb{Z})$, so ϕ induces a homomorphism $\pi_1(B_0, b_0) \to \mathrm{GL}(3, \mathbb{Z})$ which is just μ. In fact, much of the structure $\mathcal{I}$ on B_0 is contained in a lifting of μ to a homomorphism $\pi_1(B_0, b_0) \to \mathrm{GL}(3, \mathbb{Z}) \ltimes \mathbb{R}^3$.

Thus, the integral affine structure $\mathcal{I}$ on B_0 includes the monodromy μ, so if f admits a section $s : B_0 \to X_0$ then we can reconstruct $f : X_0 \to B_0$ up to isomorphism as a smooth T^3-bundle from $(B_0, \mathcal{I})$. If s can also be chosen so that its graph in X_0 is Lagrangian, then the symplectic structure ω on X_0 can also be reconstructed from $(B_0, \mathcal{I})$, up to diffeomorphisms of X_0 commuting with f. Therefore, under mild assumptions, (X_0, ω) and $f : X_0 \to B_0$ depend only on $(B_0, \mathcal{I})$. This is our answer to (a) above.

To make progress on (b) we need to impose strong assumptions on the discriminant Δ, and the nature of the singular fibres. At the level of cohomology, Gross [133] deals with this by defining *simple* fibrations $f : X \to B$. For a simple fibration the cohomology $H^*(T^b, \mathbb{R})$ of the singular fibres, and their contribution to $H^*(X, \mathbb{R})$, are determined by the monodromy μ, and so $H^*(X, \mathbb{R})$ can be reconstructed from B, B_0 and μ. If $f : X \to B$ and $\hat{f} : \hat{X} \to B$ are simple fibrations with the same discriminant Δ and topologically dual T^3-fibrations over B_0, Gross [133, §2], [138, Th. 19.16] proves Mirror Symmetry of Hodge numbers for $X, \hat{X}$, as in §9.2.1.

To try to reconstruct X and ω, Gross [134] assumes that B is a smooth 3-manifold, and the Lagrangian fibration $f : X \to B$ is *smooth* everywhere, including at singular points of singular fibres. This is a strong assumption, with powerful consequences for Δ and the nature of the singular fibres of f; in particular, it implies the discriminant Δ is 1-dimensional. Motivated by this and by examples, Gross [135, §1] defined *well-behaved* fibrations of Calabi–Yau 3-folds, encapsulating the topological features that were expected of special Lagrangian fibrations at the time. The discriminant Δ of a well-behaved fibration is a *graph* in B with finitely many vertices.

Gross [135] constructs explicit, non-Lagrangian, well-behaved mirror fibrations $f : X \to B$, $\hat{f} : \hat{X} \to B$, and uses the classification of 6-manifolds to show that $X, \hat{X}$ are diffeomorphic to the quintic in $\mathbb{CP}^4$ and its mirror. Pursuing a different course, Ruan [290–292] studies Lagrangian fibrations of the quintic in $\mathbb{CP}^4$, and [293] generalizes this to Calabi–Yau 3-hypersurfaces in toric 4-folds. Ruan's method is to construct an initial piecewise-smooth Lagrangian fibration $f' : X \to B$ via gradient flow, whose discriminant Δ' has codimension 1, and whose singular fibres have singularities incompatible with being special Lagrangian. Then he tries to modify f to a smooth, well-behaved Lagrangian fibration $f : X \to B$ whose discriminant Δ has codimension 2.

When X is a quintic in $\mathbb{CP}^4$, or more generally a Calabi–Yau hypersurface in a toric 4-fold P, Gross and Ruan's work gives a clear picture of the topological structure expected for the SYZ fibration $f : X \to B$, and to a large extent its symplectic structure as well. In their description, B is $\mathcal{S}^3$, and Δ is a *trivalent graph* in $\mathcal{S}^3$ derived from the combinatorial data (fan) defining the toric 4-fold P; the construction of B, Δ and $(B_0, \mathcal{I})$ from the fan of P is described in an explicit, succinct way by Haase and Zharkov [146]. The monodromy μ of $f : X_0 \to B_0$ is understood, and there are reasonable guesses for the topology of all the singular fibres.

In the opinion of the author, the discriminants Δ in $B = \mathcal{S}^3$ constructed by Gross and Ruan, which are 1-dimensional trivalent graphs, are the right answer for the limiting discriminant Δ_∞ in Conjectures 9.4.2 and 9.4.3. We discuss this further in §9.4.5.

Both the closed string mirror symmetry of §9.2.2, and the homological mirror symmetry of §9.3, involve only half of the geometric structures on each side of the mirror: the complex structure on X, and the symplectic structure on $\hat{X}$. The SYZ Conjecture involves all the structure on both sides, since SL 3-folds make no sense without both complex and symplectic structures. But the Gross–Ruan strategy of studying Lagrangian fibrations $f : X \to B$ uses only the symplectic structure on X.

In a similar way, one can weaken the idea of special Lagrangian fibrations by working with only the complex structure. Let (X, J) be a complex 3-fold, and θ a holomorphic volume form on X. Define a *special fibration* $f : X \to B$ to be a fibration whose nonsingular fibres T^b for $b \in B_0 = B \setminus \Delta$ are totally real 3-tori with $\mathrm{Im}\, \theta|_{T^b} \equiv 0$.

Much of the story above for Lagrangian fibrations has an analogue for special fibrations. Using $\mathrm{Im}\, \theta$ and $H_2(T^b, \mathbb{Z})$ instead of ω and $H_1(T^b, \mathbb{Z})$, we can define an integral affine structure $\mathcal{I}'$ on B_0. From $(B_0, \mathcal{I}')$, under mild conditions, we can reconstruct the T^3-bundle $f : X_0 \to B_0$ up to isomorphism, and we can define a complex structure J_0 and holomorphic 3-form θ_0 on X_0.

However, there is an important difference with the Lagrangian case: J_0, θ_0 are *not* expected to be isomorphic to J, θ on X_0, but are only first approximations to them. Conjectures by Fukaya [117] indicate how to construct J, θ near the large complex structure limit by making a series of corrections to J_0, θ_0, which depend on counts of holomorphic discs in the mirror $\hat{X}$. Thus, recovering X, J, θ from fibration data is much harder than recovering X, ω, and has not been pursued as far.

9.4.4 More sophisticated limiting conjectures related to SYZ

We now review conjectures by a number of authors which elaborate on the SYZ Conjecture, and are the fruit of several years of work and reflection upon it. They all concern the limiting behaviour as $t \to \infty$ of the geometric structures of mirror Calabi–Yau 3-folds $X_t, \hat{X}_t$ approaching large radius and complex structure limits, as in Conjectures 9.4.2 and 9.4.3. They apparently have little to do with special Lagrangian fibrations.

Some similar conjectures are expressed by Gross and Wilson [141, Conj. 6.2] and Gross [136, §5], and independently by Kontsevich and Soibelman [220, §3]. They are proved for $K3$ surfaces by Gross and Wilson [141]. Think of the Riemannian manifold (X_t, g_t) as a metric space with fibration $f_t : X_t \to B$. For large t, the diameters of the fibres (T_t^b, g_t) are expected to be small compared to the diameter of the base space B. Let us choose the parameter t and the scaling of the g_t up to homothety so that the fibres (T_t^b, g_t) have diameter $O(1)$, and the base space B has diameter $O(t)$. We can also consider the rescaled metric $(X_t, t^{-2}g_t)$; in this the fibres $(T_t^b, t^{-2}g_t)$ have diameter $O(t^{-1})$, and the base space B has diameter $O(1)$.

There is a notion of limits of compact metric spaces called *Gromov–Hausdorff convergence*. Using this, Gross–Wilson and Kontsevich–Soibelman suggest that for $b \in B \setminus \Delta$, the Gromov–Hausdorff limit of (T_t^b, g_t) as $t \to \infty$ should be a flat Riemannian 3-torus (T_∞^b, g_∞), as in Conjectures 9.4.2(c) and 9.4.3(c). Also, the Gromov–Hausdorff limit of $(X_t, t^{-2}g_t)$ as $t \to \infty$ should be a metric space (B, d). On $B_0 = B \setminus \Delta$ this d comes from a nonsingular Riemannian metric h, which is singular along Δ. That is, a family of Riemannian 6-manifolds converge to a (singular) Riemannian 3-manifold; this kind of change of dimension is allowed in Gromov–Hausdorff limits.

Furthermore, B_0 carries two integral affine structures $\mathcal{I}, \mathcal{I}'$, coming from ω_t and $\operatorname{Im} \theta_t$ as $t \to \infty$. These structures satisfy a compatibility called the *real Monge–Ampère equation*, and h is derived from $\mathcal{I}, \mathcal{I}'$. In this context, mirror symmetry says that the Gromov–Hausdorff limits $(T_\infty^b, g_\infty), (\hat{T}_\infty^b, \hat{g}_\infty)$ of the fibres of $f, \hat{f}$ for $b \in B_0$ should be dual flat Riemannian 3-tori, and the Gromov–Hausdorff limits of $(X_t, t^{-2}g_t)$ and $(\hat{X}_t, t^{-2}\hat{g}_t)$ should be the same metric space (B, d), with the same metric h on B_0, but the integral affine structures $\hat{\mathcal{I}}, \hat{\mathcal{I}}'$ coming from $\hat{X}_t$ satisfy $\hat{\mathcal{I}} = \mathcal{I}'$ and $\hat{\mathcal{I}}' = \mathcal{I}$.

Fukaya [117] takes the ideas of Gross–Wilson and Kontsevich–Soibelman further. As in §9.4.3, the symplectic structure ω_t on $f_t^{-1}(B_0) \subset X_t$ is determined exactly by an integral affine structure on B_0. Fukaya explains how to write down a conjectural asymptotic expansion for g_t, J_t on $f_t^{-1}(B_0)$ as $t \to \infty$, compatible with ω_t, where the leading term is the Gross–Wilson–Kontsevich–Soibelman picture. This specifies the Fourier modes of g_t, J_t on the T^3 fibres of f_t asymptotically as $t \to \infty$, in terms of counts of trees in B whose edges are gradient flow lines of certain functions, or equivalently, in terms of counts of holomorphic discs in $\hat{X}_t$.

Another paper of Kontsevich and Soibelman [221] sets out a programme similar to Fukaya's, but works with analytic spaces over the non-Archimedean field $\mathbb{C}((t))$ instead of asymptotic expansions. An analytic space over $\mathbb{C}((t))$ is roughly speaking a family of complex analytic varieties X_t for $t \in \mathbb{C}$ with $0 < |t| \ll 1$, which the authors want to approach a large complex structure limit as $t \to 0$. Let $B, \Delta, B_0, \mathcal{I}$ be as in §9.4.3. Then from the integral affine manifold $(B_0, \mathcal{I})$, Kontsevich and Soibelman construct a noncompact analytic space over $\mathbb{C}((t))$. The problem is to compactify this over Δ. To make this possible, the analytic space over B_0 must first be modified by terms analogous to Fukaya's corrections. The authors carry out the compactification for $K3$ surfaces.

Gross and Siebert [137, 139, 140] have begun an exciting programme which transforms the SYZ Conjecture into algebraic geometry. Let X_t for $t \in \mathbb{C}$ with $|t| > 1$ be a family of complex 3-folds with trivial canonical bundle. There is a well-defined algebro-geometric notion of when the family X_t approaches a *large complex structure limit* as $t \to \infty$, which is crucial in mirror symmetry. Gross and Siebert define when such a family X_t is a *toric degeneration*, which implies that it is a large complex structure limit. This includes many known examples of large complex structure limits, such as the Batyrev–Borisov construction.

Given a toric degeneration $(X_t)_{|t|>1}$, Gross and Siebert construct a limit X_∞ as $t \to \infty$, a singular scheme built out of toric varieties. They would like to recover the family $(X_t)_{|t|>1}$ from X_∞, uniquely up to deformation, by smoothing X_∞. However, there may be many different ways to smooth X_∞, and to select one they must put a *logarithmic structure* on X_∞, making it into a *log scheme* $X_\infty^\dagger$. Log geometry is difficult to explain, and we shall not attempt it.

Now mirror symmetry exchanges complex and symplectic geometry, so to formulate a wholly algebro-geometric version of mirror symmetry, we must somehow include symplectic structures in the algebraic geometry. Gross and Siebert do this by including a family of *ample line bundles*, or *polarizations*, $\mathcal{L}_t$ on X_t. The curvature of an appropriate connection on $\mathcal{L}_t$ is then the symplectic form ω_t on X_t. Taking the limit $t \to \infty$ gives a polarization $\mathcal{L}_\infty$ on $X_\infty^\dagger$.

Gross and Siebert then express mirror symmetry between families $(X_t)_{|t|>1}$ and $(\hat{X}_t)_{|t|>1}$ undergoing toric degenerations as a correspondence between the limiting polarized log schemes $(X_\infty^\dagger, \mathcal{L}_\infty)$ and $(\hat{X}_\infty^\dagger, \hat{\mathcal{L}}_\infty)$. This correspondence goes via singular affine 3-manifolds, in a similar way to §9.4.3. From the log scheme $X_\infty^\dagger$ one constructs a 3-manifold B, a singular set $\Delta \subset B$, an integral affine structure $\mathcal{I}$ on $B_0 = B \setminus \Delta$, and a polyhedral decomposition $\mathcal{P}$ of B. The polarization $\mathcal{L}_\infty$ gives also a multivalued piecewise-linear function φ on B. Then $(X_\infty^\dagger, \mathcal{L}_\infty)$ is equivalent to $(B, \Delta, \mathcal{I}, \mathcal{P}, \varphi)$, and the correspondence between $(X_\infty^\dagger, \mathcal{L}_\infty)$ and $(\hat{X}_\infty^\dagger, \hat{\mathcal{L}}_\infty)$ works via a *discrete Legendre transform* explicitly relating the quintuples $(B, \Delta, \mathcal{I}, \mathcal{P}, \varphi)$ and $(\hat{B}, \hat{\Delta}, \hat{\mathcal{I}}, \hat{\mathcal{P}}, \hat{\varphi})$.

9.4.5   U(1)-**invariant SL fibrations in** $\mathbb{C}^3$ **and the SYZ Conjecture**

In §8.2.6 we described the work of the author in [198, 205–207] on SL 3-folds N in $\mathbb{C}^3$ invariant under the U(1)-action

$$\mathrm{e}^{i\psi} : (z_1, z_2, z_3) \mapsto (\mathrm{e}^{i\psi} z_1, \mathrm{e}^{-i\psi} z_2, z_3) \quad \text{for } \mathrm{e}^{i\psi} \in \mathrm{U}(1). \tag{9.2}$$

We now explain how this is applied in [197, 198] and [207, §8] to construct U(1)-invariant SL fibrations of subsets of $\mathbb{C}^3$, and draw conclusions about SL fibrations of general Calabi–Yau 3-folds and the SYZ Conjecture. Following [207, Def. 8.1], define:

Definition 9.4.4   Let S be a strictly convex domain in $\mathbb{R}^2$ invariant under $(x, y) \mapsto (x, -y)$, let U be an open set in $\mathbb{R}^3$, and $\alpha \in (0, 1)$. Suppose $\Phi : U \to C^{3,\alpha}(\partial S)$ is a continuous map such that if $(a, b, c) \neq (a, b', c')$ in U then $\Phi(a, b, c) - \Phi(a, b', c')$ has exactly one local maximum and one local minimum in ∂S.

For $\boldsymbol{\alpha} = (a, b, c) \in U$, let $f_{\boldsymbol{\alpha}} \in C^{3,\alpha}(S)$ or $C^1(S)$ be the unique (weak) solution of (8.8) with $f_{\boldsymbol{\alpha}}|_{\partial S} = \Phi(\boldsymbol{\alpha})$, which exists by Theorem 8.2.3. Define $u_{\boldsymbol{\alpha}} = \frac{\partial f_{\boldsymbol{\alpha}}}{\partial y}$ and $v_{\boldsymbol{\alpha}} = \frac{\partial f_{\boldsymbol{\alpha}}}{\partial x}$. Then $(u_{\boldsymbol{\alpha}}, v_{\boldsymbol{\alpha}})$ is a solution of (8.7) in $C^{2,\alpha}(S)$ if $a \neq 0$, and a weak solution of (8.6) in $C^0(S)$ if $a = 0$. Also $u_{\boldsymbol{\alpha}}, v_{\boldsymbol{\alpha}}$ depend continuously on $\boldsymbol{\alpha} \in U$ in $C^0(S)$, by Theorem 8.2.3. For each $\boldsymbol{\alpha} = (a, b, c)$ in U, define $N_{\boldsymbol{\alpha}}$ in $\mathbb{C}^3$ by

$$N_{\boldsymbol{\alpha}} = \big\{ (z_1, z_2, z_3) \in \mathbb{C}^3 : z_1 z_2 = v_{\boldsymbol{\alpha}}(x, y) + iy, \quad z_3 = x + i u_{\boldsymbol{\alpha}}(x, y),$$
$$|z_1|^2 - |z_2|^2 = 2a, \quad (x, y) \in S^\circ \big\}. \tag{9.3}$$

Then N_α is a noncompact SL 3-fold without boundary in $\mathbb{C}^3$, invariant under (9.2), which is nonsingular if $a \neq 0$, by Proposition 8.2.1.

In [207, Th. 8.2] we show that the N_α are the fibres of an *SL fibration*.

Theorem 9.4.5 *In the situation of Definition 9.4.4, if $\alpha \neq \alpha'$ in U then $N_\alpha \cap N_{\alpha'} = \emptyset$. There exists an open set $V \subset \mathbb{C}^3$ and a continuous, surjective map $F : V \to U$ such that $F^{-1}(\alpha) = N_\alpha$ for all $\alpha \in U$. Thus, F is a special Lagrangian fibration of $V \subset \mathbb{C}^3$, which may include singular fibres.*

The key point here is that the SL 3-folds N_α for α in U are *disjoint*. It is fairly obvious that a 3-dimensional family of disjoint SL 3-folds in $\mathbb{C}^3$ will locally form an SL fibration. Here is how we prove the N_α are disjoint. Roughly speaking, U(1)-invariant SL 3-folds in $\mathbb{C}^3$ are equivalent to SL 2-folds in a *Kähler quotient* $\mathbb{C}^3 /\!/ \mathrm{U}(1)$ of $\mathbb{C}^3$ by U(1). But as in §8.1.1, SL 2-folds are holomorphic curves in disguise, and so they always intersect with *positive intersection number*. Thus, two SL 2-folds which intersect trivially in homology must be disjoint. Using this, the boundary conditions in Definition 9.4.4 ensure that $N_\alpha / \mathrm{U}(1)$ and $N_{\alpha'} / \mathrm{U}(1)$ are disjoint for all $\alpha \neq \alpha'$ in U.

It is easy to find families Φ satisfying Definition 9.4.4. For example [207, Ex. 8.3], given any $\phi \in C^{3,\alpha}(\partial S)$ we may define $U = \mathbb{R}^3$ and $\Phi : \mathbb{R}^3 \to C^{3,\alpha}(\partial S)$ by $\Phi(a, b, c) = \phi + bx + cy$. So this construction produces very large families of U(1)-invariant SL fibrations, including singular fibres, which can have any *multiplicity* and *type* in the sense of §8.2.6.

Here is a simple, explicit example. Define $F : \mathbb{C}^3 \to \mathbb{R} \times \mathbb{C}$ by

$$F(z_1, z_2, z_3) = (a, b), \quad \text{where} \quad 2a = |z_1|^2 - |z_2|^2$$

$$\text{and} \quad b = \begin{cases} z_3, & a = z_1 = z_2 = 0, \\ z_3 + \bar{z}_1 \bar{z}_2 / |z_1|, & a \geqslant 0, \ z_1 \neq 0, \\ z_3 + \bar{z}_1 \bar{z}_2 / |z_2|, & a < 0. \end{cases} \tag{9.4}$$

This is a piecewise-smooth SL fibration of $\mathbb{C}^3$. It is not smooth on $|z_1| = |z_2|$.

The fibres $F^{-1}(a, b)$ are T^2-cones singular at $(0, 0, b)$ when $a = 0$, and nonsingular $\mathcal{S}^1 \times \mathbb{R}^2$ when $a \neq 0$. They are isomorphic to the SL 3-folds of Example 8.2.4 under transformations of $\mathbb{C}^3$, but they are assembled to make a fibration in a novel way.

As a goes from positive to negative the fibres undergo a surgery, a Dehn twist on $\mathcal{S}^1$. The reason why the fibration is only piecewise-smooth, rather than smooth, is really this topological transition, rather than the singularities themselves. The fibration is not differentiable at every point of a singular fibre, rather than just at singular points, and this is because we are jumping from one moduli space of SL 3-folds to another at the singular fibres.

The author conjectures that (9.4) is the local model for codimension one singularities of SL fibrations of *generic* almost Calabi–Yau 3-folds, if such fibrations exist. The justification for this is that the T^2-cone singularities have 'index one' in the sense of §8.3, and so should occur in codimension one in families of SL 3-folds in generic almost Calabi–Yau 3-folds. Since they occur in codimension one in this family, the singular behaviour should be stable under small perturbations of the almost Calabi–Yau structure.

In [197, §7] we also describe a $U(1)$-invariant local model for codimension two singularities of SL fibrations, in which two of the codimension one T^2-cones come together to give a multiplicity two singularity in the sense of §8.2.6. The author expects this to be a typical codimension two singular behaviour in SL fibrations of generic almost Calabi–Yau 3-folds. However, the author expects that SL fibrations of Calabi–Yau 3-folds must include singular behaviour in codimension three that cannot be locally modelled on $U(1)$-invariant SL fibrations, so this approach does not provide all the ingredients necessary to understand SL fibrations of Calabi–Yau 3-folds.

Suppose $f : X \to B$ is an SL fibration of a generic (almost) Calabi–Yau 3-fold, with discriminant Δ. Based on the ideas above, we predict:

- f is smooth on $f^{-1}(B \setminus \Delta)$, but is only continuous and not smooth on $f^{-1}(\Delta)$.
- The discriminant Δ is of dimension 2, and is typically composed of intersecting 'ribbons', that is, closed 2-submanifolds with boundary in B.
- Every singular fibre $T^b = f^{-1}(b)$ for $b \in \Delta$ has only finitely many singular points.
- In Conjecture 9.4.2, if this holds, the discriminants $\Delta_t, \hat{\Delta}_t$ should be composed of 2-dimensional ribbons as above, but the limiting discriminant Δ_∞ should be a 1-dimensional graph in B, as in the Gross–Ruan picture. The convergence $\Delta_t, \hat{\Delta}_t \to \Delta_\infty$ as $t \to \infty$ happens by the ribbons contracting to zero width, via a local model described in [197, §7].

In [197, §8] the author used this to argue that the naïve form Conjecture 9.4.1 of the SYZ Conjecture does not hold, since if $f : X \to B$ and $\hat{f} : \hat{X} \to B$ are SL fibrations of general mirror Calabi–Yau 3-folds, then $f, \hat{f}$ should have *different* discriminants $\Delta, \hat{\Delta}$ in B. In Conjecture 9.4.2, if Δ_∞ is a Gross–Ruan graph and $|t| \gg 0$ so that $\Delta_t, \hat{\Delta}_t$ are composed of ribbons thickening Δ_∞, then near a trivalent vertex v of Δ_∞ in B we expect Δ_t to look like a Y-shaped ribbon lying in a single plane in B, but $\hat{\Delta}_t$ to look like three ribbons with ends lying in three different planes in B and intersecting in an interval. So $\Delta_t, \hat{\Delta}_t$ are locally nonhomeomorphic, and cannot be identified.

The ideas of §8.5.5 on the *index* of special Lagrangian singularities may be helpful in studying SL fibrations $f : X \to B$ of generic almost Calabi–Yau 3-folds X. Such fibrations come from compactifications $\overline{\mathcal{M}}_{T^3}$ of 3-dimensional moduli spaces $\mathcal{M}_{T^3}$ of SL 3-tori in X, so the SL 3-folds in $\partial \mathcal{M}_{T^3} = \overline{\mathcal{M}}_{T^3} \setminus \mathcal{M}_{T^3}$ must have singularities of index 1, 2 or 3. If we could classify SL singularities with index 1, 2 and 3, as in Problem 8.5.18, then we would understand the local models for all possible singular fibres of SL fibrations of generic almost Calabi–Yau 3-folds.

10
Hyperkähler and quaternionic Kähler manifolds

We now discuss *hyperkähler* and *quaternionic Kähler* manifolds, that is, Riemannian $4m$-manifolds (X, g) whose holonomy lies in $\mathrm{Sp}(m)$ and $\mathrm{Sp}(m)\,\mathrm{Sp}(1)$. As $\mathrm{Sp}(m) \subseteq \mathrm{SU}(2m)$, a metric g with holonomy in $\mathrm{Sp}(m)$ is Kähler and Ricci-flat, and closely related to the Calabi–Yau metrics of Chapter 7. But since $\mathrm{Sp}(m)$ preserves three complex structures J_1, J_2, J_3 on $\mathbb{R}^{4m}$ there are corresponding constant complex structures J_1, J_2, J_3 on X. Furthermore, if $a_1, a_2, a_3 \in \mathbb{R}$ with $a_1^2 + a_2^2 + a_3^2 = 1$ then $a_1 J_1 + a_2 J_2 + a_3 J_3$ is a complex structure on X, and g is Kähler with respect to it.

Thus, a metric g with $\mathrm{Hol}(g) \subseteq \mathrm{Sp}(m)$ is Kähler with respect to a whole 2-sphere S^2 of complex structures. We call g a *hyperkähler metric*, (J_1, J_2, J_3, g) a *hyperkähler structure*, and (X, J_1, J_2, J_3, g) a *hyperkähler manifold*. Hyperkähler geometry should be understood in terms of the *quaternions* $\mathbb{H}$. The complex structures J_1, J_2, J_3 on X make each tangent space into a left $\mathbb{H}$-module isomorphic to $\mathbb{H}^m$.

We begin in §10.1 with an introduction to hyperkähler geometry. Sections 10.2 and 10.3 describe examples of noncompact hyperkähler 4-manifolds, *hyperkähler ALE spaces*, and compact hyperkähler 4-manifolds, $K3$ *surfaces*. These are very well understood, by treating them as complex surfaces and using complex algebraic geometry, and we give a very precise description of the moduli space of hyperkähler structures on $K3$. Then §10.4 explains the theory of compact hyperkähler manifolds in higher dimensions. In comparison to Calabi–Yau manifolds, rather few examples are known.

Section 10.5 discusses *quaternionic Kähler manifolds* (X, g), with holonomy in $\mathrm{Sp}(m)\,\mathrm{Sp}(1)$ for $m \geqslant 2$. They are Einstein, with positive or negative scalar curvature. The name is unfortunate, as quaternionic Kähler manifolds are never Kähler. However, each quaternionic Kähler manifold X has a *twistor space*, an S^2-bundle over X, which is a complex manifold, and Kähler–Einstein if X has positive scalar curvature. Rather few examples are known, and it is conjectured that the only compact quaternionic Kähler manifolds with positive scalar curvature are symmetric spaces. We close in §10.6 with brief discussions of several other topics in quaternionic geometry.

10.1 An introduction to hyperkähler geometry

We now introduce the holonomy groups $\mathrm{Sp}(m)$ in $4m$ dimensions. If (X, g) is a Riemannian $4m$-manifold and $\mathrm{Hol}(g) \subseteq \mathrm{Sp}(m)$, then g is Kähler with respect to complex structures J_1, J_2, J_3 on X with $J_1 J_2 = J_3$. If $a_1, a_2, a_3 \in \mathbb{R}$ and $a_1^2 + a_2^2 + a_3^2 = 1$

then $a_1 J_1 + a_2 J_2 + a_3 J_3$ is also a complex structure on X, which makes g Kähler. Thus g is Kähler with respect to a whole 2-sphere $\mathcal{S}^2$ of complex structures. We call (J_1, J_2, J_3, g) a *hyperkähler structure* on X.

It is convenient to define the Lie groups $\mathrm{Sp}(m)$ and hyperkähler manifolds using the *quaternions* $\mathbb{H}$. We shall also discuss *twistor spaces* of hyperkähler manifolds, a device for translating hyperkähler geometry in $4m$ real dimensions into holomorphic geometry in $2m + 1$ complex dimensions. Some references for this section are the foundational paper of Hitchin, Karlhede, Lindström and Roček [164], and Salamon [296, Ch.s 8–9]. A good survey on the material of §10.1–§10.4 is Dancer [92].

10.1.1 The Lie groups $\mathrm{Sp}(m)$

The quaternions $\mathbb{H}$ are the associative, nonabelian real algebra

$$\mathbb{H} = \{x_0 + x_1 i_1 + x_2 i_2 + x_3 i_3 : x_j \in \mathbb{R}\} \cong \mathbb{R}^4.$$

The imaginary quaternions are $\mathrm{Im}\,\mathbb{H} = \langle i_1, i_2, i_3 \rangle \cong \mathbb{R}^3$. Multiplication is given by

$$i_1 i_2 = -i_2 i_1 = i_3, \quad i_2 i_3 = -i_3 i_2 = i_1, \quad i_3 i_1 = -i_1 i_3 = i_2, \quad i_1^2 = i_2^2 = i_3^2 = -1.$$

When $x = x_0 + x_1 i_1 + x_2 i_2 + x_3 i_3$, we define $\bar{x}$ and $|x|$ by

$$\bar{x} = x_0 - x_1 i_1 - x_2 i_2 - x_3 i_3 \quad \text{and} \quad |x|^2 = x_0^2 + x_1^2 + x_2^2 + x_3^2.$$

These satisfy $\overline{(pq)} = \bar{q}\,\bar{p}$, $|p| = |\bar{p}|$ and $|pq| = |p||q|$ for all $p, q \in \mathbb{H}$.

Let $\mathbb{H}^m$ have coordinates $(q^1, \ldots, q^m)$, with $q^j = x_0^j + x_1^j i_1 + x_2^j i_2 + x_3^j i_3 \in \mathbb{H}$ and $x_k^j \in \mathbb{R}$. Define a metric g and 2-forms $\omega_1, \omega_2, \omega_3$ on $\mathbb{H}^m$ by

$$
\begin{aligned}
g &= \sum_{j=1}^m \sum_{k=0}^3 (\mathrm{d}x_k^j)^2, & \omega_1 &= \sum_{j=1}^m \mathrm{d}x_0^j \wedge \mathrm{d}x_1^j + \mathrm{d}x_2^j \wedge \mathrm{d}x_3^j, \\
\omega_2 &= \sum_{j=1}^m \mathrm{d}x_0^j \wedge \mathrm{d}x_2^j - \mathrm{d}x_1^j \wedge \mathrm{d}x_3^j, & \omega_3 &= \sum_{j=1}^m \mathrm{d}x_0^j \wedge \mathrm{d}x_3^j + \mathrm{d}x_1^j \wedge \mathrm{d}x_2^j.
\end{aligned}
\tag{10.1}
$$

We can also express $g, \omega_1, \omega_2, \omega_3$ neatly as $g + \omega_1 i_1 + \omega_2 i_2 + \omega_3 i_3 = \sum_{j=1}^m \mathrm{d}\bar{q}^j \otimes \mathrm{d}q^j$, using multiplication in $\mathbb{H}$ to interpret $\mathrm{d}\bar{q}^j \otimes \mathrm{d}q^j$ as an $\mathbb{H}$-valued tensor.

Identify $\mathbb{H}^m$ with $\mathbb{R}^{4m}$. Then g is the Euclidean metric on $\mathbb{R}^{4m}$. Let J_1, J_2 and J_3 be the complex structures on $\mathbb{R}^{4m}$ corresponding to left multiplication by i_1, i_2 and i_3 in $\mathbb{H}^m$ respectively. Then g is Kähler with respect to each J_j, with Kähler form ω_j. Furthermore, if $a_1, a_2, a_3 \in \mathbb{R}$ with $a_1^2 + a_2^2 + a_3^2 = 1$ then $a_1 J_1 + a_2 J_2 + a_3 J_3$ is a complex structure on $\mathbb{R}^{4m}$, and g is Kähler with respect to it, with Kähler form $a_1 \omega_1 + a_2 \omega_2 + a_3 \omega_3$.

The subgroup of $\mathrm{GL}(4m, \mathbb{R})$ preserving g, ω_1, ω_2 and ω_3 is $\mathrm{Sp}(m)$. It is a compact, connected, simply-connected, semisimple Lie group of dimension $2m^2 + m$, a subgroup of $\mathrm{SO}(4m)$. Since any of g, ω_j and J_j can be written in terms of the other two, $\mathrm{Sp}(m)$ also preserves J_1, J_2 and J_3.

We can write $\mathrm{Sp}(m)$ as a group of $m \times m$ matrices over $\mathbb{H}$ by

$$\mathrm{Sp}(m) \cong \{A \in M_m(\mathbb{H}) : A\bar{A}^t = I\}.\tag{10.2}$$

To understand the action of $\mathrm{Sp}(m)$ on $\mathbb{H}^m$ we think of $\mathbb{H}^m$ as row matrices over $\mathbb{H}$, and then $A \in \mathrm{Sp}(m)$ acts by $(q^1\, q^2 \cdots q^m) \mapsto (q^1\, q^2 \cdots q^m)\bar{A}^t$. The point here is that J_1, J_2, J_3 are defined by left multiplication by i_2, i_2, i_3, so to commute with this A must act on the right; but as by convention mappings act on the left we right-multiply by $\bar{A}^t$, to preserve the order of multiplication.

We can also identify $\mathbb{H}^m$ with $\mathbb{C}^{2m}$. Define complex coordinates $(z_1, \ldots, z_{2m})$ on $\mathbb{R}^{4m}$ by $z_{2j-1} = x_0^j + ix_1^j$ and $z_{2j} = x_2^j + ix_3^j$ for $j = 1, \ldots, m$. Then $g, \omega_1, \omega_2, \omega_3$ satisfy

$$g = \sum_{j=1}^{2m} |\mathrm{d}z_j|^2, \quad \omega_1 = \frac{i}{2} \sum_{j=1}^{2m} \mathrm{d}z_j \wedge \mathrm{d}\bar{z}_j \text{ and } \omega_2 + i\omega_3 = \sum_{j=1}^{m} \mathrm{d}z_{2j-1} \wedge \mathrm{d}z_{2j}.$$

That is, g and ω_1 are the standard Hermitian metric and Hermitian form on $\mathbb{C}^{2m}$, and the (2,0)-form $\omega_2 + i\omega_3$ is a *complex symplectic form* on $\mathbb{C}^{2m}$.

Observe that $\frac{1}{m!}(\omega_2 + i\omega_3)^m = \mathrm{d}z_1 \wedge \cdots \wedge \mathrm{d}z_{2m}$, which is the usual holomorphic volume form θ on $\mathbb{C}^{2m}$. As $\mathrm{Sp}(m)$ preserves ω_2 and ω_3 it preserves θ. So $\mathrm{Sp}(m)$ preserves the metric g, Hermitian form ω_1 and complex volume form θ on $\mathbb{C}^{2m}$. From §7.1, this means that $\mathrm{Sp}(m)$ is a subgroup of $\mathrm{SU}(2m)$, the subgroup fixing the complex symplectic form $\omega_2 + i\omega_3$.

Now $\dim \mathrm{Sp}(m) = 2m^2 + m$ and $\dim \mathrm{SU}(2m) = 4m^2 - 1$. Thus, for all $m > 1$ we have $\dim \mathrm{Sp}(m) < \dim \mathrm{SU}(2m)$, and $\mathrm{Sp}(m)$ is a proper subgroup of $\mathrm{SU}(2m)$. However, when $m = 1$ we have $\dim \mathrm{Sp}(1) = \dim \mathrm{SU}(2) = 3$, and in fact $\mathrm{SU}(2) = \mathrm{Sp}(1)$, so the holonomy groups $\mathrm{SU}(2)$ and $\mathrm{Sp}(1)$ are the same.

10.1.2 The holonomy groups $\mathrm{Sp}(m)$ and hyperkähler structures

Let (X, g) be a Riemannian $4m$-manifold with $\mathrm{Hol}(g) \subseteq \mathrm{Sp}(m)$. By Proposition 2.5.2, each $\mathrm{Sp}(m)$-invariant tensor on $\mathbb{R}^{4m}$ corresponds to a tensor on X constant under the Levi-Civita connection ∇ of g. So from above there exist almost complex structures J_1, J_2, J_3 and 2-forms $\omega_1, \omega_2, \omega_3$ on X, each constant under ∇, and isomorphic to the standard models on $\mathbb{R}^{4m}$ at each point of X.

As $\nabla J_j = 0$, each J_j is an integrable complex structure, and g is Kähler with respect to J_j, with Kähler form ω_j. Similarly, if $a_1, a_2, a_3 \in \mathbb{R}$ with $a_1^2 + a_2^2 + a_3^2 = 1$ then $a_1 J_1 + a_2 J_2 + a_3 J_3$ is a complex structure on X, and g is Kähler with respect to it, with Kähler form $a_1 \omega_1 + a_2 \omega_2 + a_3 \omega_3$. Therefore g is Kähler in lots of different ways, with respect to a whole 2-sphere of complex structures. Because of this, we call g *hyperkähler*.

Definition 10.1.1 Let X be a $4m$-manifold. An *almost hyperkähler structure* on X is a quadruple (J_1, J_2, J_3, g), where J_j are almost complex structures on X with $J_1 J_2 = J_3$, and g is a Riemannian metric on X which is Hermitian with respect to J_1, J_2 and J_3.

We call (J_1, J_2, J_3, g) a *hyperkähler structure* on X if in addition $\nabla J_j = 0$ for $j = 1, 2, 3$, where ∇ is the Levi-Civita connection of g. Then (X, J_1, J_2, J_3, g) is a *hyperkähler manifold*, and g a *hyperkähler metric*. Each J_j is integrable, and g is Kähler with respect to J_j. We refer to the Kähler forms $\omega_1, \omega_2, \omega_3$ of J_1, J_2, J_3 as the *hyperkähler 2-forms* of X.

An almost hyperkähler structure is equivalent to an $\mathrm{Sp}(m)$-structure, and a hyperkähler structure to a torsion-free $\mathrm{Sp}(m)$-structure. The next proposition comes from [296, Lem. 8.4], except the last part, which follows from Proposition 7.1.1 using the inclusion $\mathrm{Sp}(m) \subseteq \mathrm{SU}(2m)$.

Proposition 10.1.2 *Suppose* X *is a* $4m$-*manifold and* (J_1, J_2, J_3, g) *an almost hyperkähler structure on* X, *and let* $\omega_1, \omega_2, \omega_3$ *be the Hermitian forms of* J_1, J_2 *and* J_3. *Then the following conditions are equivalent:*

(i) (J_1, J_2, J_3, g) *is a hyperkähler structure,*

(ii) $d\omega_1 = d\omega_2 = d\omega_3 = 0$,

(iii) $\nabla\omega_1 = \nabla\omega_2 = \nabla\omega_3 = 0$, *and*

(iv) $\mathrm{Hol}(g) \subseteq \mathrm{Sp}(m)$, *and* J_1, J_2, J_3 *are the induced complex structures.*

All hyperkähler metrics are Ricci-flat.

10.1.3 Twistor spaces of hyperkähler manifolds

Suppose (X, J_1, J_2, J_3, g) is a hyperkähler $4m$-manifold. Set $Z = \mathbb{CP}^1 \times X$, where

$$\mathbb{CP}^1 \cong \mathcal{S}^2 = \{a_1 J_1 + a_2 J_2 + a_3 J_3 : a_j \in \mathbb{R}, \ a_1^2 + a_2^2 + a_3^2 = 1\}$$

is the natural 2-sphere of complex structures on X. Then $\mathbb{CP}^1$ has a natural complex structure J_0, say. Each point $z \in Z$ is of the form (J, x) for $J \in \mathbb{CP}^1$ and $x \in X$, and the tangent space $T_z Z$ is $T_z Z = T_J \mathbb{CP}^1 \oplus T_x X$.

Now J_0 is a complex structure on $T_J \mathbb{CP}^1$, and J is a complex structure on $T_x X$. Thus $J_Z = J_0 \oplus J$ is a complex structure on $T_z Z = T_J \mathbb{CP}^1 \oplus T_x X$. This defines an almost complex structure J_Z on Z, which turns out to be integrable. Hence (Z, J_Z) is a complex $(2m+1)$-manifold, called the *twistor space* of X.

Let $p : Z \to \mathbb{CP}^1$ and $\pi : Z \to X$ be the natural projections. Then p is holomorphic, and the hypersurface $p^{-1}(J)$ is isomorphic to X with complex structure J, for each $J \in \mathbb{CP}^1$. Define $\sigma : Z \to Z$ by $\sigma : (J, x) \mapsto (-J, x)$. Then σ is a free antiholomorphic involution of Z. For each $x \in X$, the fibre $\Sigma_x = \pi^{-1}(x)$ of π is a holomorphic curve in Z isomorphic to $\mathbb{CP}^1$, with normal bundle $2m\mathcal{O}(1)$, which is preserved by σ.

There is one other piece of holomorphic data on Z given by the hyperkähler structure on X. Let $\mathcal{D}$ be the kernel of $dp : TZ \to T\mathbb{CP}^1$. Then $\mathcal{D}$ is a vector subbundle of TZ, which is holomorphic as p is holomorphic; $\mathcal{D}$ is the bundle of tangent spaces to the fibres of p. Then one can construct a nondegenerate holomorphic section ω of the holomorphic vector bundle $p^*(\mathcal{O}(2)) \otimes \Lambda^2 \mathcal{D}^*$ over Z.

Effectively this amounts to a choice of complex symplectic form on each of the fibres $p^{-1}(J) \cong (X, J)$ of p. For instance, the complex 2-form $\omega_2 + i\omega_3$ on X is holomorphic with respect to J_1, and on $p^{-1}(J_1)$ we take ω to be $\omega_2 + i\omega_3$, multiplied by some fixed element in the fibre of $\mathcal{O}(2)$ over J_1 in $\mathbb{CP}^1$. We summarize these ideas in the following theorem, taken from [164, §3(F)].

Theorem 10.1.3 *Let* (X, J_1, J_2, J_3, g) *be a hyperkähler* $4m$-*manifold. Then the twistor space* (Z, J_Z) *of* X *is a complex* $(2m+1)$-*manifold diffeomorphic to* $\mathbb{CP}^1 \times X$. *Let* $p : Z \to \mathbb{CP}^1$ *and* $\pi : Z \to X$ *be the projections. Then* p *is holomorphic, and there exists a holomorphic section* ω *of* $p^*(\mathcal{O}(2)) \otimes \Lambda^2 \mathcal{D}^*$ *which is symplectic on the fibres of the kernel* $\mathcal{D}$ *of* $dp : TZ \to T\mathbb{CP}^1$. *There is also a natural, free antiholomorphic involution* $\sigma : Z \to Z$ *satisfying* $\sigma^*(\omega) = \omega$, $\pi \circ \sigma = \pi$ *and* $p \circ \sigma = \sigma' \circ p$, *where* $\sigma' : \mathbb{CP}^1 \to \mathbb{CP}^1$ *is the antipodal map.*

It turns out that the holomorphic data Z, p, ω and σ is sufficient to reconstruct (X, J_1, J_2, J_3, g). We express this in our next result, deduced from [164, Th. 3.3].

Theorem 10.1.4 *Suppose* (Z, J_Z) *is a complex* $(2m+1)$-*manifold equipped with*

(a) *a holomorphic projection* $p : Z \to \mathbb{CP}^1$,

(b) *a holomorphic section* ω *of* $p^*(\mathcal{O}(2)) \otimes \Lambda^2 \mathcal{D}^*$ *which is symplectic on the fibres of* $\mathcal{D}$, *where* $\mathcal{D}$ *is the kernel of* $\mathrm{d}p : TZ \to T\mathbb{CP}^1$, *and*

(c) *a free antiholomorphic involution* $\sigma : Z \to Z$ *satisfying* $\sigma^*(\omega) = \omega$ *and* $p \circ \sigma = \sigma' \circ p$, *where* $\sigma' : \mathbb{CP}^1 \to \mathbb{CP}^1$ *is the antipodal map.*

Define X' *to be the set of holomorphic curves* C *in* Z *isomorphic to* $\mathbb{CP}^1$, *with normal bundle* $2m\mathcal{O}(1)$ *and* $\sigma(C) = C$. *Then* X' *is a hypercomplex* $4m$-*manifold, equipped with a natural pseudo-hyperkähler metric* g. *If* g *is positive definite, then* X' *is hyperkähler. Let* Z' *be the twistor space of* X'. *Then there is a natural, locally biholomorphic map* $\iota : Z' \to Z$, *which identifies* p, ω *and* σ *with their analogues on* Z'.

A *pseudo-hyperkähler metric* is a pseudo-Riemannian metric g of type $(4k, 4m - 4k)$ with $\mathrm{Hol}(g) \subseteq \mathrm{Sp}(k, m-k)$. It is Riemannian if $k = m$. One moral of this theorem is that hyperkähler manifolds can be written solely in terms of *holomorphic* data, and so they can be studied and explicit examples found using *complex algebraic geometry*.

10.2 Hyperkähler ALE spaces

The *ALE spaces* are a special class of noncompact hyperkähler 4-manifolds.

Definition 10.2.1 Let G be a finite subgroup of $\mathrm{Sp}(1)$, and let $(\hat{J}_1, \hat{J}_2, \hat{J}_3, \hat{g})$ be the Euclidean hyperkähler structure and $r : \mathbb{H}/G \to [0, \infty)$ the radius function on $\mathbb{H}/G$. We say that a hyperkähler 4-manifold (X, J_1, J_2, J_3, g) is *asymptotically locally Euclidean*, or *ALE*, and asymptotic to $\mathbb{H}/G$, if there exists a compact subset $S \subset X$ and a map $\pi : X \setminus S \to \mathbb{H}/G$ that is a diffeomorphism between $X \setminus S$ and $\{x \in \mathbb{H}/G : r(x) > R\}$ for some $R > 0$, such that

$$\hat{\nabla}^k(\pi_*(g) - \hat{g}) = O(r^{-4-k}) \quad \text{and} \quad \hat{\nabla}^k(\pi_*(J_j) - \hat{J}_j) = O(r^{-4-k}) \qquad (10.3)$$

as $r \to \infty$, for $j = 1, 2, 3$ and $k \geqslant 0$, where $\hat{\nabla}$ is the Levi-Civita connection of $\hat{g}$.

Hyperkähler ALE spaces are called *gravitational instantons* by physicists. What this definition means is that a hyperkähler ALE space is a noncompact hyperkähler 4-manifold X with one end which at infinity resembles $\mathbb{H}/G$, and the hyperkähler structure on X and its derivatives are required to approximate the Euclidean hyperkähler structure on $\mathbb{H}/G$ with a prescribed rate of decay.

One reason hyperkähler ALE spaces are interesting is that they give a local model for how to desingularize hyperkähler 4-orbifolds to give hyperkähler 4-manifolds, as we will see when we discuss the Kummer construction of the $K3$ surface in §10.4. They are also the simplest examples of ALE manifolds with holonomy $\mathrm{SU}(m)$.

The first examples of hyperkähler ALE spaces, asymptotic to $\mathbb{H}/\{\pm 1\}$, were written down by Eguchi and Hanson [100] and are called *Eguchi–Hanson spaces*.

Example 10.2.2 Consider $\mathbb{C}^2$ with complex coordinates (z_1, z_2), acted upon by the involution $-1 : (z_1, z_2) \mapsto (-z_1, -z_2)$. Let (X, π) be the blow-up of $\mathbb{C}^2/\{\pm 1\}$ at

0. Then X is a crepant resolution of $\mathbb{C}^2/\{\pm 1\}$. It is biholomorphic to $T^*\mathbb{CP}^1$, with $\pi_1(X) = \{1\}$ and $H^2(X, \mathbb{R}) = \mathbb{R}$. Define $f : X \setminus \pi^{-1}(0) \to \mathbb{R}$ by

$$f = \sqrt{r^4 + 1} + 2\log r - \log\left(\sqrt{r^4 + 1} + 1\right), \tag{10.4}$$

where $r = \left(|z_1|^2 + |z_2|^2\right)^{1/2}$ is the radius function on X.

Define a 2-form ω_1 on $X \setminus \pi^{-1}(0)$ by $\omega_1 = i\partial\bar\partial f$. Then ω_1 extends smoothly and uniquely to X. Furthermore, ω_1 is a closed, real, positive (1,1)-form, and is thus the Kähler form of a Kähler metric g on X. This is the *Eguchi–Hanson* metric on X, which has holonomy $\mathrm{SU}(2)$. It extends to a hyperkähler structure (J_1, J_2, J_3, g) on X, where J_1 is the natural complex structure on X, and the Kähler forms ω_2, ω_3 of J_2, J_3 satisfy $\omega_2 + i\omega_3 = \pi^*(dz_1 \wedge dz_2)$.

For large r we have $f = r^2 + O(r^{-2})$, so that $\omega_1 = i\partial\bar\partial(r^2) + O(r^{-4})$. But $i\partial\bar\partial(r^2)$ is the Kähler form of the Euclidean metric $\hat g$ on $\mathbb{C}^2/\{\pm 1\}$, so that $g = \hat g + O(r^{-4})$ for large r, as in the first equation of (10.3). In the same way we can show that (X, J_1, J_2, J_3, g) is a *hyperkähler ALE space* asymptotic to $\mathbb{H}/\{\pm 1\}$.

There is a 3-dimensional family of hyperkähler ALE spaces $(X', J_1', J_2', J_3', g')$ asymptotic to $\mathbb{H}/\{\pm 1\}$, which are called *Eguchi–Hanson spaces*, and are all isomorphic to the example (X, J_1, J_2, J_3, g) above under homotheties and rotations of the 2-sphere $\mathcal{S}^2$ of complex structures. That is, if $(X', J_1', J_2', J_3', g')$ is any hyperkähler ALE space asymptotic to $\mathbb{H}/\{\pm 1\}$ then X' is diffeomorphic to X, and we can choose a diffeomorphism $\phi : X \to X'$ such that $\phi^*(g') = t^2 g$ for some $t > 0$, and $\phi^*(J_j') = \sum_{k=1}^3 a_{jk} J_k$ for some 3×3 matrix (a_{jk}) in $\mathrm{SO}(3)$.

The Eguchi–Hanson spaces were soon generalized by Gibbons and Hawking [124], who gave explicit examples of hyperkähler ALE spaces asymptotic to $\mathbb{H}/\mathbb{Z}_k$ for all $k \geqslant 2$. Hitchin [160] constructed the same spaces using twistor methods.

Eventually, a complete construction and classification of all hyperkähler ALE spaces was achieved by Kronheimer [226, 227]. The construction makes use of the *McKay correspondence* of §7.3.1, which links the Kleinian singularities $\mathbb{C}^2/G$, their crepant resolutions, and the Dynkin diagrams of type A_r ($r \geqslant 1$), D_r ($r \geqslant 4$), E_6, E_7 and E_8. Here is a statement of Kronheimer's results, following from [226, 227].

Theorem 10.2.3 *Let G be a nontrivial finite subgroup of $\mathrm{SU}(2)$ with Dynkin diagram Γ, and let $\mathfrak{h}_{\mathbb{R}}$ be the real vector space with basis the set of nontrivial irreducible representations of G. Let Δ be the set of roots and W the Weyl group of Γ. Then $\mathrm{Aut}(\Gamma) \ltimes W$ acts naturally on $\mathfrak{h}_{\mathbb{R}}$, and Δ is a finite subset of $\mathfrak{h}_{\mathbb{R}}^*$ preserved by $\mathrm{Aut}(\Gamma) \ltimes W$. Define $U \subset \mathfrak{h}_{\mathbb{R}} \otimes \mathbb{R}^3$ by*

$$U = \big\{(\alpha_1, \alpha_2, \alpha_3) \in \mathfrak{h}_{\mathbb{R}} \otimes \mathbb{R}^3 : \text{for each } \delta \in \Delta,$$
$$\delta(\alpha_1), \delta(\alpha_2), \delta(\alpha_3) \text{ are not all zero}\big\}. \tag{10.5}$$

There is a continuous family of noncompact hyperkähler 4-manifolds and 4-orbifolds X_α parametrized by $\alpha = (\alpha_1, \alpha_2, \alpha_3)$ in $\mathfrak{h}_{\mathbb{R}} \otimes \mathbb{R}^3$, which can be written down explicitly using the hyperkähler quotient construction, satisfying the following conditions:

(a) *If $\alpha \in U$ then X_α is a hyperkähler ALE space asymptotic to $\mathbb{H}/G$, diffeomorphic to the crepant resolution of $\mathbb{C}^2/G$, with a natural isomorphism $H^2(X_\alpha, \mathbb{R}) \cong \mathfrak{h}_{\mathbb{R}}$.*

(b) If $\alpha \notin U$ then X_α is a singular orbifold asymptotic to $\mathbb{H}/G$, with $X_0 \cong \mathbb{H}/G$.

(c) Let $\alpha \in U$. Then the isomorphism $H^2(X_\alpha, \mathbb{R}) \cong \mathfrak{h}_\mathbb{R}$ identifies $[\omega_j] \in H^2(X_\alpha, \mathbb{R})$ and $\alpha_j \in \mathfrak{h}_\mathbb{R}$ for $j = 1, 2, 3$.

(d) Let $\alpha, \beta \in \mathfrak{h}_\mathbb{R} \otimes \mathbb{R}^3$. Then X_α and X_β are isomorphic as hyperkähler manifolds if and only if $\alpha = (\gamma, w) \cdot \beta$ for some $(\gamma, w) \in \mathrm{Aut}(\Gamma) \ltimes W$.

(e) Suppose X' is a hyperkähler ALE space asymptotic to $\mathbb{H}/G$. Then there exists $\alpha \in U$ such that $X' \cong X_\alpha$.

Here the *hyperkähler quotient* is a method of producing hyperkähler manifolds due to Hitchin et al. [164, §3], described in §10.6.4. Given a hyperkähler $4m$-manifold (X, J_1, J_2, J_3, g) and a suitable k-dimensional Lie group G of automorphisms of it, the hyperkähler quotient of X by G is a new hyperkähler manifold of dimension $4(m - k)$.

Kronheimer's proof of this theorem comes in two parts. First, in [226], for each finite group $G \subset \mathrm{SU}(2)$ he uses the Dynkin diagram Γ of G to write down an explicit hyperkähler quotient of $\mathbb{H}^n$ by a product of unitary groups $\mathrm{U}(k)$, and shows that for suitable values of the moment map, this quotient is a hyperkähler ALE space asymptotic to $\mathbb{H}/G$. Then in [227] he shows that every hyperkähler ALE space X asymptotic to $\mathbb{H}/G$ arises from this construction. The proof uses the twistor space of X, and facts about the deformations of $\mathbb{C}^2/G$ taken from Slodowy [315].

Here is the reason for the condition defining U in (10.5). Each element $\delta \in \Delta$ corresponds to a 2-sphere $\mathcal{S}^2$ in X, with self-intersection -2. It turns out that the volume of the corresponding $\mathcal{S}^2$ in X_α is $\left(\sum_{j=1}^3 \delta(\alpha_j)^2 \right)^{1/2}$. Thus if $\delta(\alpha_j) = 0$ for $j = 1, 2, 3$ then this $\mathcal{S}^2$ collapses to a point, and X_α becomes singular.

We shall also explain the rôle of the group $\mathrm{Aut}(\Gamma) \ltimes W$ in part (d). Here W is the Weyl group, and $\mathrm{Aut}(\Gamma)$ the automorphism group of the graph Γ, given by

$$\mathrm{Aut}(\Gamma) = \begin{cases} \{1\} & \text{if } \Gamma = A_1, E_7 \text{ or } E_8, \\ \mathbb{Z}_2 & \text{if } \Gamma = A_k \ (k \geqslant 2), D_k \ (k \geqslant 5) \text{ or } E_6, \\ S_3 & \text{if } \Gamma = D_4. \end{cases}$$

There is a natural, surjective group homomorphism $\rho : \mathrm{Aut}(\mathbb{H}/G) \to \mathrm{Aut}(\Gamma)$, such that $\mathrm{Ker}\,\rho$ is the identity component of $\mathrm{Aut}(\mathbb{H}/G)$. Hence $\mathrm{Aut}(\Gamma)$ is the group of isotopy classes of automorphisms of $\mathbb{H}/G$.

Let X be the crepant resolution of $\mathbb{C}^2/G$, regarded as a real 4-manifold. Isotopy (continuous deformation) is an equivalence relation on the diffeomorphism group of X. It turns out that $\mathrm{Aut}(\Gamma) \ltimes W$ acts on X as a group of *isotopy classes of diffeomorphisms*. For each $(\gamma, w) \in \mathrm{Aut}(\Gamma) \ltimes W$ there is an isotopy class of diffeomorphisms $\Phi : X \to X$, such that $\Phi_* : H^2(X, \mathbb{C}) \to H^2(X, \mathbb{C})$ coincides with $(\gamma, w) : \mathfrak{h}_\mathbb{C} \to \mathfrak{h}_\mathbb{C}$ under the isomorphism $H^2(X, \mathbb{C}) \cong \mathfrak{h}_\mathbb{C}$.

If $\gamma = 1$, so that $(\gamma, w) \in W$, then we can choose Φ to be the identity outside a compact subset in X. More generally, we can choose Φ to be asymptotic to any $\tau \in \mathrm{Aut}(\mathbb{H}/G)$ with $\rho(\tau) = \gamma$. Thus $\mathrm{Aut}(\Gamma) \ltimes W$ is a kind of symmetry group of the topology of X.

10.3 $K3$ surfaces

A *K3 surface* is defined to be a compact, complex surface (X, J) with $h^{1,0}(X) = 0$ and trivial canonical bundle. $K3$ surfaces occupy a special place in Kodaira's classification of complex surfaces. They are also important in Riemannian holonomy, as they are the only compact 4-manifolds carrying metrics with holonomy $\mathrm{SU}(2) = \mathrm{Sp}(1)$.

Thus $K3$ surfaces are the lowest-dimensional examples of both Calabi–Yau manifolds (with holonomy $\mathrm{SU}(m)$), and compact hyperkähler manifolds (with holonomy $\mathrm{Sp}(m)$). But because their behaviour is more typical of the hyperkähler than the Calabi–Yau case, we cover them here and not in Chapter 7. We begin with some examples, and then discuss $K3$ surfaces first as complex surfaces, and then as hyperkähler 4-manifolds.

10.3.1 Examples of $K3$ surfaces

Example 10.3.1 Define S to be the *Fermat quartic*

$$S = \big\{ [z_0, \dots, z_3] \in \mathbb{CP}^3 : z_0^4 + z_1^4 + z_2^4 + z_3^4 = 0 \big\}.$$

The *adjunction formula* [132, p. 147] shows that $K_S = (K_{\mathbb{CP}^3} \otimes L_S)|_S$, where L_S is the line bundle over $\mathbb{CP}^3$ associated to the divisor S. But $K_{\mathbb{CP}^3} = \mathcal{O}(-4)$ and $L_S = \mathcal{O}(4)$ as S is a quartic. So $K_{\mathbb{CP}^3} \otimes L_S = \mathcal{O}(0)$, and the canonical bundle K_S of S is trivial. Theorem 5.10.4 shows that $H^k(S, \mathbb{C}) \cong H^k(\mathbb{CP}^3, \mathbb{C})$ and $\pi_k(S) \cong \pi_k(\mathbb{CP}^3)$ for $k = 0, 1$, so S is connected and simply-connected. Thus $h^{1,0}(S) = 0$ and K_S is trivial, so that S is a $K3$ surface, by definition.

We shall work out the Hodge and Betti numbers $h^{p,q}$, b^k of S. The Riemann–Roch formula shows that $\chi(S) = 24$, and as $b^0 = 1$ and $b^1 = 0$ we have $b^2 = 22$. Also $h^{2,0} = 1$ as K_S is trivial, so $h^{0,2} = 1$ and $h^{1,1} = 20$. But the signature $\tau(S) = b_+^2 - b_-^2$ satisfies $\tau(S) = \sum_{p,q=0}^2 (-1)^p h^{p,q} = -16$. Hence $b_+^2 = 3$ and $b_-^2 = 19$.

More generally, using §7.6 we find that the following are all $K3$ surfaces:

- Any nonsingular quartic in $\mathbb{CP}^3$.
- A complete intersection of a cubic and a quadric in $\mathbb{CP}^4$.
- A complete intersection of 3 quadrics in $\mathbb{CP}^5$.

But these are all projective varieties. Here are some *non-algebraic* $K3$ surfaces.

Example 10.3.2 Let Λ be a *lattice* in $\mathbb{C}^2$, so that $\Lambda \cong \mathbb{Z}^4$. Then $\mathbb{C}^2/\Lambda$ is a complex 4-torus T^4. Define a map $\sigma : T^4 \to T^4$ by $\sigma : (z_1, z_2) + \Lambda \mapsto (-z_1, -z_2) + \Lambda$. Then σ fixes the 16 points

$$\big\{ (z_1, z_2) + \Lambda : (z_1, z_2) \in \tfrac{1}{2}\Lambda \big\}.$$

Thus $T^4/\langle \sigma \rangle$ is a *complex orbifold*, with 16 singular points modelled on $\mathbb{C}^2/\{\pm 1\}$.

Let S be the blow-up of $T^4/\langle \sigma \rangle$ at the 16 singular points. Then S is a *crepant resolution* of $T^4/\langle \sigma \rangle$, and is a $K3$ surface. We call this the *Kummer construction*, and S a *Kummer surface*. For a generic choice of lattice Λ, the torus T^4 is not an algebraic surface, and neither is the Kummer surface S. Thus there exist non-algebraic $K3$ surfaces.

Now $T^4/\langle \sigma \rangle$ is simply-connected, and $H^k(T^4/\langle \sigma \rangle, \mathbb{C})$ is the σ-invariant part of $H^k(T^4, \mathbb{C})$. Thus we find that $b_\pm^2(T^4/\langle \sigma \rangle) = 3$. The blow-up replaces each singular point with a copy of $\mathbb{CP}^1$, with self-intersection -2. This leaves π_1 and b_+^2 unchanged, and adds one to b^2_- for each of the 16 points. Hence S is simply-connected and has $b^2_+ = 3$ and $b^2_- = 19$, as in Example 10.3.1.

We shall return to the Kummer construction in Example 10.3.14.

10.3.2 $K3$ surfaces as complex surfaces

The theory of complex surfaces (including non-algebraic surfaces) is an old and very well understood branch of complex algebraic geometry. We now discuss what is known about $K3$ surfaces from this point of view. We begin with a result on the topology of $K3$ surfaces.

Theorem 10.3.3 *Let* (X, J) *be a* $K3$ *surface. Then* X *is simply-connected, with Betti numbers* $b^2 = 22$, $b^2_+ = 3$, *and* $b^2_- = 19$. *Also* X *is Kähler, and has Hodge numbers* $h^{2,0} = h^{0,2} = 1$ *and* $h^{1,1} = 20$. *All* $K3$ *surfaces are diffeomorphic.*

Proof Kodaira [216, Th. 13] showed that every $K3$ surface is a deformation of a non-singular quartic surface in $\mathbb{CP}^3$. Thus all $K3$ surfaces are diffeomorphic to the surface S of Example 10.3.1. Hence $\pi_1(X) = \{1\}$, $b^2 = 22$, $b^2_+ = 3$, and $b^2_- = 19$, from above. Todorov [328, Th. 2] and Siu [314] prove that every $K3$ surface is Kähler. (This is not obvious.) As K_X is trivial we have $h^{2,0} = 1$, so $h^{0,2} = 1$ as X is Kähler, and this leaves $h^{1,1} = 20$ as $b^2 = 22$. $\square$

Work on $K3$ surfaces has focussed on two main areas: firstly, the study of *algebraic* $K3$ surfaces, and secondly, the description of the moduli space of *all* $K3$ surfaces, including the non-algebraic ones. We shall explain the principal results in this second area. Some good general references on the following material are Beauville et al. [26], Barth et al. [19, Ch. VIII] and Besse [30, p. 365–368].

Here are some important tools for studying the moduli space of $K3$ surfaces.

Definition 10.3.4 Let Λ be a lattice isomorphic to $\mathbb{Z}^{22}$, with an even, unimodular quadratic form $q_\Lambda : \Lambda \rightarrow \mathbb{Z}$ of signature $(3, 19)$. All such lattices are isomorphic. A *marked* $K3$ *surface* (X, J, ϕ) is a $K3$ surface (X, J) with an isomorphism $\phi : H^2(X, \mathbb{Z}) \rightarrow \Lambda$ identifying the intersection form q_X on $H^2(X, \mathbb{Z})$ with the quadratic form q_Λ on Λ. Let $\mathcal{M}_{K3}$ be the *moduli space of marked* $K3$ *surfaces*. It is locally a complex manifold of dimension 20, by Theorem 7.7.1. However, note that globally, $\mathcal{M}_{K3}$ is not Hausdorff.

Write $\Lambda_\mathbb{R} = \Lambda \otimes_\mathbb{Z} \mathbb{R}$ and $\Lambda_\mathbb{C} = \Lambda \otimes_\mathbb{Z} \mathbb{C}$. Then ϕ induces isomorphisms $\phi_\mathbb{R} : H^2(X, \mathbb{R}) \rightarrow \Lambda_\mathbb{R}$ and $\phi_\mathbb{C} : H^2(X, \mathbb{C}) \rightarrow \Lambda_\mathbb{C}$. Now $H^{2,0}(X) \cong \mathbb{C}$. We define the *period* of (X, J, ϕ) to be $\phi_\mathbb{C}(H^{2,0}(X))$, which we regard as a point in the complex projective space $P(\Lambda_\mathbb{C})$. Define the *period map* $\mathcal{P} : \mathcal{M}_{K3} \rightarrow P(\Lambda_\mathbb{C})$ by $\mathcal{P} : (X, J, \phi) \mapsto \phi_\mathbb{C}(H^{2,0}(X))$. Then $\mathcal{P}$ is holomorphic.

Now $\mathcal{M}_{K3}$ is a (non-Hausdorff) complex manifold of dimension 20, and $P(\Lambda_\mathbb{C}) \cong \mathbb{CP}^{21}$. Thus we expect $\mathcal{P}(\mathcal{M}_{K3})$ to be a complex hypersurface in $P(\Lambda_\mathbb{C})$. To identify which hypersurface, let (X, J, ϕ) be a marked $K3$ surface, and choose a holomorphic volume form $\omega_\mathbb{C}$ on X. Then $[\omega_\mathbb{C}] \in H^{2,0}(X)$. Define $\lambda_X = \phi_\mathbb{C}([\omega_\mathbb{C}]) \in \Lambda_\mathbb{C}$. As $\omega_\mathbb{C}$ is a $(2,0)$-form we have $\omega_\mathbb{C} \wedge \omega_\mathbb{C} = 0$. Thus

$$q_\Lambda(\lambda_X) = q_X([\omega_c]) = \int_X \omega_c \wedge \omega_c = 0.$$

Near each point in X we can choose holomorphic coordinates (z_1, z_2) such that $\omega_c = dz_1 \wedge dz_2$. Then

$$(\omega_c + \bar{\omega}_c)^2 = (dz_1 \wedge dz_2 + d\bar{z}_1 \wedge d\bar{z}_2)^2 = 2 dz_1 \wedge dz_2 \wedge d\bar{z}_1 \wedge d\bar{z}_2,$$

which is a *positive* 4-form. Hence

$$q_\Lambda(\lambda_X + \bar{\lambda}_X) = q_X([\omega_c] + [\bar{\omega}_c]) = \int_X (\omega_c + \bar{\omega}_c)^2 > 0.$$

We have shown that $q_\Lambda(\lambda_X) = 0$ and $q_\Lambda(\lambda_X + \bar{\lambda}_X) > 0$. But the period of X is $\langle \lambda_X \rangle \in P(\Lambda_\mathbb{C})$. Thus we have found two conditions on the period of (X, J, ϕ).

Definition 10.3.5 Define the *period domain* Q by

$$Q = \{ [\lambda] \in P(\Lambda_\mathbb{C}) : \lambda \in \Lambda_\mathbb{C} \setminus \{0\}, \; q_\Lambda(\lambda) = 0, \; q_\Lambda(\lambda + \bar{\lambda}) > 0 \}. \tag{10.6}$$

From above, if (X, J, ϕ) is a marked $K3$ surface then the period of X lies in Q, and the period map $\mathcal{P}$ maps $\mathscr{M}_{K3} \to Q$.

We now state a series of important results on the period map. The first was published by Kodaira [216, Th. 17], who attributes it to Weil and Andreotti.

Theorem 10.3.6. (Local Torelli Theorem) *The period map* $\mathcal{P} : \mathscr{M}_{K3} \to Q$ *is a local isomorphism of complex manifolds.*

The second is due to Todorov [328, Th. 1] (see also Looijenga [246]).

Theorem 10.3.7 *The period map* $\mathcal{P} : \mathscr{M}_{K3} \to Q$ *is surjective.*

The third is due to Burns and Rapoport [65, Th. 1], and known as the Global Torelli Theorem. We state it in a weak form.

Theorem 10.3.8. (Weak Torelli Theorem) *Let* (X, J, ϕ) *and* (X', J', ϕ') *be marked* $K3$ *surfaces with the same period. Then* $(X, J) \cong (X', J')$.

Notice what is *not* said here: the theorem does not claim that the isomorphism $X \cong X'$ identifies ϕ and ϕ', and this is not in general true. To explain why, we must discuss the *Kähler cones* of $K3$ surfaces.

Definition 10.3.9 Let $\langle \, , \, \rangle_\mathbb{R}$ be the indefinite inner product on $\Lambda_\mathbb{R}$ induced by q_Λ, and $\langle \, , \, \rangle_\mathbb{C}$ the complex inner product on $\Lambda_\mathbb{C}$ induced by q_Λ. Let $\Pi \in P(\Lambda_\mathbb{C})$. Define the *root system* Δ_Π of Π by

$$\Delta_\Pi = \{ \lambda \in \Lambda : q_\Lambda(\lambda) = -2, \; \langle \lambda, p \rangle_\mathbb{C} = 0 \text{ for all } p \in \Pi \}.$$

Define the *Kähler chambers* of Π to be the connected components of

$$\{ \omega \in \Lambda_\mathbb{R} : q_\Lambda(\omega) > 0, \; \langle \omega, p \rangle_\mathbb{C} = 0 \text{ for all } p \in \Pi, \; \langle \omega, \lambda \rangle_\mathbb{R} \neq 0 \text{ for all } \lambda \in \Delta_\Pi \}.$$

It can be shown that the group G of automorphisms of Λ preserving q_Λ and Π acts transitively on the set of Kähler chambers of Π, so the Kähler chambers are really all isomorphic. Now Looijenga [246] proves

Theorem 10.3.10 *Let* (X, J, ϕ) *be a marked* $K3$ *surface with period* Π *and Kähler cone* $\mathcal{K}_X$. *Then* $\phi_\mathbb{R}(\mathcal{K}_X)$ *is one of the Kähler chambers of* Π.

Motivated by this, we define the *augmented period domain* and *map*:

Definition 10.3.11 Define the *augmented period domain* $\tilde{Q}$ by

$$\tilde{Q} = \big\{ (\Pi, C) : \Pi \in Q \text{ and } C \subset \Lambda_\mathbb{R} \text{ is a Kähler chamber of } \Pi \big\},$$

where Q is given in (10.6). Define the *augmented period map* $\tilde{\mathcal{P}} : \mathcal{M}_{K3} \to \tilde{Q}$ by

$$\tilde{\mathcal{P}} : (X, J, \phi) \mapsto \big(\phi_{\mathrm{c}}(H^{2,0}(X)), \phi_\mathbb{R}(\mathcal{K}_X) \big),$$

where $\mathcal{K}_X$ is the Kähler cone of (X, J).

Observe that $\tilde{\mathcal{P}}\big((X, J, \phi)\big)$ lies in $\tilde{Q}$ by Theorem 10.3.7, so $\tilde{\mathcal{P}}$ does map $\mathcal{M}_{K3}$ to $\tilde{Q}$. Looijenga's main result [246] may then be written:

Theorem 10.3.12 *This map* $\tilde{\mathcal{P}} : \mathcal{M}_{K3} \to \tilde{Q}$ *is a 1-1 correspondence.*

Thus we have a very precise description of the moduli space $\mathcal{M}_{K3}$ of marked $K3$ surfaces. Note that $\mathcal{M}_{K3}$ is not Hausdorff. However, the moduli space of *unmarked* $K3$ surfaces is a Hausdorff complex orbifold isomorphic to $Q/\operatorname{Aut}(\Lambda)$.

10.3.3 $K3$ surfaces as hyperkähler 4-manifolds

We are primarily interested in $K3$ surfaces not as complex surfaces (X, J), but as Riemannian 4-manifolds (X, g) with holonomy $\mathrm{Sp}(1)$, or as hyperkähler 4-manifolds (X, J_1, J_2, J_3, g). By combining the above results with Yau's solution of the Calabi Conjecture, we can deduce a great deal about metrics with holonomy $\mathrm{Sp}(1)$ and hyperkähler structures on the $K3$ 4-manifold.

Theorem 10.3.13 *Let* (X, J) *be a* $K3$ *surface. Then each Kähler class on* X *contains a unique metric with holonomy* $\mathrm{SU}(2)$. *Conversely, any compact Riemannian 4-manifold* (X, g) *with* $\mathrm{Hol}(g) = \mathrm{SU}(2) = \mathrm{Sp}(1)$ *admits a constant complex structure* J *such that* (X, J) *is a* $K3$ *surface.*

Proof As X is Kähler and $c_1(X) = 0$ there is a unique Ricci-flat Kähler metric g in every Kähler class on X by Theorem 7.1.2, and $\mathrm{Hol}^0(g) \subseteq \mathrm{SU}(2)$ by Proposition 7.1.1. From Berger's Theorem, the only possibilities for $\mathrm{Hol}^0(g)$ are $\mathrm{SU}(2)$ and $\{1\}$. But X is compact and simply-connected, so $\mathrm{Hol}^0(g) = \mathrm{Hol}(g)$, and $\mathrm{Hol}^0(g) = \{1\}$ is not possible. Hence $\mathrm{Hol}(g) = \mathrm{SU}(2)$.

Now let (X, g) be a compact Riemannian 4-manifold with $\mathrm{Hol}(g) = \mathrm{SU}(2)$. As $\mathrm{SU}(2)$ preserves the complex structure J and $(2,0)$-form $\mathrm{d}z_1 \wedge \mathrm{d}z_2$ on $\mathbb{C}^2$, there exists a constant complex structure J and a constant $(2,0)$-form θ on X. Proposition 7.1.4 shows that $\pi_1(X)$ is finite, so $h^{1,0}(X) = 0$, and K_X is trivial as θ is holomorphic. Thus (X, J) is a $K3$ surface. $\qquad\square$

From §10.1, if (X, g) has $\mathrm{Hol}(g) = \mathrm{Sp}(1)$ there are complex structures J_1, J_2, J_3 on X such that (X, J_1, J_2, J_3, g) is hyperkähler. In the context of the theorem, let (X, J_1, g) be a Kähler surface with $\mathrm{Hol}(g) = \mathrm{Sp}(1)$, and Kähler form ω_1. Choose a constant (2,0)-form $\omega_{\mathbb{C}}$ on X, scaled so that $\omega_{\mathbb{C}} \wedge \bar{\omega}_{\mathbb{C}} = 2\omega_1 \wedge \omega_1$. Then $\omega_2 = \mathrm{Re}\,\omega_{\mathbb{C}}$ and $\omega_3 = \mathrm{Im}\,\omega_{\mathbb{C}}$ are the Kähler forms of complex structures J_2, J_3 on X, and (J_1, J_2, J_3, g) is a hyperkähler structure.

Now this theorem tells us that holonomy $\mathrm{SU}(2)$ metrics exist on any $K3$ surface, but it tells us nothing about what these metrics actually look like. No explicit formulae are known for any holonomy $\mathrm{SU}(2)$ metric on a $K3$ surface, and it seems likely that no such formulae exist; that is, that these metrics are transcendental objects that admit no exact algebraic description.

However, there is one way of getting an *approximate* description of some of these hyperkähler $K3$ metrics, using the Kummer construction of Example 10.3.2. This idea was suggested by Page [275], and made rigorous by LeBrun and Singer [240] and Topiwala [330] using twistor theory, as an alternative proof of the existence of hyperkähler structures on $K3$. We explain it in our next example.

Example 10.3.14 Let the complex torus T^4, involution $\sigma : T^4 \to T^4$, and Kummer surface S resolving $T^4/\langle\sigma\rangle$ be as in Example 10.3.2. Then S is a $K3$ surface, and there is a 20-dimensional family of Kähler metrics on S with holonomy $\mathrm{SU}(2)$, isomorphic to the Kähler cone $\mathcal{K}_S$ of S, by Theorem 10.3.13.

Now $T^4 = \mathbb{C}^2/\Lambda$, where Λ acts on $\mathbb{C}^2$ by translations. Choose a Hermitian metric g_0 on $\mathbb{C}^2$. Then g_0 is invariant under Λ and σ, and so pushes down to give a flat Kähler orbifold metric g_0 on $T^4/\langle\sigma\rangle$. Let $\pi : S \to T^4/\langle\sigma\rangle$ be the blow-up map. Then $\pi^*(g_0)$ is a *singular* Kähler metric on S, which is degenerate at the 16 $\mathbb{CP}^1$ in S introduced by blowing up the 16 singular points of $T^4/\langle\sigma\rangle$.

We can think of $\pi^*(g_0)$ as a point in the boundary $\bar{\mathcal{K}}_S \setminus \mathcal{K}_S$ of the family $\mathcal{K}_S$ of holonomy $\mathrm{SU}(2)$ metrics on S, as there exists a smooth family $\{g_t : t \in (0,1)\}$ of nonsingular holonomy $\mathrm{SU}(2)$ metrics on S such that $g_t \to \pi^*(g_0)$ as $t \to 0$, in a suitable sense. Thus $\pi^*(g_0)$ approximates g_t for small t.

However, $\pi^*(g_0)$ is not a good description of g_t near the 16 $\mathbb{CP}^1$ in S where $\pi^*(g_0)$ is singular. Instead, near each $\mathbb{CP}^1$ we can approximate g_t by an *Eguchi–Hanson metric*, as in Example 10.2.2. Near each $\mathbb{CP}^1$ we can naturally identify S with the blow-up X of $\mathbb{C}^2/\{\pm 1\}$ at 0, and when t is small g_t is close to one of the Eguchi–Hanson metrics on X. Explicitly, we can choose g_t to be close to the Eguchi–Hanson metric on X with Kähler potential

$$f_t = \sqrt{r^4 + t^4} + 2t^2 \log r - t^2 \log\left(\sqrt{r^4 + t^4} + t^2\right),$$

which is a natural 1-parameter generalization of the function f of (10.4). Thus we regard the holonomy $\mathrm{SU}(2)$ metrics g_t on S as being the result of gluing a flat metric on $T^4/\langle\sigma\rangle$ together with 16 small Eguchi–Hanson spaces. The parameter $t > 0$ is a measure of the diameter of the central $\mathcal{S}^2$ in each Eguchi–Hanson space, and the approximation is best when t is small.

This example is the motivation for more complicated 'Kummer constructions' of metrics with special holonomy in higher dimensions—in particular, those of Calabi–

Yau 3-folds described in §7.5.1, and of compact manifolds with exceptional holonomy in Chapter 11.

As we studied marked $K3$ surfaces (X, J, ϕ) above, so we can consider *marked hyperkähler $K3$ surfaces* $(X, J_1, J_2, J_3, g, \phi)$. Here are the appropriate notions of moduli space, period and period domain.

Definition 10.3.15 Let $\mathcal{M}^{\text{hk}}_{K3}$ be the moduli space of marked hyperkähler $K3$ surfaces. Define the *hyperkähler period map* $\mathcal{P}^{\text{hk}} : \mathcal{M}^{\text{hk}}_{K3} \to (\Lambda_\mathbb{R})^3$ by

$$\mathcal{P}^{\text{hk}} : (X, J_1, J_2, J_3, g, \phi) \mapsto \big(\phi_\mathbb{R}([\omega_1]), \phi_\mathbb{R}([\omega_2]), \phi_\mathbb{R}([\omega_3])\big),$$

where $\omega_1, \omega_2, \omega_3$ are the Kähler forms of J_1, J_2, J_3, and $\phi_\mathbb{R}$ maps $H^2(X, \mathbb{R}) \to \Lambda_\mathbb{R}$. Let $\langle \, , \, \rangle_\mathbb{R}$ be as in Definition 10.3.5. Define the *hyperkähler period domain* Q^{hk} by

$$Q^{\text{hk}} = \big\{ (\alpha_1, \alpha_2, \alpha_3) : \alpha_i \in \Lambda_\mathbb{R}, \langle \alpha_i, \alpha_j \rangle_\mathbb{R} = a\, \delta_{ij} \text{ for some } a > 0, \text{ and for each}$$
$$\lambda \in \Lambda \text{ with } q_\Lambda(\lambda) = -2, \text{ there exists } i = 1, 2 \text{ or } 3 \text{ with } \langle \alpha_i, \lambda \rangle_\mathbb{R} \neq 0 \big\}.$$

The reason for requiring $\langle \alpha_i, \alpha_j \rangle_\mathbb{R} = a\, \delta_{ij}$ here is that the 2-forms ω_i satisfy $\omega_i \wedge \omega_j = 2\delta_{ij} dV_g$, where dV_g is the volume form on X. Hence

$$\langle \alpha_i, \alpha_j \rangle_\mathbb{R} = [\omega_i] \cup [\omega_j] = \int_X \omega_i \wedge \omega_j = 2\delta_{ij} \int_X dV_g = 2\,\text{vol}(X)\delta_{ij},$$

so that $\langle \alpha_i, \alpha_j \rangle_\mathbb{R} = a\, \delta_{ij}$ with $a = 2\,\text{vol}(X) > 0$. As a marked $K3$ surface, (X, J_1, ϕ) has period $[\alpha_2 + i\alpha_3] \in P(\Lambda_\mathbb{C})$. So by Theorem 10.3.10, a necessary condition for $(\alpha_1, \alpha_2, \alpha_3)$ to be a hyperkähler period is that α_1 should lie in one of the Kähler chambers of $[\alpha_2 + i\alpha_3]$. By Definition 10.3.9, this implies that $\langle \alpha_i, \lambda \rangle_\mathbb{R} \neq 0$ for some $i = 1, 2, 3$ whenever $\lambda \in \Lambda$ with $q_\Lambda(\lambda) = -2$. Thus we see that the period of $(X, J_1, J_2, J_3, g, \phi)$ lies in Q^{hk}, and $\mathcal{P}^{\text{hk}}$ maps $\mathcal{M}^{\text{hk}}_{K3} \to Q^{\text{hk}}$. From Theorems 10.3.10, 10.3.12 and 10.3.13 one can prove:

Theorem 10.3.16 *This map* $\mathcal{P}^{\text{hk}} : \mathcal{M}^{\text{hk}}_{K3} \to Q^{\text{hk}}$ *is a diffeomorphism.*

This description of the hyperkähler moduli space $\mathcal{M}^{\text{hk}}_{K3}$ is simpler than that of the complex structure moduli space $\mathcal{M}_{K3}$ in §10.3.2, and $\mathcal{M}^{\text{hk}}_{K3}$ is Hausdorff although $\mathcal{M}_{K3}$ is not. One can regard $\mathcal{M}^{\text{hk}}_{K3}$ as more fundamental, and $\mathcal{M}_{K3}$ as derived from it.

Observe also that there is a strong similarity between Theorem 10.3.16, describing the moduli space of hyperkähler $K3$ surfaces, and Theorem 10.2.3, describing the moduli space of hyperkähler ALE spaces asymptotic to $\mathbb{H}/G$.

Here is a geometric interpretation of the condition that $\langle \alpha_i, \lambda \rangle_\mathbb{R} \neq 0$ for some $i = 1, 2, 3$ whenever $\lambda \in \Lambda$ with $q_\Lambda(\lambda) = -2$. Each such λ corresponds to a unique minimal 2-sphere $\mathcal{S}^2$ in X with self-intersection -2. The area A of this $\mathcal{S}^2$ is given by $A^2 = \sum_{j=1}^{3} \langle \alpha_i, \lambda \rangle_\mathbb{R}^2$. If $\langle \alpha_i, \lambda \rangle_\mathbb{R} = 0$ for $i = 1, 2, 3$ then $A = 0$, which is impossible.

That is, the periods for which $\langle \alpha_i, \lambda \rangle_\mathbb{R} = 0$ for $i = 1, 2, 3$ correspond to *singular* hyperkähler structures on X in which an $\mathcal{S}^2$ collapses down to a point, giving a hyperkähler orbifold. We can think of this process as blowing down one or more $\mathbb{CP}^1$'s in a complex surface X to get a singular complex surface X'.

10.4 Higher-dimensional compact hyperkähler manifolds

We now move on to discuss hyperkähler manifolds of dimension $4m$, for $m \geqslant 2$. There are many similarities with $K3$ surfaces. As in §10.3, we find it convenient to first explain the algebraic geometry of the underlying complex manifolds, which are called *complex symplectic manifolds*, and then translate this into results on hyperkähler manifolds. We finish with some examples. Our treatment is based on Huybrechts' papers [168–170], though many of the important ideas are originally due to Beauville [25, §6–§9]. An excellent reference on this area is Huybrechts [138, Part III].

10.4.1 Complex symplectic manifolds

Complex symplectic manifolds are higher-dimensional analogues of $K3$ surfaces.

Definition 10.4.1 A *complex symplectic manifold* $(X, J, \omega_{\mathrm{c}})$ is a compact complex $2m$-manifold (X, J) admitting Kähler metrics, with a closed $(2,0)$-form ω_{c}, such that ω_{c}^m is a nonvanishing $(2m, 0)$-form. We call $(X, J, \omega_{\mathrm{c}})$ *irreducible* if X is simply-connected and cannot be written as a product $X_1 \times X_2$ of lower-dimensional complex manifolds.

Note the assumption that (X, J) *admits Kähler metrics* here, which applies throughout the section. In contrast to the $K3$ case, there do exist non-Kähler complex symplectic $2m$-manifolds for $m \geqslant 2$, and simply-connected examples are given by Guan [143]. But they are not of interest to us, as they carry no metrics with holonomy $\mathrm{Sp}(m)$.

We shall discuss *deformations* of complex symplectic manifolds. If $(X, J, \omega_{\mathrm{c}})$ is an irreducible complex symplectic manifold, then the canonical bundle of X is trivial, so Theorem 7.7.1 shows that the local moduli space $\mathcal{M}$ of deformations of (X, J) is a complex manifold of dimension $h^{2m-1,1}(X)$. But we shall prove in (10.8) that $h^{2m-1,1}(X) = h^{1,1}(X)$, so $\dim \mathcal{M} = h^{1,1}(X)$. It can also be shown [25, §8], [168, §2.4] that small deformations of (X, J) are also irreducible complex symplectic manifolds. Thus we have:

Theorem 10.4.2 *Let $(X, J, \omega_{\mathrm{c}})$ be an irreducible complex symplectic manifold. Then the moduli space of deformations of (X, J) is locally a complex manifold of dimension $h^{1,1}(X)$. Every small deformation (X', J') of (X, J) has a $(2,0)$-form ω_{c}' such that $(X', J', \omega_{\mathrm{c}}')$ is an irreducible complex symplectic manifold.*

There is a natural quadratic form on the second cohomology of a complex symplectic manifold, due to Beauville [25, Th. 5]. The final part is from Fujiki [115, Th. 4.7].

Theorem 10.4.3 *Let $(X, J, \omega_{\mathrm{c}})$ be an irreducible complex symplectic $2m$-fold, scaled so that $\int_X \omega_{\mathrm{c}}^m \wedge \bar{\omega}_{\mathrm{c}}^m = 1$. Define a quadratic form f on $H^2(X, \mathbb{R})$ by*

$$
\begin{aligned}
f\big([\alpha]\big) = {} & \tfrac{m}{2} \int_X \omega_{\mathrm{c}}^{m-1} \wedge \bar{\omega}_{\mathrm{c}}^{m-1} \wedge \alpha^2 \\
& + (1 - m)\big(\textstyle\int_X \omega_{\mathrm{c}}^{m-1} \wedge \bar{\omega}_{\mathrm{c}}^m \wedge \alpha\big) \cdot \big(\textstyle\int_X \omega_{\mathrm{c}}^m \wedge \bar{\omega}_{\mathrm{c}}^{m-1} \wedge \alpha\big).
\end{aligned}
$$

Then there exists a unique constant $c > 0$ such that $q_X = c \cdot f$ is a primitive integral quadratic form on $H^2(X, \mathbb{Z})$, of index $\big(3, b^2(X) - 3\big)$. Furthermore, there is a rational number $C_X > 0$ such that $\int_X \alpha^{2m} = C_X q_X([\alpha])^m$ for all closed 2-forms α on X.

The form q_X is called *Beauville's form*, and the constant C_X *Fujiki's constant.* When $m = 1$, so that X is a $K3$ surface, q_X is just the intersection form on $H^2(X, \mathbb{Z})$. As q_X is integer-valued on $H^2(X, \mathbb{Z})$ it is invariant under continuous deformations of the complex symplectic structure on X, even though the definition appears to depend on the cohomology class $[\omega_c] \in H^2(X, \mathbb{C})$ for $m > 1$.

Definition 10.4.4 Let Λ be a lattice, so that $\Lambda \cong \mathbb{Z}^d$, and let q_Λ be a primitive integral quadratic form on Λ of index $(3, d - 3)$. Let (X, J, ω_c) be an irreducible complex symplectic manifold with $b^2(X) = d$. A *marking* of X is an isomorphism $\phi : H^2(X, \mathbb{Z}) \to \Lambda$ identifying the quadratic forms q_X on $H^2(X, \mathbb{Z})$ and q_Λ on Λ. We call (X, J, ω_c, ϕ) a *marked irreducible complex symplectic manifold.*

We would like to understand the moduli space of all marked irreducible complex symplectic manifolds (X, J, ω_c, ϕ). In fact it is more convenient to study the underlying marked complex manifolds (X, J, ϕ).

Definition 10.4.5 Let (X, J, ω_c, ϕ) be a marked irreducible complex symplectic manifold. Define $\mathscr{M}_X$ to be the moduli space of *marked complex deformations* (X', J', ϕ') of (X, J, ϕ). That is, (X', J') is a deformation of (X, J) admitting a $(2,0)$-form ω'_c making (X', J', ω'_c) into an irreducible complex symplectic manifold, and ϕ' is a marking of (X', J', ω'_c). Then $\mathscr{M}_X$ is locally a complex manifold of dimension $h^{1,1}(X) = d - 2$, by Theorem 10.4.2. However, globally the topology of $\mathscr{M}_X$ may not be Hausdorff; that is, $\mathscr{M}_X$ is a *non-Hausdorff complex manifold.*

An important tool in studying the moduli space $\mathscr{M}_X$ is the *period map.*

Definition 10.4.6 Let (X, J, ω_c, ϕ) be a marked irreducible complex symplectic manifold, with lattice Λ. Write $\Lambda_{\mathbb{R}} = \Lambda \otimes_{\mathbb{Z}} \mathbb{R}$ and $\Lambda_{\mathbb{C}} = \Lambda \otimes_{\mathbb{Z}} \mathbb{C}$. Then ϕ induces isomorphisms $\phi_{\mathbb{R}} : H^2(X, \mathbb{R}) \to \Lambda_{\mathbb{R}}$ and $\phi_{\mathbb{C}} : H^2(X, \mathbb{C}) \to \Lambda_{\mathbb{C}}$. Now $H^{2,0}(X) \cong \mathbb{C}$. We define the *period* of (X, J, ω_c, ϕ) to be $\phi_{\mathbb{C}}\big(H^{2,0}(X)\big)$, which we regard as a point in the complex projective space $P(\Lambda_{\mathbb{C}})$.

Observe that the period depends only on (X, J, ϕ), and not on ω_c. Let $\mathscr{M}_X$ be as above, and define the *period map* $\mathcal{P} : \mathscr{M}_X \to P(\Lambda_{\mathbb{C}})$ by $\mathcal{P} : (X', J', \phi') \mapsto \phi'_{\mathbb{C}}\big(H^{2,0}(X')\big)$. It is easy to show that $\mathcal{P}$ is holomorphic. From the definition of q_X in Theorem 10.4.3, we find that $q_X(\omega_c) = 0$, and $q_X(\omega_c + \bar{\omega}_c) > 0$. Therefore, if as in (10.6) we define $Q \subset P(\Lambda_{\mathbb{C}})$ by

$$Q = \big\{ [\lambda] \in P(\Lambda_{\mathbb{C}}) : \lambda \in \Lambda_{\mathbb{C}} \setminus \{0\},\ q_\Lambda(\lambda) = 0,\ q_\Lambda(\lambda + \bar{\lambda}) > 0 \big\}, \qquad (10.7)$$

then $\mathcal{P}$ maps $\mathscr{M}_X \to Q$. We call Q the *period domain.*

Analogues of Theorems 10.3.6 and 10.3.7 (the Local Torelli Theorem and the Surjectivity Theorem for $K3$ surfaces) were proved by Beauville [25, Th. 5(b)] and Huybrechts [168, Th. 8.1], and we give them in our next result.

Theorem 10.4.7 *The period map $\mathcal{P} : \mathscr{M}_X \to Q$ is a local isomorphism of complex manifolds. Let $\mathscr{M}_X^0$ be a nonempty connected component of $\mathscr{M}_X$. Then $\mathcal{P} : \mathscr{M}_X^0 \to Q$ is surjective.*

No good analogue of the Global Torelli Theorem ([65, Th. 1], partially stated in Theorem 10.3.8) is currently known. In fact Theorem 10.3.8 is false for complex symplectic manifolds in complex dimension 4 and above, as Debarre has found examples of complex symplectic manifolds with the same period, which are birational but not biholomorphic. For more information see Huybrechts [138, §25.5]. Here is a result of Huybrechts [170, §2] on birational complex symplectic manifolds.

Theorem 10.4.8 *Suppose* $(X, J, \omega_c), (X', J', \omega'_c)$ *are birational irreducible complex symplectic manifolds. Then* X, X' *are deformation equivalent, so they are diffeomorphic and have the same Betti and Hodge numbers, and the Hodge structures and period maps of* X, X' *coincide.*

As for $K3$ surfaces in Theorem 10.3.10, Huybrechts [170, §3] and Boucksom [41] identify the *Kähler cone* of an irreducible complex symplectic manifold.

Theorem 10.4.9 *Let* (X, J, ω_c) *be an irreducible complex symplectic manifold with Beauville form* q_X *and Kähler cone* $\mathcal{K}_X$. *Define* $\mathcal{C}_X$ *to be the connected component of* $\{\alpha \in H^{1,1}(X, \mathbb{R}) : q_X(\alpha) > 0\}$ *containing* $\mathcal{K}_X$. *Then* $\mathcal{K}_X$ *is the set of* $\alpha \in \mathcal{C}_X$ *with* $\int_C \alpha > 0$ *for all smooth rational curves* C *in* X.

If X has no rational curves, which holds for generic X, this simplifies to $\mathcal{K}_X = \mathcal{C}_X$.

10.4.2 General theory of compact hyperkähler manifolds

Here are three topological results on compact manifolds with holonomy $\mathrm{Sp}(m)$.

Proposition 10.4.10 *Let* (X, g) *be a compact* $4m$-*manifold with* $\mathrm{Hol}(g) = \mathrm{Sp}(m)$. *Then* X *is simply-connected and has* $\hat{A}$-*genus* $\hat{A}(X) = m + 1$.

Proof Theorem 3.6.5 shows that $\hat{A}(X) = m+1$, and Corollary 3.5.6 shows that $\pi_1(X)$ is finite. Let $(\tilde{X}, \tilde{g})$ be the universal cover of (X, g), and $d = |\pi_1(X)|$ the degree of the cover. Then $\tilde{X}$ is compact, as $\pi_1(X)$ is finite, and $\mathrm{Hol}(\tilde{g}) = \mathrm{Hol}^0(g) = \mathrm{Sp}(m)$. Thus $\hat{A}(\tilde{X}) = m + 1$ as above. But $\hat{A}(\tilde{X}) = d \cdot \hat{A}(X)$ by properties of characteristic classes. So $m + 1 = d(m + 1)$, giving $d = 1$, and X is simply-connected. □

By analogy with Proposition 7.1.7 we have:

Proposition 10.4.11 *Let* (X, J, g) *be a compact Kähler manifold of dimension* $2m$ *with* $\mathrm{Hol}(g) = \mathrm{Sp}(m)$, *and let* $h^{p,q}$ *be the Hodge numbers of* X. *Then*

$$h^{2k,0} = 1 \text{ for } 0 \leqslant k \leqslant m, \quad h^{2k+1,0} = 0 \text{ for all } k, \text{ and } h^{p,q} = h^{2m-p,q}. \quad (10.8)$$

Proof From §10.1.1, $\mathrm{Sp}(m)$ is the subgroup of $\mathrm{SU}(2m)$ fixing a complex symplectic form $\omega_2 + i\omega_3$ in $\Lambda^{2,0}\mathbb{C}^{2m}$. Thus $\mathrm{Sp}(m)$ also fixes the powers $(\omega_2 + i\omega_3)^k$ in $\Lambda^{2k,0}\mathbb{C}^{2m}$, for $k = 0, 1, \ldots, m$. Any form in $\Lambda^{p,0}\mathbb{C}^{2m}$ fixed by $\mathrm{Sp}(m)$ is proportional to some $(\omega_2 + i\omega_3)^k$. Therefore by Proposition 7.1.5, if $\mathrm{Hol}(g) = \mathrm{Sp}(m)$ then $H^{2k,0}(X)$ is $\mathbb{C}$ if $k = 0, \ldots, m$ and $H^{2k+1,0}(X) = 0$.

Suppose that $0 \leqslant p \leqslant m$, as otherwise we may replace p by $2m - p$. Define a linear map $\phi : \Lambda^{p,q}\mathbb{C}^{2m} \to \Lambda^{2m-p,q}\mathbb{C}^{2m}$ by $\phi(\alpha) = \alpha \wedge (\omega_2 + i\omega_3)^{m-p}$. Then ϕ is a vector space isomorphism, and as $\mathrm{Sp}(m)$ preserves $\omega_2 + i\omega_3$, it is also an isomorphism of $\mathrm{Sp}(m)$-representations. Thus Theorem 3.5.3 shows that $H^{p,q}(X) \cong H^{2m-p,q}(X)$, and so $h^{p,q} = h^{2m-p,q}$. □

Combining the last part of the proposition with (5.10), we find that

$$h^{p,q} = h^{2m-p,q} = h^{p,2m-q} = h^{2m-p,2m-q} = h^{q,p} = h^{2m-q,p} = h^{q,2m-p} = h^{2m-q,2m-p}.$$

A great deal is known about the cohomology of compact hyperkähler manifolds, from work of Beauville, Bogomolov, Fujiki, Guan, Looijenga and Lunts, Salamon, Verbitsky, Wakakuwa and others; details and references can be found in Huybrechts [138, §24]. For example, the odd Betti numbers of a compact hyperkähler manifold are divisible by 4, and for a compact hyperkähler 8-manifold we have $3 \leqslant b^2 \leqslant 8$ or $b^2 = 23$, and $0 \leqslant 2b^3 \leqslant (23 - b^2)(b^2 + 4)$, and $b^4 = 10b^2 - b^3 + 46$, so there are only finitely many possibilities for the Betti numbers in 8 dimensions. For results suggesting there are only finitely many deformation types of compact hyperkähler manifolds in any given dimension, see Huybrechts [138, §26.5].

Next we discuss the relation between compact manifolds with holonomy $\mathrm{Sp}(m)$ and irreducible complex symplectic manifolds.

Theorem 10.4.12 *Let (X, J, ω_c) be an irreducible complex symplectic $2m$-fold. Then there is a unique metric with holonomy $\mathrm{Sp}(m)$ in each Kähler class on X. Conversely, if (X, g) is a compact Riemannian $4m$-manifold with holonomy $\mathrm{Sp}(m)$ then there exist a constant complex structure J and $(2, 0)$-form ω_c on X such that (X, J, ω_c) is an irreducible complex symplectic $2m$-fold.*

Proof Observe that ω_c^m is a nonvanishing holomorphic section of the canonical bundle K_X of X. Thus K_X is trivial, and $c_1(X) = 0$. So by Theorem 7.1.2, each Kähler class κ contains a unique Ricci-flat Kähler metric g. Let ∇ be the Levi-Civita connection of g. Then Proposition 7.1.5 shows that $\nabla\omega_c = 0$, as ω_c is a closed $(2,0)$-form.

Therefore $\nabla g = \nabla J = \nabla\omega_c = 0$, and $\mathrm{Hol}(g)$ preserves a metric, complex structure and complex symplectic form on $\mathbb{C}^{2m}$. This forces $\mathrm{Hol}(g) \subseteq \mathrm{Sp}(m)$. But g is irreducible, as X is irreducible, and from the classification of holonomy groups we see that $\mathrm{Hol}(g) = \mathrm{Sp}(m)$. Any holonomy $\mathrm{Sp}(m)$ metric is Ricci-flat, so g is the only metric in κ with holonomy $\mathrm{Sp}(m)$, by Theorem 7.1.2.

Now let (X, g) be a compact Riemannian $4m$-manifold with holonomy $\mathrm{Sp}(m)$. Then from §10.1 there exist a constant complex structure $J = J_1$ and $(2,0)$-form $\omega_c = \omega_2 + i\omega_3$ on X. Clearly (X, J, ω_c) is a complex symplectic manifold. Also $\pi_1(X) = \{1\}$ by Proposition 10.4.10, and X cannot be written as a product of lower-dimensional complex manifolds, as this would force g to be reducible. Hence (X, J, ω_c) is irreducible. □

Now in §10.3.3 we applied results on the moduli of $K3$ surfaces to describe the moduli space of hyperkähler structures on $K3$. In a similar way, we can apply Theorem 10.4.7 and other results in [168–170] to describe the moduli space of hyperkähler structures on X. But as we lack both a Global Torelli Theorem and a simple picture of the Kähler cone, our results will not be as strong. Following Definition 10.3.15, we define:

Definition 10.4.13 Let $\mathcal{M}_X^{\mathrm{hk}}$ be the moduli space of marked hyperkähler structures on X. Define the *hyperkähler period map* $\mathcal{P}^{\mathrm{hk}} : \mathcal{M}_X^{\mathrm{hk}} \to (\Lambda_{\mathbb{R}})^3$ by

$$\mathcal{P}^{\mathrm{hk}} : (X, J_1, J_2, J_3, g, \phi) \mapsto \big(\phi_{\mathbb{R}}([\omega_1]), \phi_{\mathbb{R}}([\omega_2]), \phi_{\mathbb{R}}([\omega_3])\big),$$

where $\omega_1, \omega_2, \omega_3$ are the Kähler forms of J_1, J_2, J_3, and $\phi_{\mathbb{R}}$ maps $H^2(X, \mathbb{R}) \to \Lambda_{\mathbb{R}}$. Let $\langle \, , \, \rangle_{\mathbb{R}}$ be as in Definition 10.3.5. Define the *hyperkähler period domain* Q_X^{hk} by

$$Q_X^{\mathrm{hk}} = \big\{(\alpha_1, \alpha_2, \alpha_3) : \alpha_i \in \Lambda_{\mathbb{R}}, \ \langle \alpha_i, \alpha_j \rangle_{\mathbb{R}} = a\,\delta_{ij} \text{ for some } a > 0\big\}.$$

Here is a rough analogue of Theorem 10.3.16.

Theorem 10.4.14 $\mathcal{P}^{\mathrm{hk}} : \mathcal{M}_X^{\mathrm{hk}} \to Q_X^{\mathrm{hk}}$ *is a local diffeomorphism, with image a dense open set in* Q_X^{hk}.

It is not yet known whether $\mathcal{P}^{\mathrm{hk}}$ is injective, nor what its image is in Q_X^{hk}.

10.4.3   Examples of compact hyperkähler manifolds

In comparison to Calabi–Yau manifolds, examples of compact hyperkähler manifolds are difficult to find, and only a few are known in each dimension. The first examples were two series of manifolds due to Beauville [25, §6–§7] and [24, §2], which generalize an example of Fujiki [114] in real dimension 8. We now explain Beauville's examples.

Let X be a compact complex surface. For $m > 1$, define X^m to be the product $X \times X \times \cdots \times X$ of m copies of X, and $X^{(m)}$ to be the mth symmetric product of X, that is, $X^{(m)} = X^m / S_m$, where S_m is the symmetric group acting on X^m by permutations. Then $X^{(m)}$ is a *complex orbifold*, of dimension $2m$.

Let $X^{[m]}$ be the *Hilbert scheme* of zero-dimensional subspaces $(Z, \mathcal{O}_Z)$ of X of length $\dim_{\mathbb{C}} \mathcal{O}_Z = m$. Then $X^{[m]}$ is a compact, nonsingular complex $2m$-manifold, with a projection $\pi : X^{[m]} \to X^{(m)}$ which is a *crepant resolution*. By results of Varouchas [332], $X^{[m]}$ is Kähler whenever X is Kähler.

Now suppose that X is a $K3$ surface or a complex torus T^4, so that X is complex symplectic. Then X^m and $X^{(m)}$ both have complex symplectic forms. As $X^{[m]}$ is a crepant resolution of $X^{(m)}$, it also has a complex symplectic form, and it admits Kähler metrics from above, so $X^{[m]}$ is a complex symplectic manifold.

If X is a $K3$ surface then Beauville [25, Prop. 6] shows that $X^{[m]}$ is also irreducible, with $b^2 = 23$. Applying Theorems 10.4.12 and 10.4.14, we find:

Theorem 10.4.15 Let X be a $K3$ surface and $m \geqslant 2$. Then the Hilbert scheme $X^{[m]}$ is an irreducible complex symplectic $2m$-manifold, with $b^2(X^{[m]}) = 23$. There exists a 61-dimensional family of metrics g on $X^{[m]}$ with holonomy $\mathrm{Sp}(m)$, and a 64-dimensional family of hyperkähler structures (J_1, J_2, J_3, g).

These were the first known examples of compact Riemannian manifolds with holonomy $\mathrm{Sp}(m)$ for $m \geqslant 2$. If Y is a complex torus T^4 then $Y^{[m]}$ is complex symplectic, but not irreducible. Instead, $Y^{[m]}$ has a finite cover isomorphic to $K^{m-1}(Y) \times Y$, where $K^{m-1}(Y)$ is an irreducible complex symplectic $(2m-2)$-manifold. To define it explicitly, regard Y as an abelian Lie group, so there is a natural map $\Sigma : Y^{(m)} \to$

Y given by summing the m points. Then $K^{m-1}(Y)$ is the kernel of the composition $Y^{[m]} \xrightarrow{\pi} Y^{(m)} \xrightarrow{\Sigma} Y$.

Now $K^1(Y)$ is a $K3$ surface, got from Y by the Kummer construction as in Example 10.3.2. But for $m \geqslant 2$, Beauville [25, Prop. 8] shows that $K^m(Y)$ is an irreducible complex symplectic manifold with $b^2 = 7$. So as above we get:

Theorem 10.4.16 *Let Y be a complex torus T^4 and $m \geqslant 2$. Then $K^m(Y)$ is an irreducible complex symplectic $2m$-manifold, with $b^2(K^m(Y)) = 7$. There exists a smooth 13-dimensional family of metrics g on $K^m(Y)$ with holonomy $\mathrm{Sp}(m)$, and a smooth 16-dimensional family of hyperkähler structures (J_1, J_2, J_3, g).*

The Betti and Hodge numbers of the $X^{[m]}$ and $K^m(Y)$ are known by work of Göttsche and Sorgel [130]. As compact real manifolds, $X^{[m]}$ and $K^m(Y)$ are independent of the choice of $K3$ surface X and complex torus Y. Thus Theorems 10.4.15 and 10.4.16 give only two distinct compact $4m$-manifolds with holonomy $\mathrm{Sp}(m)$, for each $m \geqslant 2$. Several other constructions of such manifolds are described in [168, §2], but they are all deformation equivalent (and hence diffeomorphic) to $X^{[m]}$ or $K^m(Y)$.

Two further examples of compact hyperkähler manifolds have been constructed by O'Grady [271, 272]. The first [271] is an irreducible complex symplectic 10-manifold $\widetilde{\mathcal{M}}$ constructed as a crepant resolution of a moduli space $\mathcal{M}$ of a certain kind of sheaf on a $K3$ surface. Thus $\widetilde{\mathcal{M}}$ carries metrics with holonomy $\mathrm{Sp}(5)$, by Theorem 10.4.12. O'Grady proved that $b^2(\widetilde{\mathcal{M}}) \geqslant 24$, so that $\widetilde{\mathcal{M}}$ is not diffeomorphic to $X^{[5]}$ or $K^5(Y)$.

The second [272] is an irreducible complex symplectic 6-manifold $\widehat{\mathcal{M}}$ constructed in a similar way from a moduli space of sheaves on an abelian surface, and admits metrics with holonomy $\mathrm{Sp}(3)$. O'Grady proved that $b^2(\widehat{\mathcal{M}}) = 8$, so that $\widehat{\mathcal{M}}$ is not diffeomorphic to either $X^{[3]}$ or $K^3(Y)$. Rapagnetta [283] studies this example further, and shows that $\chi(\widehat{\mathcal{M}}) = 1920$. A related, conjectural approach to finding new examples of compact hyperkähler manifolds is described by Verbitsky in [334].

10.5 Quaternionic Kähler manifolds

Next we consider the holonomy group $\mathrm{Sp}(m)\,\mathrm{Sp}(1)$ in $\mathrm{SO}(4m)$. Some good references are Salamon [294, 297]. First we explain how $\mathrm{Sp}(m)\,\mathrm{Sp}(1)$ acts on $\mathbb{H}^m$. Regard $\mathrm{Sp}(m)$ as a group of $m \times m$ matrices over the quaternions $\mathbb{H}$ as in (10.2), and $\mathbb{H}^m$ as row matrices over $\mathbb{H}$. Then $(A, q) \in \mathrm{Sp}(m) \times \mathrm{Sp}(1)$ acts on $\mathbb{H}^m$ by $(q^1\, q^2 \cdots q^m) \mapsto q(q^1\, q^2 \cdots q^m)\bar{A}^t$, generalizing the $\mathrm{Sp}(m)$ action in §10.1.1. Write I_m for the $m \times m$ identity matrix. Then $(-I_m, -1) \in \mathrm{Sp}(m) \times \mathrm{Sp}(1)$ acts trivially on $\mathbb{H}^m$, so the action pushes down to an action of $\mathrm{Sp}(m)\,\mathrm{Sp}(1) = \big(\mathrm{Sp}(m) \times \mathrm{Sp}(1)\big)/\{\pm(I_m, 1)\}$.

Note that as multiplication in $\mathbb{H}$ is not commutative, left multiplying $(q^1 \cdots q^m)$ by q is not the same as right multiplying by $q I_m$, so the $\mathrm{Sp}(1)$ action is not part of the $\mathrm{Sp}(m)$ action. Since $\mathrm{Sp}(1)\,\mathrm{Sp}(1) = \mathrm{SO}(4)$, which is uninteresting as a holonomy group, we restrict for the moment to $m \geqslant 2$.

Definition 10.5.1 A *quaternionic Kähler manifold* of dimension $4m \geqslant 8$ is a Riemannian $4m$-manifold (X, g) with holonomy group $\mathrm{Hol}(g) \subseteq \mathrm{Sp}(m)\,\mathrm{Sp}(1)$.

Quaternionic Kähler manifolds are *not* in general Kähler, as $\mathrm{Sp}(m)\,\mathrm{Sp}(1)$ is not a subgroup of $\mathrm{U}(2m)$. From [296, Cor. 9.4] we have:

Proposition 10.5.2 *Any quaternionic Kähler manifold* (X, g) *is Einstein, and if* g *is Ricci-flat then* $\mathrm{Hol}^0(g) \subseteq \mathrm{Sp}(m)$.

Thus a quaternionic Kähler manifold (X, g) has constant scalar curvature s. The cases $s > 0$ and $s < 0$ are called *positive* and *negative quaternionic Kähler manifolds* respectively. We do not consider $s = 0$, as then g is (locally) hyperkähler. Define 2-forms $\omega_1, \omega_2, \omega_3$ on $\mathbb{H}^m = \mathbb{R}^{4m}$ as in (10.1). Then $\mathrm{Sp}(m)\,\mathrm{Sp}(1)$ preserves the 4-form

$$\Omega_0 = \omega_1 \wedge \omega_1 + \omega_2 \wedge \omega_2 + \omega_3 \wedge \omega_3$$

on $\mathbb{R}^{4m}$. So by Proposition 2.5.2, every quaternionic Kähler manifold (X, g) has a constant 4-form Ω isomorphic to Ω_0 at each point. The stabilizer group of Ω_0 in $\mathrm{GL}(4m, \mathbb{R})$ is $\mathrm{Sp}(m)\,\mathrm{Sp}(1)$, so $\mathrm{Sp}(m)\,\mathrm{Sp}(1)$-structures on a $4m$-manifold X are equivalent to 4-forms Ω on X isomorphic to Ω_0 at each point. By a result of Swann, when $m \geqslant 3$ the $\mathrm{Sp}(m)\,\mathrm{Sp}(1)$-structure is torsion-free if and only if $\mathrm{d}\Omega = 0$, that is, $\mathrm{d}\Omega = 0$ is equivalent to $\nabla\Omega = 0$. But $\mathrm{d}\Omega = 0$ is weaker than $\nabla\Omega = 0$ when $m = 2$.

As for hyperkähler manifolds in §10.1.3, quaternionic Kähler manifolds have complex *twistor spaces*. Let (X, g) be a quaternionic Kähler $4m$-manifold. Then X has a $\mathrm{Sp}(m)\,\mathrm{Sp}(1)$-*structure*, a principal $\mathrm{Sp}(m)\,\mathrm{Sp}(1)$-bundle P which is a subbundle of the frame bundle of X. Define $Z = P/\mathrm{Sp}(m)\mathrm{U}(1)$, the quotient by the subgroup $\mathrm{Sp}(m)\mathrm{U}(1)$ in $\mathrm{Sp}(m)\,\mathrm{Sp}(1)$. Then Z is a bundle over X with projection $\pi : Z \to X$ and fibre $\mathrm{Sp}(m)\,\mathrm{Sp}(1)/\mathrm{Sp}(m)\mathrm{U}(1) \cong \mathrm{Sp}(1)/\mathrm{U}(1) \cong \mathcal{S}^2$.

Points $z \in Z$ may be identified with complex structures on tangent spaces $T_{\pi(z)}X$ in the following way. If $x \in X$ then there is a nonunique identification $T_xX \cong \mathbb{R}^{4m}$ identifying the metric g and 4-form Ω on T_xX with the Euclidean versions g_0, Ω_0 on $\mathbb{R}^{4m}$. Then the fibre $\pi^{-1}(x)$ is naturally identified with the family of complex structures $a_1J_1 + a_2J_2 + a_3J_3$ on $\mathbb{R}^{4m}$ for $(a_1, a_2, a_3) \in \mathcal{S}^2 \subset \mathbb{R}^3$, where J_1, J_2, J_3 are as in §10.1.1. Alternative choices of identification $T_xX \cong \mathbb{R}^{4m}$ are related by the action of $\mathrm{Sp}(m)\,\mathrm{Sp}(1)$ on $\mathbb{R}^{4m}$, and as $\mathrm{Sp}(m)\,\mathrm{Sp}(1)$ preserves the family of complex structures $a_1J_1 + a_2J_2 + a_3J_3$ the construction is well-defined.

The fundamental theorem on twistor spaces is due to Salamon [294].

Theorem 10.5.3 *Let* (X, g) *be a quaternionic Kähler* $4m$-*manifold, and* Z, π *be as above. Then* Z *has the following natural geometric structures:*

 (i) *An integrable complex structure* J *making* Z *into a complex* $(2m + 1)$-*manifold, such that for each* $x \in X$ *the fibre* $\pi^{-1}(x)$ *is a complex curve* $\mathbb{CP}^1$ *in* Z *with normal bundle* $2m\mathcal{O}(1)$.

 (ii) *An antiholomorphic involution* $\sigma : Z \to Z$ *with* $\pi \circ \sigma = \pi$, *acting as* $\sigma : J \mapsto -J$ *under the identification of points* $z \in Z$ *with complex structures* J *on* $T_{\pi(z)}X$.

 (iii) *If* g *has nonzero scalar curvature then* Z *carries a holomorphic contact structure* τ.

 (iv) *If* g *has positive scalar curvature then* Z *has a Kähler–Einstein metric* h *with positive scalar curvature.*

This should be compared with Theorem 10.1.3. As for Theorem 10.1.4, the quaternionic Kähler structure on X can be recovered uniquely (up to homothety) from the holomorphic data Z, J, σ, τ. Thus twistor spaces are powerful tools for studying quaternionic Kähler manifolds, particularly in the positive case. Theorem 10.5.3 also suggests the appropriate definition of quaternionic Kähler manifolds in dimension 4. By Atiyah, Hitchin and Singer [12], self-dual Einstein metrics on 4-manifolds have twistor spaces Z with these properties. So we define:

Definition 10.5.4 A *quaternionic Kähler manifold* of dimension 4 is an oriented Riemannian 4-manifold (X, g) which is Einstein with self-dual Weyl curvature.

Then Theorem 10.5.3 holds in dimension 4. Other aspects of quaternionic Kähler geometry are also nicely compatible with this definition, for instance, the *quaternionic Kähler quotient construction* of Galicki and Lawson [121] produces quaternionic Kähler manifolds in dimension $4m \geqslant 8$ and self-dual Einstein manifolds in dimension 4.

The most basic examples of positive quaternionic Kähler manifolds are the *Wolf spaces*, a family of compact quaternionic Kähler symmetric spaces described in 1965 by Wolf [341], who also discussed their twistor spaces. There is one Wolf space for each compact simple Lie group. In dimension $4m$ there are three families:

$$\mathbb{HP}^m = \frac{\mathrm{Sp}(m+1)}{\mathrm{Sp}(m) \times \mathrm{Sp}(1)}, \quad \mathrm{Gr}(2, \mathbb{C}^{m+2}) = \frac{\mathrm{U}(m+2)}{\mathrm{U}(m) \times \mathrm{U}(2)}, \quad \mathrm{Gr}(4, \mathbb{R}^{m+4}) = \frac{\mathrm{SO}(m+4)}{\mathrm{SO}(m) \times \mathrm{SO}(4)}.$$

These satisfy $\mathbb{HP}^1 \cong \mathcal{S}^4 \cong \mathrm{Gr}(4, \mathbb{R}^5)$ when $m = 1$ and $\mathrm{Gr}(2, \mathbb{C}^4) \cong \mathrm{Gr}(4, \mathbb{R}^7)$ when $m = 2$. In addition, for $m = 2, 7, 10, 16, 28$ there are the exceptional spaces

$$\frac{G_2}{\mathrm{SO}(4)}, \quad \frac{F_4}{\mathrm{Sp}(3)\,\mathrm{Sp}(1)}, \quad \frac{E_6}{\mathrm{SU}(6)\,\mathrm{Sp}(1)}, \quad \frac{E_7}{\mathrm{Spin}(12)\,\mathrm{Sp}(1)}, \quad \frac{E_8}{E_7\,\mathrm{Sp}(1)}.$$

Each Wolf space also has a noncompact dual, which is a negative quaternionic Kähler manifold. Note that the holonomy groups of the Wolf spaces are of the form $K\,\mathrm{Sp}(1)$ for various Lie subgroups $K \subseteq \mathrm{Sp}(m)$; this does not contradict Theorem 3.4.1, as the Wolf spaces are symmetric.

By a result of Myers [30, §6.E] on positive Ricci curvature, any *complete* positive quaternionic Kähler manifold is compact, with finite fundamental group. Theorem 10.5.3 shows that the twistor space Z of a compact, positive quaternionic Kähler manifold (X, g) is a compact Kähler–Einstein manifold with positive scalar curvature. Thus it has ample anticanonical bundle K_Z^{-1}, and so is projective, and is by definition a *Fano variety*, and also a complex contact manifold. Such Z are very special objects in algebraic geometry, and a lot is known about them.

It is conjectured that the Wolf spaces are the only compact positive quaternionic Kähler manifolds, that is, that every compact positive quaternionic Kähler manifold is symmetric. The following theorem collects the state of play at the time of writing.

Theorem 10.5.5 *A compact positive quaternionic Kähler manifold* (X, g) *of dimension* $4m$ *is a Wolf space if one of*: (i) $m \leqslant 3$, (ii) $b^2 \geqslant 1$, *or* (iii) $m = 4$ *and* $b^4 = 1$.

Any compact positive quaternionic Kähler manifold has odd Betti numbers zero. Up to homothety there are only finitely many compact positive quaternionic Kähler manifolds of dimension $4m$, *for each* $m \geqslant 1$.

Proof Part (i) follows from Hitchin [161] when $m = 1$, Poon and Salamon [282] when $m = 2$ and Herrera and Herrera [158] when $m = 3$, using twistor space methods and index theory. Part (ii) follows from LeBrun and Salamon [239, Th. 0.3], and (iii) from Galicki and Salamon [122, Th. 5.1]. The last two parts come from Salamon [294, Th. 6.6] and LeBrun and Salamon [239]. □

There exist many examples of nonsymmetric, singular positive or negative quaternionic Kähler manifolds. In particular, the *quaternionic Kähler quotient construction* of Galicki and Lawson [121] produces examples of compact, positive quaternionic Kähler *orbifolds*, with only quotient singularities.

In the negative case, Alekseevsky [8] found a large class of complete, nonsymmetric, negative quaternionic Kähler manifolds, which are homogeneous metrics on solvable Lie groups. It is not known whether these admit compact quotients. LeBrun [237] found an infinite-dimensional family of complete negative quaternionic Kähler metrics on $\mathbb{R}^{4m}$, which arise as deformations of quaternionic hyperbolic space $\mathbb{H}\mathcal{H}^{4m}$.

It is known that all the noncompact duals of the Wolf spaces admit compact quotients, but the author is unaware of any other examples of compact negative quaternionic Kähler manifolds. Semmelmann and Weingart [308] use eigenvalue estimates for Dirac operators to prove vanishing theorems and inequalities on the Betti numbers of compact negative quaternionic Kähler $4m$-manifolds, including $b_{2k+1} = 0$ for $2k + 1 < m$.

10.6   Other topics in quaternionic geometry

We finish by giving brief discussions, with references, of other interesting areas in quaternionic geometry that we have not space to explain at length.

10.6.1   3-Sasakian manifolds

A Riemannian manifold (S, g) of dimension $4m + 3$ is called 3-*Sasakian* if the cone $\left(S \times (0, \infty), \mathrm{d}r^2 + r^2 g\right)$ on S is hyperkähler, that is, if $\mathrm{Hol}(\mathrm{d}r^2 + r^2 g)$ is contained in $\mathrm{Sp}(m + 1)$. 3-Sasakian manifolds are quaternionic analogues of *Sasakian manifolds*, which are Riemannian manifolds of dimension $2m + 1$ whose cones are Kähler. 3-Sasakian manifolds were introduced by Kuo in the 1970's and studied for a few years by Japanese geometers, but then fell into obscurity until revived in the 1990's by Boyer, Galicki, Mann and others. A good survey on them is Boyer and Galicki [44].

Let (S, g) be a 3-Sasakian $(4m+3)$-manifold. Then g is Einstein with positive scalar curvature $2(2m+1)(4m+3)$. Thus, as for positive quaternionic Kähler manifolds, complete 3-Sasakian manifolds are necessarily compact, with finite fundamental group. The metric $\mathrm{d}r^2 + r^2 g$ on $S \times (0, \infty)$ extends to a hyperkähler structure $(J_1, J_2, J_3, \mathrm{d}r^2 + r^2 g)$. Define orthonormal vector fields v_1, v_2, v_3 on S by $v_a = J_a(r\frac{\partial}{\partial r})$ on $S \times \{1\} \cong S$. They satisfy the $\mathfrak{su}(2)$ Lie algebra relations $[v_1, v_2] = v_3$, $[v_2, v_3] = v_1$, $[v_3, v_1] = v_2$, so if g is complete then v_1, v_2, v_3 exponentiate to an isometric $\mathrm{SU}(2)$ action on S.

By [44, Th. 3.3.3], the quotient $X = S/\mathrm{SU}(2)$ is an *orbifold* of dimension $4m$, which turns out to be *quaternionic Kähler* with positive scalar curvature $16m(m + 2)$. Conversely, if X is a positive quaternionic Kähler $4m$-manifold (orbifold) with associated $\mathrm{Sp}(m)\,\mathrm{Sp}(1)$-bundle P then $S = P/\mathrm{Sp}(m)$ is a 3-Sasakian $(4m + 3)$-manifold (orbifold), and the projection $\pi : S \to X$ has fibre $\mathrm{Sp}(m)\,\mathrm{Sp}(1)/\mathrm{Sp}(m) \cong$

$\mathrm{Sp}(1)/\mathbb{Z}_2 \cong \mathrm{SO}(3)$. Note that the twistor space Z of X is $S/\,\mathrm{U}(1)$, so that S is the total space of a $\mathrm{U}(1)$-bundle over Z.

Note too the connection between hyperkähler and quaternionic Kähler geometry that this implies. If (X, g) is a positive quaternionic Kähler manifold then the cone $\left(S \times (0, \infty),\, \mathrm{d}r^2 + r^2 h\right)$ on the associated 3-Sasakian manifold (S, h) is a hyperkähler manifold known as the *Swann bundle*, which fibres over X with fibre $(\mathbb{H} \setminus \{0\})/\{\pm 1\}$.

As we saw in §10.5, the only known compact positive quaternionic Kähler manifolds are symmetric spaces, so one might expect compact 3-Sasakian manifolds to be equally rare. But this is not so. It is a nice observation of Boyer, Galicki and Mann [45] that if X is a quaternionic Kähler orbifold, the associated 3-Sasakian manifold S may be nonsingular; this happens provided each orbifold point of X is modelled on $\mathbb{R}^{4m}/G$ for some finite $G \subset \mathrm{Sp}(m)\,\mathrm{Sp}(1)$ acting freely on $\mathrm{Sp}(m)\,\mathrm{Sp}(1)/\,\mathrm{Sp}(m) \cong \mathrm{SO}(3)$.

This spawned a small industry in constructing and studying examples of compact 3-Sasakian manifolds. Some important contributions are Boyer et al. [45] who produced infinitely many nonhomotopic examples of compact 3-Sasakian $(4m+3)$-manifolds for each $m \geqslant 1$ using quotients as in §10.6.4, [46] who found compact 3-Sasakian 7-manifolds with b^2 arbitrary, and Pedersen and Poon [279] who show compact 3-Sasakian manifolds are *rigid*, that is, admit no nontrivial deformations. For more references see Boyer and Galicki [44].

10.6.2 Hypercomplex manifolds

We call (X, J_1, J_2, J_3) a *hypercomplex manifold*, and (J_1, J_2, J_3) a *hypercomplex structure* on X, if X is a manifold of dimension $4m$ and J_1, J_2, J_3 are integrable complex structures on X such that $J_1 J_2 = J_3$. If (X, J_1, J_2, J_3, g) is hyperkähler then (X, J_1, J_2, J_3) is hypercomplex. However, not all hypercomplex manifolds admit hyperkähler metrics, even locally. General references on hypercomplex manifolds are Salamon [295, §6] and [296, §9].

A hypercomplex manifold (X, J_1, J_2, J_3) carries a unique torsion-free connection ∇ on TX with $\nabla J_j = 0$ called the *Obata connection*, whose holonomy is a subgroup of $\mathrm{GL}(m, \mathbb{H})$. In the language of §2.6, a hypercomplex structure on X is equivalent to a *torsion-free* $\mathrm{GL}(m, \mathbb{H})$-*structure* on X. Note that if ∇ is not flat then (X, J_1, J_2, J_3) is not locally isomorphic to $\mathbb{H}^m$ with its flat hypercomplex structure. This is an important difference between complex and hypercomplex geometry, since all complex manifolds are locally trivial.

If (X, J_1, J_2, J_3) is hypercomplex and $a_1, a_2, a_3 \in \mathbb{R}$ with $a_1^2 + a_2^2 + a_3^2 = 1$, then $a_1 J_2 + a_2 J_2 + a_3 J_3$ is a complex structure on X. Thus a hypercomplex manifold has a 2-sphere $\mathcal{S}^2$ of complex structures. As for the hyperkähler case of §10.1.3, there is a *twistor construction* for hypercomplex manifolds; the difference is that in Theorem 10.1.3 there need be no holomorphic section ω of $p^*(\mathcal{O}(2)) \otimes \Lambda^2 \mathcal{D}^*$.

Compact hypercomplex 4-manifolds are classified by Boyer [43], and are either $K3$ surfaces or flat examples such as 4-tori and Hopf surfaces. The first known compact hypercomplex manifolds of dimension $4m > 4$ which are not locally hyperkähler were hypercomplex structures on Lie groups, found by Spindel et al. [316]. They were generalized by the author [180] to hypercomplex structures on homogeneous spaces. In [182] this is extended to inhomogeneous hypercomplex structures on *biquotients* $A\backslash B/C$ of

Lie groups. The author also gave in [180] a second way of constructing compact, inhomogeneous hypercomplex manifolds, using the idea of 'twisting by an instanton'.

The *deformations* of compact hypercomplex manifolds are studied by Pedersen and Poon [277], who use their theory to prove that the homogeneous hypercomplex manifolds of [180, 316] admit inhomogeneous deformations [278]. The author has proposed a theory of *hypercomplex algebraic geometry* [186], a quaternionic analogue of complex algebraic geometry, in which one studies (noncompact) hypercomplex manifolds using an 'algebra' of $\mathbb{H}$-valued 'q-holomorphic functions' upon them.

10.6.3 Quaternionic manifolds

Underlying a hyperkähler structure (J_1, J_2, J_3, g) there is a hypercomplex structure (J_1, J_2, J_3), which is a weaker structure not involving a metric, and can be studied on its own. In the same way, under any quaternionic Kähler structure is a weaker, metric-free geometry called a *quaternionic structure*, which is studied by Salamon [295].

In $4m$ dimensions for $m \geqslant 2$, a quaternionic structure on X is a *torsion-free* $GL(m, \mathbb{H}) GL(1, \mathbb{H})$-*structure* Q on X, where $GL(m, \mathbb{H}) GL(1, \mathbb{H})$ is a Lie subgroup of $GL(4m, \mathbb{R})$. That is, Q is a $GL(m, \mathbb{H}) GL(1, \mathbb{H})$-structure on X, and there exists a torsion-free connection ∇ on TX preserving Q. However, this connection ∇ is not unique, and not part of the quaternionic structure. We call (X, Q) a *quaternionic manifold*. Following Definition 10.5.4, a *quaternionic 4-manifold* is an oriented conformal 4-manifold $(X, [g])$ with self-dual Weyl curvature.

Hyperkähler manifolds, quaternionic Kähler manifolds and hypercomplex manifolds all have an underlying quaternionic structure. As for the other three quaternionic geometries, the twistor construction works for quaternionic manifolds [295, §7]. Let (X, Q) be a quaternionic manifold, so that Q is a principal $GL(m, \mathbb{H}) GL(1, \mathbb{H})$-bundle over X, and define the *twistor space* $Z = Q/ GL(m, \mathbb{H})\mathbb{C}^*$. It has a natural projection $\pi : Z \rightarrow X$, with fibre $GL(1, \mathbb{H})/\mathbb{C}^* \cong \mathbb{CP}^1$, and Theorem 10.5.3(i),(ii) hold, but in general Z has neither a complex contact structure nor a Kähler metric, as in parts (iii),(iv) for the quaternionic Kähler case. We can reconstruct (X, Q) from holomorphic data on Z. Note that quaternionic manifolds are the weakest geometry for which the twistor transform works.

Explicit examples of compact quaternionic 4-manifolds (self-dual 4-manifolds) are given by LeBrun [235, 236, 238] and the author [181]. Analytic existence theorems for compact self-dual 4-manifolds are given by Donaldson and Friedman [97] and Taubes [319, 320]. Compact quaternionic manifolds of dimension $4m > 4$ which are neither locally quaternionic Kähler nor hypercomplex were constructed by the author [180] on homogeneous spaces, and by 'twisting by an instanton'.

10.6.4 Quotient constructions

There are *quotient constructions* for hyperkähler, quaternionic Kähler, 3-Sasakian, hypercomplex and quaternionic manifolds, which given such a manifold X acted on by a Lie group G preserving the geometric structure, produce a new manifold Y with the same structure but of smaller dimension $\dim Y = \dim X - 4 \dim G$. This is a fertile source of examples. We explain the hyperkähler case first, as it is simplest.

Suppose (X, J_1, J_2, J_3, g) is a hyperkähler $4m$-manifold, and $\omega_1, \omega_2, \omega_3$ are the

Kähler forms of J_1, J_2, J_3. Let G be a Lie group of dimension k with Lie algebra $\mathfrak{g}$ which acts freely on X preserving the hyperkähler structure. Write $\phi : \mathfrak{g} \to C^\infty(TX)$ for the corresponding Lie algebra action. Then for each $v \in \mathfrak{g}$ and $a = 1, 2, 3$ the Lie derivative $\mathcal{L}_{\phi(v)}\omega_a = 0$, so $\phi(v) \cdot \omega_a$ is a closed 1-form on X. If $H^1(X, \mathbb{R}) = 0$ there is a smooth map $\mu_a : X \to \mathfrak{g}^*$ for $a = 1, 2, 3$, unique up addition of a constant in $\mathfrak{g}^*$, satisfying $\mathrm{d}(v \cdot \mu_a) = \phi(v) \cdot \omega_a$ for all $v \in \mathfrak{g}$.

Under mild conditions we can choose μ_a to be equivariant under the actions of G on X and $\mathfrak{g}^*$, and then μ_a is unique up to addition of an element of the centre $Z(\mathfrak{g}^*)$, and is a *moment map* for the action of G on the Kähler manifold (X, J_a, g). We call $\mu = (\mu_1, \mu_2, \mu_3) : X \to \mathfrak{g}^* \otimes \mathbb{R}^3$ the *hyperkähler moment map*. Hitchin et al. [164, §3] show that where it is nonsingular, $Y = \mu^{-1}(0)/G$ is a new hyperkähler manifold of dimension $4(m - k)$, called the *hyperkähler quotient* of X by G.

A similar quotient construction for quaternionic Kähler manifolds X was found by Galicki and Lawson [120, 121]. In this case, the quaternionic Kähler moment map μ is a section of $\mathfrak{g}^* \otimes E$, where E is a vector bundle over X with fibre $\mathbb{R}^3$, and it is defined uniquely rather than up to addition of a constant. Boyer, Galicki and Mann [45] wrote down a 3-Sasakian version of this by lifting the quaternionic Kähler quotient up to the associated 3-Sasakian SO(3)-bundles.

Quotient constructions for hypercomplex and quaternionic manifolds were found by the author [179], and work on a slightly different principle. In the hypercomplex case, we start from the observation that the hyperkähler moment maps (μ_1, μ_2, μ_3) above satisfy the p.d.e. $J_1(\mathrm{d}\mu_1) = J_1(\mathrm{d}\mu_2) = J_3(\mathrm{d}\mu_3)$, which also makes sense in the hypercomplex setting. So given a hypercomplex manifold (X, J_1, J_2, J_3) invariant under a Lie group G, we define a *hypercomplex moment map* to be a smooth G-equivariant map $\mu : X \to \mathfrak{g}^* \otimes \mathbb{R}^3$ satisfying $J_1(\mathrm{d}\mu_1) = J_1(\mathrm{d}\mu_2) = J_3(\mathrm{d}\mu_3)$. Nontrivial moment maps μ need not exist, and if they do they may not be unique. If $\mathrm{d}\mu : TX \to \mathfrak{g}^* \otimes \mathbb{R}^3$ is injective along $\mu^{-1}(0)$ then $Y = \mu^{-1}(0)/G$ is hypercomplex where it is nonsingular.

10.6.5 Hyperkähler moduli spaces

Let (X, J_1, J_2, J_3, g) be a hyperkähler 4-manifold such as T^4, $K3$ or $\mathbb{H}$, which should be either compact, or complete and well-behaved at infinity. Let $P \to X$ be a principal bundle over X with fibre G, a compact Lie group. An *instanton* is a connection on P with anti-self-dual curvature. The moduli space $\mathcal{M}$ of (finite energy) instantons on P modulo gauge transformations will in general be a finite-dimensional hyperkähler manifold, which may be singular, and may be noncompact. Instanton moduli spaces are important in the study of 4-manifolds. For more information, see Donaldson and Kronheimer [98].

Here is the rough reason why $\mathcal{M}$ is hyperkähler, following Atiyah and Hitchin [11, §4]. Write $\mathcal{A}$ for the family of connections on the fixed principal bundle P. This is an infinite-dimensional affine space modelled on the vector space $C^\infty(\mathrm{Ad}(P) \otimes T^*X)$. Here $\mathrm{Ad}(P)$ is the vector bundle over X with fibre $\mathfrak{g}$, the Lie algebra of G, induced by the adjoint representation of G on $\mathfrak{g}$. Write $\mathcal{G}$ for the gauge group of P, the infinite-dimensional Lie group of smooth maps $X \to G$. Then $\mathcal{M} = \{A \in \mathcal{A} : F_A^+ = 0\}/\mathcal{G}$.

As $\mathcal{A}$ has tangent spaces $C^\infty(\mathrm{ad}(P) \otimes T^*X)$ the action J_1, J_2, J_3 on T^*X induce complex structures $\hat{J}_1, \hat{J}_2, \hat{J}_3$ on $\mathcal{A}$, and an $\mathrm{Ad}(G)$-invariant metric on $\mathfrak{g}$ and the L^2-

norm w.r.t. g on $C^\infty(\mathrm{ad}(P) \otimes T^*X)$ induce a metric $\hat{g}$ on $\mathscr{A}$, making $(\hat{J}_1, \hat{J}_2, \hat{J}_3, \hat{g})$ into a flat infinite-dimensional hyperkähler structure on $\mathscr{A}$. It turns out that $\mathscr{G}$ preserves this hyperkähler structure, and the instanton equation $F_A^+ = 0$ can be interpreted as the moment map equation $\mu = 0$ for this $\mathscr{G}$ action. Hence $\mathscr{M}$ is constructed as an *infinite-dimensional hyperkähler quotient* as in §10.6.4, and so is hyperkähler.

A related problem is the study of *magnetic monopoles* on noncompact 3-manifolds. Monopole moduli spaces on $\mathbb{R}^3$ turn out to be hyperkähler, for similar reasons to the instanton case; in fact, monopoles on $\mathbb{R}^3$ can be interpreted as instantons on $\mathbb{R}^3 \times \mathbb{R}$ invariant under translations in $\mathbb{R}$. Atiyah and Hitchin [11] use these hyperkähler metrics to resolve questions about scattering of magnetic monopoles.

Moduli space techniques can be used to find interesting examples of hyperkähler manifolds, in particular metrics on *coadjoint orbits*. Let G be a compact, semisimple Lie group, with complexification G^c, and complexified Lie algebra $\mathfrak{g}^c$. Then the coadjoint orbits of G^c in $(\mathfrak{g}^c)^*$ are all noncompact complex symplectic manifolds. Each such orbit admits G-invariant Kähler metrics making it into a hyperkähler manifold. These metrics were constructed by Kronheimer [228, 229] for the highest-dimensional orbits in $(\mathfrak{g}^c)^*$, and extended to general orbits by Biquard [31] and Kovalev [222].

Kronheimer's construction was analytic: he obtained the coadjoint orbits as moduli spaces of $\mathrm{SU}(2)$-invariant G-instantons on $\mathbb{R}^4 \setminus \{0\}$, and used the fact that instanton moduli spaces on $\mathbb{R}^4$ are hyperkähler. The author considers the metrics from a more algebraic point of view in [186, §11–§12].

11
The exceptional holonomy groups

We now discuss the *exceptional holonomy groups* G_2 in 7 dimensions, and $\mathrm{Spin}(7)$ in 8 dimensions. Metrics with these holonomy groups are Ricci-flat, and on compact manifolds come in smooth moduli spaces of known dimension. Sections 11.1–11.3 define G_2, study the topology of compact Riemannian 7-manifolds (M, g) with holonomy G_2, and describe constructions of such manifolds. Sections 11.4–11.6 give a similar treatment of $\mathrm{Spin}(7)$. Section 11.7 gives a reading list on exceptional holonomy. For much more detail on compact manifolds with exceptional holonomy and many examples, see the author's monograph [188, Chs 10–15].

Since the introduction of *M-theory* in string theory, G_2-manifolds have been of particular interest to theoretical physicists. We saw in §9.1 that supersymmetric string theory requires the universe to have 10 dimensions, which are supposed to locally resemble $\mathbb{R}^4 \times M$, where $\mathbb{R}^4$ is Minkowski 4-space and M is a Calabi–Yau 3-fold with diameter of order the Planck length, 10^{-33}cm. M-theory is a variant of string theory in which the universe has 11 dimensions, and locally resembles $\mathbb{R}^4 \times M$ for M a compact G_2-manifold with diameter of order the Planck length; though to achieve realistic physics, M must have singularities [5]. We shall not discuss M-theory as it is beyond the competence of the author, but we note that it has been the driving force behind much recent research in exceptional holonomy.

11.1 The holonomy group G_2

Here is a definition of G_2 as a subgroup of $\mathrm{GL}(7, \mathbb{R})$.

Definition 11.1.1 Let $(x_1, \ldots, x_7)$ be coordinates on $\mathbb{R}^7$. Write $\mathrm{d}x_{ij\ldots l}$ for the exterior form $\mathrm{d}x_i \wedge \mathrm{d}x_j \wedge \cdots \wedge \mathrm{d}x_l$ on $\mathbb{R}^7$. Define a 3-form φ_0 on $\mathbb{R}^7$ by

$$\varphi_0 = \mathrm{d}x_{123} + \mathrm{d}x_{145} + \mathrm{d}x_{167} + \mathrm{d}x_{246} - \mathrm{d}x_{257} - \mathrm{d}x_{347} - \mathrm{d}x_{356}. \tag{11.1}$$

The subgroup of $\mathrm{GL}(7, \mathbb{R})$ preserving φ_0 is the exceptional Lie group G_2. It is compact, connected, simply-connected, semisimple and 14-dimensional, and it fixes the 4-form

$$*\varphi_0 = \mathrm{d}x_{4567} + \mathrm{d}x_{2367} + \mathrm{d}x_{2345} + \mathrm{d}x_{1357} - \mathrm{d}x_{1346} - \mathrm{d}x_{1256} - \mathrm{d}x_{1247}, \tag{11.2}$$

the Euclidean metric $g_0 = \mathrm{d}x_1^2 + \cdots + \mathrm{d}x_7^2$, and the orientation on $\mathbb{R}^7$. Note that φ_0 and $*\varphi_0$ are related by the Hodge star.

The forms φ_0 and $*\varphi_0$ above are those given by Bryant [56, p. 539] and used in [188]. They differ from those used in the author's papers [183, 184] and by Harvey and Lawson [151, p. 113] and McLean [259, p. 733], but are equivalent to them under permutations of $x_1, \ldots, x_7$ and changes of sign of φ_0 and $*\varphi_0$.

Let M be an oriented 7-manifold. For each $p \in M$, define $\mathscr{P}_p^3 M$ to be the subset of 3-forms $\varphi \in \Lambda^3 T_p^* M$ for which there exists an oriented isomorphism between $T_p M$ and $\mathbb{R}^7$ identifying φ and the 3-form φ_0 of (11.1). Then $\mathscr{P}_p^3 M$ is isomorphic to $\mathrm{GL}_+(7, \mathbb{R})/G_2$, since φ_0 has symmetry group G_2.

Now $\dim \mathrm{GL}_+(7, \mathbb{R}) = 49$ and $\dim G_2 = 14$, so $\mathrm{GL}_+(7, \mathbb{R})/G_2$ has dimension $49 - 14 = 35$. But $\Lambda^3 T_p^* M$ also has dimension $\binom{7}{3} = 35$, and thus $\mathscr{P}_p^3 M$ is an *open subset* of $\Lambda^3 T_p^* M$. Let $\mathscr{P}^3 M$ be the bundle over M with fibre $\mathscr{P}_p^3 M$ at each $p \in M$. Then $\mathscr{P}^3 M$ is an open subbundle of $\Lambda^3 T^* M$ with fibre $\mathrm{GL}_+(7, \mathbb{R})/G_2$. Note that $\mathscr{P}^3 M$ is *not* a vector subbundle of $\Lambda^3 T^* M$. We say that a 3-form φ on M is *positive* if $\varphi|_p \in \mathscr{P}_p^3 M$ for each $p \in M$.

Similarly, define $\mathscr{P}_p^4 M$ to be the subset of 4-forms $\psi \in \Lambda^4 T_p^* M$ with an oriented isomorphism between $T_p M$ and $\mathbb{R}^7$ identifying ψ and the 4-form $*\varphi_0$ of (11.2), and let $\mathscr{P}^4 M$ have fibre $\mathscr{P}_p^4 M$ at each $p \in M$. Then $\mathscr{P}^4 M$ is an open subbundle of $\Lambda^4 T^* M$ with fibre $\mathrm{GL}_+(7, \mathbb{R})/G_2$, and sections of $\mathscr{P}^4 M$ are called *positive 4-forms*.

The *frame bundle* F of M is the bundle over M whose fibre at $p \in M$ is the set of isomorphisms between $T_p M$ and $\mathbb{R}^7$. Let φ be a positive 3-form on M, and let Q be the subset of F consisting of isomorphisms between $T_p M$ and $\mathbb{R}^7$ which identify $\varphi|_p$ and φ_0 of (11.1). It is easy to show that Q is a principal subbundle of F, with fibre G_2. That is, Q is a G_2-*structure*, as in Definition 2.6.1.

Conversely, if Q is a G_2-structure on M then, as φ_0, $*\varphi_0$ and g_0 are G_2-invariant, we can use Q to define a 3-form φ, a 4-form $*\varphi$ and a metric g on M corresponding to φ_0, $*\varphi_0$ and g_0. This 3-form φ will be positive if and only if Q is an *oriented G_2-structure*, that is, a G_2-structure which induces the given orientation on M.

Thus we have found a 1-1 correspondence between positive 3-forms φ and oriented G_2-structures Q on M. Furthermore, to any positive 3-form φ on M we can associate a unique positive 4-form $*\varphi$ and metric g, such that φ, $*\varphi$ and g are identified with φ_0, $*\varphi_0$ and g_0 under an isomorphism between $T_p M$ and $\mathbb{R}^7$, for each $p \in M$. We will call g and $*\varphi$ the *metric* and *4-form associated to* φ.

Definition 11.1.2 Let M be an oriented 7-manifold, φ a positive 3-form on M, and g the associated metric. For the rest of the book, we will adopt the following abuse of notation: we shall refer to the pair (φ, g) as a G_2-*structure*. Of course (φ, g) is not, exactly, a G_2-structure, but it does at least define a unique G_2-structure.

Define a map $\Theta : \mathscr{P}^3 M \to \mathscr{P}^4 M$ by $\Theta(\varphi) = *\varphi$. That is, if φ is a positive 3-form, then $\Theta(\varphi)$ is the associated 4-form $*\varphi$. It is important to note that Θ depends *solely* on M and its orientation, and also that Θ is a *nonlinear* map. Although we define $\Theta(\varphi) = *\varphi$ and the Hodge star $*$ is linear, actually $*$ depends on the metric g, which itself depends on φ, so $\Theta(\varphi)$ is not linear in φ.

Let M be a 7-manifold, (φ, g) a G_2-structure on M, and ∇ the Levi-Civita connection of g. We call $\nabla \varphi$ the *torsion* of (φ, g). If $\nabla \varphi = 0$ then (φ, g) is called *torsion-free*. We define a G_2-*manifold* to be a triple (M, φ, g), where M is a 7-manifold, and (φ, g)

a torsion-free G_2-structure on M.

The next proposition follows from [296, Lem. 11.5].

Proposition 11.1.3 *Let M be a 7-manifold and (φ, g) a G_2-structure on M. Then the following are equivalent:* (i) (φ, g) *is torsion-free,*

(ii) $\mathrm{Hol}(g) \subseteq G_2$, *and φ is the induced 3-form,*

(iii) $\nabla \varphi = 0$ *on M, where ∇ is the Levi-Civita connection of g,*

(iv) $\mathrm{d}\varphi = \mathrm{d}^* \varphi = 0$ *on M, and*

(v) $\mathrm{d}\varphi = \mathrm{d}\,\Theta(\varphi) = 0$ *on M.*

Torsion-free G_2-structures will play an essential rôle in our construction of compact 7-manifolds with holonomy G_2. The basic idea is to find a torsion-free G_2-structure (φ, g) on M, and then show that $\mathrm{Hol}(g) = G_2$ provided $\pi_1(M)$ is finite.

The condition that (φ, g) be torsion-free is a *nonlinear p.d.e.* on the positive 3-form φ. This is most clearly seen in part (v) of Proposition 11.1.3, as Θ is a nonlinear map. Although in parts (iii) and (iv) the conditions $\nabla \varphi = 0$ and $\mathrm{d}\varphi = \mathrm{d}^* \varphi = 0$ appear linear in φ, in fact the operators ∇ and d^* depend on g, which depends on φ, so the equations $\nabla \varphi = 0$ and $\mathrm{d}^* \varphi = 0$ should be considered nonlinear in φ.

In §3.5.1 we explained that a G-structure on M induces a splitting of the bundles of tensors on M into irreducible components. Here is the decomposition of the exterior forms on a 7-manifold with a G_2-structure, which follows from [296, Lem. 11.4].

Proposition 11.1.4 *Let M be a 7-manifold and (φ, g) a G_2-structure on M. Then $\Lambda^k T^* M$ splits orthogonally into components as follows, where Λ_l^k corresponds to an irreducible representation of G_2 of dimension l:*

(i) $\Lambda^1 T^* M = \Lambda_7^1,$ $\qquad\qquad$ (ii) $\Lambda^2 T^* M = \Lambda_7^2 \oplus \Lambda_{14}^2,$

(iii) $\Lambda^3 T^* M = \Lambda_1^3 \oplus \Lambda_7^3 \oplus \Lambda_{27}^3,$ $\qquad$ (iv) $\Lambda^4 T^* M = \Lambda_1^4 \oplus \Lambda_7^4 \oplus \Lambda_{27}^4,$

(v) $\Lambda^5 T^* M = \Lambda_7^5 \oplus \Lambda_{14}^5,$ $\qquad$ *and* $\qquad$ (vi) $\Lambda^6 T^* M = \Lambda_7^6.$

The Hodge star $*$ of g gives an isometry between Λ_l^k and Λ_l^{7-k}. Also $\Lambda_1^3 = \langle \varphi \rangle$ and $\Lambda_1^4 = \langle *\varphi \rangle$, and the spaces Λ_7^k for $k = 1, 2, \ldots, 6$ are canonically isomorphic.

Let the orthogonal projection from $\Lambda^k T^* M$ to Λ_l^k be denoted π_l. So, for instance, if $\xi \in C^\infty(\Lambda^2 T^* M)$, then $\xi = \pi_7(\xi) + \pi_{14}(\xi)$. We saw in Theorem 3.1.7 that the holonomy group of a Riemannian metric g constrains its Riemann curvature. Using this, Salamon [296, Lem. 11.8] shows:

Proposition 11.1.5 *Let (M, g) be a Riemannian 7-manifold. If $\mathrm{Hol}(g) \subseteq G_2$, then g is Ricci-flat.*

Since G_2 is a simply-connected subgroup of $\mathrm{SO}(7)$, any 7-manifold M with a G_2-structure must be a *spin manifold*, by Proposition 3.6.2. Furthermore, from Theorem 3.6.1 the natural representation of G_2 on the spinors on $\mathbb{R}^7$ fixes a nonzero spinor, and so a torsion-free G_2-structure has a corresponding *parallel spinor*. In fact if S is the spin bundle of M, then there is a natural isomorphism $S \cong \Lambda_1^0 \oplus \Lambda_7^1$. Thus we have:

Proposition 11.1.6 *Let* (φ, g) *be a* G_2-*structure on a* 7-*manifold* M. *Then* M *is spin, with a preferred spin structure. If* (φ, g) *is torsion-free, then* (M, g) *has a nonzero parallel spinor.*

From the classification of Riemannian holonomy groups in §3.4 we deduce:

Theorem 11.1.7 *Write* $\mathbb{R}^7 \cong \mathbb{R}^3 \oplus \mathbb{C}^2$, *and let* $\mathrm{SU}(2)$ *act on* $\mathbb{R}^7$ *trivially on* $\mathbb{R}^3$ *and as usual on* $\mathbb{C}^2$. *Similarly, write* $\mathbb{R}^7 \cong \mathbb{R} \oplus \mathbb{C}^3$, *and let* $\mathrm{SU}(3)$ *act on* $\mathbb{R}^7$ *trivially on* $\mathbb{R}$ *and as usual on* $\mathbb{C}^3$. *Then* $\mathrm{SU}(2) \subset \mathrm{SU}(3) \subset G_2 \subset \mathrm{SO}(7)$.

The only connected Lie subgroups of G_2 *which can be the holonomy group of a Riemannian metric on a* 7-*manifold are* $\{1\}$, $\mathrm{SU}(2)$, $\mathrm{SU}(3)$ *and* G_2, *where the subgroups* $\mathrm{SU}(2)$ *and* $\mathrm{SU}(3)$ *are defined above. Hence, if* (φ, g) *is a torsion-free* G_2-*structure on a* 7-*manifold, then* $\mathrm{Hol}^0(g)$ *is one of* $\{1\}$, $\mathrm{SU}(2)$, $\mathrm{SU}(3)$ *or* G_2.

These inclusions $\mathrm{SU}(2) \subset G_2$ and $\mathrm{SU}(3) \subset G_2$ imply that from each Calabi–Yau 2- or 3-fold with holonomy $\mathrm{SU}(2)$ or $\mathrm{SU}(3)$ we can make a G_2-manifold. Here is how to do this explicitly in terms of differential forms, following [188, Prop.s 11.1.1 & 11.1.2].

Proposition 11.1.8 *Let* (M, J, h, θ) *be a Calabi–Yau* 2-*fold, with Kähler form* ω. *Let* (x_1, x_2, x_3) *be coordinates on* $\mathbb{R}^3$ *or* T^3. *Define a metric* $g = \mathrm{d}x_1^2 + \mathrm{d}x_2^2 + \mathrm{d}x_3^2 + h$ *and a* 3-*form* φ *on* $\mathbb{R}^3 \times M$ *or* $T^3 \times M$ *by*

$$\varphi = \mathrm{d}x_1 \wedge \mathrm{d}x_2 \wedge \mathrm{d}x_3 + \mathrm{d}x_1 \wedge \omega + \mathrm{d}x_2 \wedge \mathrm{Re}\,\theta - \mathrm{d}x_3 \wedge \mathrm{Im}\,\theta.$$

Then (φ, g) *is a torsion-free* G_2-*structure on* $\mathbb{R}^3 \times M$ *or* $T^3 \times M$, *and*

$$*\varphi = \tfrac{1}{2}\omega \wedge \omega + \mathrm{d}x_2 \wedge \mathrm{d}x_3 \wedge \omega - \mathrm{d}x_1 \wedge \mathrm{d}x_3 \wedge \mathrm{Re}\,\theta - \mathrm{d}x_1 \wedge \mathrm{d}x_2 \wedge \mathrm{Im}\,\theta.$$

Proposition 11.1.9 *Let* (M, J, h, θ) *be a Calabi–Yau* 3-*fold, with Kähler form* ω. *Let* x *be a coordinate on* $\mathbb{R}$ *or* $\mathcal{S}^1$. *Define a metric* $g = \mathrm{d}x^2 + h$ *and a* 3-*form* $\varphi = \mathrm{d}x \wedge \omega + \mathrm{Re}\,\theta$ *on* $\mathbb{R} \times M$ *or* $\mathcal{S}^1 \times M$. *Then* (φ, g) *is a torsion-free* G_2-*structure on* $\mathbb{R} \times M$ *or* $\mathcal{S}^1 \times M$, *and* $*\varphi = \tfrac{1}{2}\omega \wedge \omega - \mathrm{d}x \wedge \mathrm{Im}\,\theta$.

11.2 Topological properties of compact G_2-manifolds

We now discuss the topology of *compact* G_2-*manifolds*, that is, compact manifolds M equipped with a torsion-free G_2-structure (φ, g). The material of this section can be found in [183, 184] and [188, Ch. 10]. First we prove:

Proposition 11.2.1 *Let* (M, φ, g) *be a compact* G_2-*manifold. Then* $\mathrm{Hol}(g) = G_2$ *if and only if* $\pi_1(M)$ *is finite.*

Proof Since M is compact and g is Ricci-flat, Theorem 3.5.5 shows that M has a finite cover isometric to $T^k \times N$, where N is simply-connected. Thus $\pi_1(M) \cong F \ltimes \mathbb{Z}^k$, where F is a finite group. Clearly, $\pi_1(M)$ is finite if and only if $k = 0$.

Now $\mathrm{Hol}(g) \subseteq G_2$ as (φ, g) is torsion-free, so Theorem 11.1.7 shows that $\mathrm{Hol}^0(g)$ is $\{1\}$, $\mathrm{SU}(2)$, $\mathrm{SU}(3)$ or G_2. Clearly $k = 7$ when $\mathrm{Hol}^0(g) = \{1\}$, $k = 3$ when $\mathrm{Hol}^0(g) = \mathrm{SU}(2)$, $k = 1$ when $\mathrm{Hol}^0(g) = \mathrm{SU}(3)$, and $k = 0$ when $\mathrm{Hol}^0(g) = G_2$. Thus $\mathrm{Hol}(g) = G_2$ if and only if $\pi_1(M)$ is finite. $\qquad\square$

Next we apply the ideas of §3.5.2 to compact G_2-manifolds.

Definition 11.2.2 Let (M, φ, g) be a compact G_2-manifold. For each of the subbundles Λ_l^k of $\Lambda^k T^* M$ defined in Proposition 11.1.4, write

$$\mathscr{H}_l^k = \{\eta \in C^\infty(\Lambda_l^k) : \mathrm{d}\eta = \mathrm{d}^*\eta = 0\},$$

and let $H_l^k(M, \mathbb{R})$ be the vector subspace of $H^k(M, \mathbb{R})$ with representatives in $\mathscr{H}_l^k$. Define the *refined Betti numbers* b_l^k of M by $b_l^k = \dim H_l^k(M, \mathbb{R})$.

Combining Proposition 11.1.4 and Theorem 3.5.3 we find:

Theorem 11.2.3 Let (M, φ, g) be a compact G_2-manifold. Then

$$H^2(M, \mathbb{R}) = H_7^2(M, \mathbb{R}) \oplus H_{14}^2(M, \mathbb{R}),$$
$$H^3(M, \mathbb{R}) = H_1^3(M, \mathbb{R}) \oplus H_7^3(M, \mathbb{R}) \oplus H_{27}^3(M, \mathbb{R}),$$
$$H^4(M, \mathbb{R}) = H_1^4(M, \mathbb{R}) \oplus H_7^4(M, \mathbb{R}) \oplus H_{27}^4(M, \mathbb{R}),$$
$$\text{and} \quad H^5(M, \mathbb{R}) = H_7^5(M, \mathbb{R}) \oplus H_{14}^5(M, \mathbb{R}).$$

Here $H_1^3(M, \mathbb{R}) = \langle [\varphi] \rangle$, $H_1^4(M, \mathbb{R}) = \langle [*\varphi] \rangle$ *and* $H_l^k(M, \mathbb{R}) \cong H_l^{7-k}(M, \mathbb{R})$, *so that* $b_1^3 = b_1^4 = 1$ *and* $b_l^k = b_l^{7-k}$. *Also, if* $\mathrm{Hol}(g) = G_2$ *then* $H_7^k(M, \mathbb{R}) = \{0\}$ *for* $k = 1, \ldots, 6$.

This shows that if (M, g) is a compact Riemannian 7-manifold with holonomy G_2 then M has only two independent, nontrivial refined Betti numbers, b_{14}^2 and b_{27}^3, with $b^2 = b_{14}^2$ and $b^3 = b_{27}^3 + 1$. Thus all the refined Betti numbers of M are given by b^2 and b^3. To prove the last part of the theorem, observe that $\pi_1(M)$ is finite by Proposition 11.2.1, and so $H^1(M, \mathbb{R}) = \{0\}$. But $H_7^k(M, \mathbb{R}) \cong H_7^1(M, \mathbb{R})$ for $k = 1, \ldots, 6$ by Theorem 3.5.3, as the G_2 representations associated to Λ_7^k and Λ_7^1 are isomorphic.

The following lemma is easily proved by calculating in coordinates.

Lemma 11.2.4 Let (φ, g) be a G_2-structure on a 7-manifold M, and η a 2-form on M. Then $\eta \wedge \varphi = 2 * \pi_7(\eta) - *\pi_{14}(\eta)$ and

$$\eta \wedge \eta \wedge \varphi = \left(2|\pi_7(\eta)|^2 - |\pi_{14}(\eta)|^2\right)\mathrm{d}V_g,$$

where $\mathrm{d}V_g$ *is the volume form of* g.

We use this to prove two results on the cohomology of M.

Proposition 11.2.5 Let (M, φ, g) be a compact G_2-manifold, with $\mathrm{Hol}(g) = G_2$. Then $\langle \sigma \cup \sigma \cup [\varphi], [M] \rangle < 0$ for each nonzero $\sigma \in H^2(M, \mathbb{R})$.

Proof Theorem 11.2.3 gives $H_7^2(M, \mathbb{R}) = \{0\}$, so that $H^2(M, \mathbb{R}) = H_{14}^2(M, \mathbb{R})$. Thus each $\sigma \in H^2(M, \mathbb{R})$ is represented by a unique $\eta \in \mathscr{H}_{14}^2$. So

$$\langle \sigma \cup \sigma \cup [\varphi], [M] \rangle = \int_M \eta \wedge \eta \wedge \varphi = -\int_M |\eta|^2 \mathrm{d}V_g,$$

since $\eta \wedge \eta \wedge \varphi = -|\eta|^2 \mathrm{d}V_g$ by Lemma 11.2.4. Therefore $\langle \sigma \cup \sigma \cup [\varphi], [M] \rangle < 0$ when $\sigma \neq 0$, as $\eta \neq 0$. $\qquad \square$

Proposition 11.2.6 *Let (M, φ, g) be a compact G_2-manifold, R the Riemann curvature of g, and $p_1(M) \in H^4(M, \mathbb{Z})$ the first Pontryagin class. Then*

$$\langle p_1(M) \cup [\varphi], [M] \rangle = -\frac{1}{8\pi^2} \int_M |R|^2 \mathrm{d}V_g.$$

If $\mathrm{Hol}(g) = G_2$ *then g is not flat, so* $\langle p_1(M) \cup [\varphi], [M] \rangle < 0$, *and* $p_1(M) \neq 0$.

Proof From Chern–Weil theory, $p_1(M)$ is represented by $\frac{1}{8\pi^2} \mathrm{Tr}(R \wedge R)$, a closed 4-form. But the 2-form part of R lies in Λ_{14}^2. Using this and the equation $\eta \wedge \eta \wedge \varphi = -|\eta|^2 \mathrm{d}V_g$ for $\eta \in \Lambda_{14}^2$, we find that $\mathrm{Tr}(R \wedge R) \wedge \varphi = -|R|^2 \mathrm{d}V_g$. Integrating this over M, the result follows. $\qquad\qquad\square$

We summarize Propositions 11.1.6, 11.2.1 and 11.2.6 in the following theorem, which gives topological restrictions on compact 7-manifolds M with holonomy G_2.

Theorem 11.2.7 *Suppose M is a compact 7-manifold admitting metrics with holonomy G_2. Then M is orientable and spin, $\pi_1(M)$ is finite and $p_1(M) \neq 0$.*

For further interesting results on the cohomology of compact G_2-manifolds, which goes some way towards proving that compact G_2-manifolds are *formal* in the sense of rational homotopy theory, see Verbitsky [333].

We finish with two results on moduli spaces of metrics with holonomy G_2. Let M be a compact, oriented 7-manifold. Then as in §11.1, we can identify the set of all oriented G_2-structures on M with $C^\infty(\mathscr{P}^3 M)$, the set of positive 3-forms on M. Let $\mathscr{X}$ be the set of positive 3-forms corresponding to oriented, torsion-free G_2-structures. That is,

$$\mathscr{X} = \big\{ \varphi \in C^\infty(\mathscr{P}^3 M) : \mathrm{d}\varphi = \mathrm{d}\,\Theta(\varphi) = 0 \big\}.$$

Let $\mathscr{D}$ be the group of all diffeomorphisms Ψ of M isotopic to the identity. Then $\mathscr{D}$ acts naturally on $C^\infty(\mathscr{P}^3 M)$ and $\mathscr{X}$ by $\varphi \xmapsto{\ \Psi\ } \Psi_*(\varphi)$. We define the *moduli space of torsion-free G_2-structures* on M to be $\mathscr{M} = \mathscr{X}/\mathscr{D}$. Then we have:

Theorem 11.2.8 *Let M be a compact 7-manifold, and $\mathscr{M} = \mathscr{X}/\mathscr{D}$ the moduli space of torsion-free G_2-structures on M defined above. Then $\mathscr{M}$ is a smooth manifold of dimension $b^3(M)$, and the natural projection $\pi : \mathscr{M} \to H^3(M, \mathbb{R})$ given by $\pi(\varphi \mathscr{D}) = [\varphi]$ is a local diffeomorphism.*

This is proved in [183, Th. C], and also in [188, Th. 10.4.4]. Note that Theorem 11.2.8 is an entirely *local* result, and it gives little information about the global structure of $\mathscr{M}$. For instance, we do not know whether $\mathscr{M}$ is nonempty, or if it has one connected component or many, or whether π is injective, or what its image is.

In [184, Lem. 1.1.3] and [188, Prop. 10.4.5] we show that the image of $\mathscr{M}$ in $H^3(M, \mathbb{R}) \times H^4(M, \mathbb{R})$ is a *Lagrangian submanifold*.

Proposition 11.2.9 *Let M be a compact, oriented 7-manifold. By Poincaré duality $H^4(M, \mathbb{R}) \cong H^3(M, \mathbb{R})^*$, so $H^3(M, \mathbb{R}) \times H^4(M, \mathbb{R})$ has a natural symplectic structure. Define a subset L in $H^3(M, \mathbb{R}) \times H^4(M, \mathbb{R})$ by*

$$L = \big\{ ([\varphi], [*\varphi]) : (\varphi, g) \text{ is a torsion-free } G_2\text{-structure on } M \big\}.$$

Then L is a Lagrangian submanifold of $H^3(M, \mathbb{R}) \times H^4(M, \mathbb{R})$.

11.3 Constructing compact G_2-manifolds

We now explain the method used in [183,184] and [188, §11–§12] to construct examples of compact 7-manifolds with holonomy G_2. It is based on the *Kummer construction* for Calabi–Yau metrics on the $K3$ surface discussed in Examples 10.3.2 and 10.3.14, and may be divided into four steps.

Step 1. Let T^7 be the 7-torus and (φ_0, g_0) a flat G_2-structure on T^7. Choose a finite group Γ of isometries of T^7 preserving (φ_0, g_0). Then the quotient T^7/Γ is a singular, compact 7-manifold, an *orbifold*.

Step 2. For certain special groups Γ there is a method to resolve the singularities of T^7/Γ in a natural way, using complex geometry. We get a nonsingular, compact 7-manifold M, together with a map $\pi : M \to T^7/\Gamma$, the resolving map.

Step 3. On M, we explicitly write down a 1-parameter family of G_2-structures (φ_t, g_t) depending on $t \in (0, \epsilon)$. They are not torsion-free, but have small torsion when t is small. As $t \to 0$, the G_2-structure (φ_t, g_t) converges to the singular G_2-structure $\pi^*(\varphi_0, g_0)$.

Step 4. We prove using analysis that for sufficiently small t, the G_2-structure (φ_t, g_t) on M, with small torsion, can be deformed to a G_2-structure $(\tilde{\varphi}_t, \tilde{g}_t)$, with zero torsion. Finally, we show that $\tilde{g}_t$ is a metric with holonomy G_2 on the compact 7-manifold M.

We discuss each step in greater detail.

11.3.1 Step 1: Choosing an orbifold

Let (φ_0, g_0) be the Euclidean G_2-structure on $\mathbb{R}^7$ from Definition 11.1.1. Suppose Λ is a *lattice* in $\mathbb{R}^7$, that is, a discrete additive subgroup isomorphic to $\mathbb{Z}^7$. Then $\mathbb{R}^7/\Lambda$ is the torus T^7, and (φ_0, g_0) pushes down to a torsion-free G_2-structure on T^7. We must choose a finite group Γ acting on T^7 preserving (φ_0, g_0). That is, the elements of Γ are the push-forwards to T^7/Λ of affine transformations of $\mathbb{R}^7$ which fix (φ_0, g_0), and take Λ to itself under conjugation.

Here is an example of a suitable group Γ, taken from [188, §12.2].

Example 11.3.1 Let $(x_1, \ldots, x_7)$ be coordinates on $T^7 = \mathbb{R}^7/\mathbb{Z}^7$, where $x_i \in \mathbb{R}/\mathbb{Z}$. Let (φ_0, g_0) be the flat G_2-structure on T^7 from Definition 11.1.1. Let α, β and γ be the involutions of T^7 defined by

$$\alpha : (x_1, \ldots, x_7) \mapsto (x_1, x_2, x_3, -x_4, -x_5, -x_6, -x_7), \tag{11.3}$$

$$\beta : (x_1, \ldots, x_7) \mapsto (x_1, -x_2, -x_3, x_4, x_5, \tfrac{1}{2} - x_6, -x_7), \tag{11.4}$$

$$\gamma : (x_1, \ldots, x_7) \mapsto \left(-x_1, x_2, -x_3, x_4, \tfrac{1}{2} - x_5, x_6, \tfrac{1}{2} - x_7\right). \tag{11.5}$$

By inspection, α, β and γ preserve (φ_0, g_0), because of the careful choice of exactly which signs to change. Also, $\alpha^2 = \beta^2 = \gamma^2 = 1$, and α, β and γ commute. Thus they generate a group $\Gamma = \langle \alpha, \beta, \gamma \rangle \cong \mathbb{Z}_2^3$ of isometries of T^7 preserving (φ_0, g_0).

Having chosen a lattice Λ and finite group Γ, the quotient T^7/Γ is an *orbifold*, a singular manifold with only quotient singularities. The singularities of T^7/Γ come from the fixed points of non-identity elements of Γ. We now describe the singularities in our example.

Lemma 11.3.2 *In Example* 11.3.1, $\beta\gamma, \gamma\alpha, \alpha\beta$ *and* $\alpha\beta\gamma$ *have no fixed points on* T^7. *The fixed points of* α, β, γ *are each* 16 *copies of* T^3. *The singular set* S *of* T^7/Γ *is a disjoint union of* 12 *copies of* T^3, 4 *copies from each of* α, β, γ. *Each component of* S *is a singularity modelled on that of* $T^3 \times \mathbb{C}^2/\{\pm 1\}$.

Note that the translations $\frac{1}{2}$ in (11.3)–(11.5) are carefully chosen and play an essential rôle in simplifying the group Γ and the singular set. For example, $\alpha\beta$ takes $x_6 \mapsto x_6 + \frac{1}{2}$, so $\alpha\beta$ has no fixed points, and contributes no singularities. Also, $(\alpha\beta)^2 : x_6 \mapsto x_6 + 1 = x_6$ in $\mathbb{R}/\mathbb{Z}$, so $(\alpha\beta)^2 = 1$, which would not hold if we replaced $\frac{1}{2}$ in (11.4) by any other number in $(0, 1)$.

The most important consideration in choosing Γ is that we should be able to resolve the singularities of T^7/Γ within holonomy G_2. We will explain how to do this next.

11.3.2 Step 2: Resolving the singularities

Our goal is to resolve the singular set S of T^7/Γ to get a compact 7-manifold M with holonomy G_2. How can we do this? In general we cannot, because we have no idea of how to resolve general orbifold singularities with holonomy G_2. However, suppose we can arrange that every connected component of S is locally isomorphic to either

(a) $T^3 \times \mathbb{C}^2/G$, for G a finite subgroup of $SU(2)$, or
(b) $\mathcal{S}^1 \times \mathbb{C}^3/G$, for G a finite subgroup of $SU(3)$ acting freely on $\mathbb{C}^3 \setminus \{0\}$.

One can use complex algebraic geometry to find a *crepant resolution* X of $\mathbb{C}^2/G$ or Y of $\mathbb{C}^3/G$. Then $T^3 \times X$ or $\mathcal{S}^1 \times Y$ gives a local model for how to resolve the corresponding component of S in T^7/Γ. Thus we construct a nonsingular, compact 7-manifold M by using the patches $T^3 \times X$ or $\mathcal{S}^1 \times Y$ to repair the singularities of T^7/Γ. In the case of Example 11.3.1, this means gluing 12 copies of $T^3 \times X$ into T^7/Γ, where X is the *Eguchi–Hanson space* of Example 10.2.2.

Now the point of using crepant resolutions is this. In both case (a) and (b), there exists a Calabi–Yau metric on X or Y which is asymptotic to the flat Euclidean metric on $\mathbb{C}^2/G$ or $\mathbb{C}^3/G$. Such metrics are called *asymptotically locally Euclidean (ALE)*. In case (a), these are the hyperkähler ALE spaces described in §10.2, and exist for all finite $G \subset SU(2)$. In case (b), crepant resolutions of $\mathbb{C}^3/G$ exist for all finite $G \subset SU(3)$ by Roan [289], and the author [189], [188, §8] proved that they carry ALE Calabi–Yau metrics, using a noncompact version of the Calabi Conjecture.

By Propositions 11.1.8 and 11.1.9, we can use the Calabi–Yau metrics on X or Y to construct a torsion-free G_2-structure on $T^3 \times X$ or $\mathcal{S}^1 \times Y$. This gives a local model for how to resolve the singularity $T^3 \times \mathbb{C}^2/G$ or $\mathcal{S}^1 \times \mathbb{C}^3/G$ with holonomy G_2. So, this method gives not only a way to smooth out the singularities of T^7/Γ as a manifold, but also a family of torsion-free G_2-structures on the resolution which show how to smooth out the singularities of the G_2-structure.

The requirement above that S be divided into connected components of the form (a) and (b) is in fact unnecessarily restrictive. There is a more complicated and powerful method, described in [188, §11–§12], for resolving singularities of a more general kind. We require only that the singularities should *locally* be of the form $\mathbb{R}^3 \times \mathbb{C}^2/G$ or

$\mathbb{R} \times \mathbb{C}^3/G$, for G a finite subgroup of SU(2) or SU(3), and when $G \subset$ SU(3) we do *not* require that G act freely on $\mathbb{C}^3 \setminus \{0\}$.

If X is a crepant resolution of $\mathbb{C}^3/G$, where G does not act freely on $\mathbb{C}^3 \setminus \{0\}$, then the author shows [188, §9], [192] that X carries a family of Calabi–Yau metrics satisfying a complicated asymptotic condition at infinity, called *quasi-ALE* metrics. These yield the local models necessary to resolve singularities locally of the form $\mathbb{R} \times \mathbb{C}^3/G$ with holonomy G_2. Using this method we can resolve many orbifolds T^7/Γ, and prove the existence of large numbers of compact 7-manifolds with holonomy G_2.

11.3.3   Step 3: Finding G_2-structures with small torsion

For each resolution X of $\mathbb{C}^2/G$ in case (a), and Y of $\mathbb{C}^3/G$ in case (b) above, we can find a 1-parameter family $\{h_t : t > 0\}$ of metrics with the properties:

(a) h_t is a Kähler metric on X with $\mathrm{Hol}(h_t) = $ SU(2). Its injectivity radius satisfies $\delta(h_t) = O(t)$, its Riemann curvature satisfies $\|R(h_t)\|_{C^0} = O(t^{-2})$, and $h_t = h + O(t^4 r^{-4})$ for large r, where h is the Euclidean metric on $\mathbb{C}^2/G$, and r the distance from the origin.

(b) h_t is Kähler on Y with $\mathrm{Hol}(h_t) = $ SU(3), where $\delta(h_t) = O(t)$, $\|R(h_t)\|_{C^0} = O(t^{-2})$, and $h_t = h + O(t^6 r^{-6})$ for large r.

In fact we can choose h_t to be isometric to $t^2 h_1$, and then (a), (b) are easy to prove.

Suppose one of the components of the singular set S of T^7/Γ is locally modelled on $T^3 \times \mathbb{C}^2/G$. Then T^3 has a natural flat metric h_{T^3}. Let X be the crepant resolution of $\mathbb{C}^2/G$ and let $\{h_t : t > 0\}$ satisfy property (a). Then Proposition 11.1.8 gives a 1-parameter family of torsion-free G_2-structures $(\hat{\varphi}_t, \hat{g}_t)$ on $T^3 \times X$ with $\hat{g}_t = h_{T^3} + h_t$. Similarly, if a component of S is modelled on $\mathcal{S}^1 \times \mathbb{C}^3/G$, using Proposition 11.1.9 we get a family of torsion-free G_2-structures $(\hat{\varphi}_t, \hat{g}_t)$ on $\mathcal{S}^1 \times Y$.

The idea is to make a G_2-structure (φ_t, g_t) on M by gluing together the torsion-free G_2-structures $(\hat{\varphi}_t, \hat{g}_t)$ on the patches $T^3 \times X$ and $\mathcal{S}^1 \times Y$, and (φ_0, g_0) on T^7/Γ. The gluing is done using a partition of unity. Naturally, the first derivative of the partition of unity introduces 'errors', so that (φ_t, g_t) is not torsion-free. The size of the torsion $\nabla \varphi_t$ depends on the difference $\hat{\varphi}_t - \varphi_0$ in the region where the partition of unity changes. On the patches $T^3 \times X$, since $h_t - h = O(t^4 r^{-4})$ and the partition of unity has nonzero derivative when $r = O(1)$, we find that $\nabla \varphi_t = O(t^4)$. Similarly $\nabla \varphi_t = O(t^6)$ on the patches $\mathcal{S}^1 \times Y$, and so $\nabla \varphi_t = O(t^4)$ on M.

For small t, the dominant contributions to the injectivity radius $\delta(g_t)$ and Riemann curvature $R(g_t)$ are made by those of the metrics h_t on X and Y, so we expect $\delta(g_t) = O(t)$ and $\|R(g_t)\|_{C^0} = O(t^{-2})$ by properties (a) and (b) above. In this way we prove the following result [188, Th. 11.5.7], giving the estimates on (φ_t, g_t) that we need.

Theorem 11.3.3 *On the compact 7-manifold M described above, and on many other 7-manifolds constructed in a similar fashion, one can write down the following data explicitly in coordinates:*

- *positive constants A_1, A_2, A_3 and ϵ,*
- *a G_2-structure (φ_t, g_t) on M with $\mathrm{d}\varphi_t = 0$ for each $t \in (0, \epsilon)$, and*
- *a 3-form ψ_t on M with $\mathrm{d}^* \psi_t = \mathrm{d}^* \varphi_t$ for each $t \in (0, \epsilon)$.*

These satisfy three conditions:

(i) $\|\psi_t\|_{L^2} \leqslant A_1 t^4$, $\|\psi_t\|_{C^0} \leqslant A_1 t^3$ and $\|\mathrm{d}^*\psi_t\|_{L^{14}} \leqslant A_1 t^{16/7}$,

(ii) *the injectivity radius* $\delta(g_t)$ *satisfies* $\delta(g_t) \geqslant A_2 t$,

(iii) *the Riemann curvature* $R(g_t)$ *of* g_t *satisfies* $\|R(g_t)\|_{C^0} \leqslant A_3 t^{-2}$.

Here the operator d^* *and the norms* $\|\cdot\|_{L^2}$, $\|\cdot\|_{L^{14}}$ *and* $\|\cdot\|_{C^0}$ *depend on* g_t.

Here one should regard ψ_t as a *first integral* of the torsion $\nabla\varphi_t$ of (φ_t, g_t). Thus the norms $\|\psi_t\|_{L^2}$, $\|\psi_t\|_{C^0}$ and $\|\mathrm{d}^*\psi_t\|_{L^{14}}$ are measures of $\nabla\varphi_t$. So parts (i)–(iii) say that $\nabla\varphi_t$ is small compared to the injectivity radius and Riemann curvature of (M, g_t).

11.3.4   Step 4: Deforming to a torsion-free G_2-structure

We prove the following analysis result.

Theorem 11.3.4 *Let* A_1, A_2, A_3 *be positive constants. Then there exist positive constants* κ, K *such that whenever* $0 < t \leqslant \kappa$, *the following is true.*

Let M *be a compact 7-manifold, and* (φ, g) *a* G_2-*structure on* M *with* $\mathrm{d}\varphi = 0$. *Suppose* ψ *is a smooth 3-form on* M *with* $\mathrm{d}^*\psi = \mathrm{d}^*\varphi$, *and*

(i) $\|\psi\|_{L^2} \leqslant A_1 t^4$, $\|\psi\|_{C^0} \leqslant A_1 t^{1/2}$ *and* $\|\mathrm{d}^*\psi\|_{L^{14}} \leqslant A_1$,

(ii) *the injectivity radius* $\delta(g)$ *satisfies* $\delta(g) \geqslant A_2 t$, *and*

(iii) *the Riemann curvature* $R(g)$ *satisfies* $\|R(g)\|_{C^0} \leqslant A_3 t^{-2}$.

Then there is a smooth, torsion-free G_2-*structure* $(\tilde\varphi, \tilde g)$ *on* M *with* $\|\tilde\varphi - \varphi\|_{C^0} \leqslant K t^{1/2}$.

Basically, this says that if (φ, g) is a G_2-structure on M, and the torsion $\nabla\varphi$ is sufficiently small, then we can deform to a nearby G_2-structure $(\tilde\varphi, \tilde g)$ that is torsion-free. Here is a sketch of the proof of Theorem 11.3.4, ignoring several technical points. The proof is that given in [188, §11.6–§11.8], which improves that in [183].

We have a 3-form φ with $\mathrm{d}\varphi = 0$ and $\mathrm{d}^*\varphi = \mathrm{d}^*\psi$ for small ψ, and we wish to construct a nearby 3-form $\tilde\varphi$ with $\mathrm{d}\tilde\varphi = 0$ and $\tilde{\mathrm{d}}^*\tilde\varphi = 0$. Set $\tilde\varphi = \varphi + \mathrm{d}\eta$, where η is a small 2-form. Then η must satisfy a nonlinear p.d.e., which we write as

$$\mathrm{d}^*\mathrm{d}\eta = -\mathrm{d}^*\psi + \mathrm{d}^* F(\mathrm{d}\eta), \tag{11.6}$$

where F is nonlinear, satisfying $F(\mathrm{d}\eta) = O(|\mathrm{d}\eta|^2)$.

We solve (11.6) by iteration, introducing a sequence $\{\eta_j\}_{j=0}^{\infty}$ with $\eta_0 = 0$, satisfying the inductive equations

$$\mathrm{d}^*\mathrm{d}\eta_{j+1} = -\mathrm{d}^*\psi + \mathrm{d}^* F(\mathrm{d}\eta_j), \qquad \mathrm{d}^*\eta_{j+1} = 0. \tag{11.7}$$

If such a sequence exists and converges to η, then taking the limit in (11.7) shows that η satisfies (11.6), giving us the solution we want.

The key to proving this is an *inductive estimate* on the sequence $\{\eta_j\}_{j=0}^{\infty}$. The inductive estimate we use has three ingredients, the equations

$$\|\mathrm{d}\eta_{j+1}\|_{L^2} \leqslant \|\psi\|_{L^2} + C_1 \|\mathrm{d}\eta_j\|_{L^2} \|\mathrm{d}\eta_j\|_{C^0}, \tag{11.8}$$

$$\|\nabla\mathrm{d}\eta_{j+1}\|_{L^{14}} \leqslant C_2\big(\|\mathrm{d}^*\psi\|_{L^{14}} + \|\nabla\mathrm{d}\eta_j\|_{L^{14}} \|\mathrm{d}\eta_j\|_{C^0} + t^{-4} \|\mathrm{d}\eta_{j+1}\|_{L^2}\big), \tag{11.9}$$

$$\|\mathrm{d}\eta_j\|_{C^0} \leqslant C_3\big(t^{1/2} \|\nabla\mathrm{d}\eta_j\|_{L^{14}} + t^{-7/2} \|\mathrm{d}\eta_j\|_{L^2}\big). \tag{11.10}$$

Here C_1, C_2, C_3 are positive constants independent of t. Equation (11.8) is obtained from (11.7) by taking the L^2-inner product with η_{j+1} and integrating by parts. Using the fact that $d^*\varphi = d^*\psi$ and $\|\psi\|_{L^2} = O(t^4)$, $|\psi| = O(t^{1/2})$ we get a powerful estimate of the L^2-norm of $d\eta_{j+1}$.

Equation (11.9) is derived from an *elliptic regularity estimate* for the operator $d+d^*$ acting on 3-forms on M, as in §1.4. Equation (11.10) follows from the *Sobolev Embedding Theorem* Theorem 1.2.1, since $L_1^{14}(M) \hookrightarrow C^0(M)$. Both (11.9) and (11.10) are proved on small balls of radius $O(t)$ in M, using parts (ii) and (iii) of Theorem 11.3.3, and this is where the powers of t come from.

Using (11.8)–(11.10) and part (i) of Theorem 11.3.3 we show that if

$$\|d\eta_j\|_{L^2} \leqslant C_4 t^4, \quad \|\nabla d\eta_j\|_{L^{14}} \leqslant C_5, \quad \text{and} \quad \|d\eta_j\|_{C^0} \leqslant K t^{1/2}, \tag{11.11}$$

where C_4, C_5 and K are positive constants depending on C_1, C_2, C_3 and A_1, and if t is sufficiently small, then the same inequalities (11.11) apply to $d\eta_{j+1}$. Since $\eta_0 = 0$, by induction (11.11) applies for all j and the sequence $\{d\eta_j\}_{j=0}^\infty$ is bounded in the Banach space $L_1^{14}(\Lambda^3 T^* M)$. One can then use standard techniques in analysis to prove that this sequence converges to a smooth limit $d\eta$. This concludes the proof of Theorem 11.3.4.

From Theorems 11.3.3 and 11.3.4 we see that the compact 7-manifold M constructed in Step 2 admits torsion-free G_2-structures $(\tilde\varphi, \tilde g)$. Proposition 11.2.1 then shows that $\mathrm{Hol}(\tilde g) = G_2$ if and only if $\pi_1(M)$ is finite. In the example above M is simply-connected, and so M has metrics with holonomy G_2, as we want.

By considering different groups Γ acting on T^7, and also by finding topologically distinct resolutions $M_1, \ldots, M_k$ of the same orbifold T^7/Γ, we can construct many compact Riemannian 7-manifolds with holonomy G_2. Figure 11.1 displays the 252 different sets of Betti numbers of compact, simply-connected 7-manifolds with holonomy G_2 constructed in [183, 184] and [188, §12], together with 5 more sets from Kovalev [223]. They satisfy $0 \leqslant b^2 \leqslant 28$ and $4 \leqslant b^3 \leqslant 215$.

Examples are also known [188, §12.4] of compact 7-manifolds with holonomy G_2 with finite, nontrivial fundamental group. It seems likely to the author that the Betti numbers given in Figure 11.1 are only a small proportion of the Betti numbers of all compact, simply-connected 7-manifolds with holonomy G_2.

11.3.5 Other constructions of compact G_2-manifolds

Here are two more methods, from [188, §11.9], of constructing compact 7-manifolds with holonomy G_2. The first has been successfully applied by Kovalev [223], and is based on an idea due to Simon Donaldson.

Method 1. Let X be a projective complex 3-fold with canonical bundle K_X, and s a holomorphic section of K_X^{-1} which vanishes to order 1 on a smooth divisor D in X. Then D has trivial canonical bundle, so D is T^4 or $K3$. Suppose D is a $K3$ surface. Define $Y = X \setminus D$, and suppose Y is simply-connected.

Then Y is a noncompact complex 3-fold with K_Y trivial, and one infinite end modelled on $D \times S^1 \times [0, \infty)$. Using a version of the proof of the Calabi Conjecture for noncompact manifolds one constructs a complete Calabi–Yau metric h on Y, which is asymptotic to the product on $D \times S^1 \times [0, \infty)$ of a Calabi–Yau metric on D, and Euclidean metrics on S^1 and $[0, \infty)$. We call such metrics *asymptotically cylindrical*.

Suppose we have such a metric on Y. Define a torsion-free G_2-structure (φ, g) on $S^1 \times Y$ as in Proposition 11.1.9. Then $S^1 \times Y$ is a noncompact G_2-manifold with

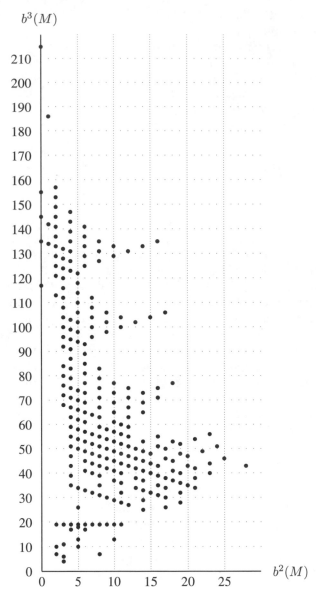

Fig. 11.1 Betti numbers (b^2, b^3) of compact, simply-connected G_2-manifolds

one end modelled on $D \times T^2 \times [0, \infty)$, whose metric is asymptotic to the product on $D \times T^2 \times [0, \infty)$ of a Calabi–Yau metric on D, and Euclidean metrics on T^2 and $[0, \infty)$.

Donaldson and Kovalev's idea is to take two such products $\mathcal{S}^1 \times Y_1$ and $\mathcal{S}^1 \times Y_2$ whose infinite ends are isomorphic in a suitable way, and glue them together to get a compact 7-manifold M with holonomy G_2. The gluing process swaps round the $\mathcal{S}^1$ factors. That is, the $\mathcal{S}^1$ factor in $\mathcal{S}^1 \times Y_1$ is identified with the asymptotic $\mathcal{S}^1$ factor in

$Y_2 \sim D_2 \times \mathcal{S}^1 \times [0, \infty)$, and vice versa.

Kovalev computes the Betti numbers (b^2, b^3) of the G_2-manifolds he constructs in [223, §8]. His examples satisfy $0 \leqslant b^2 \leqslant 9$ and $71 \leqslant b^3 \leqslant 155$, and typically have b^2 smaller than the examples of [188]. He finds 5 sets of Betti numbers not given in [188], which all have $b^2 = 0$ or 1, and are included in Figure 11.1.

The second method was outlined in [184, §4.3], but has not yet been implemented.

Method 2. Let (Y, J, h, θ) be a Calabi–Yau 3-fold, with Kähler form ω. Suppose σ : $Y \to Y$ is an involution, satisfying $\sigma^*(h) = h$, $\sigma^*(J) = -J$ and $\sigma^*(\theta) = \bar{\theta}$. We call σ a *real structure* on Y. Let N be the fixed point set of σ in Y. It is a real 3-dimensional submanifold of Y, a *special Lagrangian 3-fold*.

Let $\mathcal{S}^1 = \mathbb{R}/\mathbb{Z}$, and define a torsion-free G_2-structure (φ, g) on $\mathcal{S}^1 \times Y$ as in Proposition 11.1.9. Then $\varphi = \mathrm{d}x \wedge \omega + \mathrm{Re}\,\theta$, where $x \in \mathbb{R}/\mathbb{Z}$ is the coordinate on $\mathcal{S}^1$. Define $\hat{\sigma} : \mathcal{S}^1 \times Y \to \mathcal{S}^1 \times Y$ by $\hat{\sigma}\big((x, y)\big) = \big(-x, \sigma(y)\big)$. Then $\hat{\sigma}$ preserves (φ, g) and $\hat{\sigma}^2 = 1$. The fixed points of $\hat{\sigma}$ in $\mathcal{S}^1 \times Y$ are $\{\mathbb{Z}, \frac{1}{2} + \mathbb{Z}\} \times N$. Thus $(\mathcal{S}^1 \times Y)/\langle\hat{\sigma}\rangle$ is an orbifold. Its singular set is 2 copies of N, and each singular point is modelled on $\mathbb{R}^3 \times \mathbb{R}^4/\{\pm 1\}$.

We aim to resolve $(\mathcal{S}^1 \times Y)/\langle\hat{\sigma}\rangle$ to get a compact 7-manifold M with holonomy G_2. Locally, each singular point should be resolved like $\mathbb{R}^3 \times X$, where X is an ALE Calabi–Yau 2-fold asymptotic to $\mathbb{C}^2/\{\pm 1\}$. There is a 3-dimensional family of such X, and we need to choose one member of this family for each singular point.

Calculations by the author indicate that the data needed to do this is a closed, co-closed 1-form α on N that is nonzero at every point of N. The existence of a suitable 1-form α depends on the metric on N, which is the restriction of the metric g on Y. But g comes from the solution of the Calabi Conjecture, so we know little about it. This may make the method difficult to apply in practice.

11.4 The holonomy group $\mathrm{Spin}(7)$

We shall now discuss the holonomy group $\mathrm{Spin}(7)$, following the pattern of §11.1 for G_2. First we define $\mathrm{Spin}(7)$ as a subgroup of $\mathrm{GL}(8, \mathbb{R})$.

Definition 11.4.1 Let $\mathbb{R}^8$ have coordinates $(x_1, \dots, x_8)$. Write $\mathrm{d}x_{ijkl}$ for the 4-form $\mathrm{d}x_i \wedge \mathrm{d}x_j \wedge \mathrm{d}x_k \wedge \mathrm{d}x_l$ on $\mathbb{R}^8$. Define a 4-form Ω_0 on $\mathbb{R}^8$ by

$$
\begin{aligned}
\Omega_0 = \quad &\mathrm{d}x_{1234} + \mathrm{d}x_{1256} + \mathrm{d}x_{1278} + \mathrm{d}x_{1357} - \mathrm{d}x_{1368} - \mathrm{d}x_{1458} - \mathrm{d}x_{1467} \\
&- \mathrm{d}x_{2358} - \mathrm{d}x_{2367} - \mathrm{d}x_{2457} + \mathrm{d}x_{2468} + \mathrm{d}x_{3456} + \mathrm{d}x_{3478} + \mathrm{d}x_{5678}.
\end{aligned}
\tag{11.12}
$$

The subgroup of $\mathrm{GL}(8, \mathbb{R})$ preserving Ω_0 is the holonomy group $\mathrm{Spin}(7)$. It is a compact, connected, simply-connected, semisimple, 21-dimensional Lie group, which is isomorphic as a Lie group to the double cover of $\mathrm{SO}(7)$. This group also preserves the orientation on $\mathbb{R}^8$ and the Euclidean metric $g_0 = \mathrm{d}x_1^2 + \cdots + \mathrm{d}x_8^2$ on $\mathbb{R}^8$. We have $*\Omega_0 = \Omega_0$, where $*$ is the Hodge star on $\mathbb{R}^8$, so that Ω_0 is a self-dual 4-form.

Let M be an oriented 8-manifold. For each $p \in M$, define $\mathscr{A}_p M$ to be the subset of 4-forms $\Omega \in \Lambda^4 T_p^* M$ for which there exists an oriented isomorphism between $T_p M$ and $\mathbb{R}^8$ identifying Ω and the 4-form Ω_0 of (11.12). Then $\mathscr{A}_p M$ is isomorphic to

$GL_+(8, \mathbb{R}) / \mathrm{Spin}(7)$. Let $\mathscr{A}M$ be the bundle over M with fibre $\mathscr{A}_p M$ at each $p \in M$. Then $\mathscr{A}M$ is a subbundle of $\Lambda^4 T^* M$ with fibre $GL_+(8, \mathbb{R}) / \mathrm{Spin}(7)$. We say that a 4-form Ω on M is *admissible* if $\Omega|_p \in \mathscr{A}_p M$ for each $p \in M$.

The form Ω_0 above is given by Bryant [56, p. 545] and used in [185, 187, 188]. It differs from that used by Harvey and Lawson [151, p. 120] and McLean [259, p. 738], but is equivalent to it under changes of sign of some x_i and a change of sign of Ω_0.

Note that $\mathscr{A}M$ is *not* a vector subbundle of $\Lambda^4 T^* M$. As $\dim GL_+(8, \mathbb{R}) = 64$ and $\dim \mathrm{Spin}(7) = 21$, we see that $\mathscr{A}_p M$ has dimension $64 - 21 = 43$. But $\Lambda^4 T_p^* M$ has dimension $\binom{8}{4} = 70$, so $\mathscr{A}_p M$ has codimension 27 in $\Lambda^4 T_p^* M$. This is rather different from the G_2 case of §11.1, in which $\mathscr{P}_p^3 M$ is *open* in $\Lambda^3 T_p^* M$. As in §11.1, there is a 1-1 correspondence between oriented $\mathrm{Spin}(7)$-structures Q and admissible 4-forms $\Omega \in C^\infty(\mathscr{A}M)$ on M. Each $\mathrm{Spin}(7)$-structure Q induces a 4-form Ω on M and a metric g on M, corresponding to Ω_0 and g_0 on $\mathbb{R}^8$.

Definition 11.4.2 Let M be an oriented 8-manifold, Ω an admissible 4-form on M, and g the associated metric. As for G_2, we shall abuse notation by referring to the pair (Ω, g) as a $\mathrm{Spin}(7)$-*structure* on M. Let ∇ be the Levi-Civita connection of g. We call $\nabla \Omega$ the *torsion* of (Ω, g), and we say that (Ω, g) is *torsion-free* if $\nabla \Omega = 0$. A triple (M, Ω, g) is called a $\mathrm{Spin}(7)$-*manifold* if M is an oriented 8-manifold, and (Ω, g) a torsion-free $\mathrm{Spin}(7)$-structure on M.

The next four results are analogues of Propositions 11.1.3–11.1.6. The first three follow from Salamon [296, Lem. 12.4], [296, Prop. 12.5] and [296, Cor. 12.6] respectively, and the fourth is proved in the same way as Proposition 11.1.6.

Proposition 11.4.3 *Let M be an 8-manifold and (Ω, g) a $\mathrm{Spin}(7)$-structure on M. Then the following are equivalent:* (i) (Ω, g) *is torsion-free,*

 (ii) $\mathrm{Hol}(g) \subseteq \mathrm{Spin}(7)$, *and Ω is the induced 4-form,*

(iii) $\nabla \Omega = 0$ *on M, where ∇ is the Levi-Civita connection of g, and*

(iv) $d\Omega = 0$ *on M.*

Here $d\Omega = 0$ is a linear equation on the 4-form Ω. However, the restriction that $\Omega \in C^\infty(\mathscr{A}M)$ is nonlinear. Thus, as in the G_2 case, the condition that (Ω, g) be a torsion-free $\mathrm{Spin}(7)$-structure should be interpreted as a *nonlinear p.d.e.* upon Ω.

Proposition 11.4.4 *Let M be an oriented 8-manifold and (Ω, g) a $\mathrm{Spin}(7)$-structure on M. Then $\Lambda^k T^* M$ splits orthogonally into components, where Λ_l^k corresponds to an irreducible representation of $\mathrm{Spin}(7)$ of dimension l:*

 (i) $\Lambda^1 T^* M = \Lambda_8^1$,   (ii) $\Lambda^2 T^* M = \Lambda_7^2 \oplus \Lambda_{21}^2$,   (iii) $\Lambda^3 T^* M = \Lambda_8^3 \oplus \Lambda_{48}^3$,

(iv) $\Lambda^4 T^* M = \Lambda_+^4 T^* M \oplus \Lambda_-^4 T^* M$, $\Lambda_+^4 T^* M = \Lambda_1^4 \oplus \Lambda_7^4 \oplus \Lambda_{27}^4$, $\Lambda_-^4 T^* M = \Lambda_{35}^4$,

 (v) $\Lambda^5 T^* M = \Lambda_8^5 \oplus \Lambda_{48}^5$,   (vi) $\Lambda^6 T^* M = \Lambda_7^6 \oplus \Lambda_{21}^6$,   (vii) $\Lambda^7 T^* M = \Lambda_8^7$.

The Hodge star $*$ gives an isometry between Λ_l^k and Λ_l^{8-k}. In part (iv), $\Lambda_\pm^4 T^* M$ are the ± 1-eigenspaces of $*$ on $\Lambda^4 T^* M$. Note also that $\Lambda_1^4 = \langle \Omega \rangle$, and that there are canonical isomorphisms $\Lambda_8^1 \cong \Lambda_8^3 \cong \Lambda_8^5 \cong \Lambda_8^7$ and $\Lambda_7^2 \cong \Lambda_7^4 \cong \Lambda_7^6$.

The orthogonal projection from $\Lambda^k T^* M$ to Λ_l^k will be written π_l.

Proposition 11.4.5 *Let (M, g) be a Riemannian 8-manifold. If $\mathrm{Hol}(g)$ is a subgroup of $\mathrm{Spin}(7)$, then g is Ricci-flat.*

Proposition 11.4.6 *Let (Ω, g) be a $\mathrm{Spin}(7)$-structure on an 8-manifold M. Then M is orientable and spin, with a preferred spin structure and orientation. If (Ω, g) is torsion-free, then (M, g) has a nonzero parallel positive spinor.*

In fact, if $S = S_+ \oplus S_-$ is the spin bundle of M, then there are natural isomorphisms $S_+ \cong \Lambda_1^0 \oplus \Lambda_7^2$ and $S_- \cong \Lambda_8^1$. Here is the analogue of Theorem 11.1.7 for $\mathrm{Spin}(7)$.

Theorem 11.4.7 *The only connected Lie subgroups of $\mathrm{Spin}(7)$ which can be holonomy groups of Riemannian metrics on 8-manifolds are: (i) $\{1\}$,*

 (ii) $\mathrm{SU}(2)$, *acting on $\mathbb{R}^8 \cong \mathbb{R}^4 \oplus \mathbb{C}^2$ trivially on $\mathbb{R}^4$ and as usual on $\mathbb{C}^2$,*
(iii) $\mathrm{SU}(2) \times \mathrm{SU}(2)$, *acting on $\mathbb{R}^8 \cong \mathbb{C}^2 \oplus \mathbb{C}^2$ in the obvious way,*
 (iv) $\mathrm{SU}(3)$, *acting on $\mathbb{R}^8 \cong \mathbb{R}^2 \oplus \mathbb{C}^3$ trivially on $\mathbb{R}^2$ and as usual on $\mathbb{C}^3$,*
  (v) G_2, *acting on $\mathbb{R}^8 \cong \mathbb{R} \oplus \mathbb{R}^7$ trivially on $\mathbb{R}$ and as usual on $\mathbb{R}^7$,*
 (vi) $\mathrm{Sp}(2)$,   (vii) $\mathrm{SU}(4)$,   *and* (viii) $\mathrm{Spin}(7)$, *each acting as usual on $\mathbb{R}^8$.*

Thus, if (Ω, g) is a torsion-free $\mathrm{Spin}(7)$-structure on an 8-manifold, then $\mathrm{Hol}^0(g)$ is one of $\{1\}$, $\mathrm{SU}(2)$, $\mathrm{SU}(2) \times \mathrm{SU}(2)$, $\mathrm{SU}(3)$, G_2, $\mathrm{Sp}(2)$, $\mathrm{SU}(4)$ or $\mathrm{Spin}(7)$.

The inclusions '$\longrightarrow$' between these groups are shown below.

$$
\begin{array}{ccccccc}
\mathrm{SU}(2) & \!\!=\!\!=\!\!=\!\! & \mathrm{SU}(2) & \longrightarrow & \mathrm{SU}(3) & \longrightarrow & G_2 \\
\big\downarrow & & \big\downarrow & & \big\downarrow & & \big\downarrow \\
\mathrm{SU}(2) \times \mathrm{SU}(2) & \longrightarrow & \mathrm{Sp}(2) & \longrightarrow & \mathrm{SU}(4) & \longrightarrow & \mathrm{Spin}(7).
\end{array}
$$

As in Propositions 11.1.8 and 11.1.9, we can use these inclusions to construct $\mathrm{Spin}(7)$-manifolds out of Calabi–Yau 2-, 3- or 4-folds and G_2-manifolds. Here is how to do this explicitly in terms of differential forms, following [188, Prop.s 13.1.1–13.1.4].

Proposition 11.4.8 *Let (M, J, h, θ) be a Calabi–Yau 2-fold, with Kähler form ω. Let $(x_1, \ldots, x_4)$ be coordinates on $\mathbb{R}^4$ or T^4. Define a metric g and a 4-form Ω on $\mathbb{R}^4 \times M$ or $T^4 \times M$ by $g = \mathrm{d}x_1^2 + \cdots + \mathrm{d}x_4^2 + h$ and*

$$
\Omega = \mathrm{d}\mathbf{x}_{1234} + (\mathrm{d}\mathbf{x}_{12} + \mathrm{d}\mathbf{x}_{34}) \wedge \omega + (\mathrm{d}\mathbf{x}_{13} - \mathrm{d}\mathbf{x}_{24}) \wedge \mathrm{Re}\,\theta
$$
$$
- (\mathrm{d}\mathbf{x}_{14} + \mathrm{d}\mathbf{x}_{23}) \wedge \mathrm{Im}\,\theta + \tfrac{1}{2}\omega \wedge \omega.
$$

Then (Ω, g) is a torsion-free $\mathrm{Spin}(7)$-structure on $\mathbb{R}^4 \times M$ or $T^4 \times M$.

Proposition 11.4.9 *Let (M, J, h, θ) be a Calabi–Yau 3-fold, with Kähler form ω. Let (x_1, x_2) be coordinates on $\mathbb{R}^2$ or T^2. Define a metric g and a 4-form Ω on $\mathbb{R}^2 \times M$ or $T^2 \times M$ by $g = \mathrm{d}x_1^2 + \mathrm{d}x_2^2 + h$ and*

$$
\Omega = \mathrm{d}x_1 \wedge \mathrm{d}x_2 \wedge \omega + \mathrm{d}x_1 \wedge \mathrm{Re}\,\theta - \mathrm{d}x_2 \wedge \mathrm{Im}\,\theta + \tfrac{1}{2}\omega \wedge \omega.
$$

Then (Ω, g) is a torsion-free $\mathrm{Spin}(7)$-structure on $\mathbb{R}^2 \times M$ or $T^2 \times M$.

Proposition 11.4.10 *Let (M, φ, h) be a G_2-manifold. Define a metric g and a 4-form Ω on $\mathbb{R} \times Y$ or $S^1 \times Y$ by $g = dx^2 + h$ and $\Omega = dx \wedge \varphi + *\varphi$, where x is a coordinate on $\mathbb{R}$ or S^1. Then (Ω, g) is a torsion-free $\mathrm{Spin}(7)$-structure on $\mathbb{R} \times M$ or $S^1 \times M$.*

Proposition 11.4.11 *Let (M, J, g, θ) be a Calabi–Yau 4-fold, with Kähler form ω. Define a 4-form Ω on M by $\Omega = \frac{1}{2}\omega \wedge \omega + \mathrm{Re}\,\theta$. Then (Ω, g) is a torsion-free $\mathrm{Spin}(7)$-structure on M.*

11.5   Topological properties of compact $\mathrm{Spin}(7)$-manifolds

Next we study the topology of *compact $\mathrm{Spin}(7)$-manifolds* (M, Ω, g), that is, compact 8-manifolds M with a torsion-free $\mathrm{Spin}(7)$-structure (Ω, g). Because the dimension is divisible by four, we are able to use the *spin geometry* discussed in §3.6 to deduce important topological restrictions on M, which have no parallel in the G_2 case. Since M is oriented and spin by Proposition 11.4.6, we may consider the spin bundle $S = S_+ \oplus S_-$ and the positive Dirac operator $D_+ : C^\infty(S_+) \to C^\infty(S_-)$ on M.

As in §3.6.2, the *index* $\mathrm{ind}(D_+)$ of D_+ is determined by $\mathrm{Hol}(g)$, since g is Ricci-flat. But in §3.6.3 we explained that by the Atiyah–Singer Index Theorem, $\mathrm{ind}(D_+)$ is equal to $\hat{A}(M)$, a characteristic class of M. Thus the *geometric* invariant $\mathrm{Hol}(g)$ determines the *topological* invariant $\hat{A}(M)$. We use this idea in the following theorem, taken from [185, Th. C].

Theorem 11.5.1 *Suppose (M, Ω, g) is a compact $\mathrm{Spin}(7)$-manifold. Then the $\hat{A}$-genus $\hat{A}(M)$ of M satisfies*

$$24\hat{A}(M) = -1 + b^1 - b^2 + b^3 + b_+^4 - 2b_-^4, \qquad (11.13)$$

where b^i are the Betti numbers of M and $b_\pm^4$ the dimensions of $H_\pm^4(M, \mathbb{R})$. Moreover, if M is simply-connected then $\hat{A}(M)$ is $1, 2, 3$ or 4, and the holonomy group $\mathrm{Hol}(g)$ of g is determined by $\hat{A}(M)$ as follows:

  (i) $\mathrm{Hol}(g) = \mathrm{Spin}(7)$ *if and only if* $\hat{A}(M) = 1$,
 (ii) $\mathrm{Hol}(g) = \mathrm{SU}(4)$ *if and only if* $\hat{A}(M) = 2$,
(iii) $\mathrm{Hol}(g) = \mathrm{Sp}(2)$ *if and only if* $\hat{A}(M) = 3$, *and*
(iv) $\mathrm{Hol}(g) = \mathrm{SU}(2) \times \mathrm{SU}(2)$ *if and only if* $\hat{A}(M) = 4$.

Proof   To prove (11.13) we follow the reasoning of [294, §7]. For an 8-manifold, $\hat{A}(M)$ is given in terms of the Pontryagin classes $p_1(M)$ and $p_2(M)$ by

$$45.2^7 \hat{A}(M) = 7p_1(M)^2 - 4p_2(M).$$

From [294, eqn (7.1)], the signature $b_+^4 - b_-^4$ of M is given by

$$7p_2(M) - p_1(M)^2 = 45(b_+^4 - b_-^4),$$

and by [294, eqn (7.3)], which applies to manifolds with structure group $\mathrm{Spin}(7)$ by the remark on [294, p. 166], the Euler characteristic of M satisfies

$$4p_2(M) - p_1(M)^2 = 8(2 - 2b^1 + 2b^2 - 2b^3 + b^4).$$

Combining the last three equations gives (11.13), as we want.

Now suppose M is simply-connected. Theorem 11.4.7 shows that $\mathrm{Hol}(g)$ must be Spin(7), SU(4), Sp(2) or SU(2)×SU(2). Then Theorem 3.6.5 gives the value of $\hat{A}(M)$ for each holonomy group, which proves the 'only if' part of (i)–(iv). But as $\hat{A}(M)$ takes different values in the four cases, $\hat{A}(M)$ determines $\mathrm{Hol}(g)$. This gives the 'if' part of (i)–(iv), and completes the proof. $\qquad\square$

Next we show compact 8-manifolds with holonomy Spin(7) are simply-connected.

Proposition 11.5.2 *Suppose* (M, g) *is a compact Riemannian 8-manifold and* $\mathrm{Hol}(g)$ *is one of* Spin(7), SU(4), Sp(2) *or* SU(2) × SU(2). *Then* M *is simply-connected.*

Proof As g is Ricci-flat, $\pi_1(M)$ is finite by the argument used to prove Proposition 11.2.1. Let $\tilde{M}$ be the universal cover of M, and d the degree of the covering. Then $\tilde{M}$ is compact, and g lifts to a metric $\tilde{g}$ on $\tilde{M}$ with $\mathrm{Hol}(\tilde{g}) = \mathrm{Hol}^0(g) = \mathrm{Hol}(g)$. Thus $\hat{A}(\tilde{M}) = \hat{A}(M)$ by Theorem 11.5.1, and $\hat{A}(M) = 1, 2, 3$ or 4, depending on $\mathrm{Hol}(g)$. But $\hat{A}(\tilde{M}) = d \cdot \hat{A}(M)$, as the $\hat{A}$-genus is a characteristic class. Since $\hat{A}(M) \neq 0$, we see that $d = 1$, and so M is simply-connected. $\qquad\square$

By analogy with Definition 11.2.2 and Theorem 11.2.3, we have:

Definition 11.5.3 Let (M, Ω, g) be a compact Spin(7)-manifold. For each of the sub-bundles Λ_l^k of $\Lambda^k T^* M$ defined in Proposition 11.4.4, write

$$\mathscr{H}_l^k = \left\{ \xi \in C^\infty(\Lambda_l^k) : \mathrm{d}\xi = \mathrm{d}^*\xi = 0 \right\},$$

and let $H_l^k(M, \mathbb{R})$ be the vector subspace of $H^k(M, \mathbb{R})$ with representatives in $\mathscr{H}_l^k$. Define the *refined Betti numbers* b_l^k of M by $b_l^k = \dim H_l^k(M, \mathbb{R})$.

Theorem 11.5.4 *Let* (M, Ω, g) *be a compact* Spin(7)-*manifold. Then*

$$H^2(M, \mathbb{R}) = H_7^2(M, \mathbb{R}) \oplus H_{21}^2(M, \mathbb{R}), \quad H^3(M, \mathbb{R}) = H_8^3(M, \mathbb{R}) \oplus H_{48}^3(M, \mathbb{R}),$$
$$H_+^4(M, \mathbb{R}) = H_1^4(M, \mathbb{R}) \oplus H_7^4(M, \mathbb{R}) \oplus H_{27}^4(M, \mathbb{R}), \quad H_-^4(M, \mathbb{R}) = H_{35}^4(M, \mathbb{R}),$$
$$H^5(M, \mathbb{R}) = H_8^5(M, \mathbb{R}) \oplus H_{48}^5(M, \mathbb{R}), \quad H^6(M, \mathbb{R}) = H_7^6(M, \mathbb{R}) \oplus H_{21}^6(M, \mathbb{R}).$$

Here $H_1^4(M, \mathbb{R}) = \langle [\Omega] \rangle$ *and* $H_l^k(M, \mathbb{R}) \cong H_l^{8-k}(M, \mathbb{R})$, *so* $b_1^4 = 1$ *and* $b_l^k = b_l^{8-k}$.

If (M, Ω, g) is a compact Spin(7)-manifold, then

$$\hat{A}(M) = b_1^4 + b_7^4 - b_8^5 = 1 + b_7^2 - b^1.$$

To prove this we identify the Dirac operator $D_+ : C^\infty(S_+) \to C^\infty(S_-)$ with $\pi_8 \circ \mathrm{d} : C^\infty(\Lambda_1^4 \oplus \Lambda_7^4) \to C^\infty(\Lambda_8^5)$. This also suggests (correctly) that $\mathscr{H}_1^4$, $\mathscr{H}_7^4$ and $\mathscr{H}_8^5$ should consist of constant k-forms.

Proposition 11.5.5 *Let* (M, Ω, g) *be a compact* Spin(7)-*manifold. Then the spaces* $H_8^k(M, \mathbb{R})$ *for* $k = 1, 3, 5, 7$ *and* $H_7^k(M, \mathbb{R})$ *for* $k = 2, 4, 6$ *are represented by constant* k-*forms, and so are determined solely by* $\mathrm{Hol}(g)$. *In particular, if* $\mathrm{Hol}(g) = \mathrm{Spin}(7)$ *then these spaces are all zero, so that* $b_8^1 = b_8^3 = b_8^5 = b_8^7 = 0$ *and* $b_7^2 = b_7^4 = b_7^6 = 0$.

Proof We use the ideas of §3.5.2. Let $\xi \in \mathscr{H}_l^k$, so that $\xi \in C^\infty(\Lambda_l^k)$ and $(\mathrm{dd}^* + \mathrm{d}^*\mathrm{d})\xi = 0$. But $(\mathrm{dd}^* + \mathrm{d}^*\mathrm{d})\xi = \nabla^*\nabla\xi - 2\tilde{R}(\xi)$, by (3.13). By considering Spin(7) representations we find that if ξ lies in $\mathscr{H}_8^k$ for $k = 1, 3, 5, 7$ or $\mathscr{H}_7^k$ for $k = 2, 4, 6$ then $\tilde{R}(\xi) = 0$. Thus $\nabla^*\nabla\xi = 0$, and integrating by parts gives $\nabla\xi = 0$.

Thus $\mathscr{H}_8^k$ for $k = 1, 3, 5, 7$ and $\mathscr{H}_7^k$ for $k = 2, 4, 6$ consist of constant k-forms, and so are determined by $\mathrm{Hol}(g)$. As $\mathscr{H}_l^k \cong H_l^k(M, \mathbb{R})$ we see that $H_8^k(M, \mathbb{R})$ and $H_7^k(M, \mathbb{R})$ are determined by $\mathrm{Hol}(g)$. When $\mathrm{Hol}(g) = \mathrm{Spin}(7)$ these spaces are zero, as the corresponding Spin(7)-representations are nontrivial. $\square$

The last two results show that if (M, g) is a compact Riemannian 8-manifold with holonomy Spin(7) then M has only four nontrivial refined Betti numbers, $b_{21}^2, b_{48}^3, b_{27}^4$ and b_{35}^4. However, these are not independent, since the equation $\hat{A}(M) = 1$ implies that

$$b_{48}^3 + b_{27}^4 = b_{21}^2 + 2b_{35}^4 + 24.$$

Thus we see that there are only three independent Betti-type invariants of a compact 8-manifold with holonomy Spin(7), and we can calculate all the refined Betti numbers from b^2, b^3 and b^4.

Let (Ω, g) be a Spin(7)-structure on M. As in the G_2 case, if $\xi \in C^\infty(\Lambda_{21}^2)$ then $\xi \wedge \xi \wedge \Omega = -|\xi|^2 \mathrm{d}V_g$. So, by analogy with Propositions 11.2.5 and 11.2.6, we prove:

Proposition 11.5.6 *Suppose (M, Ω, g) is a compact Spin(7)-manifold with holonomy Spin(7). Then $\langle \sigma \cup \sigma \cup [\Omega], [M] \rangle < 0$ for each nonzero $\sigma \in H^2(M, \mathbb{R})$.*

Proposition 11.5.7 *Let (M, Ω, g) be a compact Spin(7)-manifold, R the Riemann curvature of g, and $p_1(M) \in H^4(M, \mathbb{Z})$ the first Pontryagin class. Then*

$$\langle p_1(M) \cup [\Omega], [M] \rangle = -\frac{1}{8\pi^2} \int_M |R|^2 \mathrm{d}V_g.$$

If $\mathrm{Hol}(g) = \mathrm{Spin}(7)$ then g is not flat, so $\langle p_1(M) \cup [\Omega], [M] \rangle < 0$, and $p_1(M) \neq 0$.

We summarize the results above as follows.

Theorem 11.5.8 *Suppose (M, Ω, g) is a compact Spin(7)-manifold. Then g has holonomy Spin(7) if and only if M is simply-connected and the Betti numbers of M satisfy $b^3 + b_+^4 = b^2 + 2b_-^4 + 25$. Also, if $\mathrm{Hol}(g) = \mathrm{Spin}(7)$ then M is spin and $p_1(M) \neq 0$.*

Let M be a compact, oriented 8-manifold. Then as in §11.4, we can identify the set of all oriented Spin(7)-structures on M with $C^\infty(\mathscr{A}M)$. Let $\mathscr{X} = \{\Omega \in C^\infty(\mathscr{A}M) : \mathrm{d}\Omega = 0\}$ be the set of admissible 4-forms corresponding to torsion-free Spin(7)-structures, and let $\mathscr{D}$ be the group of all diffeomorphisms of M isotopic to the identity. Then $\mathscr{D}$ acts naturally on $C^\infty(\mathscr{A}M)$ and $\mathscr{X}$. We define the *moduli space of torsion-free Spin(7)-structures* on M to be $\mathscr{M} = \mathscr{X}/\mathscr{D}$.

In a similar way to Theorem 11.2.8, our next result describes $\mathscr{M}$. It is proved in [185, Th. D] and [188, Th. 10.7.1]. Note that if M admits metrics with holonomy Spin(7) then $\hat{A}(M) = 1$ and $b^1(M) = 0$, so $\dim \mathscr{M} = 1 + b_-^4(M)$.

Theorem 11.5.9 *Let M be a compact, oriented 8-manifold, and $\mathscr{M} = \mathscr{X}/\mathscr{D}$ the moduli space of torsion-free* $\mathrm{Spin}(7)$-*structures on M. Then $\mathscr{M}$ is smooth with*

$$\dim \mathscr{M} = b_1^4 + b_7^4 + b_{35}^4 = \hat{A}(M) + b^1(M) + b_-^4(M),$$

and the projection $\pi : \mathscr{M} \to H^4(M, \mathbb{R})$ given by $\pi(\Omega\mathscr{D}) = [\Omega]$ is an immersion.

11.6 Constructing compact $\mathrm{Spin}(7)$-manifolds

There are at present two known methods for constructing examples of compact 8-manifolds with holonomy $\mathrm{Spin}(7)$. The first, due to the author in [185] and [188, §13–§14], is to resolve the singularities of torus orbifolds T^8/Γ in a similar way to the G_2 case of §11.3. It will be discussed in §11.6.1. The second, due to the author in [187] and [188, §15], is a little like Method 2 of §11.3.5 and starts with a *Calabi–Yau 4-orbifold*. It will be discussed in §11.6.2.

11.6.1 $\mathrm{Spin}(7)$-manifolds from resolving orbifolds T^8/Γ

The construction of compact 8-manifolds with holonomy $\mathrm{Spin}(7)$ from orbifolds T^8/Γ in [185] and [188, §13–§14] is very similar to the G_2 case of §11.3, so we describe it briefly using the same Steps 1–4, emphasizing only the differences with the G_2 case.

Step 1: Choosing an orbifold

The following example and lemma, taken from [185, §3.2], describe the simplest orbifold T^8/Γ the author knows that can be resolved to give a compact $\mathrm{Spin}(7)$-manifold.

Example 11.6.1 Let $(x_1, \ldots, x_8)$ be coordinates on $T^8 = \mathbb{R}^8/\mathbb{Z}^8$, where $x_i \in \mathbb{R}/\mathbb{Z}$. Define a flat $\mathrm{Spin}(7)$-structure (Ω_0, g_0) on T^8 as in Definition 11.4.1. Let α, β, γ and δ be the involutions of T^8 defined by

$$\alpha\big((x_1, \ldots, x_8)\big) = (-x_1, -x_2, -x_3, -x_4, x_5, x_6, x_7, x_8),$$
$$\beta\big((x_1, \ldots, x_8)\big) = (x_1, x_2, x_3, x_4, -x_5, -x_6, -x_7, -x_8),$$
$$\gamma\big((x_1, \ldots, x_8)\big) = (\tfrac{1}{2} - x_1, \tfrac{1}{2} - x_2, x_3, x_4, \tfrac{1}{2} - x_5, \tfrac{1}{2} - x_6, x_7, x_8),$$
$$\delta\big((x_1, \ldots, x_8)\big) = (-x_1, x_2, \tfrac{1}{2} - x_3, x_4, \tfrac{1}{2} - x_5, x_6, \tfrac{1}{2} - x_7, x_8).$$

By inspection, α, β, γ and δ preserve Ω_0 and g_0. It is easy to see that $\alpha^2 = \beta^2 = \gamma^2 = \delta^2 = 1$, and that $\alpha, \beta, \gamma, \delta$ all commute. Define Γ to be the group $\langle \alpha, \beta, \gamma, \delta \rangle$. Then $\Gamma \cong (\mathbb{Z}_2)^4$ is a group of automorphisms of T^8 preserving (Ω_0, g_0).

Lemma 11.6.2 *The nonidentity elements of Γ with fixed points in T^8 are $\alpha, \beta, \gamma, \delta$ and $\alpha\beta$. The corresponding singular sets are as follows: S_α is 4 copies of $T^4/\{\pm 1\}$, S_β is 4 copies of $T^4/\{+1\}$, S_γ is 2 copies of T^4, S_δ is 2 copies of T^4, and $S_{\alpha\beta}$ is 64 points. Here $S_\alpha \cap S_\beta = S_{\alpha\beta}$, and each $x\Gamma \in S_{\alpha\beta}$ has stabilizer $\Gamma_x = \{1, \alpha, \beta, \alpha\beta\}$.*

Thus, the singular set of T^8/Γ is rather more complex than that of Example 11.3.1.

Step 2: Resolving the singularities

Much as in §11.3.2, we can try to resolve the orbifolds T^8/Γ by gluing in *ALE spaces* with holonomy $SU(2), SU(3)$ or $SU(4)$, or more generally *Quasi-ALE spaces* with holonomy $SU(3), SU(4)$ or $Sp(2)$. Thus, we must choose T^8/Γ with all singularities of the form $\mathbb{R}^8/G$ for $G \subset Spin(7)$ conjugate in $Spin(7)$ to a subgroup of $SU(2), SU(3)$, $SU(4)$ or $Sp(2)$. However, not all such singularities can be resolved. If $G \subset SU(m)$ is a finite subgroup then $\mathbb{C}^m/G$ always admits a crepant resolution when $m = 2, 3$, but not necessarily when $m \geqslant 4$; for instance, $\mathbb{C}^4/\{\pm 1\}$ has no crepant resolution by Example 7.3.5.

This makes it more difficult to choose orbifolds T^8/Γ whose singularities can be resolved than in the G_2 case. Another problem is that in the G_2 case the singular set is made up of 1 and 3-dimensional pieces in a 7-dimensional space, so one can often arrange for the pieces to avoid each other, and resolve them independently. But in the $Spin(7)$ case the singular set is typically made up of 4-dimensional pieces in an 8-dimensional space, so they nearly always intersect. This happens even in our simple Example 11.6.1, when S_α, S_β intersect in $S_{\alpha\beta}$. The moral appears to be that when you increase the dimension, things become more difficult.

In Example 11.6.1 we can resolve T^8/Γ to get a compact 8-manifold M, using only the Eguchi–Hanson space X of Example 10.2.2. First consider S_γ, which is two disjoint copies of T^4, each modelled on the singularity of $T^4 \times \mathbb{C}^2/\{\pm 1\}$. The resolution of this singularity is $T^4 \times X$, and we may use this to resolve both S_γ and S_δ.

Now the singular sets S_α and S_β are not disjoint, but each component $T^4/\{\pm 1\}$ in S_α meets each component $T^4/\{\pm 1\}$ in S_β in 4 points, which lie in $S_{\alpha\beta}$. The singularity at each point in $S_{\alpha\beta}$ is modelled on $\mathbb{C}^2/\{\pm 1\} \times \mathbb{C}^2/\{\pm 1\}$. Here S_α corresponds locally to the subset $\mathbb{C}^2/\{\pm 1\} \times \{0\}$, and S_β corresponds locally to the subset $\{0\} \times \mathbb{C}^2/\{\pm 1\}$. Now the natural resolution of $\mathbb{C}^2/\{\pm 1\} \times \mathbb{C}^2/\{\pm 1\}$ is $X \times X$, and this is how we resolve near each point in $S_{\alpha\beta}$.

Each component of S_α and S_β is modelled locally on $T^4/\{\pm 1\} \times \mathbb{C}^2/\{\pm 1\}$, and the resolution of this is $K3 \times X$, where $T^4/\{\pm 1\}$ is resolved to give the $K3$ surface using the Eguchi-Hanson space X, as in Examples 10.3.2 and 10.3.14. Combining these resolutions gives a compact 8-manifold M with a map $\pi : M \to T^8/\Gamma$, making (M, π) a resolution of T^8/Γ. In [185, §3.2] we show M is simply-connected with Betti numbers $b^2 = 12$, $b^3 = 16$, $b^4_+ = 107$, $b^4_- = 43$ and $b^4 = 150$, so (11.13) gives $\hat{A}(M) = 1$.

Step 3: Finding $Spin(7)$*-structures with small torsion*

By a similar method to the G_2 case of §11.3.3, we rescale the Calabi–Yau structures on the resolutions of singularities of T^8/Γ by a small factor $t > 0$, and glue them to the flat $Spin(7)$-structure on T^8/Γ using a partition of unity to produce a 1-parameter family of $Spin(7)$-structures (Ω_t, g_t) on M for $t \in (0, \epsilon)$, such that (Ω_t, g_t) has *small torsion* when t is small. A convenient way to measure this torsion is to write down a small 4-form ϕ_t with $d\Omega_t + d\phi_t = 0$. We then prove the following estimates [188, Th. 13.5.8] on (Ω_t, g_t) and ϕ_t, similar to Theorem 11.3.3.

Theorem 11.6.3 *On the compact 8-manifold M described above, and on many other 8-manifolds constructed in a similar fashion, one can write down the following data* *explicitly in coordinates:*

- *positive constants* λ, μ, ν *and* ϵ,
- *a* $\mathrm{Spin}(7)$-*structure* (Ω_t, g_t) *on* M *for each* $t \in (0, \epsilon)$, *and*
- *a 4-form* ϕ_t *on* M *for each* $t \in (0, \epsilon)$ *such that* $\mathrm{d}\Omega_t + \mathrm{d}\phi_t = 0$.

These satisfy three conditions:

(i) $\|\phi_t\|_{L^2} \leqslant \lambda t^{13/3}$ *and* $\|\mathrm{d}\phi_t\|_{L^{10}} \leqslant \lambda t^{7/5}$,
(ii) *the injectivity radius* $\delta(g_t)$ *satisfies* $\delta(g_t) \geqslant \mu t$, *and*
(iii) *the Riemann curvature* $R(g_t)$ *satisfies* $\|R(g_t)\|_{C^0} \leqslant \nu t^{-2}$.

Here all norms are calculated using the metric g_t.

Step 4: Deforming to a torsion-free $\mathrm{Spin}(7)$-*structure*

We prove the following analysis result [188, Th. 13.6.1], similar to Theorem 11.3.4.

Theorem 11.6.4 *Let* λ, μ, ν *be positive constants. Then there exist positive constants* κ, K *such that whenever* $0 < t \leqslant \kappa$, *the following is true.*

Let M *be a compact 8-manifold, and* (Ω, g) *a* $\mathrm{Spin}(7)$-*structure on* M. *Suppose that* ϕ *is a smooth 4-form on* M *with* $\mathrm{d}\Omega + \mathrm{d}\phi = 0$, *and*

(i) $\|\phi\|_{L^2} \leqslant \lambda t^{13/3}$ *and* $\|\mathrm{d}\phi\|_{L^{10}} \leqslant \lambda t^{7/5}$,
(ii) *the injectivity radius* $\delta(g)$ *satisfies* $\delta(g) \geqslant \mu t$, *and*
(iii) *the Riemann curvature* $R(g)$ *satisfies* $\|R(g)\|_{C^0} \leqslant \nu t^{-2}$.

Then there exists a smooth, torsion-free $\mathrm{Spin}(7)$-*structure* $(\tilde{\Omega}, \tilde{g})$ *on* M *with* $\|\tilde{\Omega} - \Omega\|_{C^0} \leqslant K t^{1/3}$.

Here is how to interpret this. As $\nabla \Omega = 0$ if and only if $\mathrm{d}\Omega = 0$, and $\mathrm{d}\phi + \mathrm{d}\Omega = 0$, the torsion $\nabla\Omega$ is determined by $\mathrm{d}\phi$. Thus we can think of ϕ as a *first integral of the torsion* of (Ω, g). So $\|\phi\|_{L^2}$ and $\|\mathrm{d}\phi\|_{L^{10}}$ are both measures of the torsion of (Ω, g). As t is small, part (i) says that (Ω, g) has *small torsion* in a certain sense. Parts (ii) and (iii) mean that g is not too close to being singular.

Thus, the theorem as a whole says that if the torsion of (Ω, g) is small enough, and g is not too singular, then we can deform (Ω, g) to a nearby, torsion-free $\mathrm{Spin}(7)$-structure $(\tilde{\Omega}, \tilde{g})$ on M. Combining Theorems 11.6.3 and 11.6.4 gives many compact $\mathrm{Spin}(7)$-manifolds $(M, \tilde{\Omega}, \tilde{g})$. We can then use Theorem 11.5.1 to determine the holonomy group of $\tilde{g}$, which is $\mathrm{Spin}(7)$ for the case of Example 11.6.1.

In [185] and [188, §14], examples are constructed with 205 sets of Betti numbers. Using the methods of [185], Taylor [321] found 15 more examples with $b^3 = 0$, and Jang [178] (published in brief in Kim and Jang [213]) found 66 examples not given in [185]. Because of overlaps these examples realize 262 different sets of Betti numbers, which are listed in Table 11.1.

They satisfy $1 \leqslant b^2 \leqslant 50$, $0 \leqslant b^3 \leqslant 80$ and $82 \leqslant b^4 \leqslant 382$. Many of the examples also satisfy $2b^2 - 2b^3 + b^4 = 142$, but this is because [185, 213, 321] all use the same rather limited construction, which automatically produces manifolds with Euler characteristic 144. Two compact 8-manifolds with holonomy $\mathrm{Spin}(7)$ and the same Betti numbers may be distinguished by the cup products on their cohomologies (examples of this are given in [185, §3.4]), so they probably represent rather more than 262 topologically distinct 8-manifolds.

Table 11.1 Betti numbers (b^2, b^3, b^4) of compact $\mathrm{Spin}(7)$-manifolds from resolving T^8/Γ

(1,0,140)	(2,0,138)	(2,8,154)	(2,26,382)	(2,40,218)	(3,0,136)	(3,4,144)
(3,8,152)	(3,24,328)	(3,36,208)	(4,0,134)	(4,2,138)	(4,4,142)	(4,8,150)
(4,12,158)	(4,22,274)	(4,22,286)	(4,32,198)	(5,0,132)	(5,2,136)	(5,4,140)
(5,8,148)	(5,12,156)	(5,20,220)	(5,20,244)	(5,28,188)	(6,0,130)	(6,2,134)
(6,4,138)	(6,6,142)	(6,8,146)	(6,12,154)	(6,16,162)	(6,18,166)	(6,18,202)
(6,18,214)	(6,24,178)	(6,28,186)	(7,0,128)	(7,2,132)	(7,4,136)	(7,6,140)
(7,8,144)	(7,12,152)	(7,16,160)	(7,16,184)	(7,20,168)	(7,24,176)	(7,28,184)
(7,38,204)	(8,0,126)	(8,2,130)	(8,4,134)	(8,6,138)	(8,8,142)	(8,10,146)
(8,12,150)	(8,14,154)	(8,14,166)	(8,16,158)	(8,20,166)	(8,24,174)	(8,26,130)
(8,32,190)	(8,36,150)	(9,0,124)	(9,2,128)	(9,4,132)	(9,6,136)	(9,8,140)
(9,10,144)	(9,12,148)	(9,16,156)	(9,20,164)	(9,22,132)	(9,26,176)	(9,28,180)
(9,30,148)	(10,0,122)	(10,0,128)	(10,2,126)	(10,4,130)	(10,6,134)	(10,8,138)
(10,10,142)	(10,12,146)	(10,16,154)	(10,18,134)	(10,20,162)	(10,24,146)	(10,24,170)
(10,44,114)	(10,54,134)	(11,0,120)	(11,0,126)	(11,2,124)	(11,4,128)	(11,6,132)
(11,8,136)	(11,10,140)	(11,12,144)	(11,14,136)	(11,14,148)	(11,16,152)	(11,18,144)
(11,20,160)	(11,28,176)	(11,36,120)	(11,44,136)	(12,0,118)	(12,0,127)	(12,0,130)
(12,2,122)	(12,4,126)	(12,6,130)	(12,8,134)	(12,10,138)	(12,12,142)	(12,16,150)
(12,20,158)	(12,24,166)	(12,28,126)	(12,32,182)	(12,34,138)	(12,62,98)	(13,0,116)
(13,0,125)	(13,0,128)	(13,2,120)	(13,4,124)	(13,6,128)	(13,8,132)	(13,10,136)
(13,12,140)	(13,16,148)	(13,20,132)	(13,20,156)	(13,24,140)	(13,28,172)	(13,50,108)
(14,0,114)	(14,0,126)	(14,0,129)	(14,0,132)	(14,0,138)	(14,4,122)	(14,8,130)
(14,8,142)	(14,12,138)	(14,14,142)	(14,16,146)	(14,24,162)	(14,26,166)	(14,32,178)
(14,38,118)	(14,80,82)	(15,0,112)	(15,0,124)	(15,0,127)	(15,0,130)	(15,0,133)
(15,0,136)	(15,4,120)	(15,8,128)	(15,10,144)	(15,12,136)	(15,16,144)	(15,20,152)
(15,26,128)	(15,28,168)	(15,64,96)	(16,0,128)	(16,0,131)	(16,0,134)	(16,0,140)
(16,8,126)	(16,12,134)	(16,14,138)	(16,14,150)	(16,16,142)	(16,20,150)	(16,24,158)
(16,48,110)	(17,0,132)	(17,0,135)	(17,0,138)	(17,0,144)	(17,8,124)	(17,8,148)
(17,12,132)	(17,16,140)	(17,20,148)	(17,32,124)	(17,36,180)	(18,0,130)	(18,0,133)
(18,0,136)	(18,0,139)	(18,0,142)	(18,8,122)	(18,10,150)	(18,12,130)	(18,16,138)
(18,20,146)	(18,24,154)	(18,32,170)	(19,0,134)	(19,0,137)	(19,0,140)	(19,0,146)
(19,12,152)	(19,20,144)	(19,26,156)	(19,28,160)	(20,0,138)	(20,0,141)	(20,0,144)
(20,8,154)	(20,16,134)	(20,16,158)	(20,24,150)	(21,0,136)	(21,0,142)	(21,0,145)
(21,0,148)	(21,20,140)	(22,0,140)	(22,0,143)	(22,0,146)	(22,0,152)	(22,0,158)
(22,8,162)	(22,24,146)	(22,26,150)	(23,0,144)	(23,0,147)	(23,0,150)	(23,0,156)
(23,8,160)	(23,14,160)	(23,20,136)	(23,28,152)	(24,12,166)	(24,24,142)	(26,0,150)
(26,0,153)	(26,0,156)	(26,0,162)	(26,18,174)	(26,24,138)	(26,32,154)	(27,8,176)
(27,16,168)	(27,28,144)	(29,28,140)	(30,14,182)	(32,8,190)	(32,32,142)	(34,8,198)
(36,16,198)	(41,8,220)	(50,8,250)				

11.6.2 $\mathrm{Spin}(7)$-manifolds from Calabi–Yau 4-orbifolds

We now discuss a second construction of compact 8-manifolds with holonomy $\mathrm{Spin}(7)$ from [187] and [188, §15], similar to Method 2 of §11.3.5. In it we start from a *Calabi–Yau 4-orbifold* rather than from T^8. We divide the construction into five steps.

Step 1. Find a compact, complex 4-orbifold (Y, J) satisfying the conditions:

(a) Y has only finitely many singular points $p_1, \ldots, p_k$, for $k \geq 1$.

    (b) Y is modelled on $\mathbb{C}^4/\langle i \rangle$ near each p_j, where i acts on $\mathbb{C}^4$ by complex multiplication.

    (c) There exists an antiholomorphic involution $\sigma : Y \to Y$ whose fixed point set is $\{p_1, \ldots, p_k\}$.

    (d) $Y \setminus \{p_1, \ldots, p_k\}$ is simply-connected, and $h^{2,0}(Y) = 0$.

Step 2. Choose a σ-invariant Kähler class on Y. Then by Theorem 7.4.6 there exists a unique σ-invariant Ricci-flat Kähler metric g in this Kähler class. Let ω be the Kähler form of g and θ a holomorphic volume form for (Y, J, g). By multiplying θ by $e^{i\phi}$ if necessary, we can arrange that $\sigma^*(\theta) = \bar{\theta}$.

Define $\Omega = \frac{1}{2}\omega \wedge \omega + \operatorname{Re}\theta$. Then (Ω, g) is a torsion-free Spin(7)-structure on Y, by Proposition 11.4.11. Also, (Ω, g) is σ-invariant, as $\sigma^*(\omega) = -\omega$ and $\sigma^*(\theta) = \bar{\theta}$. Define $Z = Y/\langle\sigma\rangle$. Then Z is a compact real 8-orbifold with isolated singular points $p_1, \ldots, p_k$, and (Ω, g) pushes down to a torsion-free Spin(7)-structure (Ω, g) on Z.

Step 3. Z is modelled on $\mathbb{R}^8/G$ near each p_j, where G is a certain finite subgroup of Spin(7) with $|G| = 8$. We can write down two explicit, topologically distinct ALE Spin(7)-manifolds X_1, X_2 asymptotic to $\mathbb{R}^8/G$. Each carries a 1-parameter family of homothetic ALE metrics h_t for $t > 0$ with $\operatorname{Hol}(h_t) = \mathbb{Z}_2 \ltimes \operatorname{SU}(4) \subset \operatorname{Spin}(7)$.

For $j = 1, \ldots, k$ we choose $i_j = 1$ or 2, and resolve the singularities of Z by gluing in X_{i_j} at the singular point p_j for $j = 1, \ldots, k$, to get a compact, nonsingular 8-manifold M, with projection $\pi : M \to Z$.

Step 4. On M, we explicitly write down a 1-parameter family of Spin(7)-structures (Ω_t, g_t) depending on $t \in (0, \epsilon)$. They are not torsion-free, but have small torsion when t is small. As $t \to 0$, the Spin(7)-structure (Ω_t, g_t) converges to the singular Spin(7)-structure $\pi^*(\Omega_0, g_0)$.

Step 5. We prove using analysis that for sufficiently small t, the Spin(7)-structure (Ω_t, g_t) on M, with small torsion, can be deformed to a Spin(7)-structure $(\tilde{\Omega}_t, \tilde{g}_t)$, with zero torsion.

It turns out that if $i_j = 1$ for $j = 1, \ldots, k$ we have $\pi_1(M) \cong \mathbb{Z}_2$ and $\operatorname{Hol}(\tilde{g}_t) = \mathbb{Z}_2 \ltimes \operatorname{SU}(4)$, and for the other $2^k - 1$ choices of $i_1, \ldots, i_k$ we have $\pi_1(M) = \{1\}$ and $\operatorname{Hol}(\tilde{g}_t) = \operatorname{Spin}(7)$. So $\tilde{g}_t$ is a metric with holonomy Spin(7) on the compact 8-manifold M for $(i_1, \ldots, i_k) \neq (1, \ldots, 1)$.

Once we have completed Step 1, Step 2 is immediate. Steps 4 and 5 are essentially the same as Steps 3 and 4 of §11.6.1, so we discuss only Steps 1 and 3.

Step 1: An example of a suitable complex orbifold

We do Step 1 using complex algebraic geometry. The problem is that conditions (a)–(d) above are very restrictive, so it is not that easy to find *any* Y satisfying all four conditions. All the examples Y the author has found are constructed using *weighted projective spaces* $\mathbb{CP}^m_{a_0, \ldots, a_m}$, as in Definition 7.4.4.

    Here is the simplest example the author knows.

Example 11.6.5 Let Y be the hypersurface of degree 12 in $\mathbb{CP}^5_{1,1,1,1,4,4}$ given by

$$Y = \big\{ [z_0, \dots, z_5] \in \mathbb{CP}^5_{1,1,1,1,4,4} : z_0^{12} + z_1^{12} + z_2^{12} + z_3^{12} + z_4^3 + z_5^3 = 0 \big\}.$$

Calculation shows Y has singular points $p_1 = [0, 0, 0, 0, 1, -1]$, $p_2 = [0, 0, 0, 0, 1, e^{\pi i/3}]$ and $p_3 = [0, 0, 0, 0, 1, e^{-\pi i/3}]$ modelled on $\mathbb{C}^4/\langle i \rangle$, and trivial canonical bundle.

Now define a map $\sigma : Y \to Y$ by

$$\sigma : [z_0, \dots, z_5] \longmapsto [\bar z_1, -\bar z_0, \bar z_3, -\bar z_2, \bar z_5, \bar z_4].$$

Note that $\sigma^2 = 1$, though this is not immediately obvious, because of the geometry of $\mathbb{CP}^5_{1,1,1,1,4,4}$. It can be shown that conditions (a)–(d) of Step 1 above hold for Y and σ.

More suitable 4-folds Y may be found by taking hypersurfaces or complete intersections in other weighted projective spaces, possibly also dividing by a finite group, and then doing a crepant resolution to get rid of any singularities that we don't want. Examples are given in [187], [188, §15].

Step 3: Resolving $\mathbb{R}^8/G$

Define $\alpha, \beta : \mathbb{R}^8 \to \mathbb{R}^8$ by

$$\alpha : (x_1, \dots, x_8) \longmapsto (-x_2, x_1, -x_4, x_3, -x_6, x_5, -x_8, x_7),$$
$$\beta : (x_1, \dots, x_8) \longmapsto (x_3, -x_4, -x_1, x_2, x_7, -x_8, -x_5, x_6).$$

Then α, β preserve Ω_0 given in (11.12), so they lie in $\mathrm{Spin}(7)$. Also $\alpha^4 = \beta^4 = 1$, $\alpha^2 = \beta^2$ and $\alpha\beta = \beta\alpha^3$. Let $G = \langle \alpha, \beta \rangle$. Then G is a finite nonabelian subgroup of $\mathrm{Spin}(7)$ of order 8, which acts freely on $\mathbb{R}^8 \setminus \{0\}$. One can show that if Z is the compact $\mathrm{Spin}(7)$-orbifold constructed in Step 2 above, then $T_{p_j} Z$ is isomorphic to $\mathbb{R}^8/G$ for $j = 1, \dots, k$, with an isomorphism identifying the $\mathrm{Spin}(7)$-structures (Ω, g) on Z and (Ω_0, g_0) on $\mathbb{R}^8/G$, such that β corresponds to the σ-action on Y.

In the next two examples we shall construct two different ALE $\mathrm{Spin}(7)$-manifolds (X_1, Ω_1, g_1) and (X_2, Ω_2, g_2) asymptotic to $\mathbb{R}^8/G$.

Example 11.6.6 Define complex coordinates $(z_1, \dots, z_4)$ on $\mathbb{R}^8$ by

$$(z_1, z_2, z_3, z_4) = (x_1 + ix_2, x_3 + ix_4, x_5 + ix_6, x_7 + ix_8),$$

Then $g_0 = |\mathrm{d}z_1|^2 + \dots + |\mathrm{d}z_4|^2$, and $\Omega_0 = \frac{1}{2}\omega_0 \wedge \omega_0 + \mathrm{Re}\,\theta_0$, where ω_0 and θ_0 are the usual Kähler form and complex volume form on $\mathbb{C}^4$. In these coordinates, α and β are given by

$$\alpha : (z_1, \dots, z_4) \longmapsto (iz_1, iz_2, iz_3, iz_4),$$
$$\beta : (z_1, \dots, z_4) \longmapsto (\bar z_2, -\bar z_1, \bar z_4, -\bar z_3). \tag{11.14}$$

Now $\mathbb{C}^4/\langle \alpha \rangle$ is a complex singularity, as $\alpha \in \mathrm{SU}(4)$. Let (Y_1, π_1) be the blow-up of $\mathbb{C}^4/\langle \alpha \rangle$ at 0. Then Y_1 is the unique crepant resolution of $\mathbb{C}^4/\langle \alpha \rangle$. The action of β on $\mathbb{C}^4/\langle \alpha \rangle$ lifts to a *free* antiholomorphic map $\beta : Y_1 \to Y_1$ with $\beta^2 = 1$. Define $X_1 = Y_1/\langle \beta \rangle$. Then X_1 is a nonsingular 8-manifold, and the projection $\pi_1 : Y_1 \to \mathbb{C}^4/\langle \alpha \rangle$ pushes down to $\pi_1 : X_1 \to \mathbb{R}^8/G$.

There exist ALE Calabi–Yau metrics g_1 on Y_1, which were written down explicitly by Calabi [69, p. 285], and are invariant under the action of β on Y_1. Let ω_1 be the Kähler form of g_1, and $\theta_1 = \pi_1^*(\theta_0)$ the holomorphic volume form on Y_1. Define $\Omega_1 = \frac{1}{2}\omega_1 \wedge \omega_1 + \operatorname{Re}\theta_1$. Then (Ω_1, g_1) is a torsion-free Spin(7)-structure on Y_1, as in Proposition 11.4.11.

As $\beta^*(\omega_1) = -\omega_1$ and $\beta^*(\theta_1) = \bar{\theta}_1$, we see that β preserves (Ω_1, g_1). Thus (Ω_1, g_1) pushes down to a torsion-free Spin(7)-structure (Ω_1, g_1) on X_1. Then (X_1, Ω_1, g_1) is an *ALE Spin(7)-manifold* asymptotic to $\mathbb{R}^8/G$.

Example 11.6.7 Define new complex coordinates $(w_1, \ldots, w_4)$ on $\mathbb{R}^8$ by

$$(w_1, w_2, w_3, w_4) = (-x_1 + ix_3, x_2 + ix_4, -x_5 + ix_7, x_6 + ix_8).$$

Again we find that $g_0 = |\mathrm{d}w_1|^2 + \cdots + |\mathrm{d}w_4|^2$ and $\Omega_0 = \frac{1}{2}\omega_0 \wedge \omega_0 + \operatorname{Re}\theta_0$. In these coordinates, α and β are given by

$$\begin{aligned}
\alpha &: (w_1, \ldots, w_4) \mapsto (\bar{w}_2, -\bar{w}_1, \bar{w}_4, -\bar{w}_3), \\
\beta &: (w_1, \ldots, w_4) \mapsto (iw_1, iw_2, iw_3, iw_4).
\end{aligned} \tag{11.15}$$

Observe that (11.14) and (11.15) are the same, except that the rôles of α, β are reversed. Therefore we can use the ideas of Example 11.6.6 again.

Let Y_2 be the crepant resolution of $\mathbb{C}^4/\langle\beta\rangle$. The action of α on $\mathbb{C}^4/\langle\beta\rangle$ lifts to a free antiholomorphic involution of Y_2. Let $X_2 = Y_2/\langle\alpha\rangle$. Then X_2 is nonsingular, and carries a torsion-free Spin(7)-structure (Ω_2, g_2), making (X_2, Ω_2, g_2) into an ALE Spin(7)-manifold asymptotic to $\mathbb{R}^8/G$.

We can now explain the remarks on holonomy groups at the end of Step 5. The holonomy groups $\operatorname{Hol}(g_i)$ of the metrics g_1, g_2 in Examples 11.6.6 and 11.6.7 are both isomorphic to $\mathbb{Z}_2 \ltimes \mathrm{SU}(4)$, a subgroup of Spin(7). However, they are two *different* inclusions of $\mathbb{Z}_2 \ltimes \mathrm{SU}(4)$ in Spin(7), as in the first case the complex structure is α and in the second β.

The Spin(7)-structure (Ω, g) on Z also has holonomy $\operatorname{Hol}(g) = \mathbb{Z}_2 \ltimes \mathrm{SU}(4)$. Under the natural identifications we have $\operatorname{Hol}(g_1) = \operatorname{Hol}(g)$ but $\operatorname{Hol}(g_2) \neq \operatorname{Hol}(g)$ as subgroups of Spin(7). Therefore, if we choose $i_j = 1$ for all $j = 1, \ldots, k$, then Z and X_{i_j} all have the same holonomy group $\mathbb{Z}_2 \ltimes \mathrm{SU}(4)$, so they combine to give metrics $\tilde{g}_t$ on M with $\operatorname{Hol}(\tilde{g}_t) = \mathbb{Z}_2 \ltimes \mathrm{SU}(4)$.

However, if $i_j = 2$ for some j then the holonomy of g on Z and g_{i_j} on X_{i_j} are *different* $\mathbb{Z}_2 \ltimes \mathrm{SU}(4)$ subgroups of Spin(7), which together generate the whole group Spin(7). Thus they combine to give metrics $\tilde{g}_t$ on M with $\operatorname{Hol}(\tilde{g}_t) = \mathrm{Spin}(7)$.

The author was able in [187] and [188, Ch. 15] to construct compact 8-manifolds with holonomy Spin(7) realizing 14 distinct sets of Betti numbers, which are given in Table 11.2. They satisfy $0 \leqslant b^2 \leqslant 4$, $0 \leqslant b^3 \leqslant 33$ and $200 \leqslant b^4 \leqslant 11\,662$. None of them also occur in Table 11.1. Probably there are many other examples which can be produced by similar methods.

Comparing these Betti numbers with those of Table 11.1 we see that in these examples generally b^2 is smaller and b^4 much larger. Given that the two constructions of

Table 11.2 Betti numbers (b^2, b^3, b^4) of compact $\mathrm{Spin}(7)$-manifolds from resolving Calabi–Yau 4-orbifolds divided by $\mathbb{Z}_2$

$(0,0,910)$	$(0,0,1294)$	$(0,0,2446)$	$(0,0,4750)$	$(0,0,11\,662)$
$(0,6,3730)$	$(0,33,208)$	$(1,0,908)$	$(1,0,1292)$	$(1,0,2444)$
$(1,33,206)$	$(2,33,204)$	$(3,33,202)$	$(4,33,200)$	

compact 8-manifolds with holonomy $\mathrm{Spin}(7)$ that we know appear to produce sets of 8-manifolds with rather different 'geography', it is tempting to speculate that the set of all compact 8-manifolds with holonomy $\mathrm{Spin}(7)$ may be rather large, and that those constructed so far are a small sample with atypical behaviour.

11.7   Further reading on the exceptional holonomy groups

Four landmarks in the history of the exceptional holonomy groups are Berger's classification of holonomy groups in 1955, Bryant's proof of the local existence of metrics with exceptional holonomy in 1984, the author's construction of compact manifolds with exceptional holonomy in 1994, and the development of M-theory around 1995, which made string theorists very interested in G_2-manifolds, particularly from about 2001. We use these to divide our list of papers into four periods.

- **Early papers, 1955–1984.** Bonan [35] wrote down the G_2-invariant forms φ_0, $*\varphi_0$ of (11.1) and (11.2) and the $\mathrm{Spin}(7)$-invariant 4-form Ω_0 of (11.12), and showed that metrics with holonomy G_2 and $\mathrm{Spin}(7)$ are Ricci-flat.
  Fernández and Gray [106] took a G_2-structure (φ, g) on a 7-manifold, and decomposed $\nabla\varphi$ into irreducible pieces. Similarly, Fernández [103] took a $\mathrm{Spin}(7)$-structure (Ω, g) on an 8-manifold, and decomposed $\nabla\Omega$ into irreducible pieces.

- **Existence of exceptional holonomy metrics, 1984–1994.** In a very significant paper, Bryant [55, 56] used the theory of exterior differential systems to prove that there exist many metrics with holonomy G_2 and $\mathrm{Spin}(7)$ on small balls in $\mathbb{R}^7$ and $\mathbb{R}^8$. He also gave some explicit, noncomplete examples of such metrics.
  Later, Bryant and Salamon [64] wrote down explicit, complete metrics with holonomy G_2 and $\mathrm{Spin}(7)$ on noncompact manifolds, which are the total spaces of vector bundles over manifolds of dimension 3 and 4, and have large symmetry groups. The same metrics were also found by Gibbons et al. [125].
  In 1986–7 Fernández and others [77, 104, 105, 107] gave examples of compact 7-manifolds M with G_2-structures (φ, g) such that either $\mathrm{d}\varphi = 0$ or $\mathrm{d}^*\varphi = 0$, but not both. Simple examples with $\mathrm{d}^*\varphi = 0$ are also provided by real hypersurfaces in $\mathbb{O} = \mathbb{R}^8$. Bryant's paper [61] gives local results on G_2-structures which date from this period, though not published until 2003.

- **Compact manifolds with exceptional holonomy, 1994–2000.** In 1994–5 the author constructed examples of compact 7-manifolds with holonomy G_2 [183, 184], and of compact 8-manifolds with holonomy $\mathrm{Spin}(7)$ [185]. The constructions were made more powerful and many more examples constructed in the author's monograph [188]. Kovalev [223] gave a second construction of compact 7-manifolds with holonomy G_2. All this is described in §11.3 and §11.6.

At about the same time, physicists working in string theory became interested in using compact manifolds with holonomy G_2 and $\text{Spin}(7)$ as vacua for string theories. Some papers on this are Papadopoulos and Townsend [276], Shatashvili and Vafa [310, 311], Acharya [1–3], Figueroa-O'Farrill [108] and Vafa [331].

Hitchin [163] introduced an attractive point of view on G_2-structures, which has been influential in string theory, as in [96]. Given a compact oriented 7-manifold M, he considered the infinite-dimensional manifold $\mathcal{P}_\alpha$ of closed 3-forms φ on M with $[\varphi] = \alpha \in H^3(M, \mathbb{R})$ which are *positive* in the sense of §11.1, and so extend to a G_2-structure (φ, g). He defined a functional $\Phi : \mathcal{P}_\alpha \to \mathbb{R}$ taking φ to the volume of M with respect to g, and showed that φ is a stationary point of Φ if and only if (φ, g) is torsion-free. This led to an alternative proof of Theorem 11.2.8.

- **M-theory and explicit metrics with exceptional holonomy, 2001–.** String theorists constructed and studied many new examples of explicit metrics on noncompact manifolds with exceptional holonomy, generally of cohomogeneity one or two, in a similar way to Bryant and Salamon [64]. A lot of these are nonsingular and complete, and asymptotically conical or asymptotically locally conical. For examples with holonomy G_2 see Brandhuber et al. [47, 48], Chong et al. [76] and Cvetič et al. [81–83, 85, 86, 89], and for examples with holonomy $\text{Spin}(7)$ see Cvetič et al. [82, 84, 88, 89], Gukov and Sparks [144] and Kanno and Yasui [208, 209]. Cvetič et al. [87] give a review of this area.

  Some other significant string theory papers from this period are Gukov, Yau and Zaslow [145] on fibrations of G_2-manifolds by coassociative 4-folds, Atiyah and Witten [14] and Acharya and Gukov [5] on the physics of singularities of manifolds with exceptional holonomy, and Dijkgraaf et al. [96] on topological M-theory.

Here are some papers on other topics related to exceptional holonomy:

- **Gauge theory over compact manifolds with exceptional holonomy.** Let (Ω, g) be a $\text{Spin}(7)$-structure on a compact 8-manifold M, let E be a vector bundle over M, and A a connection on E with curvature F_A. We call A a $\text{Spin}(7)$ *instanton* if $\pi_7(F_A) = 0$. Such connections occur in finite-dimensional moduli spaces, and have many properties in common with instantons on 4-manifolds.

  $\text{Spin}(7)$ instantons have been studied from the mathematical point of view by Thomas [322], Lewis [243] who constructs nontrivial examples of $\text{Spin}(7)$ instantons over compact 8-manifolds with holonomy $\text{Spin}(7)$, Reyes Carrión [287], Tian [327], Tao and Tian [318], and Brendle [50], and from the string theory point of view by Acharya et al. [6] and Baulieu et al. [23].

  One can also consider connections A on vector bundles E over a compact G_2-manifold with $\pi_7(F_A) = 0$, which we call G_2 *instantons*. But the expected dimension of moduli spaces of G_2 instantons is always zero, which may make them less interesting than the $\text{Spin}(7)$ case.

- **Nearly parallel G_2-structures.** These are G_2-structures (φ, g) on M which have a *Killing spinor*, rather than a constant spinor. They satisfy $d\varphi = -8\lambda * \varphi$ and $d * \varphi = 0$, for some $\lambda \in \mathbb{R}$, and are Einstein with nonnegative scalar curvature $168\lambda^2$. Such manifolds (M, φ, g) include 3-Sasakian 7-manifolds, Einstein–Sasakian 7-manifolds, and G_2-manifolds. For more details see Friedrich et al. [113].

12

Associative, coassociative and Cayley submanifolds

In Chapter 4 we defined calibrations and calibrated submanifolds. We now apply these ideas to manifolds with exceptional holonomy. There are two types of calibrated submanifolds in 7-manifolds with holonomy G_2, called *associative 3-folds* and *coassociative 4-folds*, and one in 8-manifolds with holonomy $\mathrm{Spin}(7)$, called *Cayley 4-folds*.

Coassociative 4-folds can be defined in terms of the vanishing of a closed form. This implies that they behave rather like special Lagrangian submanifolds in Chapter 8, so that moduli spaces of compact coassociative 4-folds are smooth, for instance. Associative 3-folds and Cayley 4-folds cannot be so defined, and this gives their theories a slightly different character. We discuss associative 3-folds and coassociative 4-folds in §12.1–§12.3, and Cayley 4-folds in §12.4–§12.5.

12.1 Associative 3-folds and coassociative 4-folds in $\mathbb{R}^7$

Here is the basic definition.

Definition 12.1.1 Let $\mathbb{R}^7$ have coordinates $(x_1, \ldots, x_7)$, and as in Definition 11.1.1 define a metric $g_0 = \mathrm{d}x_1^2 + \cdots + \mathrm{d}x_7^2$, a 3-form φ_0 and a 4-form $*\varphi_0$ on $\mathbb{R}^7$ by

$$
\begin{aligned}
\varphi_0 &= \mathrm{d}\mathbf{x}_{123} + \mathrm{d}\mathbf{x}_{145} + \mathrm{d}\mathbf{x}_{167} + \mathrm{d}\mathbf{x}_{246} - \mathrm{d}\mathbf{x}_{257} - \mathrm{d}\mathbf{x}_{347} - \mathrm{d}\mathbf{x}_{356}, \\
*\varphi_0 &= \mathrm{d}\mathbf{x}_{4567} + \mathrm{d}\mathbf{x}_{2367} + \mathrm{d}\mathbf{x}_{2345} + \mathrm{d}\mathbf{x}_{1357} - \mathrm{d}\mathbf{x}_{1346} - \mathrm{d}\mathbf{x}_{1256} - \mathrm{d}\mathbf{x}_{1247},
\end{aligned}
\tag{12.1}
$$

where $\mathrm{d}\mathbf{x}_{ij\ldots l}$ is short for $\mathrm{d}x_i \wedge \mathrm{d}x_j \wedge \cdots \wedge \mathrm{d}x_l$. By Harvey and Lawson [151, Ths IV.1.4 and IV.1.16], φ_0 and $*\varphi_0$ have comass one and so are *calibrations* on $\mathbb{R}^7$. We define an *associative 3-fold* in $\mathbb{R}^7$ to be an oriented 3-dimensional submanifold of $\mathbb{R}^7$ calibrated with respect to φ_0, and a *coassociative 4-fold* in $\mathbb{R}^7$ to be an oriented 4-dimensional submanifold of $\mathbb{R}^7$ calibrated with respect to $*\varphi_0$.

As in §4.3, one of the first steps towards understanding a calibration on $\mathbb{R}^n$ is to determine the family of oriented k-planes in $\mathbb{R}^n$ that it calibrates. Define an *associative 3-plane* in $\mathbb{R}^7$ to be an oriented 3-plane U in $\mathbb{R}^7$ with $\varphi_0|_U = \mathrm{vol}_U$, and a *coassociative 4-plane* in $\mathbb{R}^7$ to be an oriented 4-plane V in $\mathbb{R}^7$ with $*\varphi_0|_V = \mathrm{vol}_V$. Write $\mathcal{F}_{\varphi_0}, \mathcal{F}_{*\varphi_0}$ for the families of associative 3- and coassociative 4-planes, which are subsets of the oriented Grassmannians $\mathrm{Gr}_+(3, \mathbb{R}^7)$ and $\mathrm{Gr}_+(4, \mathbb{R}^7)$, respectively.

We wish to describe $\mathcal{F}_{\varphi_0}$ and $\mathcal{F}_{*\varphi_0}$. Examples of an associative 3-plane U and a coassociative 4-plane V are

$$U = \{(x_1, x_2, x_3, 0, 0, 0, 0) : x_1, x_2, x_3 \in \mathbb{R}\} \subset \mathbb{R}^7, \qquad (12.2)$$

$$V = \{(0, 0, 0, x_4, x_5, x_6, x_7) : x_4, \ldots, x_7 \in \mathbb{R}\} \subset \mathbb{R}^7, \qquad (12.3)$$

with the obvious orientations. These form an orthogonal splitting $\mathbb{R}^7 = U \oplus V$, which illustrates a general principle: as in §4.3, an oriented 3-plane W in $\mathbb{R}^7$ is calibrated with respect to φ_0 if and only if the orthogonal oriented 4-plane $W^\perp$ is calibrated w.r.t. $*\varphi_0$. So taking perpendicular subspaces yields a bijection between $\mathcal{F}_{\varphi_0}$ and $\mathcal{F}_{*\varphi_0}$.

The *cross product* $\times : \mathbb{R}^7 \times \mathbb{R}^7 \to \mathbb{R}^7$ provides a good way to understand associative 3-planes. For $\mathbf{u}, \mathbf{v} \in \mathbb{R}^7$ define

$$(\mathbf{u} \times \mathbf{v})^d = \mathbf{u}^a \mathbf{v}^b (\varphi_0)_{abc} (g_0)^{cd}, \qquad (12.4)$$

using index notation, where $(g_0)^{cd}$ is the inverse of $(g_0)_{cd}$. Then $\mathbf{u} \times \mathbf{v}$ is orthogonal to $\mathbf{u}, \mathbf{v}$ with $|\mathbf{u} \times \mathbf{v}|^2 = |\mathbf{u}|^2 |\mathbf{v}|^2 - (\mathbf{u} \cdot \mathbf{v})^2$. Let W be an oriented 3-plane in $\mathbb{R}^7$, and $\mathbf{u}_1, \mathbf{u}_2, \mathbf{u}_3$ an oriented orthonormal basis of W. Then a calculation shows that $\varphi_0|_W = ((\mathbf{u}_1 \times \mathbf{u}_2) \cdot \mathbf{u}_3) \, \text{vol}_W$. Therefore W is calibrated with respect to φ_0, that is, W is associative, if and only if $\mathbf{u}_3 = \mathbf{u}_1 \times \mathbf{u}_2$.

This implies that if $\mathbf{u}, \mathbf{v}$ are linearly independent in $\mathbb{R}^7$ then the 2-plane $\langle \mathbf{u}, \mathbf{v} \rangle_\mathbb{R}$ is contained in a unique associative 3-plane, which has oriented basis $\mathbf{u}, \mathbf{v}, \mathbf{u} \times \mathbf{v}$. Now G_2 acts transitively on the Grassmannian $\text{Gr}(2, \mathbb{R}^7)$ of 2-planes in $\mathbb{R}^7$, so G_2 must act transitively on $\mathcal{F}_{\varphi_0}$. Thus $\mathcal{F}_{\varphi_0}$ is the orbit of U in (12.2) under G_2. The subgroup of G_2 fixing U is $\text{SO}(4)$. So we deduce [151, Th. IV.1.8]:

Proposition 12.1.2 *The families $\mathcal{F}_{\varphi_0}$ of associative 3-planes in $\mathbb{R}^7$ and $\mathcal{F}_{*\varphi_0}$ of coassociative 4-planes in $\mathbb{R}^7$ are both isomorphic to $G_2 / \text{SO}(4)$, with dimension 8.*

From (12.1) and (12.3) we see that $\varphi_0|_V = 0$. Therefore $\varphi_0|_W = 0$ for every coassociative 4-plane W in $\mathbb{R}^7$, since $W = \gamma \cdot V$ for some $\gamma \in G_2$, and φ_0 is G_2-invariant. It follows from [151, Cor. IV.1.20] that the converse is also true, which proves:

Proposition 12.1.3 *A 4-plane W in $\mathbb{R}^7$ is coassociative, with some unique orientation, if and only if $\varphi_0|_W \equiv 0$.*

This implies an analogue of Proposition 8.1.2 for coassociative 4-folds.

Proposition 12.1.4 *Let L be a real 4-dimensional submanifold of $\mathbb{R}^7$. Then L admits an orientation making it into a coassociative 4-fold in $\mathbb{R}^7$ if and only if $\varphi_0|_L \equiv 0$.*

This can be regarded as an *alternative definition* of coassociative 4-folds. Describing coassociative 4-folds as 4-submanifolds L in $\mathbb{R}^7$ with $\varphi_0|_L \equiv 0$ is often more useful, and easier to work with, than saying they are calibrated with respect to $*\varphi_0$. As associative and coassociative submanifolds in $\mathbb{R}^7$ are calibrated, they are minimal. Harvey and Lawson [151, Cors IV.2.5 & IV.2.10] use this to show that they are real analytic.

Theorem 12.1.5 *Let L be an associative 3-fold or a coassociative 4-fold in $\mathbb{R}^7$. Then L is real analytic wherever it is nonsingular.*

They also use exterior differential systems to prove [151, Th.s IV.4.1 & IV.4.6]:

Theorem 12.1.6 *Let P be a real analytic 2-submanifold in $\mathbb{R}^7$. Then there exists a locally unique associative 3-fold L in $\mathbb{R}^7$ containing P.*

Theorem 12.1.7 *Let P be a real analytic 3-submanifold in $\mathbb{R}^7$ with $\varphi_0|_P \equiv 0$. Then there exists a locally unique coassociative 4-fold L in $\mathbb{R}^7$ containing P.*

The condition $\varphi_0|_P \equiv 0$ is necessary by Proposition 12.1.4. One can use the same methods to show that associative 3-folds in $\mathbb{R}^7$ 'depend on 4 functions of 2 variables', and coassociative 4-folds in $\mathbb{R}^7$ 'depend on 2 functions of 3 variables', in the sense of exterior differential systems. Thus, there are very many associative and coassociative submanifolds in $\mathbb{R}^7$.

12.1.1 Associative 3-folds as graphs, and local deformations

In a similar way to SL m-folds in §8.1.2, we can write a class of associative 3-folds in $\mathbb{R}^7$ as graphs. It is convenient to do this using the quaternions $\mathbb{H}$. Let $f : \mathbb{R}^3 \to \mathbb{H}$ be a smooth function, written

$$f(x_1, x_2, x_3) = f_0(x_1, x_2, x_3) + f_1(x_1, x_2, x_3)i + f_2(x_1, x_2, x_3)j + f_3(x_1, x_2, x_3)k.$$

Define a 3-submanifold L in $\mathbb{R}^7$, the graph of f, by

$$L = \big\{ \big(x_1, x_2, x_3, f_0(x_1, x_2, x_3), \ldots, f_3(x_1, x_2, x_3)\big) : x_j \in \mathbb{R} \big\}.$$

Then Harvey and Lawson [151, §IV.2.A] calculate the conditions on f for L to be associative. Using G_2 forms as in (12.1) (different to those of [151]), the equation is

$$i\frac{\partial f}{\partial x_1} + j\frac{\partial f}{\partial x_2} - k\frac{\partial f}{\partial x_3} = C\Big(\frac{\partial f}{\partial x_1}, \frac{\partial f}{\partial x_2}, \frac{\partial f}{\partial x_3}\Big), \tag{12.5}$$

where $C : \mathbb{H} \times \mathbb{H} \times \mathbb{H} \to \mathbb{H}$ is a trilinear cross product. When f, $\mathrm{d}f$ are small, so that L approximates the associative 3-plane U of (12.2), eqn (12.5) reduces approximately to the linear equation $i\frac{\partial f}{\partial x_1} + j\frac{\partial f}{\partial x_2} - k\frac{\partial f}{\partial x_3} = 0$, which is equivalent to the *Dirac equation* on $\mathbb{R}^3$, and is elliptic. One can also show (12.5) is a nonlinear elliptic equation.

We can use this to discuss small *deformations* of associative 3-folds in $\mathbb{R}^7$, which are the local model for deformations of associative 3-folds in a G_2-manifold (M, φ, g). The family $\mathrm{Gr}_+(3, \mathbb{R}^7)$ of all oriented 3-planes in $\mathbb{R}^7$ has dimension 12, and the family $\mathcal{F}_{\varphi_0}$ of associative 3-planes has dimension 8 by Proposition 12.1.2. Thus the associative 3-planes are of codimension 4 in the set of all 3-planes. Therefore the condition for a 3-submanifold L in $\mathbb{R}^7$ to be associative is 4 real equations on each tangent space.

The freedom to vary L is the sections of its normal bundle in $\mathbb{R}^7$, which is 4 real functions. Thus, the deformation problem for associative 3-folds involves 4 *equations on 4 functions*, so it is a *determined* problem. This corresponds to the fact that (12.5) is 4 equations on 4 functions. As (12.5) is elliptic, we see that deformations of associative 3-folds are controlled by a *first-order nonlinear elliptic p.d.e.*, which makes their deformation theory fairly well behaved.

12.1.2 Coassociative 4-folds as graphs, and local deformations

In a similar way, let $f : \mathbb{H} \to \mathbb{R}^3$ be a smooth function, written

$$f(x_0 + x_1 i + x_2 j + x_3 k) = (f_1, f_2, f_3)(x_0 + x_1 i + x_2 j + x_3 k).$$

Define a 4-submanifold N in $\mathbb{R}^7$, the graph of f, by

$$N = \big\{ \big(f_1(x_0 + x_1 i + x_2 j + x_3 k), f_2(x_0 + x_1 i + x_2 j + x_3 k), \\ f_3(x_0 + x_1 i + x_2 j + x_3 k), x_0, \ldots, x_3 \big) : x_j \in \mathbb{R} \big\}.$$

Then Harvey and Lawson [151, §IV.2.B] calculate the conditions on f for N to be coassociative. Using G_2 forms as in (12.1) (different to those of [151]), the equation is

$$i \partial f_1 + j \partial f_2 - k \partial f_3 = C(\partial f_1, \partial f_2, \partial f_3), \tag{12.6}$$

where the derivatives $\partial f_j = \partial f_j(x_0 + x_1 i + x_2 j + x_3 k)$ are interpreted as functions $\mathbb{H} \to \mathbb{H}$, and C is as in (12.5). When $f, \partial f$ are small, so that L approximates the coassociative 4-plane V of (12.3), eqn (12.6) reduces approximately to the linear equation $i \partial f_1 + j \partial f_2 - k \partial f_3 = 0$. This is an *overdetermined first order linear elliptic equation*, and more generally (12.6) is an *overdetermined first order nonlinear elliptic equation*.

We now discuss *deformations* of coassociative 4-folds. The families $\mathrm{Gr}_+(4, \mathbb{R}^7)$ of oriented 4-planes in $\mathbb{R}^7$ and $\mathcal{F}_{*\varphi_0}$ of coassociative 4-planes in $\mathbb{R}^7$ have dimensions 12 and 8, so $\mathcal{F}_{*\varphi_0}$ has codimension 4 in $\mathrm{Gr}_+(4, \mathbb{R}^7)$. Thus the condition for a 4-fold L in $\mathbb{R}^7$ to be coassociative is 4 equations on each tangent space. The freedom to vary L is the sections of its normal bundle in $\mathbb{R}^7$, which is 3 real functions. Hence, the deformation problem for coassociative 4-folds involves 4 *equations on 3 functions*, as in (12.6), and is controlled by an *overdetermined first order nonlinear elliptic p.d.e.*

Now *overdetermined* elliptic equations have good regularity theory, but very poor existence theory. That is, solutions of overdetermined elliptic equations tend to be very smooth (confirming Theorem 12.1.5), but such equations tend to have few solutions or none at all, even locally. So we would expect there to exist few coassociative 4-folds in $\mathbb{R}^7$, even locally, which contradicts our claim before §12.1.1 that coassociative 4-folds in $\mathbb{R}^7$ 'depend on 2 functions of 3 variables' and are very abundant.

The explanation, which can be made precise using exterior differential systems, is that the closure $\mathrm{d}\varphi_0 = 0$ of φ_0 acts as a kind of *integrability condition* for the existence of many coassociative 4-folds locally, just as the integrability of an almost complex structure ensures the existence of many complex submanifolds of dimension $k > 1$ locally. If (φ, g) is a G_2-structure on $\mathbb{R}^7$ with $\mathrm{d}\varphi \not\equiv 0$ and L is a 4-fold in $\mathbb{R}^7$ then $\varphi|_L \equiv 0$ implies $\mathrm{d}\varphi|_L \equiv 0$, which is an extra 1 equation on each tangent space of L. But $\mathrm{d}\varphi_0|_L \equiv 0$ holds automatically as $\mathrm{d}\varphi_0 \equiv 0$, and this automatic 1 equation compensates for the difference $4 - 3$ between the 4 real equations and 3 real functions, making the problem act like a determined elliptic equation.

An important conclusion is that in any generalization of coassociative 4-folds to some class of 'almost G_2-manifolds', as we generalized special Lagrangian m-folds to almost Calabi–Yau m-folds in §8.4, we should require the G_2-structure (φ, g) to satisfy $\mathrm{d}\varphi = 0$ (or more generally $\mathrm{d}\varphi = \alpha \wedge \varphi$ for some closed 1-form α). Otherwise there will be very few coassociative 4-folds even locally, and the theory will not be interesting.

12.1.3 Cones and asymptotically conical associative and coassociative k-folds

In a similar way to the special Lagrangian case of §8.3, one can study *conical* and *asymptotically conical* associative 3-folds and coassociative 4-folds in $\mathbb{R}^7$.

Definition 12.1.8 A closed, generally singular associative 3-fold or coassociative 4-fold C in $\mathbb{R}^7$ is called a *cone* if $C = tC$ for all $t > 0$. The *vertex* 0 of C is always a singular point unless C is a linear associative $\mathbb{R}^3$ or coassociative $\mathbb{R}^4$. We call the cone C *two-sided* if $C = -C$, and *one-sided* otherwise.

Suppose $C' = C \setminus \{0\}$ is nonsingular. Write $\mathcal{S}^6$ for the unit sphere in $\mathbb{R}^7$. Then the *link* $\Sigma = C \cap \mathcal{S}^6$ of C is a compact, nonsingular submanifold of $\mathcal{S}^6$, of dimension 2 if C is associative and 3 if C is coassociative. Define $\iota : \Sigma \times (0, \infty) \to \mathbb{R}^7$ by $\iota(\sigma, r) = r\sigma$. Then $\iota : \Sigma \times (0, \infty) \to C'$ is a diffeomorphism.

A 3-dimensional cone C in $\mathbb{R}^7$ is associative if and only if its link Σ is a *pseudo-holomorphic curve* in $\mathcal{S}^6$, with respect to a certain non-integrable almost complex structure. So papers about pseudoholomorphic curves in $\mathcal{S}^6$ are effectively about associative cones. *Curvature properties* of pseudoholomorphic curves in $\mathcal{S}^6$ are studied by Hashimoto [153] and Sekigawa [307]. In the same way, a 4-dimensional cone C in $\mathbb{R}^7$ is coassociative if and only if its link Σ is *special Lagrangian* with respect to a certain $SU(3)$-structure with torsion on $\mathcal{S}^6$, but this notation is rarely used.

Definition 12.1.9 Let C be an associative or coassociative cone in $\mathbb{R}^7$ with $C \setminus \{0\}$ nonsingular, and define C', Σ, ι as above. Suppose L is a closed, nonsingular associative 3-fold or coassociative 4-fold in $\mathbb{R}^7$. We call L *Asymptotically Conical (AC)* with *rate* $\lambda < 1$ and *cone* C if there exists a compact subset $K \subset L$ and a diffeomorphism $\phi : \Sigma \times (T, \infty) \to L \setminus K$ for some $T > 0$, such that

$$\left| \nabla^k (\phi - \iota) \right| = O(r^{\lambda - k}) \quad \text{as } r \to \infty \text{ for } k = 0, 1. \tag{12.7}$$

Here $\nabla, |.|$ are computed using the cone metric $\iota^*(g_0)$ on $\Sigma \times (T, \infty)$.

This is modelled on Definition 8.3.7, but with a different convention on the rate λ: here $\lambda < 1$ and $|\phi - \iota| = O(r^\lambda)$, but in §8.3 we chose $\lambda < 2$ and $|\phi - \iota| = O(r^{\lambda - 1})$. This is because in §8.3 we worked in terms of a real function f with $\phi - \iota \approx \mathrm{d}f$ and $f = O(r^\lambda)$, but here we cannot do that.

Definition 12.1.9 is written to be as weak as possible, but it is equivalent to a stronger definition. One can show that if Definition 12.1.9 holds then making K, T larger if necessary, there exists a *unique* diffeomorphism $\phi' : \Sigma \times (T, \infty) \to L \setminus K$ with

$$(\phi' - \iota)(\sigma, r) \perp T_{\iota(\sigma, r)} C \quad \text{in } \mathbb{R}^7 \text{ for all } (\sigma, r) \in \Sigma \times (T, \infty),$$

as in (8.19), and this is the natural choice for ϕ. Furthermore, using elliptic regularity of the associative or coassociative equations and analysis on manifolds with ends as in Lockhart [245], one can show that

$$\left| \nabla^k (\phi' - \iota) \right| = O(r^{\lambda - k}) \quad \text{as } r \to \infty \text{ for all } k = 0, 1, 2, \ldots.$$

Thus it does not really matter how many derivatives we require (12.7) to hold for, any number from 1 to ∞ gives the same classes of AC (co)associative manifolds.

As in Chapter 8, (co)associative cones are interesting as local models for singularities of (co)associative manifolds in G_2-manifolds, and AC (co)associative submanifolds are interesting as local models for how to desingularize singular (co)associative submanifolds in G_2-manifolds, and also in themselves as attractive distinguished classes of calibrated submanifolds in $\mathbb{R}^7$.

Lotay [247] develops a coassociative analogue of the deformation theory for AC SL m-folds, Theorem 8.3.10 above. Here is his main result [247, Th. 7.1 & §8].

Theorem 12.1.10 *Let C be a coassociative cone in $\mathbb{R}^7$ with $C \setminus \{0\}$ nonsingular, and define Σ as above. For $\mu \in \mathbb{R}$ define D_μ to be the finite-dimensional vector space of $(\alpha, \beta) \in C^\infty(\Lambda^2 T^* \Sigma \oplus \Lambda^3 T^* \Sigma)$ satisfying $\mathrm{d}\alpha = \mu\beta$ and $\mathrm{d}(*\alpha) + \mathrm{d}^*\beta = (\mu + 2)\alpha$. Suppose $\lambda \in (-2, 1)$ with $D_\lambda = 0$, which holds for all but finitely many $\lambda \in (-2, 1)$.*

Let L be an asymptotically conical coassociative 4-fold in $\mathbb{R}^7$ with cone C and rate λ. Write $\mathcal{M}_L^\lambda$ for the moduli space of asymptotically conical coassociative 4-folds in $\mathbb{R}^7$ with cone C and rate λ isotopic to L. Then $\mathcal{M}_L^\lambda$ is a smooth manifold of computable dimension. When $\lambda \in (-2, 0)$ it satisfies

$$\dim \mathcal{M}_L^\lambda = b^0(L) - b^1(L) + b_+^2(L) + b^3(L) - b^0(\Sigma) + b^1(\Sigma) + \sum_{\mu \in (-2,\lambda): D_\mu \neq 0} \dim D_\mu. \quad (12.8)$$

Here $b^k(L), b^k(\Sigma)$ are the usual Betti numbers of L, Σ, but $b_+^2(L)$ is defined as follows. Write I for the image of the projection $H_{\mathrm{cs}}^2(L, \mathbb{R}) \to H^2(L, \mathbb{R})$ from compactly-supported to ordinary de Rham cohomology. Then the intersection form on $H_{\mathrm{cs}}^2(L, \mathbb{R})$ descends to a nondegenerate real quadratic form on I, and $b_+^2(L)$ is the dimension of a maximal subspace of I on which this form is positive definite.

This is a noncompact analogue of McLean's Theorem on deformations of compact coassociative 4-folds in G_2-manifolds, Theorem 12.3.4 below, proved using the theory of analysis on manifolds with cylindrical ends developed by Lockhart and McOwen [245]. To get a feel for Theorem 12.1.10, readers are advised to first read §12.3.1 on McLean's Theorem and its proof.

So far as the author knows, noone has yet developed a deformation theory for AC associative 3-folds in $\mathbb{R}^7$. As for compact associative 3-folds in §12.3.2 below, there will be obstruction spaces in this theory, so moduli spaces of such 3-folds will not generally be smooth. However, in contrast to the compact case, because of analytic contributions similar to the $\dim D_\mu$ terms in (12.8), the virtual dimension of these moduli spaces can be positive. So the asymptotically conical case may be more interesting than the compact case, as we expect to find moduli spaces of AC associative 3-folds which are singular, but smooth of positive, computable dimension at generic points.

Constructions are known which yield very many examples of conical and asymptotically conical associative 3-folds and coassociative 4-folds (in particular, *ruled* AC associative 3-folds and 2-*ruled* coassociative cones and AC coassociative 4-folds). These will be discussed in §12.2.

12.2 Constructing associative and coassociative k-folds in $\mathbb{R}^7$

We now review methods of constructing examples of associative 3-folds and coassociative 4-folds in $\mathbb{R}^7$, in a similar way to §8.2. Examples in other explicit noncompact

G_2-manifolds, such as those of Bryant and Salamon [64], may also be constructed using similar techniques; see for instance Karigiannis and Min-Oo [212].

12.2.1 Reduction to lower-dimensional calibrated geometries

Write $\mathbb{R}^7 = \mathbb{R} \oplus \mathbb{C}^3$, with coordinates $\big(x_1, (x_2 + ix_3, x_4 + ix_5, x_6 + ix_7)\big)$. Then as in Proposition 11.1.9 we have $\varphi_0 = dx_1 \wedge \omega + \operatorname{Re}\theta$ and $*\varphi_0 = \frac{1}{2}\omega \wedge \omega - dx_1 \wedge \operatorname{Im}\theta$, where ω, θ are the Kähler form and holomorphic volume on $\mathbb{C}^3$ given in (7.1). It follows easily that:

- If Σ is a *holomorphic curve* in $\mathbb{C}^3$ (and so calibrated with respect to ω) then $\mathbb{R} \times \Sigma$ is an associative 3-fold in $\mathbb{R}^7$.
- If $x \in \mathbb{R}$ and L is a *special Lagrangian 3-fold* in $\mathbb{C}^3$ (and so calibrated with respect to $\operatorname{Re}\Omega$) then $\{x\} \times L$ is an associative 3-fold in $\mathbb{R}^7$.
- If $x \in \mathbb{R}$ and S is a *holomorphic surface* in $\mathbb{C}^3$ (and so calibrated with respect to $\frac{1}{2}\omega \wedge \omega$) then $\{x\} \times S$ is a coassociative 4-fold in $\mathbb{R}^7$.
- If L is a *special Lagrangian 3-fold* in $\mathbb{C}^3$ with *phase* $-i$ (and so calibrated with respect to $-\operatorname{Im}\Omega$) then $\mathbb{R} \times L$ is a coassociative 4-fold in $\mathbb{R}^7$.

Enormous numbers of examples of holomorphic curves and surfaces in $\mathbb{C}^3$ may be produced using algebraic geometry. Many examples of SL 3-folds in $\mathbb{C}^3$ may be found as in §8.2. Together these give many examples of associative 3-folds and coassociative 4-folds in $\mathbb{R}^7$, including singular examples which are interesting as local models for singular associative 3-folds and coassociative 4-folds in G_2-manifolds.

These enable us to draw some useful conclusions. Roughly speaking, anything that goes wrong for holomorphic curves or surfaces or SL 3-folds will also go wrong for associative 3-folds or coassociative 4-folds. For instance, as singularities of SL 3-folds need not be real analytic [151, p. 97], singularities of associative 3-folds and coassociative 4-folds need not be real analytic either. But in general, we regard associative 3-folds and coassociative 4-folds in $\mathbb{R}^7$ as more interesting if they are not derived from lower-dimensional calibrated geometries.

12.2.2 Symmetry group and evolution equation constructions

As for the special Lagrangian case in §8.2.1, the simplest way to find nontrivial examples of associative or coassociative submanifolds L in $\mathbb{R}^7$ is to suppose L is preserved by a Lie subgroup G of the group $G_2 \ltimes \mathbb{R}^7$ of automorphisms of $\mathbb{R}^7$ preserving (φ_0, g_0). If G acts with cohomogeneity one on L then the (co)associative condition reduces to a first-order p.d.e. on G-orbits of the appropriate dimension, which can often be solved explicitly. One can also study associative or coassociative cones C in $\mathbb{R}^7$ preserved by a Lie subgroup G of G_2. This is easiest when G acts transitively or with cohomogeneity one on the link $\Sigma = C \cap \mathcal{S}^6$ of C.

Here are some papers using symmetry methods to study *associative cones* in $\mathbb{R}^7$, often expressed in terms of pseudoholomorphic curves in $\mathcal{S}^6$. Borůvka [39] finds a pseudoholomorphic $\mathcal{S}^2$ orbit of an SO(3) subgroup of G_2 acting irreducibly on $\mathbb{R}^7$. Ejiri [101, §5–§6] classifies pseudoholomorphic $\mathcal{S}^2$'s in $\mathcal{S}^6$ invariant under a U(1) subgroup of G_2. Kong, Terng and Wang [218, §6] give an integrable systems construction of all associative cones in $\mathbb{R}^7$ invariant under some U(1) subgroup of G_2. The authors do not address the periodicity question, and so determine whether there exist many $\mathrm{U}(1)$-invariant T^2-cones, but this should be a solvable problem. Lotay [250, §4] also studies $\mathrm{U}(1)$-invariant associative cones by more elementary methods; he constructs [250, Ths 4.3 and 4.4] a 4-dimensional family of very explicit $\mathrm{U}(1)$-invariant associative T^2-cones.

Constructions of non-conical associative 3-folds in $\mathbb{R}^7$ using analogues of the ideas of §8.2.1–§8.2.3 are given by Lotay [248] and [250, Th. 4.7].

For coassociative 4-folds there are fewer papers. Harvey and Lawson [151, §IV.3] give examples of coassociative 4-folds in $\mathbb{R}^7$ invariant under $\mathrm{SU}(2)$, acting on $\mathbb{R}^7 \cong \mathbb{R}^3 \oplus \mathbb{C}^2$ as $\mathrm{SO}(3) = \mathrm{SU}(2)/\{\pm 1\}$ on $\mathbb{R}^3$ and $\mathrm{SU}(2)$ on $\mathbb{C}^2$. Mashimo [255] classifies *coassociative cones* C in $\mathbb{R}^7$ with $C \cap \mathcal{S}^6$ an orbit of a Lie subgroup of G_2.

12.2.3 Integrable systems approaches to studying associative cones in $\mathbb{R}^7$

Bryant [54, §4] studies compact Riemann surfaces Σ in $\mathcal{S}^6$ pseudoholomorphic with respect to the almost complex structure J on $\mathcal{S}^6$ induced by its inclusion in $\mathrm{Im}\,\mathbb{O} \cong \mathbb{R}^7$. Then the cone on Σ is an *associative cone* on $\mathbb{R}^7$. He shows that any Σ has a *torsion* τ, a holomorphic analogue of the Serret–Frenet torsion of real curves in $\mathbb{R}^3$.

The torsion τ is a section of a holomorphic line bundle on Σ, and $\tau = 0$ if $\Sigma \cong \mathbb{CP}^1$. If $\tau = 0$ then Σ is the projection to $\mathcal{S}^6 = G_2/\mathrm{SU}(3)$ of a *holomorphic curve* $\tilde{\Sigma}$ in the *projective complex manifold* $G_2/\mathrm{U}(2)$. This reduces the problem of understanding null-torsion associative cones in $\mathbb{R}^7$ to that of finding holomorphic curves $\tilde{\Sigma}$ in $G_2/\mathrm{U}(2)$ satisfying a *horizontality condition*, which is a problem in complex algebraic geometry. In integrable systems language, null torsion curves are called *superminimal*.

Bryant also shows that *every* Riemann surface Σ may be embedded in $\mathcal{S}^6$ with null torsion in infinitely many ways, of arbitrarily high degree. This shows that there are many associative cones in $\mathbb{R}^7$, on oriented surfaces of every genus. These provide many local models for *singularities* of associative 3-folds.

Perhaps the simplest nontrivial example of a pseudoholomorphic curve Σ in $\mathcal{S}^6$ with null torsion is the *Borůvka sphere* [39], which is an $\mathcal{S}^2$ orbit of an $\mathrm{SO}(3)$ subgroup of G_2 acting irreducibly on $\mathbb{R}^7$. Other examples are given by Ejiri [101, §5–§6], who classifies pseudoholomorphic $\mathcal{S}^2$'s in $\mathcal{S}^6$ invariant under a $\mathrm{U}(1)$ subgroup of G_2, and Sekigawa [307].

Bryant's paper is one of the first steps in the study of associative cones in $\mathbb{R}^7$ using the theory of *integrable systems*. Bolton et al. [33, §6], [34] use integrable systems methods to prove important results on pseudoholomorphic curves Σ in $\mathcal{S}^6$. When Σ is a torus T^2, they show it is of *finite type* [33, Cor. 6.4], and so can be classified in terms of algebro-geometric *spectral data*, and perhaps even in principle be written down explicitly. Further results along these lines are given by Kong, Terng and Wang [218].

12.2.4 Ruled associative 3-folds

Motivated by the special Lagrangian case of §8.2.3, Lotay [248, §6] studies *ruled* associative 3-folds L in $\mathbb{R}^7$, that is, associative 3-folds L with a smooth map $\pi : L \to \Sigma$ to a 2-manifold Σ such that each fibre $\pi^{-1}(\sigma)$ is an affine line $\mathbb{R}$ in $\mathbb{R}^7$.

Actually, to allow singularities in L we need a more subtle idea: we take L to be a smooth 3-manifold with a smooth map $\iota : L \to \mathbb{R}^7$ which is generically an immersion, and whose image is associative where it is nonsingular, such that $\iota\big(\pi^{-1}(\sigma)\big)$ is an affine line in $\mathbb{R}^7$ for each $\sigma \in \Sigma$. Then $\iota(L)$ can be singular at the image of points where ι is not an immersion. But we shall neglect this issue. With this convention, any two-sided associative cone is ruled, as we shall shortly explain.

From each ruled associative 3-fold L we can construct a unique, 2-sided cone C with ruling $\pi_0 : C \to \Sigma$, such that $\pi_0^{-1}(\sigma)$ is the real line through 0 in $\mathbb{R}^7$ parallel to $\pi^{-1}(\sigma)$ for all $\sigma \in \Sigma$. It can happen that C collapses to a 1 or 2 dimensional cone. But if C is 3-dimensional then it is associative where it is nonsingular. If also L and $C \setminus \{0\}$ are nonsingular then L is *asymptotically conical*, with cone C, and rate 0.

Let C be a two-sided associative cone with isolated singularity at 0. Define $\tilde{\Sigma} = C \cap \mathcal{S}^6$. Then $\tilde{\Sigma}$ is an oriented Riemann surface. Since C is two-sided, the map $-1 : \mathcal{S}^6 \to \mathcal{S}^6$ restricts to an orientation-reversing involution $-1 : \tilde{\Sigma} \to \tilde{\Sigma}$. Define $\Sigma = \tilde{\Sigma}/\{\pm1\}$, a surface with a conformal structure. We can recover $\tilde{\Sigma}$ from Σ as the set of points (x, o) for $x \in \Sigma$ and o an orientation on $T_x\Sigma$. Let $\tilde{C}$ be the real blow-up of C at 0, with blow-up map $\iota : \tilde{C} \to C$. It is a nonsingular 3-manifold, with $\iota^{-1}(0) \cong \tilde{\Sigma}$. There is a natural smooth projection $\pi_0 : \tilde{C} \to \tilde{\Sigma}$ such that $\iota \circ \pi_0^{-1}(\{\pm\sigma\}) = \{r\sigma : r \in \mathbb{R}\}$ for all $\sigma \in \tilde{\Sigma}$. Then $\tilde{C}, \iota, \tilde{\Sigma}, \pi_0$ make up a ruled associative 3-fold.

Lotay [248, §6] shows that the ruled associative 3-folds L with cone C correspond to solutions of a linear equation on Σ, and so form a finite-dimensional vector space. These solutions can be interpreted as holomorphic sections of $T\mathcal{S}^6|_{\tilde{\Sigma}}$ which are invariant under $-1 : \tilde{\Sigma} \to \tilde{\Sigma}$, and so push down to Σ. Holomorphic vector fields on $\tilde{\Sigma}$ automatically yield holomorphic sections of $T\mathcal{S}^6|_{\tilde{\Sigma}}$. Let v be a holomorphic vector field on $\tilde{\Sigma}$ invariant under -1. Then Lotay [248, Th. 6.9] constructs a ruled associative 3-fold L_v in $\mathbb{R}^7$. For generically nonzero v this should be nonsingular, and is asymptotically conical with cone C and rate -1.

Requiring $\tilde{\Sigma}$ to have a generically nonzero holomorphic vector field forces each component of $\tilde{\Sigma}$ to be $\mathcal{S}^2$ or T^2. For L_v connected, there are four possibilities:

(a) Σ is $\mathcal{S}^2$ and $\tilde{\Sigma}$ is two $\mathcal{S}^2$ exchanged by -1, the space of -1-invariant holomorphic vector fields v on $\tilde{\Sigma}$ is $\mathbb{R}^6$, and L_v is diffeomorphic to $\mathcal{S}^2 \times \mathbb{R}$ for generic v.

(b) Σ is $\mathbb{RP}^2$ and $\tilde{\Sigma}$ is $\mathcal{S}^2$, the space of allowed v is $\mathbb{R}^3$, and L_v is diffeomorphic to the canonical bundle of $\mathbb{RP}^2$ for generic v.

(c) Σ is T^2 and $\tilde{\Sigma}$ is two T^2 exchanged by -1, the space of allowed v is $\mathbb{R}^2$, and L_v is diffeomorphic to $T^2 \times \mathbb{R}$ for generic v.

(d) Σ is the Klein bottle $\mathbb{K}$ and $\tilde{\Sigma}$ is T^2, the space of allowed v is $\mathbb{R}$, and L_v is diffeomorphic to the canonical bundle of $\mathbb{K}$ for generic v.

Using the material of [110, §6], it seems plausible that one could also prove the existence of nontrivial holomorphic sections of $T\mathcal{S}^6|_{\tilde{\Sigma}}$ for pseudoholomorphic curves $\tilde{\Sigma}$ in $\mathcal{S}^6$ of genus $g > 1$, and so construct AC associative 3-folds in $\mathbb{R}^7$ asymptotic to cones on surfaces of every genus. Combined with §12.2.3, these ideas give a way to construct many asymptotically conical associative 3-folds.

Given any 2-dimensional minimal surface M in $\mathbb{R}^4$, Ionel, Karigiannis and Min-Oo [173, §4.3] construct a vector subbundle $E \to M$ with fibre $\mathbb{R}$ of the restriction to M of the trivial bundle $\Lambda^2_-(\mathbb{R}^4) \to \mathbb{R}^4$ with fibre $\mathbb{R}^3$, such that the total space of E is a ruled associative 3-fold in the total space $\mathbb{R}^7$ of $\Lambda^2_-(\mathbb{R}^4)$.

12.2.5   2-ruled coassociative 4-folds

A similar theory to §12.2.4 for coassociative 4-folds is developed by Lotay [249] and Fox [110]. A 2-*ruled* coassociative 4-fold L is a coassociative 4-fold L in $\mathbb{R}^7$ with a smooth map $\pi : L \to \Sigma$ to a 2-manifold Σ such that each fibre $\pi^{-1}(\sigma)$ is an affine 2-plane $\mathbb{R}^2$ in $\mathbb{R}^7$. Actually, to include singularities we allow L to be a smooth 4-manifold with a smooth map $\iota : L \to \mathbb{R}^7$ which is generically an immersion, and whose image is coassociative where it is nonsingular, such that $\iota(\pi^{-1}(\sigma))$ is an affine 2-plane $\mathbb{R}^2$ in $\mathbb{R}^7$ for each $\sigma \in \Sigma$. But we neglect this point.

From each 2-ruled coassociative 4-fold L we can construct a unique, 2-ruled cone C with 2-ruling $\pi_0 : C \to \Sigma$, such that $\pi_0^{-1}(\sigma)$ is the 2-plane $\mathbb{R}^2$ through 0 in $\mathbb{R}^7$ parallel to $\pi^{-1}(\sigma)$ for all $\sigma \in \Sigma$. It can happen that C collapses to a 2 or 3 dimensional cone. But if C is 4-dimensional then it is coassociative where it is nonsingular. If also L and $C \setminus \{0\}$ are nonsingular then L is *asymptotically conical*, with cone C, and rate 0.

From [249, §4], the 2-ruled coassociative 4-folds L asymptotic to a given 2-ruled coassociative cone C with isolated singularity at 0 correspond to solutions of a linear equation on Σ, and so form a finite-dimensional vector space. Here Σ has the structure of a Riemann surface (or at least of a conformal 2-manifold, since it may not be orientable), and solutions of the linear equation can be roughly interpreted as holomorphic sections of a vector bundle over Σ with fibre $\mathbb{C}^2$. When L is in the *flat gauge* in the sense of [249, §4.1], by [249, Th. 4.10], holomorphic vector fields v on Σ give solutions of the equation. Thus if Σ is $\mathcal{S}^2$, $\mathbb{RP}^2$, T^2, or the Klein bottle $\mathbb{K}$, as in §12.2.4 we can construct 2-ruled coassociative 4-folds L_v, which should be nonsingular for generic v, and asymptotically conical with cone C and rate -1.

Fox [110] produces a beautiful correspondence between associative cones and 2-ruled coassociative cones. The basic idea is this. Let C_Σ be a 2-ruled coassociative cone with 2-ruling $\pi : C_\Sigma \to \Sigma$, for Σ an oriented 2-manifold. Construct a map $\phi : \Sigma \to \mathcal{S}^6$ as follows. For $\sigma \in \Sigma$, $\pi^{-1}(\sigma)$ is a vector space $\mathbb{R}^2 \subset \mathbb{R}^7$, with orientation determined by the orientations of C_Σ and Σ. Let $\mathbf{v}_1, \mathbf{v}_2$ be an oriented orthonormal basis of $\pi^{-1}(\sigma)$ and define $\phi(\sigma) = \mathbf{v}_1 \times \mathbf{v}_2$, using the cross product of (12.4). Then the image of the map $\Sigma \times [0, \infty) \to \mathbb{R}^7$ given by $(\sigma, r) \mapsto r\phi(\sigma)$ is an associative cone A_Σ in $\mathbb{R}^7$.

The extraordinary thing is that this map $C_\Sigma \mapsto A_\Sigma$ admits a *right inverse*. That is, for any associative cone A_Σ, Fox constructs a preferred 2-ruled coassociative cone C_Σ yielding A_Σ under the above construction. There may also be other C'_Σ yielding A_Σ, which correspond to holomorphic sections of a certain $\mathbb{CP}^2$-bundle over Σ.

Fox [110, §9] also constructs a family of non-2-ruled, non-conical coassociative 4-folds in $\mathbb{R}^7$ from each 2-ruled coassociative cone. These are generally asymptotically conical with rate $-\frac{3}{2}$, and Fox's method also yields a new, non-2-ruled coassociative cone from each 2-ruled coassociative cone. Combined with the material of §12.2.3, these ideas give ways of constructing very many coassociative cones and asymptotically conical coassociative 4-folds.

Ionel, Karigiannis and Min-Oo [173] construct 2-ruled coassociative 4-folds in $\mathbb{R}^7$ as the total spaces of vector bundles with fibre $\mathbb{R}^2$ over a minimal surface M in $\mathbb{R}^4$. Their first attempt [173, §4.2] turns out to yield only complex surfaces in some $\mathbb{C}^3 \subset \mathbb{R}^7$, as in §12.2.1. But their second construction [173, §4.5] (which was intended to find Cayley

4-folds in $\mathbb{R}^8$) produces genuine 2-ruled coassociative 4-folds which do not come from a lower-dimensional calibration.

12.2.6 Exterior differential systems

As for special Lagrangian submanifolds in §8.2.5, one can apply the theory of exterior differential systems to study coassociative 4-folds in $\mathbb{R}^7$ whose second fundamental form at a generic point has a prescribed stabilizer group. This has been done by Daniel Fox [110,111]. Several interesting distinguished classes of coassociative 4-folds emerge, including a special type of 2-ruled coassociative 4-folds, which explains Fox's interest in these. Presumably a similar analysis is possible for associative 3-folds.

12.3 Associative 3- and coassociative 4-folds in G_2-manifolds

We now discuss the calibrated submanifolds of G_2-manifolds.

Definition 12.3.1 Let (M, φ, g) be a G_2-manifold, as in §11.1. Then the 3-form φ and 4-form $*\varphi$ are *calibrations* on (M, g). We define an *associative 3-fold* in M to be an oriented 3-submanifold N of M calibrated with respect to φ, and a *coassociative 4-fold* in M to be an oriented 4-submanifold N of M calibrated with respect to $*\varphi$.

Proposition 12.1.4 implies an alternative characterization of coassociative 4-folds:

Proposition 12.3.2 *Let* (M, φ, g) *be a* G_2-*manifold, and* N *a 4-dimensional submanifold of* M. *Then* N *admits an orientation making it into a coassociative 4-fold if and only if* $\varphi|_N \equiv 0$.

In §8.4 we defined *almost Calabi–Yau m-folds*, and showed how special Lagrangian geometry in Calabi–Yau m-folds also works in almost Calabi–Yau m-folds. The following is a good analogue of this for G_2 geometry.

Definition 12.3.3 An *almost* G_2-*manifold* (M, φ, g) is a 7-manifold M equipped with a G_2-structure (φ, g) such that $\mathrm{d}\varphi \equiv 0$. Note that we do not require $\mathrm{d} * \varphi \equiv 0$.

Let (M, φ, g) be an almost G_2-manifold. Then φ is a calibration on (M, g), and we define an *associative 3-fold* in M to be an oriented 3-submanifold of M calibrated with respect to φ. Following Proposition 12.3.2, define a *coassociative 4-fold* in M to be a 4-submanifold N in M with $\varphi|_N \equiv 0$. Then N admits a unique orientation such that $*\varphi|_{T_x N} = \mathrm{vol}_{T_x N}$ for all $x \in N$. This would be the condition for N to be calibrated w.r.t. $*\varphi$, except that $*\varphi$ may not be a calibration as we have not assumed $\mathrm{d} * \varphi \equiv 0$.

The motivation for this comes from the end of §12.1.2, where we saw that $\mathrm{d}\varphi \equiv 0$ is a kind of 'integrability condition' for the existence of many coassociative 4-folds locally in a 7-manifold M with G_2-structure (φ, g). However, the condition $\mathrm{d} * \varphi \equiv 0$ is not used in an important way in associative or coassociative geometry. The main thing we lose by omitting it is that compact coassociative 4-folds need not be volume-minimizing, and their volume is not given by a (co)homological formula.

As for almost Calabi–Yau m-folds, there are two main advantages in working with almost G_2-manifolds rather than G_2-manifolds. Firstly, examples of (compact) almost G_2-manifolds can be constructed very explicitly, but no explicit compact G_2-manifolds

(M, φ, g) are known except finite quotients of (T^7, φ_0, g_0); for other examples we know only the existence of (φ, g), as in §11.3.

Secondly, almost G_2-manifolds come in *infinite-dimensional families*. If (M, φ, g) is an almost G_2-manifold and φ' is any closed 3-form with $|\varphi' - \varphi|_g$ sufficiently small then φ' has stabilizer G_2 at every point, since to have stabilizer G_2 is an open condition on 3-forms. Thus there is a unique Riemannian metric g' on M such that (M, φ', g') is also an almost G_2-manifold. For instance, we can take $\varphi' = \varphi + \mathrm{d}\eta$ for any C^1-small 2-form η on M. By contrast, Theorem 11.2.8 shows that compact G_2-manifolds come in finite-dimensional families up to diffeomorphism.

Therefore, choosing a *generic* almost G_2-manifold is a powerful thing to do. The author expects that singularities of compact associative 3-folds and coassociative 4-folds should be significantly simpler in a generic almost G_2-manifold than in an arbitrary almost G_2-manifold, in a similar way to the special Lagrangian case of Conjecture 8.5.9.

12.3.1 Deformation and obstruction theory of coassociative 4-folds

Here is the main result in the deformation theory of compact coassociative 4-folds, proved by McLean [259, Th. 4.5] in the G_2-manifold case. The extension to almost G_2-manifolds is immediate, as McLean does not use $\mathrm{d} * \varphi \equiv 0$. As our sign conventions for $\varphi_0, *\varphi_0$ in (12.1) are different to McLean's, we use self-dual 2-forms in place of McLean's anti-self-dual 2-forms.

Theorem 12.3.4 *Let (M, φ, g) be an (almost) G_2-manifold, and N a compact coassociative 4-fold in M. Then the moduli space $\mathcal{M}_N$ of coassociative 4-folds isotopic to N in M is a smooth manifold of dimension $b_+^2(N)$.*

Sketch proof Suppose for simplicity that N is an embedded submanifold. There is a natural orthogonal decomposition $TM|_N = TN \oplus \nu$, where $\nu \to N$ is the normal bundle of N in M. We shall construct a natural isomorphism $\nu \cong \Lambda_+^2 T^* N$. Let $x \in N$ and $V \in \nu_x$. Then $V \in T_x M$, so $V \cdot \varphi|_x \in \Lambda^2 T_x^* M$, and $(V \cdot \varphi|_x)|_{T_x N} \in \Lambda^2 T_x^* N$. It turns out that $(V \cdot \varphi|_x)|_{T_x N}$ actually lies in $\Lambda_+^2 T_x^* N$, the bundle of *self-dual 2-forms* on N, and that the map $V \mapsto (V \cdot \varphi|_x)|_{T_x N}$ defines an isomorphism $\nu \xrightarrow{\cong} \Lambda_+^2 T^* N$.

Let T be a small *tubular neighbourhood* of N in M. Then we can identify T with a neighbourhood of the zero section in ν, using the exponential map. The isomorphism $\nu \cong \Lambda_+^2 T^* N$ then identifies T with a neighbourhood U of the zero section in $\Lambda_+^2 T^* N$. Let $\pi : T \to N$ be the obvious projection.

Under this identification, submanifolds N' in $T \subset M$ which are C^1-close to N are identified with the *graphs* $\Gamma(\alpha)$ of small smooth sections α of $\Lambda_+^2 T^* N$ lying in U. Write $C^\infty(U)$ for the subset of the vector space of smooth self-dual 2-forms $C^\infty(\Lambda_+^2 T^* N)$ on N lying in $U \subset \Lambda_+^2 T^* N$. Then for each $\alpha \in C^\infty(U)$ the graph $\Gamma(\alpha)$ is a 4-submanifold of U, and so is identified with a 4-submanifold of T. We need to know: which 2-forms α correspond to coassociative 4-folds $\Gamma(\alpha)$ in T?

Well, N' is coassociative if $\varphi|_{N'} \equiv 0$. Now $\pi|_{N'} : N' \to N$ is a diffeomorphism, so we can push $\varphi|_{N'}$ down to N, and regard it as a function of α. That is, we define

$$P : C^\infty(U) \longrightarrow C^\infty(\Lambda^3 T^* N) \quad \text{by} \quad P(\alpha) = \pi_*(\varphi|_{\Gamma(\alpha)}). \tag{12.9}$$

Then the moduli space $\mathcal{M}_N$ is locally isomorphic near N to the set of small self-dual 2-forms α on N with $\varphi|_{\Gamma(\alpha)} \equiv 0$, that is, to a neighbourhood of 0 in $P^{-1}(0)$.

To understand the equation $P(\alpha) = 0$, note that at $x \in N$, $P(\alpha)|_x$ depends on the tangent space to $\Gamma(\alpha)$ at $\alpha|_x$, and so on $\alpha|_x$ and $\nabla\alpha|_x$. Thus the functional form of P is

$$P(\alpha)|_x = F\big(x, \alpha|_x, \nabla\alpha|_x\big) \quad \text{for } x \in N,$$

where F is a smooth function of its arguments. Hence $P(\alpha) = 0$ is a nonlinear first order p.d.e. in α. The linearization $\mathrm{d}P(0)$ of P at $\alpha = 0$ turns out to be

$$\mathrm{d}P(0)(\beta) = \lim_{\epsilon \to 0}\big(\epsilon^{-1}P(\epsilon\beta)\big) = \mathrm{d}\beta.$$

Therefore $\mathrm{Ker}(\mathrm{d}P(0))$ is the vector space $\mathcal{H}_+^2$ of closed self-dual 2-forms β on N, which by Hodge theory is a finite-dimensional vector space isomorphic to $H_+^2(N, \mathbb{R})$, with dimension $b_+^2(N)$. This is the *Zariski tangent space* of $\mathcal{M}_N$ at N, the *infinitesimal deformation space* of N as a coassociative 4-fold.

To complete the proof we must show that $\mathcal{M}_N$ is locally isomorphic to its Zariski tangent space $\mathcal{H}_+^2$, and so is a smooth manifold of dimension $b_+^2(N)$. To do this rigorously requires some technical analytic machinery, which is passed over in a few lines in [259, p. 731]. Here is one way to do it.

As $C^\infty(\Lambda_+^2 T^*N), C^\infty(\Lambda^3 T^*N)$ are not Banach spaces, we extend P in (12.9) to act on *Hölder spaces* $C^{k+1,\gamma}(\Lambda_+^2 T^*N), C^{k,\gamma}(\Lambda^3 T^*N)$ for $k \geqslant 0$ and $\gamma \in (0,1)$, giving

$$P_{k,\gamma} : C^{k+1,\gamma}(U) \longrightarrow C^{k,\gamma}(\Lambda^3 T^*N) \quad \text{defined by} \quad P_{k,\gamma}(\alpha) = \pi_*(\varphi|_{\Gamma(\alpha)}).$$

Then $P_{k,\gamma}$ is a smooth map of Banach manifolds. Let $V_{k,\gamma} \subset C^{k,\gamma}(\Lambda^3 T^*N)$ be the Banach subspace of *exact* $C^{k,\gamma}$ 3-forms on N.

As φ is closed, $\varphi|_N \equiv 0$, and $\Gamma(\alpha)$ is isotopic to N, we see that $\varphi|_{\Gamma(\alpha)}$ is an exact 3-form on $\Gamma(\alpha)$, so that $P_{k,\gamma}$ maps into $V_{k,\gamma}$. Using elliptic regularity results for $\mathrm{d} + \mathrm{d}^*$ we can show that the linearization

$$\mathrm{d}P_{k,\gamma}(0) : C^{k+1,\gamma}(\Lambda_+^2 T^*N) \longrightarrow V_{k,\gamma}, \qquad \mathrm{d}P_{k,\gamma}(0) : \beta \longmapsto \mathrm{d}\beta$$

is *surjective* as a map of Banach spaces.

Thus, $P_{k,\gamma} : C^{k+1,\gamma}(U) \to V_{k,\gamma}$ is a smooth map of Banach manifolds, with $\mathrm{d}P_{k,\gamma}(0)$ surjective. The *Implicit Mapping Theorem for Banach spaces*, Theorem 1.2.5, now implies that $P_{k,\gamma}^{-1}(0)$ is, near 0, a smooth submanifold of $C^{k+1,\gamma}(U)$, locally isomorphic to $\mathrm{Ker}(\mathrm{d}P_{k,\gamma}(0))$. But $P_{k,\gamma}(\alpha) = 0$ is an *overdetermined elliptic equation* for small α, and so elliptic regularity implies that solutions α are smooth. Therefore $P_{k,\gamma}^{-1}(0) = P^{-1}(0)$ near 0, and similarly $\mathrm{Ker}(\mathrm{d}P_{k,\gamma}(0)) = \mathrm{Ker}(\mathrm{d}P(0)) = \mathcal{H}_+^2$. This completes the proof. $\qquad\square$

The theorem shows that, as for SL m-folds in §8.4.1, the deformation theory of compact coassociative 4-folds is *unobstructed*, and the moduli space is *always* a smooth manifold with dimension given by a topological formula. This is unusual in moduli problems; a more common pattern is to find obstruction spaces and possibly singular moduli spaces, as we will see in the associative and Cayley cases of §12.3.2 and §12.5.1.

At the heart of the proof is the fact that a 4-submanifold N in M is coassociative, with the appropriate orientation, if and only if $\varphi|_N \equiv 0$, for φ a closed 3-form on M. SL m-folds are also defined by the vanishing of closed forms $\omega, \operatorname{Im}\Omega$, which is why the two theories are so similar. We saw in §12.1.2 that the condition for a 4-fold N in M to be coassociative is *overdetermined*. But Theorem 12.3.4 shows that coassociative 4-folds may often form positive-dimensional moduli spaces, which seems surprising for an overdetermined equation. Again, the reason is that $d\varphi \equiv 0$ acts as a kind of integrability condition, so that coassociative 4-folds are more abundant than one might expect.

By the same method as Proposition 8.4.6, from Proposition 12.3.2 we deduce:

Proposition 12.3.5 *Let (M, φ, g) be an (almost) G_2-manifold, and N a 4-submanifold in M. Then a necessary condition for N to be isotopic to a coassociative 4-fold N' in M is that $[\varphi|_N] = 0$ in $H^3(N, \mathbb{R})$.*

Now suppose $\{(M, \varphi_t, g_t) : t \in (-\epsilon, \epsilon)\}$ is a smooth 1-parameter family of (almost) G_2-manifolds, and N_0 a compact coassociative 4-fold in (M, φ_0, g_0). When can we extend N_0 to a smooth family of coassociative 4-folds N_t in (M, φ_t, g_t) for small t? By Proposition 12.3.5, a necessary condition is that $[\varphi_t|_{N_0}] = 0$ for all t. Our next result shows that locally, this is also sufficient.

Theorem 12.3.6 *Let $\{(M, \varphi_t, g_t) : t \in (-\epsilon, \epsilon)\}$ be a smooth 1-parameter family of (almost) G_2-manifolds, and N_0 a compact coassociative 4-fold in (M, φ_0, g_0). Suppose that $[\varphi_t|_{N_0}] = 0$ in $H^3(N_0, \mathbb{R})$ for all $t \in (-\epsilon, \epsilon)$. Then N_0 extends to a smooth 1-parameter family $\{N_t : t \in (-\delta, \delta)\}$, where $0 < \delta \leqslant \epsilon$ and N_t is a compact coassociative 4-fold in (M, φ_t, g_t).*

This can be proved using similar techniques to Theorem 12.3.4, though McLean did not prove it. It shows that coassociative 4-folds are pretty stable under small deformations of the underlying almost G_2-manifold (M, φ, g).

12.3.2 Deformation theory of associative 3-folds

Associative 3-folds cannot be defined in terms of the vanishing of closed forms, and this gives their deformation theory a different character to the coassociative case. Here is how the theory works, drawn mostly from McLean [259, §5].

Let N be a compact associative 3-fold in an almost G_2-manifold (M, φ, g). The normal bundle $\nu \to N$ of N in M has fibre $\mathbb{R}^4$, and it can be regarded as a *twisted spin bundle* on N, since there is an isomorphism $\nu \otimes_{\mathbb{R}} \mathbb{C} \cong \mathbb{S} \otimes_{\mathbb{C}} E$, where $\mathbb{S}, E$ are vector bundles over N with fibre $\mathbb{C}^2$, with $\mathbb{S}$ the spin bundle of N. Complex conjugation on $\nu \otimes_{\mathbb{R}} \mathbb{C}$ is of the form $\sigma \otimes \tilde{\sigma}$, where $\sigma : \mathbb{S} \to \mathbb{S}$ and $\tilde{\sigma} : E \to E$ are complex antilinear vector bundle isomorphisms with $\sigma^2 = -1 = \tilde{\sigma}^2$.

There is an anti-self-adjoint first order linear elliptic operator $D_N : C^\infty(\nu) \to C^\infty(\nu)$ which is a *twisted Dirac operator*, since it is derived from the usual Dirac operator $D : C^\infty(\mathbb{S}) \to C^\infty(\mathbb{S})$. The *kernel* $\operatorname{Ker} D_N$ is the set of *infinitesimal deformations* of N as an associative 3-fold or Cayley 4-fold. The *cokernel* $\operatorname{Coker} D_N$ is the *obstruction space* for these deformations. But as D_N is anti-self-adjoint, the kernel and cokernel are equal, so the infinitesimal deformation space is equal to the obstruction space. Thus the *index* of D_N is $\text{ind}(D_N) = \dim \text{Ker} \, D_N - \dim \text{Coker} \, D_N = 0$, which is obvious anyway as indices on compact odd-dimensional manifolds are always zero.

The expected, or virtual, dimension of the moduli space $\mathscr{M}_N$ of associative deformations of N is $\text{ind}(D_N) = 0$. If (M, φ, g) is a *generic* almost G_2-manifold then the obstruction spaces $\text{Coker}(D_N)$ are zero. This means the infinitesimal deformation spaces $\text{Ker}(D_N)$ are zero, and so $\mathscr{M}_N$ is a smooth 0-dimensional manifold, that is, a discrete set of points, which will probably be finite if we restrict to 3-folds in a fixed homology class in $H_3(M, \mathbb{Z})$. Furthermore, the associative 3-folds in $\mathscr{M}_N$ will persist under small deformations of (M, φ, g).

For nongeneric (M, φ, g) the deformation and obstruction spaces may be nonzero, and then $\mathscr{M}_N$ may not be smooth, or may have a larger than expected dimension; but $\mathscr{M}_N$ will reduce to a discrete set of points under small generic deformations of (φ, g). For example, the 7-torus T^7 has a 35-dimensional family of flat G_2-structures. Writing T^7 as a product $T^3 \times T^4$, it turns out that for a 31-dimensional subfamily of these G_2-structures, $T^3 \times \{p\}$ is an associative 3-fold for each $p \in T^4$. These associative T^3's deform in a moduli space of dimension 4. However, for generic flat G_2-structures there exist no associative 3-folds in T^7 at all. Thus the family of associative T^3's vanishes under small deformations of the underlying G_2-structure.

A general conclusion is that moduli spaces of associative 3-folds are fairly boring. However, it is plausible that one could define interesting invariants by counting (with some appropriate weight) associative 3-folds in a generic almost G_2-manifold (M, φ, g) in a fixed class α in $H_3(M, \mathbb{Z})$. See [194] for some similar ideas for SL 3-folds.

12.3.3   Examples of compact associative 3-folds and coassociative 4-folds

Here is a method used in [188, §10.8 and §12.6] to find examples of compact associative 3-folds in compact 7-manifolds with holonomy G_2. A *nontrivial isometric involution* of (M, g) is a diffeomorphism $\sigma : M \to M$ such that $\sigma^*(g) = g$, and $\sigma \neq \text{id}$ but $\sigma^2 = \text{id}$, where id is the identity on M.

Proposition 12.3.7 *Let (M, φ, g) be an (almost) G_2-manifold, and $\sigma : M \to M$ a nontrivial isometric involution with $\sigma^*(\varphi) = \varphi$. Then $N = \{p \in M : \sigma(p) = p\}$ is a closed, nonsingular, embedded associative 3-fold in M, which is compact if M is.*

Proof   Clearly N is a closed, nonsingular, embedded submanifold of M, and compact if M is compact. If $p \in N$ then $d\sigma : T_pM \to T_pM$ satisfies $(d\sigma)^2 = 1$, and T_pN is the subspace of T_pM fixed by $d\sigma$. If $d\sigma = 1$ then $T_pN = T_pM$ and $\dim N = 7$, so $N = M$ as M is connected. But this contradicts $\sigma \neq \text{id}$, and so $d\sigma \neq 1$. Also, $d\sigma$ preserves $\varphi|_p$. Thus, identifying T_pM with $\mathbb{R}^7$, we can regard $d\sigma : T_pM \to T_pM$ as an element of G_2. If $\gamma \in G_2$ and $\gamma \neq 1$ but $\gamma^2 = 1$, then γ is conjugate in G_2 to the map

$$(x_1, \ldots, x_7) \longmapsto (x_1, x_2, x_3, -x_4, -x_5, -x_6, -x_7).$$

The fixed set of this map is the associative 3-plane U of (12.2). Thus, if $\gamma \in G_2$ and $\gamma \neq 1$ but $\gamma^2 = 1$, then the fixed set of γ in $\mathbb{R}^7$ is an associative 3-plane. So T_pN is an associative 3-plane in T_pM. As this holds for all $p \in N$, we see N is associative. $\square$

Following [188, Ex. 12.6.1], we can use this to construct examples of compact associative 3-folds in the compact 7-manifolds with holonomy G_2 discussed in §11.3.

Example 12.3.8 Let T^7 and Γ be as in Example 11.3.1. Define $\sigma : T^7 \to T^7$ by

$$\sigma : (x_1, \ldots, x_7) \mapsto (x_1, x_2, x_3, \tfrac{1}{2} - x_4, -x_5, -x_6, -x_7).$$

Then σ preserves (φ_0, g_0) and commutes with Γ, and so its action pushes down to T^7/Γ. The fixed points of σ on T^7 are 16 copies of T^3, and $\sigma\delta$ has no fixed points in T^7 for all $\delta \neq 1$ in Γ. Thus the fixed points of σ in T^7/Γ are the image of the 16 T^3 fixed by σ in T^7. Calculation shows that these 16 T^3 do not intersect the fixed points of α, β or γ, and that Γ acts freely on the set of 16 T^3 fixed by σ. So the image of the 16 T^3 in T^7 is 2 T^3 in T^7/Γ. Now the resolution of T^7/Γ to get a compact G_2-manifold $(M, \tilde{\varphi}, \tilde{g})$ with $\mathrm{Hol}(\tilde{g}) = G_2$ described in §11.3 may be done in a σ-*equivariant* way, so that σ lifts to $\sigma : M \to M$ with $\sigma^*(\tilde{\varphi}) = \tilde{\varphi}$. The fixed points of σ in M are again 2 copies of T^3, which are *associative* 3-*folds* by Proposition 12.3.7.

The same idea works for coassociative 4-folds. If $\gamma : \mathbb{R}^7 \to \mathbb{R}^7$ is linear with $\gamma^2 = 1$ and $\gamma^*(\varphi_0) = -\varphi_0$, then either $\gamma = -1$, or γ is conjugate under G_2 to the map

$$(x_1, \ldots, x_7) \longmapsto (-x_1, -x_2, -x_3, x_4, x_5, x_6, x_7).$$

The fixed set of this is the coassociative 4-plane V of (12.3). Thus, the fixed point set of α is either $\{0\}$, or a coassociative 4-plane in $\mathbb{R}^7$. So following Proposition 12.3.7 we prove [188, Prop. 10.8.5]:

Proposition 12.3.9 Let (M, φ, g) be an (almost) G_2-manifold, and $\sigma : M \to M$ a nontrivial isometric involution with $\sigma^*(\varphi) = -\varphi$. Then each connected component of the fixed point set $\{p \in M : \sigma(p) = p\}$ of σ is either a closed, nonsingular, embedded coassociative 4-fold which is compact if M is, or a single point.

Here [188, Ex. 12.6.4] is an example in the G_2-manifold of §11.3.

Example 12.3.10 Let T^7 and Γ be as in Example 11.3.1. Define $\sigma : T^7 \to T^7$ by

$$\sigma : (x_1, \ldots, x_7) \mapsto (\tfrac{1}{2} - x_1, x_2, x_3, x_4, x_5, \tfrac{1}{2} - x_6, \tfrac{1}{2} - x_7).$$

Then σ commutes with Γ, preserves g_0 and takes φ_0 to $-\varphi_0$. The fixed points of σ in T^7 are 8 copies of T^4, and the fixed points of $\sigma\alpha\beta$ in T^7 are 128 points. If $\delta \in \Gamma$ then $\sigma\delta$ has no fixed points unless $\delta = 1, \alpha\beta$. Thus the fixed points of σ in T^7/Γ are the image of the fixed points of σ and $\sigma\alpha\beta$ in T^7.

Now Γ acts freely on the sets of 8 σ T^4's and 128 $\sigma\alpha\beta$ points. So the fixed point set of σ in T^7/Γ is the union of T^4 and 16 isolated points, none of which intersect the singular set of T^7/Γ. When we resolve T^7/Γ to get $(M, \tilde{\varphi}, \tilde{g})$ with $\mathrm{Hol}(\tilde{g}) = G_2$ in a σ-equivariant way, the action of σ on M has $\sigma^*(\tilde{\varphi}) = -\tilde{\varphi}$, and again fixes T^4 and 16 points. By Proposition 12.3.9, this T^4 is *coassociative*.

More examples of compact associative and coassociative submanifolds with different topologies are given in [188, §12.6]. Bryant [59] uses the idea of Proposition 12.3.9 to construct many *local* examples of compact coassociative 4-folds in G_2-manifolds.

Theorem 12.3.11. (Bryant [59]) *Let (N, g) be a compact, real analytic, oriented Riemannian 4-manifold whose bundle of self-dual 2-forms is trivial. Then N may be embedded isometrically as a coassociative 4-fold in a G_2-manifold (M, φ, g), as the fixed point set of an involution σ.*

Note here that M need not be *compact*, nor (M, g) *complete*. Roughly speaking, Bryant's proof constructs (φ, g) as the sum of a power series on $\Lambda^2_+ T^* N$ converging near the zero section $N \subset \Lambda^2_+ T^* N$, using the theory of *exterior differential systems*. The involution σ acts as -1 on $\Lambda^2_+ T^* N$, fixing the zero section. The requirement that $\Lambda^2_+ T^* N$ be trivial is technical, and probably not necessary. One moral of Theorem 12.3.11 is that to be coassociative places no significant geometric restrictions on a Riemannian 4-manifold, other than real analyticity and orientability; for instance, it implies no extra curvature identities on (N, g).

There is also a second method for constructing examples of compact associative or coassociative submanifolds of compact G_2-manifolds, which we now describe. It is implicit in Kovalev's proposal [224], but the author is not aware of other uses of it in the literature. We illustrate it for the T^7/Γ construction of §11.3.1–§11.3.4, but the same idea should work for the other constructions of G_2-manifolds in §11.3.5.

Suppose, as in §11.3, that we construct a 1-parameter family of torsion-free G_2-structures $(\tilde{\varphi}_t, \tilde{g}_t)$ for $t \in (0, \epsilon)$ on a compact 7-manifold M, by resolving the singularities of an orbifold T^7/Γ with flat G_2-structure (φ_0, g_0). Then $(\tilde{\varphi}_t, \tilde{g}_t) \to (\varphi_0, g_0)$ in T^7/Γ away from the singular set S of T^7/Γ as $t \to 0$.

Let N_0 be a compact coassociative 4-fold in T^7/Γ which does not intersect S, and which need not be the fixed set of any isometric involution. For instance, using the ideas of §12.2.1 we could try taking N_0 to be the image in T^7/Γ of $\{x\} \times H$ in $T^7 = \mathcal{S}^1 \times T^6$, where $x \in \mathcal{S}^1$ and H is a complex hypersurface in the complex 3-fold T^6. By a mild generalization of Theorem 12.3.6 (needed because the convergence $(\tilde{\varphi}_t, \tilde{g}_t) \to (\varphi_0, g_0)$ may not be smooth in t at $t = 0$) we can show that if $[\tilde{\varphi}_t|_{N_0}] = 0$ in $H^3(N_0, \mathbb{R})$ for $t \in (0, \epsilon)$ then for some $\delta \in (0, \epsilon]$ we can extend N_0 to a smooth family of coassociative 4-folds $\tilde{N}_t$ in $(M, \tilde{\varphi}_t, \tilde{g}_t)$ for $t \in (0, \delta)$, with $\tilde{N}_t \to N_0$ as $t \to 0$. But this condition $[\tilde{\varphi}_t|_{N_0}] = 0$ is automatic in the construction of §11.3, so $\tilde{N}_t$ exists for small t.

In a similar way, suppose N_0 is a compact associative 3-fold in T^7/Γ which does not intersect S, and which is *unobstructed* in the sense that $\operatorname{Ker} D_{N_0} = \operatorname{Coker} D_{N_0} = \{0\}$ in the notation of §12.3.2. (Note that N_0 cannot be the bijective image of an associative 3-fold in T^7, since using translations in T^7 as infinitesimal deformations we can prove $\dim \operatorname{Ker} D_{N_0} \geqslant 4$, so any such N_0 is obstructed.) Then one can show that for some $\delta \in (0, \epsilon]$ we can extend N_0 to a smooth family of associative 3-folds $\tilde{N}_t$ in $(M, \tilde{\varphi}_t, \tilde{g}_t)$ for $t \in (0, \delta)$, with $\tilde{N}_t \to N_0$ as $t \to 0$.

12.3.4 Further topics in coassociative geometry

Here are two other interesting topics in coassociative geometry.

Coassociative 4-folds with isolated conical singularities

The theory of SL m-folds with isolated conical singularities described in §8.5 should also have an analogue for coassociative 4-folds. The first part of this programme has been carried out by Jason Lotay [251]. His main result [251, Ths 7.9 and 7.13] describes the deformation and obstruction theory of coassociative 4-folds with conical singularities, in a similar way to the special Lagrangian case Theorem 8.5.5 and the asymptotically conical coassociative case Theorem 12.1.10. Lotay is also developing a coassociative analogue of the desingularization theory of §8.5.3.

One could also study associative 3-folds with isolated conical singularities, but the theory will be less interesting as the virtual dimension of their moduli spaces will always be nonpositive.

Coassociative fibrations

In §9.4 we discussed the SYZ Conjecture and fibrations of Calabi–Yau 3-folds by SL 3-folds, including singular fibres. As coassociative 4-folds behave like SL 3-folds, it is natural also to consider fibrations of G_2-manifolds by compact coassociative 4-folds, including singular fibres.

If N is a nonsingular fibre of such a fibration, then to get one coassociative 4-fold passing through each point of M near N we need the moduli space $\mathcal{M}_N$ of coassociative deformations of N to have dimension 3, so that $b_+^2(N) = 3$ by Theorem 12.3.4, and $\Lambda_+^2 T^* N$ needs to be trivial, since this is the normal bundle of N in M. The obvious 4-manifolds satisfying these two conditions are T^4 and the $K3$ surface. From [188, §12.6], examples are known of coassociative T^4's and $K3$'s in compact 7-manifolds with holonomy G_2, by the method of Proposition 12.3.9.

The idea of studying coassociative $K3$ fibrations was first suggested by McLean [259, p. 745] (written in 1990, before the SYZ Conjecture, though published in 1998). In string theory, coassociative fibrations appear in the work of Acharya [3,4] and Gukov, Yau and Zaslow [145]. Kovalev [224] has published a proposal for constructing an example of a fibration of a compact 7-manifold with holonomy G_2 by coassociative $K3$'s, including singular fibres. In the opinion of the author, the geometric idea in Kovalev's construction is beautiful, and can almost certainly be made to work; however, at the time of writing, the analytic proofs are still incomplete. To finish them will require a well-developed theory of coassociative 4-folds with isolated conical singularities, as discussed above.

Here is a sketch of Kovalev's idea. We work in the situation of the construction of Method 1 in §11.3.5. This involves a noncompact Calabi–Yau 3-fold Y with a cylindrical end modelled on $D \times S^1 \times [0, \infty)$, where D is a $K3$ surface. We find a holomorphic function $f : Y \to \mathbb{C}$ which is asymptotic on the cylindrical end to the function $(x, e^{i\theta}, t) \mapsto e^{t+i\theta}$, for $x \in D$, $e^{i\theta} \in S^1$ and $t \in [0, \infty)$. Provided Y is chosen sufficiently generically, this f can be taken to be a *holomorphic Morse function*. That is, there are finitely many stationary points of f on Y, which occur at distinct values $c \in \mathbb{C}$ of f, and near each stationary point f looks like $f = z_1^2 + z_2^2 + z_3^2 + c$ in appropriate local holomorphic coordinates (z_1, z_2, z_3) on Y.

Then $f : Y \to \mathbb{C}$ is a fibration of Y by complex surfaces, with generic fibre $K3$, and has finitely many singular fibres each of which has one isolated conical singularity modelled on a quadric cone in $\mathbb{C}^3$. From the ideas of §12.2.1, if S is a surface in a Calabi–Yau 3-fold X and $x \in S^1$ then $\{x\} \times S$ is coassociative in the natural G_2-structure on $S^1 \times X$. Therefore $\mathrm{id} \times f : S^1 \times Y \to S^1 \times \mathbb{C}$ is a *coassociative fibration* of $\mathcal{S}^1 \times Y$, with generic fibre $K3$. The family of singular fibres is finitely many copies of $\mathcal{S}^1$, and each singular fibre has one isolated conical singularity.

Now Kovalev [223] constructs a compact 7-manifold M with holonomy G_2 by gluing together two such G_2-manifolds $\mathcal{S}^1 \times Y_1, \mathcal{S}^1 \times Y_2$ along their common cylindrical end $D \times \mathcal{S}^1 \times \mathcal{S}^1 \times \mathbb{R}$. The idea of [224] is that one can take coassociative fibrations $\mathrm{id} \times f_1 : \mathcal{S}^1 \times Y_1 \to \mathcal{S}^1 \times \mathbb{C}$ and $\mathrm{id} \times f_2 : \mathcal{S}^1 \times Y_2 \to \mathcal{S}^1 \times \mathbb{C}$ of the two pieces, as above, and glue them together to make a coassociative fibration of the G_2-manifold M. The main analytic problem in doing this is ensuring that the coassociative fibrations near and including the singular fibres continue to exist under the small deformation between the G_2-structures on $\mathcal{S}^1 \times Y_i$ and M.

Associative fibrations are not really a sensible idea, since compact associative 3-folds are not expected to form 4-dimensional moduli spaces in any generic situation, though they have been considered by physicists such as Acharya [3, 4]. In the opinion of the author, a good way to interpret Acharya's ideas mathematically is to consider the formulations Conjectures 9.4.2 and 9.4.3 of the SYZ Conjecture in §9.4.2.

A G_2 analogue of Conjecture 9.4.2 involving fibrations $f_t : X_t \to B$ of G_2-manifolds X_t over a 4-manifold B, whose generic fibres $T_t^b = f_t^{-1}(b)$ are associative T^3's in X_t, is *not* plausible, as T_t^b should not admit nontrivial associative deformations. But it *would* be plausible, and in the spirit of the conjectures of §9.4.4, to make a G_2 analogue of Conjecture 9.4.3, in which the fibres T_t^b are required to be only *approximately* associative, and to converge to exactly associative, flat T^3's as $t \to \infty$.

12.4 Cayley 4-folds in $\mathbb{R}^8$

We now turn to the geometry of Cayley submanifolds in $\mathbb{R}^8$ and $\mathrm{Spin}(7)$-manifolds.

Definition 12.4.1 Let $\mathbb{R}^8$ have coordinates $(x_1, \ldots, x_8)$, and as in Definition 11.4.1 define a metric $g_0 = \mathrm{d}x_1^2 + \cdots + \mathrm{d}x_8^2$ and a 4-form Ω_0 on $\mathbb{R}^8$ by

$$
\begin{aligned}
\Omega_0 = \quad &\mathrm{d}\mathbf{x}_{1234} + \mathrm{d}\mathbf{x}_{1256} + \mathrm{d}\mathbf{x}_{1278} + \mathrm{d}\mathbf{x}_{1357} - \mathrm{d}\mathbf{x}_{1368} - \mathrm{d}\mathbf{x}_{1458} - \mathrm{d}\mathbf{x}_{1467} \\
&- \mathrm{d}\mathbf{x}_{2358} - \mathrm{d}\mathbf{x}_{2367} - \mathrm{d}\mathbf{x}_{2457} + \mathrm{d}\mathbf{x}_{2468} + \mathrm{d}\mathbf{x}_{3456} + \mathrm{d}\mathbf{x}_{3478} + \mathrm{d}\mathbf{x}_{5678},
\end{aligned}
\tag{12.10}
$$

where $\mathrm{d}\mathbf{x}_{i\ldots l}$ is short for $\mathrm{d}x_i \wedge \cdots \wedge \mathrm{d}x_l$. By Harvey and Lawson [151, Th. IV.1.24], Ω_0 has comass one and so is a *calibration* on $\mathbb{R}^8$. We define a *Cayley 4-fold* in $\mathbb{R}^8$ to be an oriented 4-dimensional submanifold of $\mathbb{R}^8$ calibrated with respect to Ω_0.

We shall describe the family $\mathcal{F}_{\Omega_0}$ of *Cayley 4-planes* in $\mathbb{R}^8$, that is, 4-dimensional oriented vector subspaces in $\mathbb{R}^8$ calibrated by Ω_0. An example of a Cayley 4-plane is

$$
V = \big\{ (x_1, x_2, x_3, x_4, 0, 0, 0, 0) : x_1, x_2, x_3, x_4 \in \mathbb{R} \big\} \subset \mathbb{R}^8,
$$

with the obvious orientation. The stabilizer subgroup of V in $\mathrm{Spin}(7)$ is isomorphic to $\mathrm{SU}(2)^3/\mathbb{Z}_2$, where $\mathbb{Z}_2$ is the subgroup $\{(1, 1, 1), (-1, -1, -1)\}$ in $\mathrm{SU}(2)^3$. By [151, Th. IV.1.38] the action of $\mathrm{Spin}(7)$ on $\mathcal{F}_{\Omega_0}$ is transitive. This proves:

Proposition 12.4.2 *The family $\mathcal{F}_{\Omega_0}$ of Cayley 4-planes in $\mathbb{R}^8$ has dimension 12 and is isomorphic to $\mathrm{Spin}(7)/K$, where $K \cong \mathrm{SU}(2)^3/\mathbb{Z}_2$ is a Lie subgroup of $\mathrm{Spin}(7)$.*

As for Theorem 12.1.5 we have:

Theorem 12.4.3 *A Cayley 4-fold L in $\mathbb{R}^8$ is real analytic wherever it is nonsingular.*

Harvey and Lawson use exterior differential systems to prove [151, Th. IV.4.3]:

Theorem 12.4.4 *Let P be a real analytic 3-submanifold in $\mathbb{R}^8$. Then there exists a locally unique Cayley 4-fold L in $\mathbb{R}^8$ containing P.*

This implies there are very many Cayley 4-folds in $\mathbb{R}^8$. In the sense of exterior differential systems, Cayley 4-folds 'depend on 4 functions of 3 variables'.

12.4.1 Cayley 4-folds as graphs, and local deformations

As in §12.1.1–§12.1.2, we can write a class of Cayley 4-folds in $\mathbb{R}^8$ as graphs. Let $f = f_0 + f_1 i + f_2 j + f_3 k = f(x_0 + x_1 i + x_2 j + x_3 k) : \mathbb{H} \to \mathbb{H}$ be smooth. Choosing signs for compatibility with (11.12), define a 4-submanifold L in $\mathbb{R}^8$ by

$$L = \big\{ \big(-x_0, x_1, x_2, x_3, f_0(x_0 + x_1 i + x_2 j + x_3 k), -f_1(x_0 + x_1 i + x_2 j + x_3 k),$$
$$-f_2(x_0 + x_1 i + x_2 j + x_3 k), f_3(x_0 + x_1 i + x_2 j + x_3 k) \big) : x_j \in \mathbb{R} \big\}.$$

Following [151, §IV.2.C], but using the $\mathrm{Spin}(7)$ form (11.12), we find L is Cayley if

$$\frac{\partial f}{\partial x_0} + i \frac{\partial f}{\partial x_1} + j \frac{\partial f}{\partial x_2} + k \frac{\partial f}{\partial x_3} = C(\partial f), \qquad (12.11)$$

for $C : \mathbb{H} \otimes_{\mathbb{R}} \mathbb{H} \to \mathbb{H}$ a homogeneous cubic polynomial. This is a first-order nonlinear elliptic p.d.e. on f. The linearization at $f = 0$ is equivalent to the *positive Dirac equation* on $\mathbb{R}^4$. More generally, first order deformations of a Cayley 4-fold L in a $\mathrm{Spin}(7)$-manifold (M, Ω, g) correspond to solutions of a *twisted positive Dirac equation* on L.

We can use this to discuss small *deformations* of Cayley 4-folds in $\mathbb{R}^8$. The families $\mathrm{Gr}_+(4, \mathbb{R}^8)$ of oriented 4-planes in $\mathbb{R}^8$ and $\mathcal{F}_{\Omega_0}$ of Cayley 4-planes in $\mathbb{R}^8$ have dimensions 16 and 12, so $\mathcal{F}_{\Omega_0}$ has codimension 4 in $\mathrm{Gr}_+(4, \mathbb{R}^8)$. Thus the condition for a 4-fold L in $\mathbb{R}^8$ to be Cayley is 4 equations on each tangent space. The freedom to vary L is the sections of its normal bundle in $\mathbb{R}^8$, which is 4 real functions. Hence, the deformation problem for Cayley 4-folds involves 4 *equations on* 4 *functions*, as in (12.11), and is controlled by a *first order nonlinear elliptic p.d.e.* This makes their deformation theory fairly well behaved.

12.4.2 Examples of Cayley 4-folds in $\mathbb{R}^8$

As in §12.2.1, we can make examples of Cayley 4-folds from simpler calibrations on $\mathbb{R}^7$ and $\mathbb{R}^8$. Write $\mathbb{R}^8 = \mathbb{C}^4$, with complex coordinates $(x_1 + ix_2, x_3 + ix_4, x_5 + ix_6, x_7 + ix_8)$. Then as in Proposition 11.4.11 we have $\Omega_0 = \frac{1}{2}\omega \wedge \omega + \mathrm{Re}\,\theta$, where ω, θ are the Kähler form and holomorphic volume on $\mathbb{C}^4$ given in (7.1). It follows easily that:

- If L is a *holomorphic surface* in $\mathbb{C}^4$ (and so calibrated with respect to $\frac{1}{2}\omega \wedge \omega$, with $\mathrm{Re}\,\theta|_L \equiv 0$) then L is Cayley in $\mathbb{R}^8$.
- If L is a *special Lagrangian 4-fold* in $\mathbb{C}^4$ (and so calibrated with respect to $\mathrm{Re}\,\Omega$, with $\frac{1}{2}\omega \wedge \omega|_L \equiv 0$) then L is Cayley in $\mathbb{R}^8$.

Similarly, write $\mathbb{R}^8 = \mathbb{R} \oplus \mathbb{R}^7$. Then as in Proposition 11.4.10 we have $\Omega_0 = \mathrm{d}x_1 \wedge \varphi_0 + *\varphi_0$, where $\varphi_0, *\varphi_0$ are the G_2-forms of (12.1) with $x_1, \dots, x_7$ replaced by $x_2, \dots, x_8$. It follows easily that:

- If L is an *associative* 3-*fold* in $\mathbb{R}^7$ then $\mathbb{R} \times L$ is Cayley in $\mathbb{R}^8$.
- If $x \in \mathbb{R}$ and L is a *coassociative* 4-*fold* in $\mathbb{R}^7$ then $\{x\} \times L$ is Cayley in $\mathbb{R}^8$.

Thus, singularities of holomorphic surfaces, SL 4-folds, associative 3-folds, and coassociative 4-folds all yield models for singularities of Cayley 4-folds. In general though, we regard examples of Cayley 4-folds in $\mathbb{R}^8$ as more interesting (as Cayley 4-folds) if they do not come from one of these simpler geometries.

There are few papers on examples of Cayley 4-folds in $\mathbb{R}^8$ at the time of writing. Lotay [249] studies 2-*ruled Cayley* 4-*folds* in $\mathbb{R}^8$, that is, Cayley 4-folds L fibred by a 2-dimensional family Σ of affine 2-planes $\mathbb{R}^2$ in $\mathbb{R}^8$, as for the coassociative case in §12.2.5. He constructs explicit families of 2-ruled Cayley 4-folds in $\mathbb{R}^8$, including some depending on an arbitrary holomorphic function $w : \mathbb{C} \to \mathbb{C}$, [249, Th. 5.1], which do not reduce to a simpler geometry as above. Lotay [250, §5] and Gu and Pries [142] both study Cayley 4-folds in $\mathbb{R}^8$ invariant under $\mathrm{SU}(2)$ subgroups in $\mathrm{Spin}(7)$; there is some overlap between their examples.

As in §12.2.5, Ionel, Karigiannis and Min-Oo [173] construct 2-ruled Cayley 4-folds in $\mathbb{R}^8$ from minimal surfaces in $\mathbb{R}^4$, but unfortunately their method yields only examples of the form $\mathbb{R} \times L$ for L associative in $\mathbb{R}^7$ or $\{x\} \times L$ for L coassociative in $\mathbb{R}^7$. In a sequel, Karigiannis and Min-Oo [212] find examples of Cayley 4-folds in the explicit noncompact 8-manifolds with holonomy $\mathrm{Spin}(7)$ of Bryant and Salamon [64].

12.5 Cayley 4-folds in $\mathrm{Spin}(7)$-manifolds

Definition 12.5.1 Let (M, Ω, g) be a $\mathrm{Spin}(7)$-manifold, as in §11.4. Then the 4-form Ω is a calibration on (M, g). We define a *Cayley* 4-*fold* in M to be an oriented 4-submanifold of M calibrated with respect to Ω.

As for the special Lagrangian case of §8.4 and the G_2 cases of §12.3, it is natural to ask whether there is some natural class of 'almost $\mathrm{Spin}(7)$-manifolds' generalizing $\mathrm{Spin}(7)$-manifolds, in which Cayley 4-folds make sense and still have good properties. Recall from §11.4 that a $\mathrm{Spin}(7)$-*manifold* is an 8-manifold M with a torsion-free $\mathrm{Spin}(7)$-structure (Ω, g), and that (Ω, g) is torsion-free if and only if $\mathrm{d}\Omega = 0$.

Now the deformation theory of Cayley 4-folds in §12.5.1 does not use $\mathrm{d}\Omega = 0$, nor is this condition very important in other parts of the theory. Thus, the appropriate notion of *almost* $\mathrm{Spin}(7)$-*manifold* (M, Ω, g) is an 8-manifold M with a $\mathrm{Spin}(7)$-structure (Ω, g), but without assuming $\mathrm{d}\Omega = 0$. We then define a Cayley 4-fold in (M, Ω, g) to be an oriented 4-submanifold N in M with $\Omega|_{T_x N} = \mathrm{vol}_{T_x N}$ for all $x \in N$, which would be the condition for N to be Ω-calibrated if Ω were a calibration.

$\mathrm{Spin}(7)$-structures on M are parametrized by arbitrary smooth sections of a bundle with fibre $\mathrm{GL}(8, \mathbb{R})/\mathrm{Spin}(7)$, of dimension 43, so almost $\mathrm{Spin}(7)$-manifolds come in infinite-dimensional families. Thus choosing a generic almost $\mathrm{Spin}(7)$-manifold is a powerful thing to do, and should simplify the singular behaviour of Cayley 4-folds.

12.5.1 Deformations of Cayley 4-folds

Cayley 4-folds cannot be defined in terms of the vanishing of closed forms, and this makes their deformation theory like that of associative 3-folds in §12.3.1, rather than the special Lagrangian and coassociative cases of §8.4.1 and §12.3.1. Here is how the theory works, drawn mostly from McLean [259, §6]. Since we use a different convention for Ω_0 in (12.10) to McLean, we exchange $\mathbb{S}_+, \mathbb{S}_-$ below compared to [259, §6].

Let N be a compact Cayley 4-fold in a Spin(7)-manifold (or almost Spin(7)-manifold) (M, Ω, g). Let $\nu \to N$ be the normal bundle of N in M. Then there is a vector bundle $E \to N$ with fibre $\mathbb{R}^4$, and a first-order elliptic operator $D_N : C^\infty(\nu) \to C^\infty(E)$ on N. The *kernel* $\operatorname{Ker} D_N$ is the set of *infinitesimal deformations* of N as a Cayley 4-fold. The *cokernel* $\operatorname{Coker} D_N$ is the *obstruction space* for these deformations. Both are finite-dimensional vector spaces, and the *index* of D_N is

$$\operatorname{ind}(D_N) = \dim \operatorname{Ker} D_N - \dim \operatorname{Coker} D_N.$$

It is a topological invariant, given in terms of characteristic classes by the *Atiyah–Singer Index Theorem*. A calculation by Christopher Lewis and the author shows that

$$\operatorname{ind}(D_N) = -\tau(N) - \tfrac{1}{2}\chi(N) - \tfrac{1}{2}[N] \cdot [N], \tag{12.12}$$

with τ the signature, χ the Euler characteristic and $[N] \cdot [N]$ the self-intersection of N. Note that the sign of $\tau(N)$ in (12.12) depends on our convention (12.10) for Ω_0; with McLean's conventions the orientation of N is reversed, giving $\tau(N)$ instead of $-\tau(N)$.

Now Cayley 4-folds need not be spin—for instance, a complex surface $\mathbb{CP}^2$ in a Calabi–Yau 4-fold is an example of a non-spin Cayley 4-fold in a Spin(7)-manifold. Suppose N is spin, with positive and negative spin bundles $\mathbb{S}_\pm$. Then in a similar way to §12.3.2, there are isomorphisms $\nu \otimes_{\mathbb{R}} \mathbb{C} \cong \mathbb{S}_+ \otimes_{\mathbb{C}} F$ and $E \otimes_{\mathbb{R}} \mathbb{C} \cong \mathbb{S}_- \otimes_{\mathbb{C}} F$, where $\mathbb{S}_\pm, F$ are vector bundles with fibre $\mathbb{C}^2$. Complex conjugation on $\nu \otimes_{\mathbb{R}} \mathbb{C}$ and $E \otimes_{\mathbb{R}} \mathbb{C}$ are of the form $\sigma_+ \otimes \sigma$ and $\sigma_- \otimes \sigma$, where $\sigma_\pm : \mathbb{S}_\pm \to \mathbb{S}_\pm$ and $\sigma : F \to F$ are complex anti-linear isomorphisms with $\sigma_\pm^2 = -1 = \sigma^2$. Then D_N is a twisted version of the *positive Dirac operator* $D_+ : C^\infty(\mathbb{S}_+) \to C^\infty(\mathbb{S}_-)$ of N.

In a *generic* situation we expect $\operatorname{Coker} D_N = 0$, and then deformations of N will be unobstructed, so that the moduli space $\mathcal{M}_N$ of Cayley deformations of N will locally be a smooth manifold of dimension $\operatorname{ind}(D_N)$. However, in nongeneric situations the obstruction space may be nonzero, and then the moduli space may not be smooth, or may have a larger than expected dimension.

This general structure is found in the deformation theory of other important mathematical objects—for instance, pseudoholomorphic curves in almost complex manifolds, and instantons and Seiberg–Witten solutions on 4-manifolds. In each case, the moduli space is only smooth with topologically determined dimension under a *genericity assumption* which forces the obstructions to vanish.

12.5.2 Examples of compact Cayley 4-folds

By the method of Propositions 12.3.7 and 12.3.9 one can prove [188, Prop. 10.8.6]:

Proposition 12.5.2 *Let* (M, Ω, g) *be a* Spin(7)-*manifold, and* $\sigma : M \to M$ *a nontrivial isometric involution with* $\sigma^*(\Omega) = \Omega$. *Then each connected component of the fixed point set* $\{p \in M : \sigma(p) = p\}$ *of* σ *is either a closed, nonsingular, embedded Cayley 4-fold which is compact if* M *is, or a single point.*

Following [188, Ex. 14.3.1], we use this to construct examples of compact Cayley 4-folds in the compact 8-manifolds with holonomy Spin(7) discussed in §11.6.

Example 12.5.3 Let T^8 and Γ be as in Example 11.6.1. Define $\sigma : T^8 \to T^8$ by

$$\sigma : (x_1, \ldots, x_8) \mapsto (-x_1, \tfrac{1}{2} - x_2, x_3, x_4, x_5, x_6, -x_7, \tfrac{1}{2} - x_8).$$

Then σ commutes with Γ and $\sigma^*(\Omega_0) = \Omega_0$. The fixed points of σ in T^8 are 16 T^4's, the fixed points of $\sigma\alpha\gamma$ are 256 points, and $\sigma\epsilon$ acts freely on T^8 for all $\epsilon \neq 1, \alpha\gamma$ in Γ. Also Γ acts freely on the 16 σ T^4's and the 256 $\sigma\alpha\gamma$ points.

Thus the fixed points of σ in T^8/Γ are the disjoint union of T^4 and 16 points, none of which intersect the singular set of T^8/Γ. When we resolve T^8/Γ in a σ-equivariant way to get M with holonomy $\mathrm{Spin}(7)$, the fixed points of σ in M are again the disjoint union of T^4 and 16 points. Proposition 12.5.2 shows this T^4 is a *Cayley* 4-*fold* in M.

Using different involutions σ, in [188, §14.3] we find examples of Cayley 4-folds in the same 8-manifold M diffeomorphic to $T^4/\mathbb{Z}_2$, $K3$, $\mathcal{S}^2 \times \mathcal{S}^2$, and with other topologies. Note too that the second method for constructing examples of associative 3-folds described after Theorem 12.3.11 will also work for Cayley 4-folds, supposing N_0 is a compact Cayley 4-fold in T^8/Γ not intersecting the singular set S of T^8/Γ, which is *unobstructed* in the sense that $\mathrm{Coker}(D_N) = 0$ in the notation of §12.5.1.

12.5.3 Further topics in Cayley geometry

Here are three other topics to do with Cayley 4-folds.

Cayley 4-folds with isolated conical singularities

As for SL m-folds in §8.5 and coassociative 4-folds in §12.3.4, it would be interesting to develop a theory of Cayley 4-folds with *isolated conical singularities* in $\mathrm{Spin}(7)$-manifolds. So far as the author knows, no work has yet been done on this.

Cayley fibrations

As for special Lagrangians in §9.4 and coassociative 4-folds in §12.3.4, it is natural to ask whether a compact 8-manifold with holonomy $\mathrm{Spin}(7)$ can be fibred by Cayley 4-folds N with some singular fibres. In String Theory, Cayley fibrations appear in the work of Acharya [3, 4], with generic fibre T^4. Here are some general considerations.

Suppose (M, Ω, g) is a compact $\mathrm{Spin}(7)$-manifold, $f : M \to B$ is a Cayley fibration, and $N = f^{-1}(b)$ is a nonsingular fibre of f. Then $\mathrm{d}f|_N$ induces an isomorphism of the normal bundle ν of N with the trivial bundle $N \times T_b B$. As ν is trivial, the self-intersection $[N] \cdot [N]$ appearing in (12.12) is zero. In §12.5.1 we saw that $\nu \otimes_{\mathbb{R}} \mathbb{C} \cong \mathbb{S}_+ \otimes_{\mathbb{C}} F$ if N is spin, and this holds locally in N even if N is not spin. Using this one can show that N is spin, with a unique spin structure such that under $\nu \otimes_{\mathbb{R}} \mathbb{C} \cong \mathbb{S}_+ \otimes_{\mathbb{C}} F$ the trivialization of ν is compatible with trivializations of $\mathbb{S}_+$ and F. Thus N is spin with trivial positive spin bundle $\mathbb{S}_+$, implying that N admits an $\mathrm{SU}(2)$-structure. This suggests N should be T^4 or $K3$. Using characteristic classes one can show that $\mathbb{S}_+$ trivial implies $3\tau(N) + 2\chi(N) = 0$.

If Cayley fibrations are to exist in any generic situation, we expect deformations of N to be *unobstructed* in the sense of §12.5.1, and clearly we would like the moduli space $\mathcal{M}_N$ of Cayley deformations of N to be smooth of dimension 4, so that it locally parametrizes the fibres of f. Thus as $[N] \cdot [N] = 0$, from (12.12) we want $-\tau(N) -$

$\frac{1}{2}\chi(N) = 4$. Combined with $3\tau(N) + 2\chi(N) = 0$ this forces $\tau(N) = -16$ and $\chi(N) = 24$. These hold for $N = K3$, but not for $N = T^4$ as $\tau(T^4) = \chi(T^4) = 0$. So we can rule out T^4 as a plausible nonsingular fibre for generic Cayley fibrations, despite [3,4]. But Cayley fibrations with nonsingular fibre $K3$ may well be a workable idea.

As for associative fibrations at the end of §12.3.4, the author suggests that Acharya's proposal [3,4] involving Cayley T^4-fibrations should be interpreted via a Spin(7) analogue of Conjecture 9.4.3, involving 1-parameter families of fibrations $f_t : X_t \to B$ of Spin(7)-manifolds X_t, whose fibres T_t^b are required to be only *approximately* Cayley in X_t, and to converge to exactly Cayley, flat T^4's as $t \to \infty$.

The author has an idea for a method to construct a Cayley fibration, including singular fibres, of the compact 8-manifold M with holonomy Spin(7) constructed from a Calabi–Yau 4-orbifold in §11.6.2. It is a little similar to Kovalev's proposal [224] for the G_2 case, and will be described in a future paper.

Cayley 4-folds as the bubbling loci of Spin(7) instantons

In §11.7 we briefly discussed Spin(7) *instantons*, that is, connections A on a vector or principal bundle E over a compact Spin(7)-manifold (M, Ω, g) whose curvature F_A satisfies $\pi_7(F_A) = 0$. An important result of Tian [327] shows that for a sequence $A_1, A_2, \ldots$ of such connections on a fixed bundle E, there is a sequence of gauge transformations $\gamma_1, \gamma_2, \ldots$ of E such that $\gamma_i(A_i)$ converges to a nonsingular Spin(7) instanton on M outside a closed singular set S, the 'blow-up locus'.

Furthermore, S can be given the structure of a *closed integral current* in M, in the sense of geometric measure theory, which is calibrated with respect to Ω. That is, S is a (possibly very singular) *Cayley 4-fold* in M. So Cayley 4-folds appear naturally in gauge theory as the bubbling loci of Spin(7) instantons. Christopher Lewis [243] constructed nontrivial examples of families of Spin(7) instantons with gauge group SU(2) on the compact 8-manifold M with holonomy Spin(7) described in §11.6.1, which bubble along two Cayley $K3$'s in M.

We can now make the following very speculative argument, which is probably nonsense, but may contain a grain of truth. Let (M, Ω, g) be a generic compact almost Spin(7)-manifold. Then we can form moduli spaces $\mathcal{M}_E$ of Spin(7) instantons on a bundle E on M. By Tian's work $\mathcal{M}_E$ has a compactification $\overline{\mathcal{M}}_E$, whose boundary $\partial\mathcal{M}_E = \overline{\mathcal{M}}_E \setminus \mathcal{M}_E$ is (very) roughly a moduli space of Cayley integral currents.

Therefore we expect moduli spaces of Cayley integral currents in a generic compact (almost) Spin(7)-manifold (or at least, those that are suitable to be bubbling loci of Spin(7) instantons, which may not be all Cayley integral currents) to be compact and *without boundary*, since the boundary of a boundary is empty. This suggests that the singular behaviour of Cayley 4-folds that can happen in real codimension 1 may be special, and more well-controlled than one might expect. If so, perhaps interesting invariants 'counting' Cayley 4-folds in a given homology class can be defined.

References

[1] B.S. Acharya, '$N = 1$ heterotic/M-theory and Joyce manifolds,' Nuclear Physics B475 (1996), 579–596. hep-th/9603033.

[2] B.S. Acharya, 'Dirichlet Joyce manifolds, discrete torsion and duality,' Nuclear Physics B492 (1997), 591–606. hep-th/9611036.

[3] B.S. Acharya, 'On mirror symmetry for manifolds of exceptional holonomy,' Nuclear Physics B524 (1998), 269–282. hep-th/9707186.

[4] B.S. Acharya, 'Exceptional mirror symmetry'. In C. Vafa and S.-T. Yau, editors, 'Winter school on Mirror Symmetry, vector bundles and Lagrangian submanifolds,' AMS/IP Studies in Advanced Mathematics, vol. 23, pages 1–14, A.M.S., Providence, RI, 2001.

[5] B.S. Acharya and S. Gukov, 'M-theory and singularities of exceptional holonomy manifolds,' Physics Reports 392 (2004), 121–189. hep-th/0409191.

[6] B.S. Acharya, M. O'Loughlin, and B. Spence, 'Higher-dimensional analogues of Donaldson–Witten theory,' Nuclear Physics B503 (1997), 657–674. hep-th/9705138.

[7] D.V. Alekseevsky, 'Riemannian spaces with exceptional holonomy,' Functional Analysis and its Applications 2 (1968), 97–105.

[8] D.V. Alekseevsky, 'Classification of quaternionic spaces with a transitive solvable group of motions,' Mathematics of the USSR. Izvestija 9 (1975), 297–339.

[9] F.J. Almgren, 'Almgren's big regularity paper,' World Scientific, River Edge, NJ, 2000. Edited by V. Scheffer and J.E. Taylor.

[10] W. Ambrose and I.M. Singer, 'A theorem on holonomy,' Transactions of the American Mathematical Society 75 (1953), 428–443.

[11] M.F. Atiyah and N.J. Hitchin, 'The geometry and dynamics of magnetic monopoles,' Princeton University Press, Princeton, 1988.

[12] M.F. Atiyah, N.J. Hitchin, and I.M Singer, 'Self-duality in four-dimensional Riemannian geometry,' Proceedings of the Royal Society of London Series A 362 (1978), 425–461.

[13] M.F. Atiyah and I.M. Singer, 'The index of elliptic operators. III,' Annals of Mathematics 87 (1968), 546–604.

[14] M. Atiyah and E. Witten, 'M-theory dynamics on a manifold of G_2 holonomy,' Advances in Theoretical and Mathematical Physics 6 (2003), 1–106. hep-th/0107177.

[15] T. Aubin, 'Métriques riemanniennes et courbure,' Journal of Differential Geometry 4 (1970), 383–424.

[16] T. Aubin, '*Some nonlinear problems in Riemannian geometry*,' Springer Monographs in Mathematics, Springer-Verlag, Berlin, 1998.

[17] D. Auroux, L. Katzarkov, and D. Orlov, '*Mirror symmetry for weighted projective planes and their noncommutative deformations*,' math.AG/0404281, 2004.

[18] D. Auroux, L. Katzarkov, and D. Orlov, '*Mirror symmetry for Del Pezzo surfaces: vanishing cycles and coherent sheaves*,' math.AG/0506166, 2005.

[19] W. Barth, C. Peters, and A. Van de Ven, '*Compact complex surfaces*,' Ergebnisse der Mathematik und ihre Grenzgebiete, vol. 4, Springer-Verlag, New York, 1984.

[20] V.V. Batyrev, '*Dual polyhedra and mirror symmetry for Calabi–Yau hypersurfaces in toric varieties*,' Journal of Algebraic Geometry 3 (1994), 493–535. alg-geom/9310003.

[21] V.V. Batyrev, '*Non-Archimedean integrals and stringy Euler numbers of log-terminal pairs*,' Journal of the European Mathematical Society 1 (1999), 5–33. math.AG/9803071.

[22] V.V. Batyrev and D.I. Dais, '*Strong McKay correspondence, string-theoretic Hodge numbers and mirror symmetry*,' Topology 35 (1996), 901–929. alg-geom/9410001.

[23] L. Baulieu, H. Kanno, and I.M. Singer, '*Special quantum field theories in eight and other dimensions*,' Communications in Mathematical Physics 194 (1998), 149–175. hep-th/9704167.

[24] A. Beauville, '*Some remarks on Kähler manifolds with $c_1 = 0$*'. In K. Ueno, editor, '*Classification of algebraic and analytic manifolds, 1982*,' Progress in Mathematics, vol. 39, pages 1–26, Birkhäuser, 1983.

[25] A. Beauville, '*Variétés Kählériennes dont la première classe de Chern est nulle*,' Journal of Differential Geometry 18 (1983), 755–782.

[26] A. Beauville et al., '*Géometrie des surfaces $K3$: modules et périodes, Séminaire Palaiseau 1981–2*,' Astérisque 126 (1985).

[27] M. Berger, '*Sur les groupes d'holonomie homogène des variétés à connexion affines et des variétés riemanniennes*,' Bulletin de la Société Mathématique de France 83 (1955), 279–330.

[28] M. Berger, '*A panoramic view of Riemannian geometry*,' Springer-Verlag, Berlin, 2003.

[29] M. Bershadsky, S. Cecotti, H. Ooguri, and C. Vafa, '*Kodaira–Spencer theory of gravity and exact results for quantum string amplitudes*,' Communications in Mathematical Physics 165 (1994), 311–427. hep-th/9309140.

[30] A.L. Besse, '*Einstein manifolds*,' Springer-Verlag, New York, 1987.

[31] O. Biquard, '*Sur les équations de Nahm et la structure de Poisson des algèbres de Lie semi-simples complexes*,' Mathematische Annalen 304 (1996), 253–276.

[32] S. Bochner, '*Vector fields and Ricci curvature*,' Bulletin of the American Mathematical Society 52 (1946), 776–797.

[33] J. Bolton, F. Pedit, and L.M. Woodward, '*Minimal surfaces and the affine Toda field model*,' Journal für die Reine und Angewandte Mathematik 459 (1995), 119–150.

[34] J. Bolton, L. Vrancken, and L.M. Woodward, '*On almost complex curves in the nearly Kähler 6-sphere,*' Oxford Quarterly Journal of Mathematics 45 (1994), 407–427.

[35] E. Bonan, '*Sur les variétés riemanniennes à groupe d'holonomie G_2 ou* Spin(7),' Comptes Rendus de l'Académie des Sciences. Série A, Sciences Mathematiques 262 (1966), 127–129.

[36] A.I. Bondal and M.M. Kapranov, '*Enhanced triangulated categories,*' Math. USSR Sbornik 70 (1991), 93–107.

[37] A. Bondal and D. Orlov, '*Reconstruction of a variety from the derived category and groups of autoequivalences,*' Compositio Math. 125 (2001), 327–344. alggeom/9712029.

[38] A. Borel, '*Some remarks about Lie groups transitive on spheres and tori,*' Bulletin of the American Mathematical Society 55 (1949), 580–587.

[39] O. Borůvka, '*Sur les surfaces represéntées par les fonctions sphériques de prèmiere espèce,*' Journal de Mathématiques Pures et Appliquées 12 (1933), 337–383.

[40] R. Bott and L.W. Tu, '*Differential Forms in Algebraic Topology,*' Graduate Texts in Mathematics, vol. 82, Springer-Verlag, New York, 1995.

[41] S. Boucksom, '*Le cône kählérien d'une variété hyperkählérienne,*' Comptes Rendus de l'Académie des Sciences. Série A, Sciences Mathematiques 333 (2001), 935–938.

[42] J.-P. Bourguignon et al., '*Première classe de Chern et courbure de Ricci: preuve de la conjecture de Calabi, Séminaire Palaiseau 1978,*' Astérisque 58 (1978).

[43] C.P. Boyer, '*A note on hyper-Hermitian four-manifolds,*' Proceedings of the American Mathematical Society 102 (1988), 157–164.

[44] C.P. Boyer and K. Galicki, '*3-Sasakian manifolds*'. In C. LeBrun and M. Wang, editors, '*Essays on Einstein manifolds,*' Surveys in Differential Geometry, vol. V, pages 123–184, International Press, 1999. hep-th/9810250.

[45] C.P. Boyer, K. Galicki, and B.M. Mann, '*The geometry and topology of 3-Sasakian manifolds,*' Journal für die Reine und Angewandte Mathematik 455 (1994), 183–220.

[46] C.P. Boyer, K. Galicki, B.M. Mann, and E. Rees, '*Compact 3-Sasakian manifolds with arbitrary second Betti number,*' Inventiones mathematicae 131 (1998), 321–344.

[47] A. Brandhuber, '*G_2 holonomy spaces from invariant three-forms,*' Nuclear Physics B629 (2002), 393–416. hep-th/0112113.

[48] A. Brandhuber, J. Gomis, S.S. Gubser, and S. Gukov, '*Gauge theory at large N and new G_2 holonomy metrics,*' Nuclear Physics B611 (2001), 179–204. hep-th/0106034.

[49] G.E. Bredon, '*Topology and Geometry,*' Graduate Texts in Mathematics, vol. 139, Springer-Verlag, Berlin, 1993.

[50] S. Brendle, '*Complex anti-self-dual instantons and Cayley submanifolds,*' math.DG/0302094, 2003.

[51] T. Bridgeland, '*Stability conditions on triangulated categories*,' to appear in Annals of Mathematics. math.AG/0212237, 2002.

[52] T. Bridgeland, '*Derived categories of coherent sheaves*,' math.AG/0602129, 2006.

[53] R. Brown and A. Gray, '*Riemannian manifolds with holonomy group* $\mathrm{Spin}(9)$'. In S. Kobayashi et al., editors, '*Differential geometry (in honour of Kentaro Yano)*,' pages 41–59, Kinokuniya, Tokyo, 1972.

[54] R.L. Bryant, '*Submanifolds and special structures on the octonians*,' Journal of Differential Geometry 17 (1982), 185–232.

[55] R.L. Bryant, '*Metrics with holonomy G_2 or* $\mathrm{Spin}(7)$'. In '*Arbeitstagung Bonn 1984*,' Lecture Notes in Mathematics, vol. 1111, pages 269–277, Springer-Verlag, 1985.

[56] R.L. Bryant, '*Metrics with exceptional holonomy*,' Annals of Mathematics 126 (1987), 525–576.

[57] R.L. Bryant, '*Classical, exceptional, and exotic holonomies: a status report*'. In A.L. Besse, editor, '*Actes de la Table Ronde de Géométrie Différentielle (Luminy, 1992)*,' Seminaires et Congres, vol. 1, pages 93–165, Société Mathematique de France, Paris, 1996.

[58] R.L. Bryant, '*Some examples of special Lagrangian tori*,' Advances in Theoretical and Mathematical Physics 3 (1999), 83–90. math.DG/9902076.

[59] R.L. Bryant, '*Calibrated embeddings: the special Lagrangian and coassociative cases*,' Annals of Global Analysis and Geometry 18 (2000), 405–435. math.DG/9912246.

[60] R.L. Bryant, '*Second order families of special Lagrangian 3-folds*'. In V. Apostolov, A. Dancer, N.J. Hitchin, and M. Wang, editors, '*Perspectives in Riemannian geometry*,' CRM Proceedings and Lecture Notes, vol. 40, pages 63–98, A.M.S., Providence, RI, 2006. math.DG/0007128.

[61] R.L. Bryant, '*Some remarks on G_2-structures*'. In S. Akbulut, T. Önder, and R.J. Stern, editors, '*Proceedings of Gökova Geometry–Topology Conference 2005*,' pages 75–109, International Press, Somerville, MA, 2006. math.DG/0305124.

[62] R.L. Bryant, '$\mathrm{SO}(n)$-*invariant special Lagrangian submanifolds of* $\mathbb{C}^{n+1}$ *with fixed loci*,' Chinese Annals of Mathematics Series B 27 (2006), 95–112. math.DG/0402201.

[63] R.L. Bryant, S.S. Chern, R.B. Gardner, H.L. Goldschmidt, and P.A. Griffiths, '*Exterior differential systems*,' M.S.R.I. Publications, vol. 18, Springer-Verlag, New York, 1991.

[64] R.L. Bryant and S.M. Salamon, '*On the construction of some complete metrics with exceptional holonomy*,' Duke Mathematical Journal 58 (1989), 829–850.

[65] D. Burns and M. Rapoport, '*On the Torelli problem for Kählerian $K3$ surfaces*,' Annales scientifiques de l'École Normale Superieure 8 (1975), 235–274.

[66] A. Butscher, '*Regularizing a singular special Lagrangian variety*,' Communications in Analysis and Geometry 12 (2004), 733–791. math.DG/0110053.

[67] E. Calabi, '*The space of Kähler metrics*'. In '*Proceedings of the International Congress of Mathematicians, Amsterdam, 1954*,' vol. 2, pages 206–207, North–Holland, Amsterdam, 1956.

[68] E. Calabi, '*On Kähler manifolds with vanishing canonical class*'. In R.H. Fox, D.C. Spencer, and A.W. Tucker, editors, '*Algebraic geometry and topology, a symposium in honour of S. Lefschetz*,' pages 78–89, Princeton University Press, Princeton, 1957.

[69] E. Calabi, '*Métriques kählériennes et fibrés holomorphes*,' Annales scientifiques de l'École Normale Superieure 12 (1979), 269–294.

[70] P. Candelas, X.C. de la Ossa, P.S. Green, and L. Parkes, '*A pair of Calabi–Yau manifolds as an exactly soluble superconformal theory*,' Nuclear Physics B359 (1991), 21–74.

[71] P. Candelas, M. Lynker, and R. Schimmrigk, '*Calabi–Yau manifolds in weighted $\mathbb{P}_4^*$*,' Nuclear Physics B341 (1990), 383–402.

[72] E. Carberry and I. McIntosh, '*Minimal Lagrangian 2-tori in $\mathbb{CP}^2$ come in real families of every dimension*,' Journal of the London Mathematical Society 69 (2004), 531–544. math.DG/0308031.

[73] I. Castro and F. Urbano, '*New examples of minimal Lagrangian tori in the complex projective plane*,' Manuscripta Mathematica 85 (1994), 265–281.

[74] I. Castro and F. Urbano, '*On a new construction of special Lagrangian immersions in complex Euclidean space*,' Oxford Quarterly Journal of Mathematics 55 (2004), 253–265. math.DG/0403108.

[75] G. Cheeger and M. Gromoll, '*The splitting theorem for manifolds of nonnegative Ricci curvature*,' Journal of Differential Geometry 6 (1971), 119–128.

[76] Z.W. Chong, M. Cvetič, G.W. Gibbons, H. Lü, C.N. Pope, and P. Wagner, '*General metrics of G_2 holonomy and contraction limits*,' Nuclear Physics B638 (2002), 459–482. hep-th/0204064.

[77] L.A. Cordero and M. Fernández, '*Some compact solvmanifolds of G_2 coassociative type*,' Analele Universitatii Bucuresti Matematica 36 (1987), 16–19.

[78] K. Costello, '*Topological conformal field theories and Calabi–Yau categories*,' math.QA/0412149, 2004.

[79] K. Costello, '*The Gromov–Witten potential associated to a TCFT*,' math.QA/0509264, 2005.

[80] D.A. Cox and S. Katz, '*Mirror symmetry and algebraic geometry*,' Mathematical Surveys and Monographs, vol. 68, A.M.S., Providence, RI, 1999.

[81] M. Cvetič, G.W. Gibbons, H. Lü, and C.N. Pope, '*A G_2 unification of the deformed and resolved conifolds*,' Physics Letters B534 (2002), 172–180. hep-th/0112138.

[82] M. Cvetič, G.W. Gibbons, H. Lü, and C.N. Pope, '*Cohomogeneity one manifolds of $\mathrm{Spin}(7)$ and G_2 holonomy*,' Physical Review D65 (2002), 106004. hep-th/0108245.

[83] M. Cvetič, G.W. Gibbons, H. Lü, and C.N. Pope, '*M-theory conifolds*,' Physics Review Letters 88 (2002), 121602. hep-th/0112098.

[84] M. Cvetič, G.W. Gibbons, H. Lü, and C.N. Pope, '*New complete non-compact* $\mathrm{Spin}(7)$ *manifolds*,' Nuclear Physics B620 (2002), 29–54. hep-th/0103155.

[85] M. Cvetič, G.W. Gibbons, H. Lü, and C.N. Pope, '*Supersymmetric M3-branes and G_2-manifolds*,' Nuclear Physics B620 (2002), 3–28. hep-th/0106026.

[86] M. Cvetič, G.W. Gibbons, H. Lü, and C.N. Pope, '*Bianchi IX self-dual Einstein metrics and singular G_2 manifolds*,' Classical and Quantum Gravity 20 (2003), 4239–4268. hep-th/0206151.

[87] M. Cvetič, G.W. Gibbons, H. Lü, and C.N. Pope, '*Special holonomy spaces and M-theory*'. In C. Bachas, A. Bilal, M. Douglas, N. Nekrasov, and F. David, editors, '*Unity from duality: gravity, gauge theory and strings*,' pages 523–545, Springer-Verlag, Berlin, 2003. hep-th/0206154.

[88] M. Cvetič, G.W. Gibbons, H. Lü, and C.N. Pope, '*New cohomogeneity one metrics with* $\mathrm{Spin}(7)$ *holonomy*,' Journal of Geometry and Physics 49 (2004), 350–365. math.DG/0105119.

[89] M. Cvetič, G.W. Gibbons, H. Lü, and C.N. Pope, '*Orbifolds and slumps in G_2 and* $\mathrm{Spin}(7)$ *metrics*,' Annals of Physics 310 (2004), 265–301. hep-th/0111096.

[90] J. Dadok and R. Harvey, '*Calibrations on* $\mathbb{R}^6$,' Duke Mathematical Journal 50 (1983), 1231–1243.

[91] J. Dadok, R. Harvey, and F. Morgan, '*Calibrations on* $\mathbb{R}^8$,' Transactions of the American Mathematical Society 307 (1988), 1–40.

[92] A.S. Dancer, '*Hyper-Kähler manifolds*'. In C. LeBrun and M. Wang, editors, '*Essays on Einstein manifolds*,' Surveys in Differential Geometry, vol. V, pages 15–38, International Press, 1999.

[93] G. De Rham, '*Sur la réductibilité d'un espace de Riemann*,' Commentarii Mathematici Helvetici 26 (1952), 328–344.

[94] J. Denef and F. Loeser, '*Motivic integration, quotient singularities and the McKay correspondence*,' Composition Mathematica 131 (2002), 267–290. math.AG/9903187.

[95] D.-E. Diaconescu, '*Enhanced D-brane categories from string field theory*,' Journal of High Energy Physics 06 (2001), 016. hep-th/0104200.

[96] R. Dijkgraaf, S. Gukov, A. Neitzke, and C. Vafa, '*Topological M-theory as unification of form theories of gravity*,' hep-th/0411073, 2004.

[97] S.K. Donaldson and R. Friedman, '*Connected sums of self-dual manifolds and deformations of singular spaces*,' Nonlinearity 2 (1989), 197–239.

[98] S.K. Donaldson and P.B. Kronheimer, '*The Geometry of Four-Manifolds*,' OUP, Oxford, 1990.

[99] M.R. Douglas, '*The Geometry of String Theory*'. In M. Douglas, J. Gauntlett, and M. Gross, editors, '*Strings and Geometry*,' Clay Mathematics Proceedings, vol. 3, pages 1–30, American Mathematical Society, Providence, RI, 2004.

[100] T. Eguchi and A.J. Hanson, '*Asymptotically flat solutions to Euclidean gravity*,' Physics Letters 74B (1978), 249–251.

[101] N. Ejiri, '*Equivariant minimal immersions of S^2 into $S^{2m}(1)$*,' Transactions of the American Mathematical Society 297 (1986), 105–124.

[102] H. Federer, '*Geometric Measure Theory*,' Grundlehren der mathematischen Wissenschaften, vol. 153, Springer-Verlag, Berlin, 1969.

[103] M. Fernández, 'A *classification of Riemannian manifolds with structure group* $\mathrm{Spin}(7)$,' Annali di matematica pura ed applicata 143 (1986), 101–122.

[104] M. Fernández, '*An example of a compact calibrated manifold associated with the exceptional Lie group* G_2,' Journal of Differential Geometry 26 (1987), 367–370.

[105] M. Fernández, 'A *family of compact solvable* G_2-*calibrated manifolds*,' Tohoku Mathematical Journal 39 (1987), 287–289.

[106] M. Fernández and A. Gray, '*Riemannian manifolds with structure group* G_2,' Annali di matematica pura ed applicata 132 (1982), 19–45.

[107] M. Fernández and T. Iglesias, '*New examples of Riemannian manifolds with structure group* G_2,' Rendiconti del Circolo Matematico di Palermo 35 (1986), 276–290.

[108] J.M. Figueroa-O'Farrill, '*Extended superconformal algebras associated with manifolds of exceptional holonomy*,' Physics Letters B392 (1997), 77–84.

[109] A.P. Fordy and J.C. Wood, editors, '*Harmonic maps and integrable systems*,' Aspects of Mathematics, vol. E23, Vieweg, Wiesbaden, 1994.

[110] D. Fox, '*Coassociative cones that are ruled by 2-planes*,' math.DG/0511458, 2005.

[111] D. Fox, '*Second order families of coassociative 4-folds*,' Ph.D. thesis, Duke University, 2005.

[112] R. Friedman, '*Simultaneous resolution of threefold double points*,' Mathematische Annalen 274 (1986), 671–689.

[113] Th. Friedrich, I. Kath, A. Moroianu, and U. Semmelmann, '*On nearly parallel* G_2-*structures*,' Journal of Geometry and Physics 23 (1997), 259–286.

[114] A. Fujiki, '*On primitively symplectic Kähler V-manifolds of dimension 4*'. In K. Ueno, editor, '*Classification of algebraic and analytic manifolds, 1982*,' Progress in Mathematics, vol. 39, pages 71–250, Birkhäuser, 1983.

[115] A. Fujiki, '*On the de Rham cohomology group of a compact Kähler symplectic manifold*'. In T. Oda, editor, '*Algebraic Geometry, Sendai, 1985*,' Advanced Studies in Pure Mathematics, vol. 10, pages 105–165, Kinokuniya, Tokyo, 1987.

[116] K. Fukaya, '*Morse homotopy, A_∞-category, and Floer homologies*'. In H.-J. Kim, editor, '*Proceedings of GARC Workshop on Geometry and Topology, Seoul, 1993*,' Lecture Notes Series, vol. 18, pages 1–102, Seoul National University, 1993.

[117] K. Fukaya, '*Multivalued Morse theory, asymptotic analysis and mirror symmetry*'. In M. Lyubich and L. Takhtajan, editors, '*Graphs and patterns in mathematics and theoretical physics*,' Proceedings of Symposia in Pure Mathematics, vol. 73, pages 205–278, A.M.S., Providence, RI, 2005.

[118] K. Fukaya, Y.-G. Oh, H. Ohta, and K. Ono, '*Lagrangian intersection Floer homology – anomaly and obstruction*,' in preparation, 2006.

[119] W. Fulton, '*Introduction to Toric Varieties*,' Annals of Mathematics Studies, no. 131, Princeton University Press, Princeton, 1993.

[120] K. Galicki, 'A generalization of the momentum mapping construction for quaternionic Kähler manifolds,' Communications in Mathematical Physics 108 (1987), 117–138.

[121] K. Galicki and H.B. Lawson, 'Quaternionic reduction and quaternionic orbifolds,' Mathematische Annalen 282 (1988), 1–21.

[122] K. Galicki and S. Salamon, 'Betti numbers of 3-Sasakian manifolds,' Geometriae Dedicata 63 (1996), 45–68.

[123] S.I. Gelfand and Y.I. Manin, 'Methods of Homological Algebra,' second edition, Springer Monographs in Mathematics, Springer-Verlag, Berlin, 2003.

[124] G.W. Gibbons and S.W. Hawking, 'Gravitational multi-instantons,' Physics Letters 78B (1978), 430–432.

[125] G.W. Gibbons, D.N. Page, and C.N. Pope, 'Einstein metrics on S^3, $\mathbb{R}^3$ and $\mathbb{R}^4$ bundles,' Communications in Mathematical Physics 127 (1990), 529–553.

[126] D. Gilbarg and N.S. Trudinger, 'Elliptic partial differential equations of second order,' Grundlehren der mathematischen Wissenschaften, vol. 224, Springer-Verlag, Berlin, 1977.

[127] E. Goldstein, 'A construction of new families of minimal Lagrangian submanifolds via torus actions,' Journal of Differential Geometry 58 (2001), 233–261. math.DG/0007135.

[128] E. Goldstein, 'Calibrated fibrations,' Communications in Analysis and Geometry 10 (2002), 127–150. math.DG/9911093.

[129] M. Goresky and R. MacPherson, 'Stratified Morse theory,' Ergebnisse der Mathematik und ihrer Grenzgebiete, vol. 14, Springer-Verlag, Berlin, 1988.

[130] L. Göttsche, 'Hilbert schemes of zero-dimensional subschemes of smooth varieties,' Springer Lecture Notes in Mathematics, vol. 1572, Springer-Verlag, Berlin, 1994.

[131] P. Green and T. Hübsch, 'Calabi–Yau manifolds as complete intersections in products of complex projective spaces,' Communications in Mathematical Physics 109 (1987), 99–108.

[132] P. Griffiths and J. Harris, 'Principles of Algebraic Geometry,' Wiley, New York, 1978.

[133] M. Gross, 'Special Lagrangian fibrations I: Topology'. In M.-H. Saito, Y. Shimizu, and K. Ueno, editors, 'Integrable Systems and Algebraic Geometry,' pages 156–193, World Scientific, Singapore, 1998. alg-geom/9710006.

[134] M. Gross, 'Special Lagrangian fibrations II: Geometry'. In 'Differential Geometry inspired by String Theory,' Surveys in Differential Geometry, vol. 5, pages 341–403, International Press, Boston, MA, 1999. math.AG/9809072.

[135] M. Gross, 'Topological mirror symmetry,' Inventiones mathematicae 144 (2001), 75–137. math.AG/9909015.

[136] M. Gross, 'Examples of special Lagrangian fibrations'. In K. Fukaya, Y.-G. Oh, K. Ono, and G. Tian, editors, 'Symplectic geometry and mirror symmetry, proceedings of the 4th KIAS annual international conference,' pages 81–109, World Scientific, Singapore, 2001. math.AG/0012002.

[137] M. Gross, '*Toric degenerations and Batyrev–Borisov duality*,' Mathematische Annalen 333 (2005), 645–688. math.AG/0406171.

[138] M. Gross, D. Huybrechts, and D. Joyce, '*Calabi–Yau manifolds and related geometries*,' Universitext, Springer-Verlag, Berlin, 2003.

[139] M. Gross and B. Siebert, '*Affine manifolds, log structures, and mirror symmetry*,' Turkish Journal of Mathematics 27 (2003), 33–60. math.AG/0211094.

[140] M. Gross and B. Siebert, '*Mirror symmetry via logarithmic degeneration data. I*,' Journal of Differential Geometry 72 (2006), 169–338. math.AG/0309070.

[141] M. Gross and P.M.H. Wilson, '*Large complex structure limits of $K3$ surfaces*,' Journal of Differential Geometry 55 (2000), 475–546. math.DG/0008018.

[142] W. Gu and C. Pries, '*Examples of Cayley 4-manifolds*,' Houston J. Math. 30 (2004), 55–87.

[143] D. Guan, '*Examples of compact holomorphic symplectic manifolds which are not Kählerian. II*,' Inventiones mathematicae 121 (1995), 135–145.

[144] S. Gukov and J. Sparks, '*M-theory on* $\mathrm{Spin}(7)$ *manifolds*,' Nuclear Physics B625 (2002), 3–39. hep-th/0109025.

[145] S. Gukov, S.-T. Yau, and E. Zaslow, '*Duality and fibrations on G_2-manifolds*,' Turkish Journal of Mathematics 27 (2003), 61–97. hep-th/0203217.

[146] C. Haase and I. Zharkov, '*Integral affine structures on spheres and torus fibrations of Calabi–Yau toric hypersurfaces. I*,' math.AG/0205321, 2002.

[147] H.A. Hamm, '*Lefschetz theorems for singular varieties*'. In P. Orlik, editor, '*Singularities*,' Proceedings of Symposia in Pure Mathematics, vol. 40, part 1, pages 547–557, American Mathematical Society, Rhode Island, 1983.

[148] J. Harris, '*Algebraic Geometry, A First Course*,' Graduate Texts in Mathematics, vol. 133, Springer-Verlag, New York, 1992.

[149] R. Hartshorne, '*Algebraic Geometry*,' Graduate Texts in Mathematics, vol. 52, Springer-Verlag, New York, 1977.

[150] R. Harvey, '*Spinors and calibrations*,' Perspectives in Mathematics, vol. 9, Academic Press, San Diego, 1990.

[151] R. Harvey and H.B. Lawson, '*Calibrated geometries*,' Acta Mathematica 148 (1982), 47–157.

[152] R. Harvey and F. Morgan, '*The faces of the Grassmannian of three-planes in $\mathbb{R}^7$ (Calibrated geometries on $\mathbb{R}^7$)*,' Inventiones mathematicae 83 (1986), 191–228.

[153] H. Hashimoto, '*J-holomorphic curves of a 6-dimensional sphere*,' Tokyo Mathematics Journal 23 (2000), 137–159.

[154] M. Haskins, '*The geometric complexity of special Lagrangian T^2-cones*,' Inventiones mathematicae 157 (2004), 11–70. math.DG/0307129.

[155] M. Haskins, '*Special Lagrangian cones*,' American Journal of Mathematics 126 (2004), 845–871. math.DG/0005164.

[156] M. Haskins and N. Kapouleas, '*Special Lagrangian cones with higher genus links*,' to appear in Inventiones mathematicae. math.DG/0512178, 2005.

[157] S. Helgason, '*Differential Geometry and Symmetric Spaces*,' Academic Press, New York, 1962.

[158] H. Herrera and R. Herrera, '$\hat{A}$-genus on non-spin manifolds with S^1 actions and the classification of positive quaternion-Kähler 12-manifolds,' Journal of Differential Geometry 61 (2002), 341–364.

[159] H. Hironaka, 'On resolution of singularities (characteristic zero)'. In 'Proceedings of the International Congress of Mathematicians, Stockholm, 1962,' pages 507–521, Mittag–Leffler, 1963.

[160] N.J. Hitchin, 'Polygons and gravitons,' Mathematical Proceedings of the Cambridge Philosphical Society 85 (1979), 465–476.

[161] N.J. Hitchin, 'Kählerian twistor spaces,' Proceedings of the London Mathematical Society 43 (1981), 133–150.

[162] N.J. Hitchin, 'The moduli space of special Lagrangian submanifolds,' Annali della Scuola Normale Superiore di Pisa. Classe di Scienze 25 (1997), 503–515. dg-ga/9711002.

[163] N.J. Hitchin, 'The geometry of three-forms in 6 and 7 dimensions,' Journal of Differential Geometry 55 (2000), 547–576. math.AG/0010054.

[164] N.J. Hitchin, A. Karlhede, U. Lindström, and M. Roček, 'Hyperkähler metrics and supersymmetry,' Communications in Mathematical Physics 108 (1987), 535–589.

[165] K. Hori, A. Iqbal, and C. Vafa, 'D-branes and Mirror Symmetry,' hep-th/0005247, 2000.

[166] K. Hori, S. Katz, A. Klemm, R. Pandharipande, R. Thomas, C. Vafa, R. Vakil, and E. Zaslow, 'Mirror Symmetry,' Clay Mathematics Monographs, vol. 1, A.M.S., Providence, RI, 2003.

[167] T. Hübsch, 'Calabi–Yau manifolds, a bestiary for physicists,' World Scientific, Singapore, 1992.

[168] D. Huybrechts, 'Compact hyperkähler manifolds: basic results,' Inventiones mathematicae 135 (1999), 63–113. alg-geom/9705025.

[169] D. Huybrechts, 'Erratum: "Compact hyperkähler manifolds: basic results",' Inventiones mathematicae 152 (2003), 209–212.

[170] D. Huybrechts, 'The Kähler cone of a compact hyperkähler manifold,' Mathematische Annalen 326 (2003), 499–513. math/9909109.

[171] S. Iitaka, 'Algebraic Geometry, an Introduction to Birational Geometry of Algebraic Varieties,' Graduate Texts in Mathematics, vol. 76, Springer-Verlag, New York, 1982.

[172] M. Ionel, 'Second order families of special Lagrangian submanifolds in $\mathbb{C}^4$,' Journal of Differential Geometry 65 (2003), 211–272. math.DG/0303082.

[173] M. Ionel, S. Karigiannis, and M. Min-Oo, 'Bundle constructions of calibrated submanifolds in $\mathbb{R}^7$ and $\mathbb{R}^8$,' Mathematical Research Letters 12 (2005), 493–512. math.DG/0408005.

[174] Y. Ito, 'Crepant resolution of trihedral singularities,' Proceedings of the Japan Academy. Series A, Mathematical Sciences 70 (1994), 131–136.

[175] Y. Ito, 'Crepant resolution of trihedral singularities and the orbifold Euler characteristic,' International Journal of Mathematics 6 (1995), 33–43.

[176] Y. Ito and M. Reid, '*The McKay correspondence for finite subgroups of* $SL(3, \mathbb{C})$'. In M. Andreatta et al., editors, '*Higher dimensional complex varieties (Trento, June 1994)*,' pages 221–240, de Gruyter, 1996.

[177] T.A. Ivey and J.M. Landsberg, '*Cartan for beginners: differential geometry via moving frames and exterior differential systems,*' Graduate Studies in Math., vol. 61, A.M.S., Providence, RI, 2003.

[178] G. Jang, '*Compact manifolds with holonomy group* $\mathrm{Spin}(7)$,' Ph.D. thesis, Seoul National University, 1999.

[179] D.D. Joyce, '*The hypercomplex quotient and the quaternionic quotient,*' Mathematische Annalen 290 (1991), 323–340.

[180] D.D. Joyce, '*Compact hypercomplex and quaternionic manifolds,*' Journal of Differential Geometry 35 (1992), 743–761.

[181] D.D. Joyce, '*Explicit construction of self-dual 4-manifolds,*' Duke Mathematical Journal 77 (1995), 519–552.

[182] D.D. Joyce, '*Manifolds with many complex structures,*' Oxford Quarterly Journal of Mathematics 46 (1995), 169–184.

[183] D.D. Joyce, '*Compact Riemannian 7-manifolds with holonomy G_2. I,*' Journal of Differential Geometry 43 (1996), 291–328.

[184] D.D. Joyce, '*Compact Riemannian 7-manifolds with holonomy G_2. II,*' Journal of Differential Geometry 43 (1996), 329–375.

[185] D.D. Joyce, '*Compact 8-manifolds with holonomy* $\mathrm{Spin}(7)$,' Inventiones mathematicae 123 (1996), 507–552.

[186] D.D. Joyce, '*Hypercomplex algebraic geometry,*' Oxford Quarterly Journal of Mathematics 49 (1998), 129–162.

[187] D.D. Joyce, '*A new construction of compact 8-manifolds with holonomy* $\mathrm{Spin}(7)$,' Journal of Differential Geometry 53 (1999), 89–130. math.DG/9910002.

[188] D.D. Joyce, '*Compact Manifolds with Special Holonomy,*' Oxford Mathematical Monographs, Oxford University Press, Oxford, 2000.

[189] D.D. Joyce, '*Asymptotically Locally Euclidean metrics with holonomy* $\mathrm{SU}(m)$,' Annals of Global Analysis and Geometry 19 (2001), 55–73. math.AG/9905041.

[190] D.D. Joyce, '*Constructing special Lagrangian m-folds in* $\mathbb{C}^m$ *by evolving quadrics,*' Mathematische Annalen 320 (2001), 757–797. math.DG/0008155.

[191] D.D. Joyce, '*Evolution equations for special Lagrangian 3-folds in* $\mathbb{C}^3$,' Annals of Global Analysis and Geometry 20 (2001), 345–403. math.DG/0010036.

[192] D.D. Joyce, '*Quasi-ALE metrics with holonomy* $\mathrm{SU}(m)$ *and* $\mathrm{Sp}(m)$,' Annals of Global Analysis and Geometry 19 (2001), 103–132. math.AG/9905043.

[193] D.D. Joyce, '*Special Lagrangian 3-folds and integrable systems,*' math.DG/0101249, 2001. To appear in *Surveys on Geometry and Integrable Systems*, editors M. Guest, R. Miyaoka and Y. Ohnita, Advanced Studies in Pure Mathematics series, Mathematical Society of Japan, 2001.

[194] D.D. Joyce, '*On counting special Lagrangian homology 3-spheres*'. In A.J. Berrick, M.C. Leung, and X.W. Xu, editors, '*Topology and Geometry: Commem-*

orating SISTAG,' Contemporary Mathematics, vol. 314, pages 125–151, A.M.S., 2002. hep-th/9907013.

[195] D.D. Joyce, '*Ruled special Lagrangian 3-folds in* $\mathbb{C}^3$,' Proceedings of the London Mathematical Society 85 (2002), 233–256. math.DG/0012060.

[196] D.D. Joyce, '*Special Lagrangian* m-*folds in* $\mathbb{C}^m$ *with symmetries,*' Duke Mathematical Journal 115 (2002), 1–51. math.DG/0008021.

[197] D.D. Joyce, '*Singularities of special Lagrangian fibrations and the SYZ Conjecture,*' Communications in Analysis and Geometry 11 (2003), 859–907. math.DG/0011179.

[198] D.D. Joyce, '$U(1)$-*invariant special Lagrangian 3-folds in* $\mathbb{C}^3$ *and special Lagrangian fibrations,*' Turkish Journal of Mathematics 27 (2003), 99–114. math.DG/0206016.

[199] D.D. Joyce, '*Singularities of special Lagrangian submanifolds*'. In S.K. Donaldson, Y. Eliashberg, and M. Gromov, editors, '*Different Faces of Geometry,*' International Mathematical Series, vol. 3, pages 163–198, Kluwer/Plenum, 2004.

[200] D.D. Joyce, '*Special Lagrangian submanifolds with isolated conical singularities. I. Regularity,*' Annals of Global Analysis and Geometry 25 (2004), 201–251. math.DG/0211294.

[201] D.D. Joyce, '*Special Lagrangian submanifolds with isolated conical singularities. II. Moduli spaces,*' Annals of Global Analysis and Geometry 25 (2004), 301–352. math.DG/0211295.

[202] D.D. Joyce, '*Special Lagrangian submanifolds with isolated conical singularities. III. Desingularization, the unobstructed case,*' Annals of Global Analysis and Geometry 26 (2004), 1–58. math.DG/0302355.

[203] D.D. Joyce, '*Special Lagrangian submanifolds with isolated conical singularities. IV. Desingularization, obstructions and families,*' Annals of Global Analysis and Geometry 26 (2004), 117–174. math.DG/0302356.

[204] D.D. Joyce, '*Special Lagrangian submanifolds with isolated conical singularities. V. Survey and applications,*' Journal of Differential Geometry 63 (2003), 279–347. math.DG/0303272.

[205] D.D. Joyce, '$U(1)$-*invariant special Lagrangian 3-folds. I. Nonsingular solutions,*' Advances in Mathematics 192 (2005), 35–71. math.DG/0111324.

[206] D.D. Joyce, '$U(1)$-*invariant special Lagrangian 3-folds. II. Existence of singular solutions,*' Advances in Mathematics 192 (2005), 72–134. math.DG/0111326.

[207] D.D. Joyce, '$U(1)$-*invariant special Lagrangian 3-folds. III. Properties of singular solutions,*' Advances in Mathematics 192 (2005), 135–182. math.DG/0204343.

[208] H. Kanno and Y. Yasui, '*On* $\mathrm{Spin}(7)$ *holonomy metric based on* $\mathrm{SU}(3)/U(1)$. *I,*' Journal of Geometry and Physics 43 (2002), 293–309. hep-th/0108226.

[209] H. Kanno and Y. Yasui, '*On* $\mathrm{Spin}(7)$ *holonomy metric based on* $\mathrm{SU}(3)/U(1)$ *II,*' Journal of Geometry and Physics 43 (2002), 310–326. hep-th/0111198.

[210] A.N. Kapustin and D.O. Orlov, '*Remarks on A-branes, Mirror Symmetry, and the Fukaya category,*' Journal of Geometry and Physics 48 (2003), 84–99. hep-

th/0109098.

[211] A.N. Kapustin and D.O. Orlov, '*Lectures on mirror symmetry, derived categories, and D-branes,*' Russian Math. Surveys 59 (2004), 101–134. math.AG/0308173.

[212] S. Karigiannis and M. Min-Oo, '*Calibrated subbundles in noncompact manifolds of special holonomy,*' Annals of Global Analysis and Geometry 28 (2005), 371–394. math.DG/0412312.

[213] H.-J. Kim and G. Jang, '*Riemannian manifolds with exceptional holonomy groups*'. In '*Proceedings of the fourth international workshop on Differential Geometry (Taegu, 1999),*' pages 27–31, Kyungpook National University, Taegu, 2000.

[214] S. Kobayashi and K. Nomizu, '*Foundations of Differential Geometry,*' vol. 1, Wiley, New York, 1963.

[215] S. Kobayashi and K. Nomizu, '*Foundations of Differential Geometry,*' vol. 2, Wiley, New York, 1963.

[216] K. Kodaira, '*On the structure of compact complex analytic surfaces. I,*' American Journal of Mathematics 86 (1964), 751–798.

[217] K. Kodaira, '*Complex Manifolds and Deformation of Complex Structures,*' Grundlehren der mathematischen Wissenschaften, vol. 283, Springer-Verlag, New York, 1986.

[218] S. Kong, C.-L. Terng, and E. Wang, '*Associative cones and integrable systems,*' Chinese Annals of Mathematics Series B 27 (2006), 153–168. math.DG/0602565.

[219] M. Kontsevich, '*Homological algebra of mirror symmetry*'. In '*Proceedings of the International Congress of Mathematicians (Zürich, 1994),*' pages 120–139, Birkhäuser, Basel, 1994. alg-geom/9411018.

[220] M. Kontsevich and Y. Soibelman, '*Homological mirror symmetry and torus fibrations*'. In K. Fukaya, Y.-G. Oh, K. Ono, and G. Tian, editors, '*Symplectic geometry and mirror symmetry,*' pages 203–263, World Scientific, River Edge, NJ, 2001. math.SG/0011041.

[221] M. Kontsevich and Y. Soibelman, '*Affine structures and non-Archimedean analytic spaces*'. In P. Etingof, V. Retakh, and I. M. Singer, editors, '*The unity of mathematics,*' Progress in Mathematics, vol. 244, pages 321–385, Birkhäuser, Boston, MA, 2006. math.AG/0406564.

[222] A.G. Kovalev, '*Nahm's equations and complex adjoint orbits,*' Oxford Quarterly Journal of Mathematics 47 (1996), 41–58.

[223] A.G. Kovalev, '*Twisted connected sums and special Riemannian holonomy,*' Journal für die Reine und Angewandte Mathematik 565 (2003), 125–160. math.DG/0012189.

[224] A.G. Kovalev, '*Coassociative $K3$ fibrations of compact G_2-manifolds,*' math.DG/0511150, 2005.

[225] M. Kreuzer and H. Skarke, '*Complete classification of reflexive polyhedra in four dimensions,*' Advances in Theoretical and Mathematical Physics 4 (2000), 1209–1230. hep-th/0002240.

[226] P.B. Kronheimer, '*The construction of ALE spaces as hyperkähler quotients,*' Journal of Differential Geometry 29 (1989), 665–683.

[227] P.B. Kronheimer, '*A Torelli-type theorem for gravitational instantons,*' Journal of Differential Geometry 29 (1989), 685–697.

[228] P.B. Kronheimer, '*A hyper-kählerian structure on coadjoint orbits of a semisimple complex group,*' Journal of the London Mathematical Society 42 (1990), 193–208.

[229] P.B. Kronheimer, '*Instantons and the geometry of the nilpotent variety,*' Journal of Differential Geometry 32 (1990), 473–490.

[230] S. Lang, '*Real Analysis,*' second edition, Addison Wesley, Reading, Mass., 1983.

[231] G. Lawlor, '*The angle criterion,*' Inventiones mathematicae 95 (1989), 437–446.

[232] H.B. Lawson, '*Lectures on Minimal Submanifolds,*' vol. 1, Publish or Perish, Wilmington, Delaware, 1980.

[233] H.B. Lawson and M.L. Michelson, '*Spin Geometry,*' Princeton University Press, Princeton, 1989.

[234] C.I. Lazaroiu, '*Graded Lagrangians, exotic topological D-branes and enhanced triangulated categories,*' Journal of High Energy Physics 06 (2001), 064. hep-th/0105063.

[235] C. LeBrun, '*Anti-self-dual Hermitian metrics on blown up Hopf surfaces,*' Mathematische Annalen 289 (1991), 383–392.

[236] C. LeBrun, '*Explicit self-dual metrics on $\mathbb{CP}^2 \# \cdots \# \mathbb{CP}^2$,*' Journal of Differential Geometry 34 (1991), 223–253.

[237] C. LeBrun, '*On complete quaternion-Kähler manifolds,*' Duke Mathematical Journal 63 (1991), 723–743.

[238] C. LeBrun, '*Scalar-flat Kähler metrics on blown-up ruled surfaces,*' Journal für die Reine und Angewandte Mathematik 420 (1991), 161–177.

[239] C. LeBrun and S. Salamon, '*Strong rigidity of positive quaternion-Kähler manifolds,*' Inventiones mathematicae 118 (1994), 109–132.

[240] C. LeBrun and M. Singer, '*A Kummer-type construction of self-dual 4-manifolds,*' Mathematische Annalen 300 (1994), 165–180.

[241] D.A. Lee, '*Connected sums of special Lagrangian submanifolds,*' Communications in Analysis and Geometry 12 (2004), 553–579. math.DG/0303224.

[242] Y.-I. Lee, '*Embedded special Lagrangian submanifolds in Calabi–Yau manifolds,*' Communications in Analysis and Geometry 11 (2003), 391–423.

[243] C. Lewis, '$\mathrm{Spin}(7)$ *instantons,*' Ph.D. thesis, Oxford University, 1998.

[244] A. Lichnerowicz, '*Spineurs harmoniques,*' Comptes Rendus de l'Académie des Sciences. Série A, Sciences Mathematiques 257 (1963), 7–9.

[245] R.B. Lockhart, '*Fredholm, Hodge and Liouville theorems on noncompact manifolds,*' Transactions of the American Mathematical Society 301 (1987), 1–35.

[246] E. Looijenga, '*A Torelli theorem for Kähler–Einstein $K3$ surfaces*'. In '*Geometry Symposium, Utrecht, 1980,*' Lecture Notes in Mathematics, vol. 894, pages 107–112, Springer-Verlag, 1982.

[247] J. Lotay, '*Deformation theory of asymptotically conical coassociative 4-folds,*' math.DG/0411116, 2004.

[248] J. Lotay, '*Constructing associative 3-folds by evolution equations,*' Communications in Analysis and Geometry 13 (2005), 999–1037. math.DG/0401123.

[249] J. Lotay, '*2-ruled calibrated 4-folds in* $\mathbb{R}^7$ *and* $\mathbb{R}^8$,' Journal of the London Mathematical Society 74 (2006), 219–243. math.DG/0401125.

[250] J. Lotay, '*Calibrated submanifolds of* $\mathbb{R}^7$ *and* $\mathbb{R}^8$ *with symmetries,*' to appear in the Oxford Quarterly Journal of Mathematics. math.DG/0601764, 2006.

[251] J. Lotay, '*Coassociative 4-folds with conical singularities,*' math.DG/0601762, 2006.

[252] H. Ma and Y. Ma, '*Totally real minimal tori in* $\mathbb{CP}^2$,' Mathematische Zeitschrift 249 (2005), 241–267. math.DG/0106141.

[253] D.G. Markushevich, M.A. Olshanetsky, and A.M. Perelomov, '*Description of a class of superstring compactifications related to semi-simple Lie algebras,*' Communications in Mathematical Physics 111 (1987), 247–274.

[254] S.P. Marshall, '*Deformations of special Lagrangian submanifolds,*' Ph.D. thesis, Oxford University, 2002.

[255] K. Mashimo, '*Homogeneous totally real submanifolds of* S^6,' Tsukaba J. Math. 9 (1985), 185–202.

[256] D. McDuff and D. Salamon, '*Introduction to Symplectic Topology,*' second edition, Oxford University Press, Oxford, 1998.

[257] I. McIntosh, '*Special Lagrangian cones in* $\mathbb{C}^3$ *and primitive harmonic maps,*' Journal of the London Mathematical Society 67 (2003), 769–789. math.DG/0201157.

[258] J. McKay, '*Graphs, singularities, and finite groups,*' Proceedings of Symposia in Pure Mathematics 37 (1980), 183–186.

[259] R.C. McLean, '*Deformations of calibrated submanifolds,*' Communications in Analysis and Geometry 6 (1998), 705–747.

[260] S. Merkulov and L. Schwachhöfer, '*Classification of irreducible holonomies of torsion-free affine connections,*' Annals of Mathematics 150 (1999), 77–149. math.DG/9907206.

[261] J. Milnor and J. Stasheff, '*Characteristic classes,*' Annals of Mathematics Studies, no. 76, Princeton University Press, Princeton, 1974.

[262] D. Montgomery and H. Samelson, '*Transformation groups of spheres,*' Annals of Mathematics 44 (1943), 454–470.

[263] F. Morgan, '*The exterior algebra* $\Lambda^k \mathbb{R}^n$ *and area minimization,*' Linear Algebra and its Applications 66 (1985), 1–28.

[264] F. Morgan, '*Area-minimizing surfaces, faces of Grassmannians, and calibrations,*' American Mathematical Monthly 95 (1988), 813–822.

[265] F. Morgan, '*Geometric Measure Theory, a Beginner's Guide,*' Academic Press, San Diego, 1995.

[266] C.B. Morrey, '*Second order elliptic systems of partial differential equations*'. In L. Bers, S. Bochner, and F. John, editors, '*Contributions to the theory of Partial*

Differential Equations,' Annals of Mathematics Studies, vol. 33, pages 101–159, Princeton University Press, Princeton, 1954.

[267] C.B. Morrey, '*Multiple Integrals in the Calculus of Variations*,' Grundlehren der mathematischen Wissenschaften, vol. 130, Springer-Verlag, Berlin, 1966.

[268] D.R. Morrison, '*Mirror symmetry and rational curves on quintic 3-folds: a guide for mathematicians*,' Journal of the American Mathematical Society 6 (1993), 223–247. alg-geom/9202004.

[269] D.R. Morrison and G. Stevens, '*Terminal quotient singularities in dimensions three and four*,' Proceedings of the American Mathematical Society 90 (1984), 15–20.

[270] T. Oda, '*Convex Bodies and Algebraic Geometry*,' Springer-Verlag, New York, 1988.

[271] K. O'Grady, '*Desingularized moduli spaces of sheaves on a $K3$*,' Journal für die Reine und Angewandte Mathematik 512 (1999), 49–117. alg-geom/9708009 and alg-geom/9805099.

[272] K. O'Grady, '*A new six-dimensional irreducible symplectic variety*,' Journal of Algebraic Geometry 12 (2003), 435–505. math.AG/0010187.

[273] Y. Ohnita, '*Stability and rigidity of special Lagrangian cones over certain minimal Legendrian orbits*,' preprint, 2006.

[274] T. Pacini, '*Deformations of asymptotically conical special Lagrangian submanifolds*,' Pacific Journal of Mathematics 215 (2004), 151–181. math.DG/0207144.

[275] D.N. Page, '*A physical picture of the K3 gravitational instanton*,' Physics Letters 80B (1978), 55–57.

[276] G. Papadopoulos and P. Townsend, '*Compactification of $D = 11$ supergravity on spaces of exceptional holonomy*,' Physics Letters B357 (1995), 300–306.

[277] H. Pedersen and Y.S. Poon, '*Deformations of hypercomplex structures*,' Journal für die Reine und Angewandte Mathematik 499 (1998), 81–99.

[278] H. Pedersen and Y.S. Poon, '*Inhomogeneous hypercomplex structures on homogeneous manifolds*,' Journal für die Reine und Angewandte Mathematik 516 (1999), 159–181.

[279] H. Pedersen and Y.S. Poon, '*A note on rigidity of 3-Sasakian manifolds*,' Proceedings of the American Mathematical Society 127 (1999), 3027–3034.

[280] A. Polishchuk, '*Homological mirror symmetry with higher products*'. In C. Vafa and S.-T. Yau, editors, '*Winter school on mirror symmetry, vector bundles and Lagrangian submanifolds*,' AMS/IP Studies in Advanced Mathematics, vol. 23, pages 247–259, A.M.S., Providence, RI, 2001. math.AG/9901025.

[281] A. Polishchuk and E. Zaslow, '*Categorical mirror symmetry: the elliptic curve*,' Advances in Theoretical and Mathematical Physics 2 (1998), 443–470. math.AG/9801119.

[282] Y.S. Poon and S.M. Salamon, '*Quaternionic Kähler 8-manifolds with positive scalar curvature*,' Journal of Differential Geometry 33 (1991), 363–378.

[283] A. Rapagnetta, '*Topological invariants of O'Grady's six-dimensional irreducible symplectic variety*,' math.AG/0406026, 2004.

[284] M. Reid, '*Minimal models of canonical 3-folds*'. In S. Iitaka, editor, '*Algebraic Varieties and Analytic Varieties*,' Advanced Studies in Pure Mathematics, no. 1, pages 131–180, North Holland, Amsterdam, 1983.

[285] M. Reid, '*Young person's guide to canonical singularities*'. In '*Algebraic Geometry, Bowdoin 1985*,' Proceedings of Symposia in Pure Mathematics, no. 46, pages 345–416, American Mathematical Society, 1987.

[286] M. Reid, '*Le correspondance de McKay*,' Astérisque 276 (2002), 53–72. math.AG/9911165.

[287] R. Reyes Carrión, '*A generalization of the notion of instanton*,' Differential Geometry and its Applications 8 (1998), 1–20.

[288] S.-S. Roan, '*On the generalization of Kummer surfaces*,' Journal of Differential Geometry 30 (1989), 523–537.

[289] S.-S. Roan, '*Minimal resolution of Gorenstein orbifolds*,' Topology 35 (1996), 489–508.

[290] W.-D. Ruan, '*Lagrangian torus fibration of quintic Calabi–Yau hypersurfaces I: Fermat quintic case*'. In C. Vafa and S.-T. Yau, editors, '*Winter school on Mirror Symmetry, vector bundles and Lagrangian submanifolds*,' AMS/IP Studies in Advanced Mathematics, vol. 23, pages 297–332, A.M.S., Providence, RI, 2001. math.DG/9904012.

[291] W.-D. Ruan, '*Lagrangian torus fibration of quintic Calabi–Yau hypersurfaces II: technical results on gradient flow construction*,' Journal of Symplectic Geometry 1 (2002), 435–521. math.SG/0411264.

[292] W.-D. Ruan, '*Lagrangian torus fibration of quintic Calabi–Yau hypersurfaces III: symplectic topological SYZ mirror construction for general quintics*,' Journal of Differential Geometry 63 (2003), 171–229. math.DG/9909126.

[293] W.-D. Ruan, '*Lagrangian torus fibration and mirror symmetry of Calabi–Yau hypersurface in toric variety*,' math.DG/0007028, 2000.

[294] S.M. Salamon, '*Quaternionic Kähler manifolds*,' Inventiones mathematicae 67 (1982), 143–171.

[295] S.M. Salamon, '*Differential geometry of quaternionic manifolds*,' Annales scientifiques de l'École Normale Superieure 19 (1986), 31–55.

[296] S.M. Salamon, '*Riemannian geometry and holonomy groups*,' Pitman Research Notes in Mathematics, vol. 201, Longman, Harlow, 1989.

[297] S.M. Salamon, '*Quaternion-Kähler geometry*'. In C. LeBrun and M. Wang, editors, '*Essays on Einstein Manifolds*,' Surveys in Differential Geometry, vol. V, pages 83–121, International Press, 1999.

[298] I. Satake, '*On a generalization of the notion of manifold*,' Proceedings of the National Academy of Sciences of the U.S.A. 42 (1956), 359–363.

[299] M. Schlessinger, '*Rigidity of quotient singularities*,' Inventiones mathematicae 14 (1971), 17–26.

[300] R. Schoen and J. Wolfson, '*Minimizing volume among Lagrangian submanifolds*'. In M. Giaquinta, J. Shatah, and S.R.S. Varadhan, editors, '*Differen-*

tial equations: La Pietra 1996,' Proceedings of Symposia in Pure Mathematics, vol. 65, pages 181–199, A.M.S., Providence, RI, 1999.

[301] R. Schoen and J. Wolfson, '*Minimizing area among Lagrangian surfaces: the mapping problem*,' Journal of Differential Geometry 58 (2001), 1–86. math.DG/0008244.

[302] P. Seidel, '*Graded Lagrangian submanifolds*,' Bull. Soc. Math. France 128 (2000), 103–149. math.SG/9903049.

[303] P. Seidel, '*Vanishing cycles and mutation*'. In C. Casacuberta, R.M. Miró-Roig, J. Verdera, and S. Xambó-Descamps, editors, '*Proceedings of the 3rd European Congress of Mathematics, Barcelona, 2000, vol. II*,' Progress in Mathematics, vol. 202, pages 65–85, Birkhäuser, Basel, 2001. math.SG/0007115.

[304] P. Seidel, '*More about vanishing cycles and mutation*'. In K. Fukaya, Y.-G. Oh, K. Ono, and G. Tian, editors, '*Symplectic geometry and mirror symmetry*,' pages 429–465, World Scientific, River Edge, NJ, 2001. math.SG/0010032.

[305] P. Seidel, '*Homological mirror symmetry for the quartic surface*,' math.SG/0310414, 2003.

[306] P. Seidel, '*Fukaya categories and Picard–Lefschetz theory*,' book in preparation, available at www.math.uchicago.edu/~seidel/, 2006.

[307] K. Sekigawa, '*Almost complex submanifolds of a 6-dimensional sphere*,' Kodai Mathematics Journal 6 (1983), 174–185.

[308] U. Semmelmann and G. Weingart, '*Vanishing theorems for quaternionic Kähler manifolds*,' Journal für die Reine und Angewandte Mathematik 544 (2002), 111–132. math.DG/0001061.

[309] R.A. Sharipov, '*Minimal tori in the five-dimensional sphere in $\mathbb{C}^3$*,' Theor. Math. Phys. 87 (1991), 363–369. Revised version: math.DG/0204253.

[310] S.L. Shatashvili and C. Vafa, '*Exceptional magic*,' Nuclear Physics B Proceedings Supplement 41 (1995), 345–356.

[311] S.L. Shatashvili and C. Vafa, '*Superstrings and manifolds of exceptional holonomy*,' Selecta Mathematica 1 (1995), 347–381. hep-th/9407025.

[312] L. Simon, '*Isolated singularities of extrema of geometric variational problems*'. In E. Giusti, editor, '*Harmonic mappings and minimal immersions*,' Springer Lecture Notes in Mathematics, vol. 1161, pages 206–277, Springer, Berlin, 1985.

[313] J. Simons, '*On the transitivity of holonomy systems*,' Annals of Mathematics 76 (1962), 213–234.

[314] Y.T. Siu, '*Every $K3$ surface is Kähler*,' Inventiones mathematicae 73 (1983), 139–150.

[315] P. Slodowy, '*Simple singularities and simple algebraic groups*,' Lecture Notes in Mathematics, vol. 815, Springer-Verlag, Berlin, 1980.

[316] Ph. Spindel, A. Sevrin, W. Troost, and A. Van Proeyen, '*Extended supersymmetric σ-models on group manifolds*,' Nuclear Physics B308 (1988), 662–698.

[317] A. Strominger, S.-T. Yau, and E. Zaslow, '*Mirror symmetry is T-duality*,' Nuclear Physics B479 (1996), 243–259. hep-th/9606040.

[318] T. Tao and G. Tian, '*A singularity removal theorem for Yang–Mills fields in higher dimensions*,' Journal of the American Mathematical Society 17 (2004), 557–593. math.DG/0209352.

[319] C.H. Taubes, '*The existence of anti-self-dual conformal structures*,' Journal of Differential Geometry 36 (1992), 163–253.

[320] C.H. Taubes, '*Metrics, connections and gluing theorems*,' Regional Conference Series in Mathematics, vol. 89, American Mathematical Society, Rhode Island, 1996.

[321] C. Taylor, '*New examples of compact 8-manifolds of holonomy* $\mathrm{Spin}(7)$,' Mathematical Research Letters 6 (1999), 557–561.

[322] R.P. Thomas, '*Gauge theory on Calabi–Yau manifolds*,' Ph.D. thesis, Oxford University, 1997.

[323] R.P. Thomas, '*Derived categories for the working mathematician*'. In C. Vafa and S.-T. Yau, editors, '*Winter school on Mirror Symmetry, vector bundles and Lagrangian submanifolds*,' AMS/IP Studies in Advanced Mathematics, vol. 23, pages 349–361, A.M.S., Providence, RI, 2001. math.AG/0001045.

[324] R.P. Thomas, '*Moment maps, monodromy and mirror manifolds*'. In K. Fukaya, Y.-G. Oh, K. Ono, and G. Tian, editors, '*Symplectic geometry and mirror symmetry (Seoul, 2000)*,' pages 467–498, World Scientific, River Edge, NJ, 2001. math.DG/0104196.

[325] R.P. Thomas and S.-T. Yau, '*Special Lagrangians, stable bundles and mean curvature flow*,' Communications in Analysis and Geometry 10 (2002), 1075–1113. math.DG/0104197.

[326] G. Tian, '*Smoothness of the universal deformation space of compact Calabi–Yau manifolds and its Peterson–Weil metric*'. In S.-T. Yau, editor, '*Mathematical aspects of String Theory*,' Advanced series in Mathematical Physics, vol. 1, pages 629–646, World Scientific, 1987.

[327] G. Tian, '*Gauge theory and calibrated geometry. I,*' Annals of Mathematics 151 (2000), 193–268. math.DG/0010015.

[328] A.N. Todorov, '*Applications of Kähler–Einstein–Calabi–Yau metrics to moduli of $K3$ surfaces*,' Inventiones mathematicae 61 (1980), 251–265.

[329] A.N. Todorov, '*The Weil–Petersson geometry of the moduli space of* $\mathrm{SU}(n \geqslant 3)$ *(Calabi–Yau) manifolds I*,' Communications in Mathematical Physics 126 (1989), 325–346.

[330] P. Topiwala, '*A new proof of the existence of Kähler–Einstein metrics on $K3$. I*,' Inventiones mathematicae 89 (1987), 425–448.

[331] C. Vafa, '*Modular invariance and discrete torsion on orbifolds*,' Nuclear Physics B273 (1986), 592–606.

[332] J. Varouchas, '*Kähler spaces and proper open morphisms*,' Mathematische Annalen 283 (1989), 13–52.

[333] M. Verbitsky, '*Manifolds with parallel differential forms and Kähler identities for G_2-manifolds*,' math.DG/0502540, 2005.

[334] M. Verbitsky and D. Kaledin, '*Hyperkähler manifolds*,' International Press, Somerville, MA, 1999. alg-geom/9712012 and alg-geom/9710026.

[335] C. Voisin, '*Mirror Symmetry*,' SMF/AMS Texts and Monographs, vol. 1, A.M.S., Providence, RI, 1999.

[336] M.Y. Wang, '*Parallel spinors and parallel forms*,' Annals of Global Analysis and Geometry 7 (1989), 59–68.

[337] S.H. Wang, '*Compact special Legendrian surfaces in S^5*,' math.DG/0211439, 2002.

[338] F.W. Warner, '*Foundations of differentiable manifolds and Lie groups*,' Scott, Foresman and Co., Illinois, 1971.

[339] P.M.H. Wilson, '*Calabi–Yau manifolds with large Picard number*,' Inventiones mathematicae 98 (1989), 139–155.

[340] P.M.H. Wilson, '*The Kähler cone on Calabi–Yau threefolds*,' Inventiones mathematicae 107 (1992), 561–583.

[341] J. Wolf, '*Complex homogeneous contact manifolds and quaternionic symmetric spaces*,' Journal of Mathematics and Mechanics 14 (1965), 1033–1047.

[342] Y. Xin, '*Minimal submanifolds and related topics*,' Nankai Trancts in Mathematics, vol. 8, World Scientific, Singapore, 2003.

[343] H. Yamabe, '*On an arcwise connected subgroup of a Lie group*,' Osaka Journal of Mathematics 2 (1950), 13–14.

[344] S.-T. Yau, '*On Calabi's conjecture and some new results in algebraic geometry*,' Proceedings of the National Academy of Sciences of the U.S.A. 74 (1977), 1798–1799.

[345] S.-T. Yau, '*On the Ricci curvature of a compact Kähler manifold and the complex Monge–Ampère equations. I*,' Communications on pure and applied mathematics 31 (1978), 339–411.

[346] Y. Yuan, '*Global solutions to special Lagrangian equations*,' Proceedings of the American Mathematical Society 134 (2006), 1355–1358. math.AP/0501456.

Index